Lecture Notes in Artificial In

Subseries of Lecture Notes in Computer Science

LNAI Series Editors

Randy Goebel
University of Alberta, Edmonton, Canada
Yuzuru Tanaka
Hokkaido University, Sapporo, Japan
Wolfgang Wahlster
DFKI and Saarland University, Saarbrücken, Germany

LNAI Founding Series Editor

Joerg Siekmann
DFKI and Saarland University, Saarbrücken, Germany

Ivan Habernal   Václav Matoušek (Eds.)

Text, Speech, and Dialogue

16th International Conference, TSD 2013
Pilsen, Czech Republic, September 1-5, 2013
Proceedings

 Springer

Volume Editors

Ivan Habernal
University of West Bohemia
306 14 Pilsen, Czech Republic
E-mail: habernal@kiv.zcu.cz

Václav Matoušek
University of West Bohemia
306 14 Pilsen, Czech Republic
E-mail: matousek@kiv.zcu.cz

ISSN 0302-9743   e-ISSN 1611-3349
ISBN 978-3-642-40584-6   e-ISBN 978-3-642-40585-3
DOI 10.1007/978-3-642-40585-3
Springer Heidelberg New York Dordrecht London

Library of Congress Control Number: 2013946087

CR Subject Classification (1998): I.2, H.3-5, J.1, H.2, I.5, F.1

LNCS Sublibrary: SL 7 – Artificial Intelligence

Typesetting: Camera-ready by author, data conversion by Scientific Publishing Services, Chennai, India

Printed on acid-free paper

Springer is part of Springer Science+Business Media (www.springer.com)

Preface

The annual Text, Speech and Dialogue Conference (TSD), which originated in 1998, constitutes a recognized platform for the presentation and discussion of state-of-the-art technology and recent achievements in natural language processing. It has become an interdisciplinary forum, interweaving the themes of speech technology and language processing. The conference attracts researchers not only from Central and Eastern Europe, but also from other parts of the world. Indeed, one of its goals has always been to bring together NLP researchers with different interests from all over world and to promote their mutual cooperation. One of the ambitions of the conference is, as its title says, not only to deal with dialogue systems, but also to improve dialogue between researchers in two areas of NLP, i.e., between "dialogue" and "speech" people. Moreover, the TSD 2013 conference was organized in parallel with the SPECOM conference. The SPECOM conference focuses on the processing of Slavic languages and brought to Plzeň (Pilsen) many specialists from Russia and former Soviet Union countries. Several informal discussions between both groups of participants were organized.

The TSD 2013 Conference was the 16th event in the series of International Conferences on Text, Speech and Dialogue supported by the International Speech Communication Association (ISCA) and Czech Society for Cybernetics and Informatics (ČSKI). The conference was held in the Conference Center of the Vienna International Hotel Angelo, during September 1–5, 2013, and was organized by the University of West Bohemia in Plzeň in cooperation with Masaryk University, Brno, Czech Republic. Like its predecessors, TSD 2013 highlighted to both the academic and scientific world the importance of text and speech processing and its most recent breakthroughs in current applications. Both experienced researchers and professionals as well as newcomers to text and speech processing, interested in designing or evaluating interactive software, developing new interaction technologies, or investigating overarching theories of text and speech processing, found in the TSD conference a forum to communicate with people sharing similar interests.

This year's conference was partially oriented toward challenges for highly inflected languages and applications of large language corpora, which had been chosen as the main topic of the conference. All invited speakers – Hynek Hermansky (Johns Hopkins University, Baltimore), Torbjörn Lager (University of Gothenburg, Sweden), Ron Cole (President of Boulder Language Technologies), Ralf Steinberger (Language Technology Project Manager at the European Commission's Joint Research Centre in Ispra) and Victor Zakharov (Saint-Petersburg State University) – gave nice invited talks on the latest achievements in the relatively broad and still unexplored area of highly inflected languages and their processing. Many interesting questions on the newest speech applications were answered in these talks; the convergence of Western and Eastern countries evident in the implementation of several speech applications can be seen as one of the most significant results of the conference.

This volume contains a collection of submitted papers presented at the conference, which were thoroughly reviewed by three members of the conference reviewing team consisting of more than 60 top specialists in the conference topic areas. A total of 65 accepted papers out of 148 submissions, altogether contributed by 171 authors and co-authors, were selected by the Program Committee for presentation at the conference and for inclusion in this book. Theoretical and more general contributions were presented in common (plenary) sessions. Problem-oriented sessions as well as panel discussions brought together specialists in limited problem areas with the aim of exchanging knowledge and skills resulting from research projects of all kinds.

Last but not least, we would like to express our gratitude to the authors for providing their papers on time, to the members of the conference reviewing team and Program Committee for their careful reviews and paper selection, and to the editors for their hard work in preparing this volume. Special thanks are due to the members of the Local Organizing Committee for their tireless effort and enthusiasm during the conference organization. We hope that you will benefit from the work presented in these proceedings.

June 2013

Václav Matoušek

Organization

TSD 2013 was organized by the Faculty of Applied Sciences, University of West Bohemia in Plzeň (Pilsen), in cooperation with the Faculty of Informatics, Masaryk University in Brno, Czech Republic. The conference website is located at: `http://www.kiv.zcu.cz/tsd2013/` or `http://www.tsdconference.org`.

Program Committee

Hynek Heřmanský (USA), *Chair*
Eneko Agirre (Spain)
Geneviève Baudoin (France)
Paul Cook (Australia)
Jan Černocký (Czech Republic)
Simon Dobrisek (Slovenia)
Karina Evgrafova (Russia)
Darja Fišer (Slovenia)
Radovan Garabík (Slovakia)
Alexander Gelbukh (Mexico)
Louise Guthrie (UK)
Jan Hajič (Czech Republic)
Eva Hajičová (Czech Republic)
Yannis Haralambous (France)
Ludwig Hitzenberger (Germany)
Jaroslava Hlaváčová
    (Czech Republic)
Aleš Horák (Czech Republic)
Eduard Hovy (USA)
Maria Khokhlova (Russia)
Daniil Kocharov (Russia)
Ivan Kopeček (Czech Republic)
Valia Kordoni (Germany)
Steven Krauwer (The Netherlands)
Siegfried Kunzmann (Germany)
Natalija Loukachevitch (Russia)
Václav Matoušek (Czech Republic)
Diana McCarthy (UK)
France Mihelić (Slovenia)
Hermann Ney (Germany)

Elmar Nöth (Germany)
Karel Oliva (Czech Republic)
Karel Pala (Czech Republic)
Nikola Pavešić, (Slovenia)
Vladimír Petkevič (Czech Republic)
Fabio Pianesi (Italy)
Maciej Piasecki (Poland)
Jan Pomikálek (Czech Republic)
Adam Przepiorkowski, (Poland)
Josef Psutka (Czech Republic)
James Pustejovsky (USA)
German Rigau (Spain)
Léon J. M. Rothkrantz
    (The Netherlands)
Anna Rumshishky (USA)
Milan Rusko (Slovakia)
Mykola Sazhok (Ukraine)
Stefan Steidl (Germany)
Pavel Skrelin (Russia)
Pavel Smrž (Czech Republic)
Petr Sojka (Czech Republic)
Georg Stemmer (Germany)
Marko Tadić (Croatia)
Tamás Varadi (Hungary)
Zygmunt Vetulani (Poland)
Pascal Wiggers (The Netherlands)
Yorick Wilks (UK)
Marcin Woliński (Poland)
Victor Zakharov (Russia)

Local Organizing Committee

Václav Matoušek *(Chair)*
Tomáš Brychcín
Kamil Ekštein
Ivan Habernal
Michal Konkol

Miloslav Konopík
Roman Mouček
Tomáš Ptáček
Anna Habernalová *(Secretary)*

Acknowledgements

Special thanks to the following reviewers who devoted their valuable time to review more than five papers each and thus helped to keep the high quality of the conference review process.

Alexander Gelbukh
Tino Haderlein
Yannis Haralambous
Hynek Hermansky
Maria Khokhlova
Daniil Kocharov
Ulrich Kordon
Natalia Loukachevitch
Karel Oliva

Maciej Piasecki
Josef Psutka
Adam Przepiórkowski
Leon Rothkrantz
Milan Rusko
Pavel Skrelin
Marcin Woliński
Victor Zakharov

Sponsoring Institutions

International Speech Communication Association (ISCA)
Czech Society for Cybernetics and Informatics (CSKI)

About Plzeň (Pilsen)

The new town of Pilsen was founded at the confluence of four rivers – Radbuza, Mže, Úhlava and Úslava – following a decree issued by the Czech king Wenceslas II. He did so in 1295. From the very beginning, the town was a busy trade center located at the crossroads of two important trade routes. These linked the Czech lands with the German cities of Nuremberg and Regensburg.

In the fourteenth century, Pilsen was the third largest town after Prague and Kutna Hora. It comprised 290 houses on an area of 20 ha. Its population was 3,000 inhabitants. In the sixteenth century, after several fires that damaged the inner center of the town, Italian architects and builders contributed significantly to the changing character of the city. The most renowned among them was Giovanni de Statia. The Holy Roman Emperor, the Czech king Rudolf II, resided in Pilsen twice between 1599 and 1600. It was at the time of the Estates Revolt. He fell in love with the city and even bought two houses neighboring the town hall and had them reconstructed according to his taste.

Later, in 1618, Pilsen was besieged and captured by Count Mansfeld's army. Many Baroque-style buildings dating to the end of the seventeenth century were designed by Jakub Auguston. Sculptures were made by Kristian Widman. The historical heart of the city – almost identical to the original Gothic layout – was declared a protected historic city reserve in 1989.

Pilsen experienced a tremendous growth in the first half of the nineteenth century. The City Brewery was founded in 1842 and the Skoda Works in 1859. With a population of 175,038 inhabitants, Pilsen prides itself on being the seat of the University of West Bohemia and Bishopric.

The historical core of the city of Pilsen is limited by the line of the former town fortification walls. These gave way, in the middle of the nineteenth century, to a green belt of town parks. Entering the grounds of the historical center, you walk through streets that still respect the original Gothic urban layout, i.e., the unique developed chess ground plan.

You will certainly admire the architectonic dominant features of the city. These are mainly the Church of St. Bartholomew, the loftiness of which is accentuated by its slim church spire. The spire was reconstructed into its modern shape after a fire in 1835, when it was hit by a lightning bolt during a night storm.

The placement of the church within the grounds of the city square was also rather unique for its time. The church stands to the right of the city hall. The latter is a Renaissance building decorated with graffiti in 1908–1912. You will certainly also notice the Baroque spire of the Franciscan monastery.

All architecture lovers can also find more hidden jewels, objects appreciated for their artistic and historic value. These are burgher houses built by our ancestors in the styles of the Gothic, Renaissance, or Baroque periods. The architecture of these sights was successfully modeled by the construction whirl of the end of the nineteenth century and the beginning of the twentieth century.

Thanks to the generosity of the Gothic builders, the town of Pilsen was predestined for free architectonic development since its very coming into existence. The town has therefore become an example of a harmonious coexistence of architecture both historical and historicizing.

Table of Contents

Invited Talks

Conference Papers

Corpora of the Russian Language*

Victor Zakharov

Saint-Petersburg State University, Saint-Petersburg, Russia
vz1311@yandex.ru

Abstract. The paper describes corpora of the Russian language and the state of
the art of Russian corpus linguistics. The main attention is paid to the Russian
National Corpus and to specialized corpora.

Keywords: Russian corpus linguistics, corpora, the Russian language, specialized corpora.   -

1 Introduction

1.1 Prehistory of Russian Corpus Linguistics

In recent years creation of different text corpora became one of the cutting edge directions in the applied linguistics. In Western countries the corpus linguistics shaped itself
as a separate linguistic universe in early 90's, even though the concept of the corpus and
the first physical corpora had been known long before.

The earliest Russian corpus was built in 1980s at the University of Uppsala (Sweden). But another project influenced this direction in Russia so much that it should
be mentioned. In 1960–70s the **Frequency Dictionary of Russian** was created by
L.N. Zasorina (a printed version, 1977 [1]). Text database for the dictionary counted
about 1 million tokens. During its compilation a huge number of notorious issues
of corpus linguistics was discussed: representativeness, tokenization, normalization,
lemmatization. So it was the earliest computerized corpus of Russian that doesn't exist
nowadays.

In 1980s the **Computer Fund of the Russian Language** project started. The idea belonged to the academician Andrei Yershov. It was formulated in his paper "On methodology of constructing dialogue systems: the phenomenon of business prose" [2]. The
idea was stated as follows: "Any progress in the field of constructing models and algorithms will remain a purely academic exercise, unless a most important problem of creating a Computer fund of the Russian language is solved. It is to be hoped that creation
of such a Computer fund by linguists, qualified for the task, will precede construction of large systems for application purposes. This would minimize labor costs and
simultaneously would protect the 'tissues' of the Russian language from arbitrary and
incompetent intervention". The Fund was to include the following databases: 1) general lexicon of Russian, 2) databases of various dictionaries, 3) terminology database,
4) information system for the Russian grammar, 5) other subsystems (phonetics, dialectology, diachronic lexis etc), 6) and last but not least, collection of texts, i.e. corpus.
Unfortunately, the bulk of the accumulated results was either abandoned or lost.

---

* Invited talk.

I. Habernal and V. Matousek (Eds.): TSD 2013, LNAI 8082, pp. 1–13, 2013.

1.2 History

The most renowned Russian corpus for many years was the **Uppsala Corpus of Russian Texts**. By now its linguistic material is neither up to date in terms of the volume (one million word occurrences), nor complies with modern conceptions of a national corpus at all. The Uppsala corpus has 600 texts, its volume is 1 million tokens, equally divided between specialized texts and fiction. The aim of the corpus was to represent literary language, and thus the collection doesn't cover spoken language. Full specialized texts from 1985 till 1989 were selected for the corpus and fiction from 1960 through 1988. Texts were presented in Latin alphabet.

The Uppsala corpus belongs to so called **Tübingen Russian corpora** that were created during the project "Linguistische Datenstrukturen. Theoretische und empirische Grundlagen der Grammatikforschung" (SFB 441) of the Tübingen University in 1990-2000s with online access [3].

Russian newspaper corpus was built at the Department of Philology of the Moscow State University in 2000-2002 at the Laboratory of General and Computational Lexicology and Lexicography (1 million tokens in total, online version is limited to 200 thousand tokens). Texts and text items are automatically or semi-automatically marked by various tags: the source, text volume, genre, date of the publication etc. (for texts); grammatical, lexical, morphemic or other categories (for words) [3].

2   Modern Corpora of Russian

The National Corpus of the Russian Language, hereinafter referred to as the Russian National Corpus, is the most popular one among linguists for both being the most well known and the opportunities which it presents. However, being unable to go into a deeper analysis within the framework of this paper, we will zero on its general characteristics together with its most unique features. Also, to show the state of the art in modern Russian corpus linguistics we will touch in greater detail upon other corpora that are not so much known but are worth mentioning.

2.1 The Russian National Corpus

The Russian National Corpus (RNC)[1] includes primarily original prose representing standard Russian but also, albeit in smaller volumes, translated works (parallel with the original texts) and poetry, as well as texts, representing the non-standard forms of modern Russian [4,5]. It was started in 2003 and from April, 2004 is accessible via Internet. The corpus size in total is about 500 million tokens (March 2013).

The corpus allows us to study the variability and volatility of linguistic phenomena frequencies, as well as to obtain reliable results in the following areas: 1) the study of morphological variants of words and their evolution; 2) the study of word-formation options and related issues; 3) the study of changes in syntactic relations; 4) the research of changes in the system of Russian accent; 5) a study of lexical variation, in particular, changes in synonym series and lexical groups, as well as semantic relations in them.

---

[1] http://ruscorpora.ru

Within the main corpus the RNC includes the following subcorpora:

1) **The Main Corpus.** It includes texts representing standard Russian and may be subdivided into 2 parts: modern written texts (from the 1950s to the present day) and early texts (from the middle of the 18th to the middle of the 20th centuries; pre-1918 texts are given in modern orthography). The main corpus counts in total 230 million tokens. The search is carried out in both groups. It is possible to choose one of them and add search parameters on the *Customize your corpus* page.

The part of modern texts is the largest one of the subcorpora. Texts are represented in proportion to their share in real-life usage. For example, the share of fiction does not exceed 40%.

Every text included in the main corpus is subject to metatagging and morphological tagging. Morphological tagging is carried out automatically. In a small part of the main corpus (around 6 mln tokens) grammatical homonyms are disambiguated by hand, and results of automated morphological analysis are corrected. This part is the model morphological corpus and serves as a testing ground for various search algorithms and programs of morphological analysis and automated processing. Disambiguated texts are automatically supplied with indicators of stress. Stress annotation may be turned off for printing or saving the search results.

2) **The Corpus of Spoken Russian.** It represents real-life Russian speech and includes the recordings of public and spontaneous spoken Russian and the transcripts of the Russian movies. To record the spoken specimens the standard spelling was used. The corpus contains the patterns of different genres/types and of different geographic origins. The corpus covers the time frame from 1930 to 2007.

3) **Deeply Annotated Corpus (treebank).** This corpus contains texts augmented with morphosyntactic annotation. Besides the morphological information, every sentence has its syntax structure (disambiguated). The corpus uses dependency trees as its annotation formalism (Fig. 1).

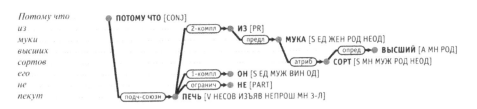

Fig. 1. Dependency tree of a Russian sentence

Nodes in such a tree are words of the sentence, while its edges are labeled with names of syntax relationships. The repertory of relationships for the corpus, as well as other specific linguistic decisions has been developed in the Laboratory for Computational Linguistics, Institute for Information Transmission, Russian Academy of Sciences that compiled the corpus.

4) **Parallel Text Corpus.** The site contains the parallel text corpora for the following languages: English, German, French, Spanish, Italian, Polish, Ukrainian, Belorussian, and multilingual.

5) Dialectal Corpus. The dialectal corpus contains recordings of dialectal speech (presented in loosely standardized orthography) from different regions of Russia. The corpus employs special tags for specifically dialectal morphological features (including those absent in standard language); moreover, purely dialectal lexemes are supplied with commentary.

6) Poetry Corpus. At the moment the poetry corpus covers the time frame between 1750 and 1890s, but also includes some poets of the 20th century. Apart from the usual morphological tagging, there is a number of tags adapted for poetry.

7) The poetic part of the **Accentological Subcorpus** includes Russian poetry of 18–21 centuries with the marked arses (potentially stressed syllables). It gives a user the possibility to figure out the real stress of a word-form according to the simple set of rules. For example, it is possible to search for texts written in various poetic meters.

8) Educational Corpus. The educational corpus is a small disambiguated corpus adapted for the Russian school educational program.

9) Newspaper Corpus. It covers articles from the media of the 2000s.

10) Multimodal/Multimedia Corpus (see below).

2.2   Semantic Annotation in the Russian National Corpus

The main corpus of the RNC contains semantic annotation, too [6]. Semantic annotation in the main corpus is a unique feature of RNC that makes it distinct from other national corpora.

Semantic and derivational parameters involved are *person*, *substance*, *space*, *movement*, *diminutive*, etc. There are three groups of tags assigned to words to reflect lexical and semantic information: class, lexical and semantic features, derivational features. The set of semantic and lexical parameters is different for different parts of speech. Moreover, nouns are divided into three subclasses (concrete nouns, abstract nouns, and proper names), each with its own hierarchy of tags.

Lexical and semantic tags are grouped as follows: taxonomy, mereology, topology, causation, auxiliary status, evaluation. A word in the semantic dictionary is assigned a set of characteristics along many other parameters.

The meta-language of tags is based on English notation; it is, however, possible to make a search using traditional Russian category names in the search "semantic features" form. The following are some tags from an inventory of available tags with examples in parenthesis.

Nouns: categories: r:concr – concrete nouns, r:abstr – abstract nouns, r:propn – proper names. Some tags for concrete nouns:

Taxonomy: t:hum – person (*человек* (human), *учитель* (teacher)), t:hum:etn – ethnonyms (*эфиоп* (Ethiopian), *итальянка* (Italian)), t:hum:kin – kinship terms (*брат* (brother), *бабушка* (grandmother)), t:animal – animals (*корова* (cow), *сорока* (magpie)), etc.

Mereology: pt:part – parts (*верхушка* (top)), pt:part& pc:plant – parts of plants (*ветка* (limb), *корень* (root)), pt:part& pc:constr – parts of buildings and constructions (*комната* (room), *дверь* (door)), etc.

Topology: top:contain – containers (*комната* (room), *озеро* (lake)), top:horiz – horizontal surfaces (*пол* (floor), *площадка* (ground, area)), etc.

Evaluation: ev – evaluation (neither positive nor negative) (*озорник* (mischief-maker)), ev:posit – positive evaluation (*умница* (clever man or woman)), ev:neg – negative evaluation (*негодяй* (scoundrel))).

Some tags for verbs:

t:move – movement (*бежать* (run), *бросить* (throw))

t:put – placement (*положить* (put), *спрятать* (hide))

t:impact – physical impact (*бить* (beat), *колоть* (prick))

t:be:exist – existence (*жить* (live), *происходить* (happen))

t:be:appear – start of existence (*возникнуть* (arise), *создать* (create))

t:be:disapp – end of existence (*убить* (kill), *улетучиться* (disappear))

t:loc – location (*лежать* (lie), *стоять* (stand)).

And these are just a few examples of 200 tags available in the corpus that are structured in a hierarchical way.

2.3   Search in the Russian National Corpus

For search RNC uses the search engine Yandex-Server which has been specially adapted for corpus needs. One can search by an exact form, by a set phrase, by lexico-grammatical and semantic features, by additional features such as a specified position (before or after punctuation marks, in the beginning or in the end of a sentence, capitalization, etc.).

Words we are searching for could be combined with logical operators «AND», «OR» and «NOT». For compound searches parenthesis are used. For example, the query S & (nom|acc) yields nouns in nominative or accusative. It can be used with both left or right truncation. Distance between words could be set from minimum to maximum. The distance between words next to each other is 1 word; the distance of 0 is interpreted as concurrence of word-forms. For lexico-grammatical search, we can input a sequence of lexemes and/or word-forms with certain grammatical and/or semantic features. We can combine them in any way.

A simpler way to search for certain grammatical features is to use a selection window. The selection window contains a list of appropriate features, subdivided by categories: i.e., part of speech, case, gender, voice, number for morphology, etc. To invert selection within a category, one uses the equivalent of the "NO" operator.

The **semantic features** field allows for listing the semantic and derivational features of the lexeme. As a rule, semantic features have a hierarchy. In semantic search we must remember that words tagged as belonging to a category may often not belong to a subcategory: for example, verbs belonging to the "physical impact" category include verbs not belonging to the subcategories "creation" and "destruction", such as verbs of processing like *вымыть* (wash). There exists a capability to uncheck the boxes next to all subcategories.

By default all the tagged meanings of a given word are searchable. For instance, the parameter Human qualities selected in the Semantic features field will yield both *умный* 'intelligent', *верный* 'faithful', *коварный* 'perfidious' (where the parameter is present in its basic meaning), as well as *мягкий* 'soft' or *холодный* 'cold' that apply to human beings only metaphorically.

To refine the scope of the search, we could select one or two parameters:

"sem" — only the first meaning given in dictionaries is searched (thus human qualities will yield words like 'intelligent', 'faithful' or 'perfidious', but not those like 'soft' or 'cold'); "sem2" — the meanings other than the first ones are searched (thus only words like 'soft' or 'cold' will be found).

The search can be limited to a subcorpus which is chosen as one of the above mentioned subcorpora or as a combination of metadata features.

The types of annotation which are specific for the special corpora (MURCO, poetical, dialectological, etc.) define the peculiarities of the appropriate interface in comparison with the interface of the RNC proper.

Search results can be presented twofold: a horizontal text (a broader context) and a concordance (Fig 2). In both cases grammatical and semantic features of any word can be checked out (Fig 2 shows that for the word *женщина* (women)).

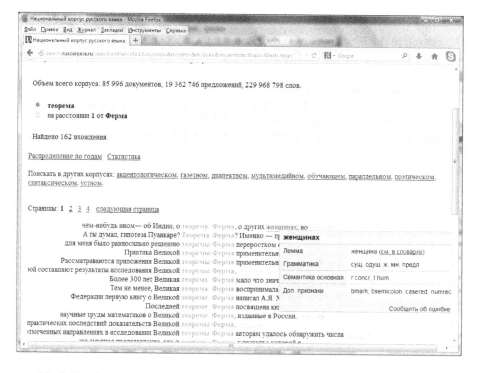

Fig. 2. Search results for the word combination *теорема Ферма* (Fermat's theorem)

From the search page one can get to *Графики* service (Charts) (*Распределение по годам* link (chronological distribution)).

2.4   Charts

In terms of functionality Charts service of RNC is similar to Google Books Ngram Viewer. It shows chronological distribution of lexical units (text forms, phrases), found in the main corpus of RNC. You can get to this link from the search results page of RNC, Fig 2, as well as from the main menu, Fig 3.

You can set time limits too, e.g. from 1930 through 1960. Clicking the button *По-строить* (Draw), we will get a chart, Fig 4, where each object we compare is shown in its own color with legend located in the top right corner (here *Черчилль* (Churchill), *Рузвельт* (Roosevelt), and *Франко* (Franco) are shown in comparison).

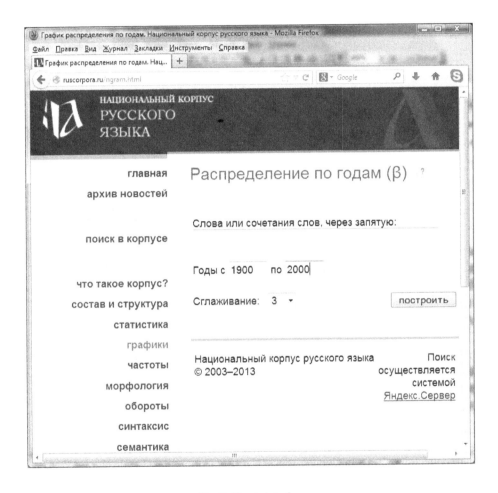

Fig. 3. Charts interface

The vertical axis is for relative usage frequency of a lexical unit. Mouse over any point on the curve, and you would see the relative usage frequency (in ipm), for a respective year. Smoothing the charts allows the general trend to be seen beyond random frequency volatility. Thus, with the 10 years smoothing the word frequency is averaged over 5 prior and 5 consecutive years, i.e. the average for 11 years is taken for any given year.

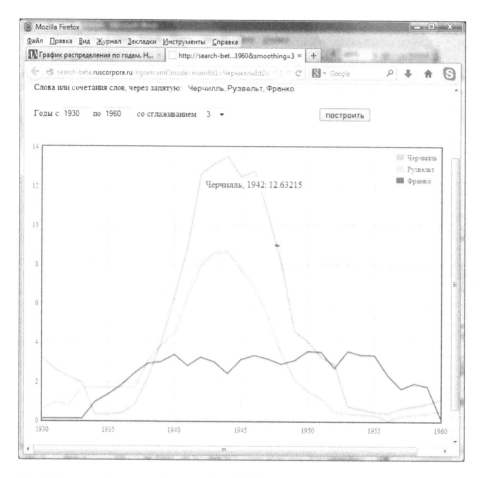

Fig. 4. Occurrence frequency of names Churchill, Roosevelt and Franco in the texts published from 1930 through 1960

There is a possibility to display tables with absolute frequency of occurrence for any year. Links from the tables allow to show examples from the corpus for each year.

From Charts one can easily jump to Google Ngram Viewer working on a Russian language collection of texts in Google Books. National Corpus and Google Ngram Viewer while having similar ideology, use different formulae to calculate relative frequencies.

2.5   Other Text Corpora of Russian

Helsinki Annotated Corpus (HANCO). The HANCO Corpus project has been running since 2001 in the Department of Slavonic and Baltic Languages and Literatures at the University of Helsinki. The corpus was envisaged to include morphological, syntactic and functional (semantic) information about approximately 100000 running words, extracted from a modern Russian magazines and representing the contemporary Russian language.

The main principles of creation are as follows [7]:

1. Targeting a wider audience. Potential users are not only a narrow circle of experts, but also students and teachers of Russian.
2. Focus on the accuracy of the grammatical description.
3. Multilevel grammatical information combined in the process of searching.
4. Possibility of alternative interpretations of linguistic facts.

For now the following types of linguistic information are to be included in the HANCO.

- Morphological information. In the HANCO, complete morphological description of every running word is given. The morphological analysis and the subsequent disambiguation procedure have been carried out automatically with further manual processing.
- Syntactic information. Syntactic information is given at three levels: word collocations, clauses, and sentences. The full description of units for every level is given according to the Academy Grammar of Russian.

Leeds University Corpora. In the 2000s a large number of corpora for different languages (including a Russian one) were created at the Leeds University by S. Sharoff.[2] Among them there is a version of the Russian National Corpus amounting to 116 mln words. The new frequency dictionary of Russian was created on its base. The interface to Russian corpora is supported by CQP IMS Corpus Workbench. It uses a special query language including Regular Expression language and allows for lexical and grammatical search. There are options to set the output interface. It is also possible to receive collocation lists calculated on the base of association measures such as MI, T-score, Log-likelihood. On this site[3] there is a collection of various software tools to process text corpus data.

Moshkov's Library Corpus. There is a big corpus of Russian texts, 680 mln words, on the site of NLP group created by A. Sokirko.[4] Using a powerful query language of the DDC corpus manager, it allows to search for lexical units taking into account parts of speech and morphological characteristics. At the same site there is a search service of bigrams (54 mln), calculated from a corpus by MI measure.

Sketch Engine Corpora. The English linguistic company Lexical Computing Ltd. (A. Kilgarriff) provides on a commercial basis an access to more than 140 corpora of various languages. Among them there is a few corpora of Russian, and primarily, the ruTenTen corpus created from Internet texts with the Wacky technology, totaling about 20 billion word usage. English researchers together with the Czech developers from the Masaryk University in Brno developed a special corpus manager, the Sketch Engine[5] [8]. The manager possesses many unique opportunities. Besides standard search with concordance output it issues lists of collocations based on individual syntactic models

---

[2] http://corpus.leeds.ac.uk/ruscorpora.html

[3] http://corpus.leeds.ac.uk/tools

[4] http://aot.ru/search1.html

[5] http://sketchengine.co.uk

(word sketches), forms a word frequency list, groups lexical units into lexico-semantic fields with internal clustering, and shows the strength of syntagmatic relations between lexemes.

2.6 Speech Corpora of Russian

Oral speech, and especially, the nonpublic oral improvised speech, according to many scientists, is the most important version of language, the closest to its "kernel", and showing the most characteristic models of language. Therefore it is important to dwell on Russian speech corpora.

The Corpus of Spoken Russian is the collections of transcripts of the spoken texts of different types (private speech, public speech, and movie speech-tracks). Its volume just now is around 10,5 million tokens. These transcripts are annotated morphologically and semantically by the RNC annotation system. In addition, the corpus has its own annotation: the accentological and the sociological one. The accentological annotation presupposes that in every word-form the real rather than normative stress is marked. Therefore, a user can investigate the history of Russian accentological system and its normative requirements, which are specific for any particular period. The sociological annotation means that to every text the information on the sex, the age and the name of a speaker is assigned, so a user can form his own subcorpora according to all these parameters and their combinations.

The Spoken corpus of the RNC gives a user various possibilities, but all these tasks must not be connected or based on the real phonation. Therefore the **Multimodal Russian corpus (MURCO)** was formed as a part of the RNC [9]. Its material are fragments of movies of the 1930s through the 2000s. The main principle of the MURCO is the alignment of the text transcripts with the parallel sound and video tracks. Consequently, when a user makes his data query he may obtain not only a written text, annotated from different points of view, but also the corresponding sound and video material. This possibility let a researcher use the obtained information at his will. He utilize his own manner of phonetic transcription or speech and intonation analyzers of his own choice; he may pose and solve all types of research tasks connected with phonetics, etc. The total volume of the movie transcripts in the RNC is around 3,5 million tokens.

The types of annotation in the MURCO are as follows:

- orthoepic annotation: combinations of sounds are marked;
- annotation of accentological structure: the word structure in regards to the stress position is defined;
- speech act annotation: the types of speech acts and vocal gestures, used in a clip are described;
- gesture annotation: the type of gesticulation in a clip is described.

Another speech corpus worth dwelling upon is the **ORD corpus** developed at Institute of Philological Researches of the St. Petersburg State University [10].The abbreviation ORD stems from Russian *Odin Rechevoj Den'*, literally translated as *one day of speech*. The main aim of creating the ORD corpus is to collect recordings of actual speech which we use in our everyday communication. For the first series of recordings a demographically balanced group of 30 persons representing various social and age strata in the population of St. Petersburg was selected. These individuals spent one day with recorders dangling around their necks and recording all their communications. So recorders were on while having breakfast at home with the family, then while preparing to go to work, on the way to work, speaking on the cell phone, then business and informal conversations at work with colleagues, lunch time, shopping, recreation, etc. In the result more than 240 hours of recording were obtained with 170 hours containing speech data quite suitable for further linguistic analysis, and more than 50 hours of recordings good enough for further phonetic analysis. The corpus was divided into 2202 communication episodes. 134 episodes are already transcribed in detail. At present, orthographic transcription of the corpus numbers more than 50000 word-forms. The corpus presents the unique linguistic material, allowing to perform fundamental research in many aspects including complex behaviour of people in real world. These utterly natural recordings are to be used for practical purposes: to verify scientific hypotheses, to make adjustments to improve systems of speech synthesis and speech recognition, etc.

2.7 Special Corpora

A special corpus is a balanced corpus, of smaller size, as a rule, meant specifically for certain research tasks helping to resolve corresponding problems of user choice.

SPbEFL LC, a Learner English corpus (English as a foreign language), started at Herzen University (Saint Petersburg, Russia) in 2009 is a multi-mother tongue (Russian, Chinese, Japanese, Korean, Thai, and Vietnamese) corpus that compiles written texts (essays and personal letters), monologues, and dialogues (in scripts). The contributors' pre-tested language proficiency is intermediate (26%) and advanced (74%). The language/text relevant criteria include medium, genre, topic, technicality and task setting. SPbEF LC is an attempt to compile a target-specific structure, a text collection in accord with essential corpus design criteria. Operated with reliable free tools, the corpus proves efficient enough in spotting and analyzing the learner language with reference to syntactically parsed texts, concordance, frequency and collocation lists.

The corpus is aimed at interlanguage studies based on the assumption that both the vocabulary and the sentence patterns presumably reflect the actual language fund that the learners subconsciously resort to in case of FL communication. SPbEFL corpus findings pinpoint "atypical" mistakes in learner interpretation and use of basic structures that address issues of both learner universals and new language learning and teaching materials [11].

A new multimodal corpus of learners' spontaneous dialogues made according to a short outline is under construction in Irkutsk State Linguistic university (Russia). It is called **UMCO** *(Uchebnyj Multimodalnyj Corpus (Learner multimodal corpus))* and belongs to a category of learner's corpora. The dialogues are made up by the students of Chinese, Russian and German. Now UMCO consists of 25 video clips lasting from 1,5 to 3 minutes. ELAN is used as a corpus manager, for its multiple advantages, including the possibility of typing Chinese characters.

The corpus contains a number of parallel subcorpora where small thematic blocks of native speakers dialogues are aligned with those produced by the learners of the same language.

Among other specialized corpora worth mentioning are **Regensburg diachronic corpus of Russian** (texts in Old Russian), **Corpus of Old Russia manuscripts** (birchbark letters), a parallel Corpus for translations of **The Tale of Igor's Campaign**, a Corpus of electronic Russian Heritage **Manuscript**, a historic **St. Petersburg Corpus of Hagiographic Texts** of XV–XVII centuries (SKAT), etc.

The demand for corpora of specialized texts can be comparable with that for national ones. Any specialized branch corpus gives a specialist the most important material: professional terms in their typical context thus providing means to monitor terminology evolution including the birth of new terms.

3   Conclusion: Corpus Oriented Researches

At the moment all corpora of the Russian language and mostly the RNC are used by both Russian and foreign researchers. The RNC has English interface and the help system in English. Its subcorpora with their special annotation provide various possibilities for linguistic studies. The RNC site has a special division called *Studiorum*. It includes some data of researches in Russian language.

The studies based upon the semantic annotation are of special interest. There are a few works which address word sense disambiguation and lexical constructions – the chains of lexical units, one of which is usually a lexical constant and others are variables [12]. The basic results obtained in the experiments have to do with revealing and classifying of different types of context markers to specify different meanings of target words. The type and degree of specification of the RNC semantic annotation could provide the rules for associating context tags of special semantic classes with different meanings.

References

1. Zasorina, L.N. (ed.): Chastotnyi slovar' russkogo yazyka. Moskva (1977)
2. Yershov, A.P.: K metodologii postroeniya dialogovykh sistem. Fenomen delovoi prozy. Novosibirsk (1979)
3. Reznikova, T.I.: Slavyanskaya korpusnaya lingvistika. In: Plungyan, V.A. (ed.) Natsionalnyi Korpus Russkogo Yazyka: 2006–2008, Saint-Petersburg, pp. 404–465 (2009)
4. Natsionalnyi korpus russkogo yazyka: 2003–2005, Moskva (2005)
5. Natsionalnyi korpus russkogo yazyka: 2006–2008, Saint-Petersburg (2009)
6. Lashevskaja, O.N., Shemanaeva, O.J.: Semantic Annotation Layer in Russian National Corpus: Lexical Classes of Nouns and Adjectives. In: Proceedings of the Sixth International Language Resources and Evaluation (LREC 2008), Marrakech, Morocco, pp. 3355–3358 (2008)
7. Kopotev, M., Mustajoki, A.: Printsipy sozdaniya Helsingskogo annotirovannogo korpusa russkikh tekstov (HANCO) v seti Internet (Principles of the Creation of the Helsinki Annotated Corpus HANCO). Nauchno-tekhnicheskaya Informatsiya 2(6), 33–37 (2003)

8. Kilgarriff, A., Rychlý, P., Smrž, P., Tugwell, D.: The Sketch Engine. In: Proceedings of the XIth Euralex International Congress, pp. 105–116. Universite de Bretagne-Sud., Lorient (2004)

9. Grishina, E.: Multimodal Russian Corpus (MURCO): General Structure and User Interface. In: Levická, J., Garabík, R. (eds.) Slovko 2009. NLP, Corpus Linguistics, Corpus Based Grammar Research, Bratislava, Slovakia, pp. 119–131 (2009)

10. Sherstinova, T.: The structure of the ORD speech corpus of Russian everyday communication. In: Matoušek, V., Mautner, P. (eds.) TSD 2009. LNCS, vol. 5729, pp. 258–265. Springer, Heidelberg (2009)

11. Kamshilova, O.: Learner Language analysis in SPbEFL Learner Corpus. In: Learner Language, Learner Corpora. LLLC 2012 Conference, October 5-6. The University of Oulu (2012), http://www.oulu.fi/hutk/sutvi/oppijankieli/LLLC/LLLC2012_abstracts.pdf

12. Lashevskaja, O., Mitrofanova, O.: Disambiguation of Taxonomy Markers in Context: Russian Nouns. In: Jokinen, K., Bick, E. (eds.) 17th Nordic Conference on Computational Linguistics (NODALIDA 2009), Odense, Denmark. NEALT Proceedings Series 2009, vol. 4, pp. 111–117 (2009)

Long, Deep and Wide Artificial Neural Nets for Dealing with Unexpected Noise in Machine Recognition of Speech⋆

Hynek Hermansky⋆⋆

Center for Language and Speech Processing
The Johns Hopkins University, Baltimore, Maryland, USA
hynek@jhu.edu

Abstract. Most emphasis in current deep learning artificial neural network based automatic recognition of speech is put on deep net architectures with multiple sequential levels of processing. . The current work argues that benefits can be also seen in expanding the nets longer in temporal direction, and wider into multiple parallel processing streams.

Keywords: artificial neural networks, machine recognition of speech, robustness to noise, unexpected distortions, parallel processing.

1 How Is Speech Recognized

1.1 Automatic Recognition of Speech

In current stochastic automatic recognition of speech (ASR), training data (both acoustic speech data and text) are used to build a set of models of a spoken language. The recognition task consists of finding a model from this set, which represents a string of words $\hat{W}$ that satisfies

$$\hat{W} = argmax_W\{p(X|W)P(W)\}, \tag{1}$$

where X represents measurements describing the speech signal, $p(X|W)$ stands for a likelihood of W given the data X and $P(W)$ represents a prior probability of the word sequence W.

Current state-of-the-art stochastic ASR systems often estimate the likelihood $p(X|W)$ by a discriminatively-trained multi-layer perceptron artificial neural network (MLP) [1]. The recent reincarnations of this technique, illustrated in the upper part of Fig. 1 introduce various alternations of the basic MLP architecture, mostly aiming for

---

⋆ Invited talk.

⋆⋆ This work was supported in parts by the DARPA RATS project D10PC0015, IARPA BABEL project W911NF12-C-0013, and by the Johns Hopkins Center of Excellence in Human Language Technologies. Any opinions, findings and conclusions or recommendations expressed in this material are those of the author and do not necessarily reflect the views of the DARPA, IARPA or JHU HLTCOE.

multiple levels of processing with research focusing on particular net configurations and on proper training [2,3].

While the resulting speech sound likelihood estimates are demonstrated to be better that the earlier used likelihoods derived by generative Gaussian Mixture Models, unexpected signal distortions that were not seen in the training data can still make the acoustic likelihoods unacceptably low. A step towards addressing the unreliable acoustic evidence might be in expanding the net architectures not only into deeper but also into longer and wider structures, where substantial temporal context attempts to cover whole coarticulation patterns of speech sounds, and multiple processing paths, attending to multiple parts of information-carrying space, attempt to capitalize on redundancies of coding of information in speech [4], possibly allowing for adaptive alleviation of corrupted processing streams. This approach, illustrated in the lowr part of the Fig. 1 has been pursued for some time [5,6,7,8,5,9], and has some support in known properties of human cognition [10].

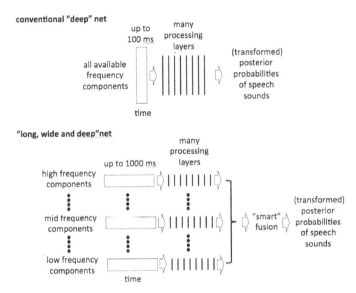

Fig. 1. Deep net, shown in the upper part of the Figure, typically imputs a single phoneme-length time-frequency pattern of speech, and in multiple sequential processing steps derives posterior probabilities of speech sounds of interest. The long, wide and deep net consists in its first processing stage of a number of parallel nets, each inputting relatively long time-frequency pattern, covering only a part of the available frequency space, derives posterior probabilities of speech sounds in each of its first-stage processing streams, and fuses these estimates to deliver the final posterior estimates, while allowing for alleviation of possibly corrupted first-stage estimates.

1.2 Human Recognition of Speech

The approach mentioned above is motivated by [11] which suggest that human speech recognition is carried out in individual frequency bands and the final error in recognition is given by

$$P(\varepsilon) = \prod P(\varepsilon_i), \tag{2}$$

is given by a product of probabilities of errors $P(\varepsilon_i)$ in the individual frequency streams.

The "product of errors" rule (Eq. 2) are satisfied for conditionally independent errors. In speech recognition this implies that conditionally independent speech processing channels exist in human speech communication process. That requires that speech message is coded and decoded in parallel redundant channels, and that human listener can correctly identify if the given processing channel yields the correct information or is in error.

To the extent that we could make such an assumption, we could explain why the "product of errors" rule holds in speech perception and, subsequently, why speech communication is relatively robust to the presence of many local signal distortions. Emulating this ability in ASR would be desirable.

2   Multi-stream ASR

One engineering technique that represents a step towards such a parallel processing system is multi-stream ASR [12,13,14,15].

The techniques was spurred by insightful interpretations of earlier works [16,17] The fundamental motivation behind multi-stream ASR is that when message cues are conflicting or corrupted in some processing streams, such a situation can be identified and a corrective action can focus on the more reliable streams that still provide enough cues to facilitate the recognition.

To build a multi-stream system, one needs to deal with

– ways of forming the appropriate processing streams, and
– optimal fusion of the information from the individual streams.

Forming the streams is by no means the solved problem. Each stream should should carry some part of desired but complementary information, and should be formed so that noises and distortions affect only some of them. Our current systems typically use relatively long temporal segments of spectral energies of band-limited signals as information carrying features [18,19].

3   Performance Monitoring Approach to Fusion of Information

While forming the streams and describing the information carried in each stream is not a trivial task, it is the fusion of information from the individual processing streams that presents the most tempting challenge. One engineering approach is to form processing streams as all possible combinations of sub-streams, each combination fused by another MLP, and to pick up at every instance the best processing stream based on outputs from all streams. The whole scheme is depicted in Fig. 2. To my knowledge, that was first introduced in [7,8].

The obvious question is how to find the "best" stream combination. [7,8] discussed several different measures of confidence in classification in different processing streams,

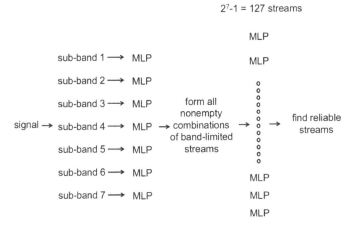

Fig. 2. Multi-stream processing [7]. In the first processing stage, 7 band-limited stream are formed and posterior probabilities of phonemes are derived in each band-limited stream. In the second processing stage, all 127 nonempty combinations of the 7 band-limited streams are formed and posterior probabilities are derived from the concatenated outputs of the combined band-limited streams.

aiming for leaving out the corrupted streams. In the current work we attempt for more principal approach, which is more in line with the notion that human listeners appear to "know when they know", and are thus able to pick up the processing streams, which yield correct answers.

Human listeners posses wast knowledge from past experience what speech should sound like. Machine "knowledge" comes only from its training data. Thus, when evaluating to what extent the machine "knows", it makes sense to evaluate its performance with respect to its performance on its training data.

The fundamental premise of the proposed approach is that the *ASR system can never work better than it does on the data on which it was trained.* In this ideal situation, the data that the system encounters during its operation come from the same distribution as do the data on which the system was trained and the system is at its best.

– First, let's postulate that the classifier performance is optimal for the data on which the classifier was trained. Then, any deviation from this ideal condition only degrades the performance.
– Further, let's propose that, even when the result is not known in advance, the classifier performance can be characterized by observing its output for a sufficient interval of time. The well trained classifier exposed to its training data should yield results that reflect statistical properties of a language (some phonemes occur much more often than others [20]), yield "reasonable" confusions (confusions within phoneme class are more likely than confusions across classes. [21]), yield properly spaced phoneme classes (after certain time span the classifier should yield different phoneme class), etc. Such properties could be represented in statistic or in parametric model of some elements of the classifier output.

- Next, comparing statistics or evaluating likelihood of a model derived from the classifier's output on the "good" training data and on any "corrupted" data.
- Finally, attempts can be made to modify the system in a way that decreases the difference between observed classifier performance on its training data and in the test. The assumption is then that the system performance also improves.

Notice that for evaluating any atypical response, there is no need for knowing what the result of the classification should be. All that is required is to observe the behavior of the machine output on its training data and on the given test data.

More recent works include [22] who forms four processing streams by compounding emulated cortical-like receptive fields that attend to different parts of the spectro-temporal acoustic space. To describe the classifier output, [22] an autocorrelation matrix, derived from the appropriately post-processed output of a neural net based classifier (posteriogram). Diagonal elements of the is matrix reflect estimated frequencies of occurrence of each phoneme and off-diagonal elements reflect co-activations of different phoneme posterior estimates. The matrix does not tell anything about correctness or incorrectness of the estimates. However, the second-order off-diagonal terms reflect possible confusions among estimates. This statistic does not use any temporal information and is based solely on instantaneous outputs from the classifier. Any additional distortion of the signal results in a change of the statistic that this matrix describes. Thus, computing a measure of similarity between the autocorrelation matrices derived from the clean signal and from the corrupted signal indicates degradation of the stream due to the distortion. More current work [23] reports on more extensive evaluations and on some alternative evaluation criteria.

Subsequent work [24] replaces search for the best processing streams by a full feedback-control system, where the fusion is adaptively modified using a particle filtering technique. Appropriate initial weights for a weighted summation of individual estimates are found by cross-correlating phoneme probability estimates with the phoneme labels on the training (cross-validation) data. The autocorrelation matrix of the final weighted probability estimates is adopted as the measure that summarizes the system performance during the operation. The particle filtering procedure changes the fusion weight of each phoneme in each stream so that the statistics of the output are more similar to the statistics derived on the training data.

The most recent work re-visits the original multi-stream concept of Sharma et al [7,8], which first creates M band-limited streams, each yielding a vector of posterior probabilities of phonemes. In a subsequent step, all nonempty combinations of these band-limited streams are formed, yielding $2^M - 1$ processing streams. Performance monitoring module selects the N-best performing streams, which are used to yield the vector of final posterior probabilities. Training of the whole system is done in stages, the firts layer of M three-layer MPLs is trained to estimate posterior probabilities of phonemes from temporal trajectories of Hilbert envelopes in frequency limited bands (each about 3 critical bands wide). All combinations of these posterior estimates from the band-limited streams are used in the second stage of the processing where $2^M - 1$ three-layer MLP are trained to again estimate posteriors of phonemes, thus resulting in $2^M - 1$ processing streams, each of which covering different set of frequencies of

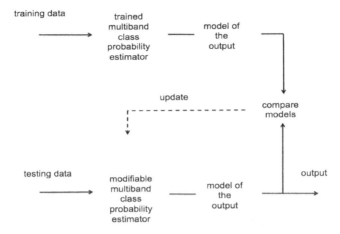

Fig. 3. Emulating human-like performance monitoring in ASR system. Deteriorated output from the classifier is identified by a large divergence between the statistic that was learned on the training data and the statistic derived on test data during the operation. Modification of the classifier is currently done by selecting the most reliable streams for final posterior fusion.

the original signal. Performance monitoring described below is then applied to select N most reliable streams for further processing.

The most successful performance monitoring technique to date is based on first order statistics of temporal elements of classifier output, computed from mean divergence between posterior probability vectors spaced Δt apart

$$M(\Delta t) = \frac{\sum_{t=0}^{T-\Delta t} D(P_t, P_{t+\Delta t})}{T - \Delta t} \tag{3}$$

where D is divergence between two feature vectors P_t and $P_{t+\Delta t}$, and T indicates time interval over which is the function evaluated. Comparison of statistics derived on training data and in test yielded 10 best processing streams that were selected for each utterance and their posterior vectors were averaged for the subsequent Viterbi search for the best phoneme sequence. When compared to the baseline single-stream conventional system, the multi streams system yielded 9% relative improvement in error rate in clean conditions and as much as 30% improvement in car noise at 0 dB signal-to-noise ratio. These automatically obtained results compare well with the best possible "cheating" results where the top reliable stream was selected by a "human-in-the-loop" who knew the correct answers in advance, and yielded 19 % improvement in clean conditions and 34% relative improvement in noise. More details can be found in [25,26].

4 Conclusions

The paper suggests that MLP-based estimation of posterior probabilities of speech sounds should be done from relatively long segments of speech signal, and in many parallel interacting streams, resulting on MLP architectures that are not only deep but also long and wide. The streams should describe the speech signal in different ways, capitalizing on the redundant way the message is coded in the signal. Given the constantly changing acoustic environment, the choice of the best streams for the final decision about the message should be done adaptively.

Acknowledgments. The multi-stream framework was developed with S. Sharma, P. Jain and M. Pavel. The performance monitoring based information fusion was proposed with N. Mesgarani, and further developed with S. Thomas, E. Variani, F. Li, V. Peddinti and T. Ogawa.

References

1. Bourlard, H., Wellekens, C.J.: Links between markov models and multilayer perceptrons. IEEE Trans. Patt. Anal. and Machine Intell. 12(12), 1167–1178 (1990)
2. Seide, F., Li, G., Yu, D.: Conversational speech transcription using context-dependent deep neural networks. In: Proc. Interspeech, pp. 437–412 (2011)
3. Morgan, N.: Deep and wide: Multiple layers in automatic speech recognition. IEEE Transactions on Audio, Speech, and Language Processing 20(1), 7–13 (2012)
4. Miller, G.: Language and Communication. McGraw-Hill Book Company (1951)
5. Hermansky, H., Sharma, S.: TRAPS, classifiers of temporal patterns. In: Proc. Int. Conf. Spoken Language Processing. I. S. C. Association, Syndey (1998)
6. Tibrewala, S., Hermansky, H.: Multi-stream approach in acoustic modeling. In: Proc. DARPA Large Vocabulary Continuous Speech Recognition Hub 5 Workshop, pp. 1255–1258 (1997)
7. Tibrewala, S., Hermansky, H.: Sub-band based recognition of noisy speech. In: Proc. Int. Conf. Spoken Language Processing. International Speech Communication Association (1997)
8. Sharma, S.: Multi-stream approach to robust speech recognition. Ph.D. dissertation, Oregon Graduate Institute of Science and Technology, Portland, Oregon (1999)
9. Jain, P., Hermansky, H.: Beyond a single critical-band in TRAP based ASR. In: Proc. Eurospeech, pp. 437–440 (2003)
10. Hermansky, H.: Multistream recognition of speech: Dealing with unknown unknowns (invited paper). Proceedings of Institute of Electriocal and Electronics Engineers 101(5), 1076–1088 (2013)
11. Fletcher, H.: Speech and Hearing in Communication. Krieger, New York (1953)
12. Duchnowski, P.: A new structure for automatic speech recognition. Ph.D. dissertation, Massachusetts Instittute of Technology, Cambridge, MA (1992)
13. Bourlard, H., Dupont, S., Hermansky, H., Morgan, N.: Towards subband-based speech recognition. In: Proc. EUSIPCO 1996, pp. 1579–1582 (1996)
14. Hermansky, H., Tibrewala, S., Pavel, M.: Towards ASR on partially corrupted speech. In: Proc. Int. Conf. Spoken Language Processing, pp. 462–465 (1996)
15. Bourlard, H., Dupont, S.: A new ASR approach based on independent processing and recombination of partial frequency bands. In: Proc. Int. Conf. Spoken Language Processing, pp. 426–429 (1996)
16. Allen, J.B.: Personal communicaton. DoD Summer Workshop at Rutgers University (1993)
17. Allen, J.B.: Articulation and Intelligibility. Morgan & Claypool (2005)
18. Hermansky, H.: History of modulation spectrum in ASR. In: Proc. IEEE Int. Conf. Acoust., Speech and Signal Processing, pp. 5458–5461 (2010)

19. Hermansky, H.: Speech recognition from spectral dynamics (invited paper). Sādhanā, Indian Academy of Sciences 36(5), 729–744 (2011)
20. Dewey, E.: Relative Frequency of English Speech Sounds. Harvard University Press, Cambridge (1923)
21. Miller, G.A., Nicely, P.: An analysis of perceptual confusions among some english consonants. J. Acoust. Soc. Amer. 27(2), 338–352 (1955)
22. Mesgarani, N., Thomas, S., Hermansky, H.: Towards optimizing stream fusion. Express Letters of the Acoustical Society of America 139(1), 14–18 (2011)
23. Variani, E., Hermansky, H.: Estimating classifier performance in unknown noise. To appear in Proc. Interspeech (2012)
24. Mesgarani, N., Thomas, S., Hermansky, H.: Adaptive stream fusion in multistream recognition of speech. In: Proc. Interspeech, pp. 2329–2332 (2011)
25. Hermansky, H., Variani, E., Peddinti, V.: Mean temporal distance: Predicting asr error from temporal properties of speech signal. JHU Center for Language and Speech Processing. Technical Report (December 2012)
26. Variani, E., Peng, L., Hermansky, H.: Multi-stream recogntion of noisy speech with performance monitoring. In: Proceedings Interspeech (2013)

Multilingual Media Monitoring and Text Analysis – Challenges for Highly Inflected Languages*

Ralf Steinberger[1], Maud Ehrmann[2], Júlia Pajzs[3], Mohamed Ebrahim[4],
Josef Steinberger[5], and Marco Turchi[6]

[1] European Commission – Joint Research Centre, IPSC-GlobeSec, Ispra (VA), Italy
Ralf.Steinberger@jrc.ec.europa.eu
[2] Sapienza University of Rome, Department of Computer Science, Rome, Italy
ehrmann@di.uniroma1.it
[3] Hungarian Academy of Sciences, Research Institute for Linguistics, Budapest, Hungary
pajzs.julia@nytud.mta.hu
[4] Cognizant SetCon, Munich, Germany
Mohamed.Ebrahim@cognizant.de
[5] University of West Bohemia, Faculty of Applied Sciences, Department of Computer Science
and Engineering, NTIS Centre, Pilsen, Czech Republic
jstein@kiv.zcu.cz
[6] Fondazione Bruno Kessler, Human Language Technology group, Trento, Italy
turchi@fbk.eu

Abstract. We present the highly multilingual news analysis system *Europe Media Monitor* (EMM), which gathers an average of 175,000 online news articles per day in tens of languages, categorises the news items and extracts named entities and various other information from them. We also give an overview of EMM's text mining tool set, focusing on the issue of how the software deals with highly inflected languages such as those of the Slavic and Finno-Ugric language families. The questions we ask are: How to adapt extraction patterns to such languages? How to de-inflect extracted named entities? And: Will document categorisation benefit from lemmatising the texts?

Keywords: multilinguality, text mining, information extraction, text classification, inflection, Slavic and Finno-Ugric languages, media monitoring.

1 Introduction

Languages differ substantially in their degree of morphological variability. The higher the degree of inflection, the higher the type-token ratio between lemmas (the uninflected dictionary forms) and their possible inflection forms. Romance languages have many more different verb forms than English. Slavic and Finno-Ugric languages additionally inflect for case. Hungarian and Turkish may add even further morphemes as suffixes to words in a productive process that makes it almost impossible to predict all possible inflected word forms. When developing Natural Language Processing tools for Information Retrieval, Information Extraction, Machine Translation and other applications,

---

* Invited talk.

I. Habernal and V. Matousek (Eds.): TSD 2013, LNAI 8082, pp. 22–33, 2013.
© Springer-Verlag Berlin Heidelberg 2013

this morphological variability in highly inflected languages must be taken care of, in one way or another. In this paper, we focus on Information Extraction and on document classification, but the observations are likely to be useful for other Language Technology applications, as well.

The most common approach to Named Entity Recognition (NER) for highly inflected languages is to make use of linguistic pre-processing tools. This is true for both rule-based systems (e.g. [7] for Polish and [5] for Turkish) and for systems based on machine learning methods (e.g. [4] for Czech and [2] for Hungarian). As EMM's NER tools cover over twenty languages and we have no morphological analysers, alternative ways of dealing with the morphological variation had to be explored.

Even when lemmatisers are available, it is not a given that this solution is satisfactory. Piskorski et al. [8] tested a number of lemmatisers on Polish named entities (as opposed to common words) and found their accuracy to be rather low, i.e. between 35% and 75%. The authors listed a number of reasons that are responsible for this low performance: (a) Declensions of Polish surnames sometimes follows a different pattern from the declension of nouns. (b) The inflection patterns of foreign names depend on the pronunciation of the name (e.g. the genitive form of the name *Wilde* pronounced in the English way is *Wilde'a* while the German pronunciation would yield *Wildego*). (c) There may be identical homographic inflection forms for different base forms so that it is difficult to know which inflection paradigm to use for the lemmatisation. An example is the name *Marka*, which could be the uninflected nominative form of the female name or the genitive form of the male names *Marek* or *Mark*.

Our challenge is thus to find ways to deal with inflection in many languages in a generic way that does not require much language-specific effort and that helps us capture at least the most frequent cases. Due to the large amounts of frequently redundant news texts in EMM, we generally favour high precision and we accept lower recall.

In the following sections, we give an overview of EMM (Section 2) and we briefly sketch the workings of its mostly language-agnostic Language Technology tool set (Section 3). We then present our current approach to NER in highly inflected languages, which involves matching uninflected dictionary and term lists against the inflected word forms found in text, and which also requires lemmatising inflected entity names (Section 4). In Section 5, we shed light on the impact of high morphological inflection on document categorisation by presenting experimental results on using lemmatisers for the classification of documents in four languages with varying degrees of morphological complexity. In Section 6, we briefly summarise our insights. Pointers to related work will be weaved into the various sections.

While we cannot claim that our insights are universally valid because of our particular highly multilingual setting and the subsequent choice of shallow text processing, we hope that fellow researchers working on inflected languages will nevertheless benefit somewhat from our descriptions.

2 The Europe Media Monitor EMM and Its Text Analysis Tools

EMM [13] is freely accessible via the URL http://emm.newsbrief.eu/overview.html. It gathers an average of 175,000 news articles per day in up to 75 languages from about

4,000 different news and information websites, which are visited up to every five minutes. The news items get fed into four different online news analysis applications. In *EMM-NewsBrief* and in the *Medical Information System MedISys* (the latter focuses on the Public Health domain), the articles downloaded during the last four hours are clustered every ten minutes, separately for each language, using hierarchical group-average clustering. The news clusters are displayed in order of their size, together with the historically collected list of related news articles. Each cluster is accompanied with automatically extracted meta-information, which includes: lists of categories the documents belong to; lists of persons, organisations and locations; quotations by and about entities. The news can also be viewed via content category pages, displaying a whole range of multilingual statistics and early warning graphs. A translation tool shows users the first few words of each text in English, and a link allows users to translate the whole news text into any language offered by Google Translate.

While the purpose of the two mentioned EMM applications is to allow monitoring the news live and to possibly issue early warning messages and updates, the *EMM-NewsExplorer* provides long-term trends by clustering the news once a day and by displaying hyperlinks to related news over longer periods of time and across its 21 languages. The hyperlinks allow users to explore the news over time and across languages. NewsExplorer also features entity pages, where it displays information about entities collected in the course of its almost ten years of multilingual news analysis: For each of its over one million entities detected automatically during this period, the entity pages display: titles mentioned with the entity name; name variant spellings found; quotations by and about this entity; news clusters in which this entity was mentioned; and other entities mentioned in the same news clusters. A major feature of NewsExplorer is its capability to automatically determine which variant spellings – within the same language and across languages and scripts – belong to the same entity. Linking these variants to the same unique entity identifier allows the application to collect and display multilingual entity information on the same page (independently of how the name was spelled) and to establish hyperlinks between related documents.

EMM-Labs features a loose collection of text analysis applications that have not yet been entirely integrated with the other three applications: a multi-document summariser; an event extraction tool that automatically fills in document scenario templates for potentially dangerous events; a social network application; a media trend monitoring tool, and more. As of May 2013, EMM-NewsBrief is also freely available as an app for Apple iPads. A version for Android will follow.

EMM users include the European institutions, the national authorities of the European Union's (EU) 27 Member States, international organisations such as various United Nations sub-organisations and the African Union, national authorities of EU partner countries such as the US, Canada and China, as well as about 30,000 anonymous internet users per day. The institutional users mostly use EMM as a means to review the media output for their field of interest and also for early warning purposes.

3   EMM's Approach to Reaching High Multilinguality

In order to allow the functionality described in Section 2, EMM has tightly integrated a whole range of Language Technology tools. The tools have all been developed at the JRC under the strict conditions that it must be possible to develop them relatively fast for tens of languages (typically adding on all functionalities for a new language in not more than three person months of effort). In order to reach this objective, the information extraction software consists mostly of a language-agnostic set of rules that applies to all languages. The effort to adapt the software applications to a new language mostly consists of creating the language-specific lexical resources, of evaluating the output and of then tuning the applications. Such lexical resources include lists of first names, surnames, titles and professions for NER; lists of category-defining words for subject domain classification; various targeted stop word lists; lists of reporting verbs for quotation recognition, etc. A variety of machine learning methods are used to speed up the lexical acquisition for the language adaptation process.

The philosophy behind our approach is described in detail in [12], together with a series of challenges and with an overview of approaches taken by other multilingual system developers. Major guidelines of the philosophy include the following: As far as possible, use language-independent rules to limit the language-specific effort; underspecify rules and the lexicon, i.e. only provide the minimal information necessary for the application; avoid the use of language pair-specific linguistic resources (there are 210 language pair combinations for the currently 21 languages of NewsExplorer!); favour lexical learning over black box classifiers to allow quality control and human intervention for tuning purposes. To make these ideas more concrete, we sketch here how EMM's NER and name variant matching system works (for details, see [9]). [12] contains descriptions of further text processing modules from the point of view of the design principles described in this section.

The recognition of person names in EMM in currently 21 languages is performed using the same language-agnostic rule set, but involving separate application-specific dictionaries for each of the languages. Individual rules are added to some of the languages, where necessary. Arabic, which does not distinguish case, has its own set of rules. The NER tool searches for groups of at least two uppercase words. These uppercase words are identified as potentially being person names if one of the two uppercase words is a known first name (our mixed-country name lists contain tens of thousands of first names) or if the uppercase words are found next to one or more *trigger words*. Trigger words can be titles (e.g. *president*), professions (e.g. *soldier, tennis player*), country or other location adjectives (*French, Bavarian*), words referring to religion or ethnic group (*Hindu, Berber*), and further expressions (e.g. *X-year-old, has deceased*). The patterns also allow combinations such as *76-year-old former Kenyan president*.

These patterns recognise one thousand or more new names every day and a name matching procedure tries to determine which of these new name forms might be a spelling variant of any of the over 1.5 million previously recognised names in the EMM database. This is done by (a) where necessary, transliterating names into the Roman script, using standard transliteration rules; (b) normalising the name strings by applying about thirty hand-crafted string operation rules inspired by empirically found name variations; (c) removing all vowels to produce a consonant signature and – if the consonant signature of a newly identified name is identical to that of an existing name – applying string similarity metrics to the pair of full names (after transliteration) and to the normalised names. If the similarity is above a certain threshold, the two names are automatically marked as being variants of each other. The name normalisation rules are the same for all 21 languages. The rules remove diacritics and name 'infixes' (like van, de la, bin, etc.), turn double consonants into single consonants, change –ow name endings (for Russian names) into –ov, etc., to name just a few. The rules thus convert a name found such as Malik al Saïdoullaïev into the normalised form Malik Saidulaiev and into the consonant signature mlk sdlv. The name spelling Malik Sayduláyev has the same consonant signature and the calculated string similarity is high enough to automatically merge these two name variants.

Many name variant spellings are due to typos or they are caused by different transliteration rules, depending on the target language. The next section addresses name variants due to morphological inflection.

4   Inflection and Information Extraction

For the NER approach described in Section 3 and considering our requirement that variant spellings belonging to the same name be linked to the same unique entity identifier, highly inflected languages pose three types of problems:

1. The lookup of hundreds of thousands of known entity names contained in geographical gazetteers (e.g. city and country names) and of the many known person and organisation names accumulated in the NewsExplorer database is a challenge because, in documents, these names may occur in an inflected form.
2. The number of lexical items (trigger words) that are part of NER patterns (such as titles, country adjectives, etc.) is much higher compared to that for less-inflected languages. It is thus more difficult to come up with a reasonably complete list of these words.
3. For any entity name newly found in text (e.g. Czech *Josefu Steinbergerovi*), we have to determine whether it is a new name or whether it is an (inflected) variant of a previously known name (in this case *Josef Steinberger*). If it is a previously unknown name, we furthermore want to avoid adding an inflected name variant to the database of known names.

Outside the context of EMM, depending on the task at hand, it may not be necessary to solve all three problems. For the task of name variant mapping (co-reference resolution) within and across languages (e.g. [3] and [8]), it is sufficient to determine whether two strings belong to the same name and it will not be necessary to identify the base form of the name. In such cases, a string similarity measure such as that proposed by Piskorski et al. [8] will be sufficient. In EMM, however, we need a solution to each of the three challenges. Without having access to a morphological analysis tool, our options for solutions are rather limited: (A) We can use wild cards such as the Kleene star (e.g. to solve problems 1 and 2); (B) we can try to build a simple replication of the functionality of a generator of morphological variants (to solve problems 1 and 2) or of a morphological analyser (to solve problem 3). We have tested these methods for NER on Hungarian and on Czech texts. Sub-sections 4.1 and 4.2 summarise our experiences.

4.1   Lookup of Known Entity Names and other Words (e.g. person titles)

For the purpose of geo-tagging, lookup procedures are unavoidable as a term found in text such as 'London' must be associated with its geographical co-ordinates (and more) if the place should be shown on a map, as it is the case in EMM. For the frequently found person and organisation names historically collected by EMM (e.g. *Mitt Romney*), it is also more efficient to use a lookup procedure rather than recognising them each time using the NER patterns. We can think of three ways to perform the lookup of the items of such name and word lists in inflected languages: (a) add Kleene star type wild cards to the search terms (possibly truncating the search term to allow for stem changes); (b) pre-generate all, or at least the most frequent, inflection variants; (c) lemmatise the text collections that are to be searched. The latter is the least practical solution as it requires lemmatisers and it is computationally heavy. In EMM, we use solutions (a) for the agglutinative language Hungarian and (b) for the Slavic languages Czech, Slovene and Russian. In their effort to match English and Russian names found in comparable corpora, Klementiev and Roth [3] used the Kleene star solution (a) by radically truncating all names found after the fifth letter. Note that some languages inflect both name parts (e.g. Polish and Czech) while others inflect only the last name part (e.g. Hungarian and German).

For Hungarian, we thus tested simply adding a wild card to the end of each entity name in order to capture inflection variants of the known names (e.g. *London%* and *Romney%*, where the % stands for 0, 1 or more characters). By doing this, the recall of the lookup mechanism rises, but (a) some unwanted words may get captured and (b) there are wanted variants that are not captured, i.e. in the case when the last letter(s) of the name get changed during inflection. Hungarian examples for case (b) are *Obamának* (for US president *Obama*), where the final '–a' gets replaced by '–á' and *Mórahalmon* (location with the base form *Mórahalom*), where the last two letters of the unmarked nominative form change. To capture case (b), we tested removing the last letter of base forms and adding the % to the stem. For names ending in '–a' and '–e', this produces very good results (e.g. see Table 1), but it also introduces more errors. For instance, this pattern erroneously recognises the Hungarian words Romnet and Romnetnek as inflection variants of Romney. The final compromise is to add the % after all names (with a minimum length requirement of four characters) and to capture the root-changing cases (Hungarian names ending in '–a' or in '–e') by changing the search string: In addition to Obama, we would thus also be searching for Obamá%. That way, the number of false positives is relatively low, but all cases similar to those in Table 1 will be captured. For fine-tuning and to avoid high numbers of false positives, it is recommended to look at a frequency list of matches in a larger corpus. The language-specific effort of this wild-card solution consists of (a) identifying the required wildcard patterns to search for (e.g. in Hungarian: % after base form except when name ends in '–a' or '–e'), and (b) to test and tune the results.

For Slovene, Russian and Czech, which have a rich morphology, but no agglutination, we took another path, i.e. that of creating inflection rules and of pre-generating inflection variants according to these patterns. For instance, correct Czech inflection forms of the name *Jana* are *Jany* and *Janou*. These can be pre-generated by a suffix replacement rule that substitutes the final letter '–a' with '–y' and '–ou'. In some cases,

Table 1. Hungarian inflection forms of the name *Obama* and their frequency in the news (search pattern: *obam%*)

Name form	Frequency
obama	31262
obamát	1439
obamának	1035
obamával	598
obamára	238
obamáék	157
obamától	136
obamáról	120
obamához	88
obamánál	53
obamában	39
obamán	28
obamáé	24
obamáné	21
obamáig	20
obamáját	17
obamáéknak	13
obamáért	12
obamáékkal	9
obamai	8
obamából	7
obamája	6
obame	4

two final letters need to be checked to find the right replacement rule: *Jitka – Jitce* (*–ka* turns to *–ce*). Thus the algorithm has to match the longest suffix when searching for the right rule (i.e. rather substitute –ka than –a). For Czech, we have identified a number of expansion rules. There are separate rules for male first names (6 expansion patterns), for female first names (10), for surnames (12), for demonyms (country adjectives; 10 patterns) and for other words that need to be looked up (titles, professions, positions, etc.; 9 patterns). Note that these rules will only cover the major inflection forms, but that there are always irregular inflections that will either be missed or that need adding separately (e.g. the male first name *Zdeněk* expands to *Zdeňka*, *Zdeňkovi*, *Zdeňkem*, a paradigm not covered by the expansion rules). Gender plays an important role in Czech as for the same suffix quite different substitution rules exist. A Czech speciality is adding the suffix –ová to female surnames, which is done even for foreign names (e.g. *Merkelová*). The suffix generator thus needs input about the entity gender. A further difficulty is that foreign names – which are frequently found in the news and which make up the majority of items on our lookup lists – may follow different inflection patterns. However, through the manually created inflection rules described here, we recognise a whole range of inflected words we would otherwise miss. This inflection pattern approach is more time-consuming to produce than the wild card approach, and it requires the dedication of a linguistically skilled specialist.

Both methods allow us to recognise inflection forms and to also know the base form, which we need in order to decide whether we have encountered a new name or a variant of a previously known name. The rules were written, tested and refined on real-world data, but we have not yet carried out a formal evaluation that would allow us to show precision and recall of the methods.

4.2   De-inflecting Newly Found Entity Names

Section 4.1 described the effort of looking up inflection forms of known names and words, but the methods described there do not help when a new name has been identified by a NER pattern like the ones described in Section 3. When recognising a new entity like the name *Borisz Paszternak* in Hungarian text, we need to know whether this is the base form or whether this is a morphological variant (e.g. of *Borisz Paszter*) or a spelling variant of a name already present in our database (e.g. *Boris Pasternak*). In order to lemmatise inflected Hungarian names, we manually produced a list of 66 frequent proper name suffixes (e.g. '–nak', '–ot', '–on', '–en', '–val', '–ről', '–képpen', '–ként') and tested whether it would be safe to simply strip these off names newly found in Hungarian text. We did not consider single-letter suffixes such as –t (e.g. *Angela Merkelt*) as we considered that these would apply too often so that stripping them would be too dangerous. We applied the suffix stripping only to new names that we had not seen during our large-scale and multiannual analysis of the news in other languages. Out of 18,414 names newly found in the Hungarian news, the suffix-stripping rules changed 287 (1.56%). Out of these, 8 names (2.8%) should not have been lemmatised (e.g. *Barbara Gordon* was erroneously changed to *Barbara Gord*). The manual evaluation allowed us to also detect 26 further NER errors (9% error rate), which allowed us to improve our NER resources.

In our EMM setting, erroneously merging two unrelated names to the same unique name identifier is highly unwanted, while having different name variants of the same name with different identifiers is a more acceptable mistake. As it still happened that the suffix-stripping procedure erroneously mapped new unknown names to base forms that were not related to the newly found names (there are 1.5 million names in our name database), we decided for the moment *not* to apply this suffix stripping method to newly identified names. In our experience, 12% of Hungarian person names appear in their uninflected nominative form (judging by looking at media VIPs such as *Viktor Orbán* and *Barack Obama*, see Table 1), which is very similar to the 13% of inflected person names reported for Polish [8]. This means that we might in principle add quite a number of inflected person names to our database. However, we can assume that most of these cases will be of relatively low frequency because the base form of frequent names will also be found so that the wild card solution described in Section 4.1 will eventually capture such inflected name variants.

We have thus not yet entirely solved the challenge of how to lemmatise entity names. We plan to explore further how to efficiently add de-inflection to our processes. One possible area for improvement is to add more conditions under which the de-inflection of names will automatically apply, in order to make the rules safer. Another option is to explore ways of learning automatic methods for de-inflection.

5 Inflection and Multi-label Document Categorisation

In addition to the news subject domain categorisation described in Section 2, news clusters in *NewsExplorer* are additionally automatically multi-label classified according to the EuroVoc thesaurus[1], which is the wide coverage classification scheme used by EU institutions and other organisations to categorise official documents into nearly 7,000 classes. NewsExplorer identifies news clusters in different languages as being about the same event or subject if they talk about similar (EuroVoc) subject domains and if they at least partially make reference to the same entities (persons, organisations and locations: see [11] for details). The multi-label classifier used was trained on tens of thousands of manually labelled almost parallel documents per language for 22 official EU languages. We tackled the challenge of selecting the most appropriate among the thousands of EuroVoc categories by treating it like a profile-based category ranking task (see [14] for details). The profile of each class (descriptor, category) consists of a ranked list of typical features for this class.

In the default setting of the application, no word normalisation (lemmatisation or stemming) was used. The comparative results for all 22 languages in [14] shows that the tool consistently performed better for the Germanic EU languages (Dutch, Danish, English, German and Swedish) than for the Slavic EU languages (Bulgarian, Czech, Polish, Slovak and Slovene). We saw as one possible explanation of this regularity that the larger feature space in the more highly inflected Slavic languages has a negative impact on the automatic classifiers. If this were indeed the case, the Slavic results should be better when applying word normalisation mechanisms such as lemmatisation or stemming. In earlier work [10], applying a lemmatiser to Spanish documents was indeed found to improve categorisation results by two points on the F-measure scale. In order to better understand the reasons for this performance difference between the language families, we carried out a number of experiments (described in [1]) where we changed the document representation to test the impact of lemmatisation and of part-of-speech tagging for four languages belonging to different language families: Czech (Slavic), English (Germanic), Estonian (Baltic) and French (Romance). Czech and Estonian are representatives of highly inflected languages, while French has no noun inflection and English very little inflection altogether. In addition to reducing the feature space through lemmatisation, we also tested adding part-of-speech information to each word feature. As we do not have our own lemmatisers and part-of-speech taggers and we could therefore not use these tools as part of our processing chain, these experiments were carried out using third-party tools, exclusively for the purpose of better understanding the problem.

The results of that experiment are repeated in Table 2. We would like to note that the results may not be entirely comparable across languages because we did not have the same amount of training documents, the amount of noise in the data may differ, and the linguistic pre-processing tools used may not perform equally well. However, in our view, the results nevertheless show a rather clear picture:

1. For all four languages, when lemmatising the training and the test data, not only the classification performance (F1-score) goes down compared to using inflected

---

[1] See http://eurovoc.europa.eu/. Site last visited on 6.6.2013.

Table 2. Experimental categorisation results (Precision, Recall and F1-Score, with standard deviation) for the EuroVoc indexing experiment in the four languages Czech, English, Estonian and French, using four different document representations each. Best results per language are highlighted in boldface.

	Czech	English	Estonian	French
1. Words (basic)				
Categories total	3338	4633	3577	4096
Categories trained	**1727**	2446	1979	2451
Precision	0.4476±0.0059	0.4620±0.0025	0.4779±0.0045	0.4786±0.0023
Recall	0.5052±0.0057	0.5362±0.0019	0.5329±0.0078	0.5577±0.0041
F1-Score	0.4747±0.0056	0.4963±0.0022	0.5039±0.0059	**0.5151±0.0025**
2. Lemmas				
Categories total	3338	4633	3577	4096
Categories trained	1663	2420	1951	2409
Precision	0.4303±0.0060	0.4589±0.0031	0.4654±0.0034	0.4723±0.0025
Recall	0.4858±0.0059	0.5330±0.0026	0.5189±0.0061	0.5502±0.0037
F1-Score	0.4564±0.0058	0.4932±0.0028	0.4907±0.0043	0.5083±0.0025
3. Lemmas + POS				
Categories total	3338	4633	3577	4096
Categories trained	1664	2472	1957	2430
Precision	0.4309±0.0063	0.4655±0.0032	0.4672±0.0035	0.4709±0.0028
Recall	0.4865±0.0060	0.5403±0.0025	0.5208±0.0065	0.5484±0.0043
F1-Score	0.4570±0.0060	0.5001±0.0029	0.4925±0.0046	0.5067±0.0030
4. Words + POS				
Categories total	3338	4633	3577	4096
Categories trained	1725	**2497**	**1856**	**2467**
Precision	0.4508±0.0058	0.4686±0.0032	0.4843±0.0039	0.4775±0.0032
Recall	0.5087±0.0057	0.5439±0.0027	0.5395±0.0077	0.5565±0.0040
F1-Score	**0.4780±0.0055**	**0.5035±0.0029**	**0.5104±0.0055**	0.5140±0.0030

word forms, but also the number of trained categories is smaller (comparing experiments two and one in Table 2). It thus seems that, while reducing data sparseness, lemmatisation reduces some of the knowledge available to the classifier, and this inflection knowledge seems to be relevant.

2. Secondly, for all languages except French, adding POS information improved the results. This is true when adding the POS to the word form (comparing experiments four and one) and when adding it to the lemma (comparing experiments three and two). We could not observe the same trend for French, but in both comparisons, the French performance difference is not statistically significant. Unlike lemmatisation, adding POS information is likely to increase the number of features used and it increases the information available to the classifier, yielding positive results.

3. The best-performing document representation for the three languages Czech, English and Estonian is the combination of inflected words and their POS (see experiment four), which uses the largest number of features. The improvement compared to the basic setting (experiment one) is statistically significant for English and Estonian. For French, the situation is inversed: the simple word forms perform best, but the comparison with the word-POS combination is not statistically significant.

The experiments thus do not explain why classifiers perform less well for the Slavic languages than for the Germanic languages, but they indicate that conflating inflection forms into lemmas has a negative impact on classification performance. This is in line with findings by Moschitti and Basili [6] for Italian and English, and by Toman et al. [15] for Czech.

6   Conclusion

The *Europe Media Monitor* (EMM) is a suite of news gathering and analysis applications whose analysis is based on language technology tools such as software for information extraction, clustering and classification, the generation of social networks based on information automatically extracted from the news, co-reference resolution, opinion mining, machine translation, and more. Due to the high number of languages in EMM – 75 for news gathering and 21 for information extraction – and the unavailability of parsers, morphological analysers, etc., EMM applies shallow processing methods. In this paper, we presented the main challenges highly inflected languages such as those from the Slavic or Finno-Ugric language families pose for EMM. These challenges are related to dictionary lookup (requiring the mapping of uninflected lists of names and terms against the inflected word forms found in text) and to the necessary lemmatisation of inflected names found in news articles. We presented two light-weight methods (and their limitations) to satisfy these requirements – using Kleene star type wild cards and using hand-crafted name inflection and name lemmatisation rules. We also presented experiments, performed on languages from four different language families, showing that inflection forms are actually beneficial for the performance of document classification software.

The solutions for dealing with name inflection in named entity recognition and name variant matching described in Section 4 are not entirely satisfactory. We therefore plan to refine these methods and to also explore further solutions, such as automatically learning de-inflection rules. The expected bottleneck for this is the lack of reliable training data, especially for the mostly foreign location and person names in our large NER word lists.

References

1. Mohamed, E., Ehrmann, M., Turchi, M., Steinberger, R.: Multi-label EuroVoc classification for Eastern and Southern EU Languages. In: Vertan, C., Hahn, W. (eds.) Multilingual Processing in Eastern and Southern EU languages - Low-resourced Technologies and Translation, pp. 370–394. Cambridge Scholars Publishing, Cambridge (2012)
2. Farkas, R., Szarvas, G., Kocsor, A.: Named entity recognition for Hungarian using various machine learning algorithms. Acta Cybernetica 17(3), 633–646 (2006)
3. Klementiev, A., Roth, D.: Weakly supervised named-entity transliteration and discovery from multilingual comparable corpora. In: Proceedings of ACL 2006 Conference (2006)
4. Konkol, M., Konopík, M.: Maximum Entropy Named Entity Recognition for Czech Language. In: Habernal, I., Matoušek, V. (eds.) TSD 2011. LNCS (LNAI), vol. 6836, pp. 203–210. Springer, Heidelberg (2011)

5. Küçük, D., Yazıcı, A.: Named Entity Recognition Experiments on Turkish Texts. In: Andreasen, T., Yager, R.R., Bulskov, H., Christiansen, H., Larsen, H.L. (eds.) FQAS 2009. LNCS, vol. 5822, pp. 524–535. Springer, Heidelberg (2009)

6. Moschitti, A., Basili, R.: Complex linguistic features for text classification: A comprehensive study. In: Proceedings of the 26th European Conference on Information Retrieval Research, Sunderland, UK (2004)

7. Piskorski, J.: Extraction of Polish Named-Entities. In: Proceedings of the Fourth International Conference on Language Resources and Evaluation (LREC), pp. 313–316 (2004)

8. Piskorski, J., Wieloch, K., Sydow, M.: On knowledge-poor methods for person name matching and lemmatization for highly inflectional languages. Inf. Retrieval 12, 275–299 (2009)

9. Pouliquen, B., Steinberger, R.: Automatic Construction of Multilingual Name Dictionaries. In: Goutte, C., Cancedda, N., Dymetman, M., Foster, G. (eds.) Learning Machine Translation. Advances in Neural Information Processing Systems Series (NIPS), pp. 59–78. MIT Press (2009)

10. Pouliquen, B., Steinberger, R., Ignat, C.: Automatic Annotation of Multilingual Text Collections with a Conceptual Thesaurus. In: Proceedings of the Workshop 'Ontologies and Information Extraction' at the EuroLan Summer School 'The Semantic Web and Language Technology' (EUROLAN 2003), Bucharest, Romania (2003)

11. Pouliquen, B., Steinberger, R., Deguernel, O.: Story tracking: linking similar news over time and across languages. In: Proceedings of the 2nd Workshop Multi-source Multilingual Information Extraction and Summarization (MMIES 2008) held at CoLing 2008, Manchester, UK (2008)

12. Steinberger, R.: A survey of methods to ease the development of highly multilingual Text Mining applications. Language Resources and Evaluation Journal 46(2), 155–176 (2012)

13. Steinberger, R., Pouliquen, B., van der Goot, E.: An Introduction to the Europe Media Monitor Family of Applications. In: Gey, F., Kando, N., Karlgren, J. (eds.) Information Access in a Multilingual World - Proceedings of the SIGIR 2009 Workshop (SIGIR-CLIR 2009), Boston, USA, pp. 1–8 (2009)

14. Steinberger, R., Ebrahim, M., Turchi, M.: JRC EuroVoc Indexer JEX - A freely available multi-label categorisation tool. In: Proceedings of the 8th International Conference on Language Resources and Evaluation (LREC 2012), Istanbul, pp. 798–805 (2012)

15. Toman, M., Tesar, R., Ježek, K.: Influence of Word Normalization on Text Classification. In: Proceedings of InSciT 2006, Merida, Spain (2006)

Spoken Dialogs with Children for Science Learning and Literacy*

Ron Cole, Wayne Ward, Daniel Bolanos, Cindy Buchenroth-Martin, and Eric Borts

Mentor InterActive Inc.
and
Boulder Language Technologies, Boulder, Colorado, USA

Abstract. Advances in human language and character animation technologies have enabled a new generation of intelligent tutoring systems that support conversational interaction between young learners and a lifelike computer character that was designed to behave like a sensitive and effective human tutor My Science Tutor is a spoken dialog system in which children learn to construct science explanations through conversations with Marni, the virtual science tutor, in multimedia environments. MyST displays illustrations, silent animations or interactive simulations to the student, while Marni asks open-ended questions like "Whats going here?". Based on MySTs analysis of the students spoken response, the system decides what the student understands about the science and what the student has not yet explained (or doesnt know), and generates a follow-on question a new prompt, and possibly a new animation, that is designed to scaffold learning and challenge the student to reason about the science. Two large scale evaluations were conducted in which third, fourth and fifth grade students received over 5 hours of tutoring during sixteen 20-minute sessions in four different areas of science. The results revealed that, relative to students who did not receive tutoring, students who used My Science Tutor achieved significant learning gains in standardized tests of science achievement, equivalent to gains achieved by students who received tutoring by expert human tutors. In recent research, we have extended the technologies used in MyST to a develop a new generation of interactive books that use text, speech and dialog technologies to help children learn to read science texts fluently, expressively, and with good comprehension. We will demonstrate these MindStars Books and present initial results of classroom testing.

---

* Invited talk.

I. Habernal and V. Matousek (Eds.): TSD 2013, LNAI 8082, p. 34, 2013.

Statecharts and SCXML for Dialogue Management[*]

Torbjörn Lager

Department of Philosophy
Linguistics and Theory of Science University of Gothenburg
Box 200, SE-405 30 Gothenburg, Sweden
torbjorn.lager@ling.gu.se

Abstract. The World Wide Web Consortium (W3C) has selected Harel State-charts, under the name of State Chart XML (SCXML), as the basis for future standards in the area of (multimodal) dialog systems. In this talk, I give a brief introduction to Statecharts and to SCXML, show what it can do (and not do) for someone wanting to use it for dialogue management, describe its relation to other dialogue system components, explain its possible use as a meta-dialogue manager (i.e. as a manager of dialogue managers), as well as some ideas for how to compile other dialogue management languages (such as VoiceXML) into SCXML. If time permits, I will also give some pointers to existing SCXML implementations.

---

[*] Invited talk.

I. Habernal and V. Matousek (Eds.): TSD 2013, LNAI 8082, p. 35, 2013.

A Comparison of Deep Neural Network Training Methods for Large Vocabulary Speech Recognition

László Tóth* and Tamás Grósz

MTA-SZTE Research Group on Artificial Intelligence
Hungarian Academy of Sciences and University of Szeged
tothl@inf.u-szeged.hu, groszt@sol.cc.u-szeged.hu

Abstract. The introduction of deep neural networks to acoustic modelling has brought significant improvements in speech recognition accuracy. However, this technology has huge computational costs, even when the algorithms are implemented on graphic processors. Hence, finding the right training algorithm that offers the best performance with the lowest training time is now an active area of research. Here, we compare three methods; namely, the unsupervised pre-training algorithm of Hinton et al., a supervised pre-training method that constructs the network layer-by-layer, and deep rectifier networks, which differ from standard nets in their activation function. We find that the three methods can achieve a similar recognition performance, but have quite different training times. Overall, for the large vocabulary speech recognition task we study here, deep rectifier networks offer the best tradeoff between accuracy and training time.

Keywords: deep neural networks, TIMIT, LVCSR.

1 Introduction

Recently there has been a renewed interest in applying neural networks (ANNs) to speech recognition, thanks to the invention of deep neural nets. As the name suggests, deep neural networks differ from conventional ones in that they consist of several hidden layers, while conventional ANN-based recognizers work with only one hidden layer. The application of a deep structure can provide significant improvements in speech recognition results compared to previously used techniques [1]. However, modifying the network architecture also requires modifications to the training algorithm, because the conventional backpropagation algorithm encounters difficulties when training many-layered feedforward networks [2]. As a solution, Hinton et al. presented a pre-training algorithm that works in an unsupervised fashion [3]. After this pre-training step, the backpropagation algorithm can find a much better local optimum. The first tests of deep networks for speech recognition were performed on the TIMIT database [4], which is much smaller than the corpora routinely used for the training of industrial-scale speech recognizers. Hence, since their invention, a lot of effort has been gone into trying to

---
* This publication is supported by the European Union and co-funded by the European Social Fund. Project title: Telemedicine-focused research activities in the fields of mathematics, informatics and medical sciences. Project number: TÁMOP-4.2.2.A-11/1/KONV-2012-0073.

I. Habernal and V. Matousek (Eds.): TSD 2013, LNAI 8082, pp. 36–43, 2013.

scale up deep networks to much larger datasets and large vocabulary tasks [5,6,7]. The main problem here is that Hinton's pre-training algorithm is very CPU-intensive, even when implemented on graphic processors (GPUs). Several solutions have been proposed to alleviate or circumvent the computational burden of pre-training, but the search for the optimal training technique is still continuing.

Here, we compare three different technologies for the training of deep networks. One is the original pre-training algorithm of Hinton et al.[3]. It treats the network as a deep belief network (DBN) built out of restricted Bolztmann machines (RBMs), and optimizes an energy-based target function using the contrastive divergence (CD) algorithm. After pre-training, the network has to be trained further using some conventional training method like backpropagation.

The second algorithm is called 'discriminative pre-training' by Seide et al. [5]. This method constructs a deep network by adding one layer at a time, and trains these sub-networks after the addition of each layer. Both the pre-training of the partial nets and the final training of the full network are performed by backpropagation, so no special training algorithm is required.

As for the third method, it is different from the two above in the sense that in this case it is not the training algorithm that is slightly modified, but the neurons themselves. Namely, the usual sigmoid activation function is replaced with the rectifier function $\max(0, x)$. These kinds of neural units have been proposed by Glorot et al., and were successfully applied to image recognition and NLP tasks [8]. Rectified linear units were also found to improve restricted Boltzmann machines [9]. It has been shown recently that a deep rectifier network can attain the same phone recognition performance as that for the pre-trained nets of Mohamed et al. [4], but without the need for any pre-training [10].

Here, we first compare the three methods on the TIMIT database, but the main goal of the paper is to obtain results for a large vocabulary recognition task. For this purpose, we trained a recognition system on a 28-hour speech corpus of Hungarian broadcast news. This recognizer is a hybrid HMM/ANN system [11] that gets the state-level posterior probability values from the neural net, while the decoder is the HDecode program, which is a part of the HTK package [12]. As Hungarian is an agglutinative language, our system runs with a relatively large dictionary of almost five hundred thousand word forms.

2   Training Algorithms for Deep Neural Networks

2.1   DBN Pre-Training

This efficient unsupervised algorithm, first described in [3], can be used for learning the connection weights of a deep belief network (DBN) consisting of several layers of restricted Boltzmann machines (RBMs). As their name implies, RBMs are a variant of Boltzmann machines, with the restriction that their neurons must form a bipartite graph. They have an input layer, representing the features of the given task, a hidden layer which has to learn some representation of the input, and each connection in an RBM must be between a visible unit and a hidden unit. RBMs can be trained using the one-step contrastive divergence (CD) algorithm described in [3]. An RBM assigns the following energy value to each configuration of visible and hidden state vectors, denoted by v and h, respectively:

$$E(v, h|\Theta) = -\sum_{i=1}^{V}\sum_{j=1}^{H} w_{ij}v_ih_j - \sum_{i=1}^{V} b_iv_i - \sum_{j=1}^{H} a_jh_j. \tag{1}$$

Derived from the gradient of the joint likelihood function of data and labels, the one-step contrastive divergence update rule for the visible-hidden weights is

$$\Delta w_{ij} \propto \langle v_ih_j\rangle_{input} - \langle v_ih_j\rangle_1, \tag{2}$$

where $\langle.\rangle_1$ represents the expectation with respect to the distribution of samples got from running a Gibbs sampler initialized on the data for one full step.

Although RBMs with the energy function of Eq. (1) are suitable for binary data, in speech recognition the acoustic input is typically represented by real-valued feature vectors. For real-valued input vectors, the Gaussian-Bernoulli restricted Boltzmann machine (GRBM) can be used, and it requires making only a minor modification of Eq. (1). The GRBM energy function is given by:

$$E(v, h|\Theta) = \sum_{i=1}^{V} \frac{(v_i - b_i)^2}{2} - \sum_{i=1}^{V}\sum_{j=1}^{H} w_{ij}v_ih_j - \sum_{j=1}^{H} a_jh_j \tag{3}$$

Hinton et al. showed that the weights resulting from the unsupervised pre-training algorithm can be used to initialize the weights of a deep, but otherwise standard, feed-forward neural network. After this initialization step, we simply use the backpropagation algorithm to fine-tune the network weights with respect to a supervised criterion.

2.2 Discriminative Pre-training

'Discriminative pre-training' (DPT) was proposed in [5] as an alternative to DBN pre-training. It is a simple algorithm where first we train a network with one hidden layer to full convergence using backpropagation; then we replace the softmax layer by another randomly initialized hidden layer and a new softmax layer on top, and we train the network again; this process is repeated until we reach the desired number of hidden layers. Seide et al. found that this method gives the best results if one performs only a few iterations of backpropagation in the pre-training phase (instead of training to full convergence) with an unusually large learn rate. In their paper, they concluded that this simple training strategy performs just as well as the much more complicated DBN pre-training method described above [5].

2.3 Deep Rectifier Neural Networks

In the case of the third method it is not the training algorithm, but the neurons that are slightly modified. Instead of the usual sigmoid activation, here we apply the rectifier function $\max(0, x)$ for all hidden neurons [2]. There are two fundamental differences between the sigmoid and the rectifier functions. One is that the output of rectifier neurons does not saturate as their activity gets higher. Glorot et al. conjecture that this is very important in explaining their good performance in deep nets: because of this linearity, there is no gradient vanishing effect [2]. The other difference is the hard saturation at 0 for negative activity values: because of this, only a subset of neurons are active for a given input. One might suppose that this could harm optimization by blocking gradient backpropagation, but the experimental results do not support this hypothesis. It seems that the hard nonlinearities do no harm as long as the gradient can propagate along some paths.

The main advantage of deep rectifier nets is that they can be trained with the standard backpropagation algorithm, without any pre-training. On the TIMIT database they were found to yield phone recognition results similar to those of sigmoid networks pre-trained with the DBN algorithm [10].

3   Experimental Setup

Here, we report the results of applying the ANN-based recognizers on two databases. The first one is the classic TIMIT database of English sentences, while the second is a corpus of Hungarian broadcast news. On TIMIT quite a lot of phone recognition results are available, so it is good for comparative purposes. However, TIMIT is quite small and usually only phone-level results are reported on it. Hence, our second group of tests on the Hungarian corpus sought to measure the large vocabulary recognition performance of the methods used.

As regards TIMIT, the training set consisted of the standard 3696 'si' and 'sx' sentences, while testing was performed on the core test set (192 sentences). A random 10% of the training set was held out for validation purposes, and this block of data will be referred to as the 'development set'. The scores reported are phone recognition error rates using a phone bigram language model.

The speech data of Hungarian broadcast news was collected from eight Hungarian TV channels. It contains about 28 hours of recordings, from which 22 hours were selected for the training set, 2 hours for the development set and 4 hours for the test set. The language model was created from texts taken from the [origo] news portal (www.origo.hu), from a corpus of about 50 million words. Hungarian is an agglutinative language with a lot of word forms, hence we limited the size of the recognition dictionary to 486982 words by keeping only those words that occurred at least twice in the corpus. The pronunciations of these words were obtained from the 'Hungarian Pronunciation Dictionary' [13]. Based on the [origo] corpus, a trigram language model was built using the language modelling tools of HTK [12].

As for the acoustic features, we applied the standard MFCC coefficients, extracted from 25 ms frames with 10 ms frame skips. We used 13 MFCC coefficients (including the zeroth one), along with the corresponding Δ and $\Delta\Delta$ values. In each case, the neural network was trained on 15 neighboring frames, so the number of inputs to the acoustic model was 585.

Neural networks require a frame-level labelling of the training data. For this purpose, we first trained a standard hidden Markov model (HMM) speech recognizer, again using the HTK toolkit. For the TIMIT dataset, monophone 3-state models were created, which resulted in 183 states. For the broadcast news dataset, triphone models were constructed, consisting of 2348 tied triphone states in total. The HMM states were then aligned to the training data using forced alignment. These labels served as training targets for the neural nets.

For the recognition process, we applied the decoders of the HTK package. We used HVite for the phone recognition experiments on TIMIT, while the HDecode routine was applied for the large vocabulary recognition tests on the broadcast news task. In both cases the acoustic modeling module of HTK required a slight modification in order to be able to work with the posterior probability values produced by the neural nets. For the TIMIT dataset, the language model weight and the insertion penalty factor were set to 1.0 and 0.0, respectively. With the broadcast news dataset, these meta-parameters were tuned on the development set. Lastly, for a fairness of comparison, the pruning beam width was set to the same value for each network.

3.1   Training Parameters for the Neural Networks

In the case of the DBN-based pre-training method (see Section 2.1), we applied stochastic gradient descent (i.e. backpropagation) training with a mini-batch size of 128. For Gaussian-binary RBMs, we ran 50 epochs with a fixed learning rate of 0.002, while for binary-binary RBMs we used 30 epochs with a learning rate of 0.02. Then, to fine-tune the pre-trained nets, again backpropagation was applied with the same mini-batch size as that used for pre-training. The initial learn rate was set to 0.01, and it was halved after each epoch when the error on the development set increased. During both the pre-training and fine-tuning phases, the learning was accelerated by using a momentum of 0.9 (except for the first epoch of fine-tuning, which did not use the momentum method).

Turning to the discriminative pre-training method (see Section 2.2), the initial learn rate was set to 0.01, and it was halved after each epoch when the error on the development set increased. The learn rate was restored to its initial value of 0.01 after the addition of each layer. Furthermore, we found that using 5 epochs of backpropagation after the introduction of each layer gave the best results. For both the pre-training and fine-tuning phases we used a batch size of 128 and momentum of 0.8 (except for the first epoch). The initial learn rate for the fine-tuning of the full network was again set to 0.01.

The training of deep rectifier nets (see Section 2.3) did not require any pre-training at all. The training of the network was performed using backpropagation with an initial learn rate of 0.001 and a batch size of 128.

4   Experimental Results: TIMIT

Fig. 1 shows the phone recognition error rates obtained on the TIMIT development set and core test set with a varying number of hidden layers, each hidden layer containing 2048 neurons. As can be seen, the three training methods performed very similarly on the test set, the only exception being the case of five hidden layers, where the rectifier net performed slightly better. It also significantly outperformed the other two methods on the development set.

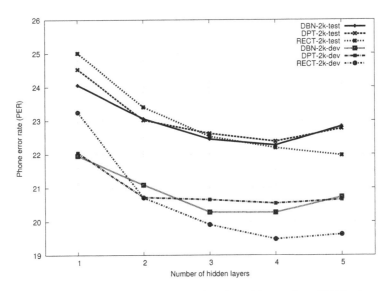

Fig. 1. Phone error rates on TIMIT as a function of the number of hidden layers

Using similar features, training labels and network sizes, Mohamed at al. reported a 22.3% error rate with DBN pre-training [4], while Tóth reported a 21.8% figure with rectifier nets [10]. As our scores fall in the same range, the results also demonstrate the soundness of our methodology.

We mention that a single hidden layer net with the same amount of weights as the best deep net yielded 23.7%. This proves that the better results are due to the deep architecture and not simply due to the increased amount of parameters.

5   Experimental Results: Hungarian Broadcast News

Fig. 2 shows the word error rates got for the large vocabulary broadcast news recognition task. Similar to the TIMIT tests, 2048 neurons were used for each hidden layer, with a varying number of hidden layers. The trends of the results are quite similar to those for the TIMIT database. The error rates seem to saturate at 4-5 hidden layers, and the curves for the three methods run parallel and have only slightly different values. The lowest error rate is attained with the five-layer rectifier network, both on the development and the test sets.

Although their recognition accuracy scores are quite similar, the three methods differ significantly in the training times required. Table 1 shows the training times we measured on an NVIDIA GTX-560 TI graphics card. Evidently, the DBN pre-training algorithm has the largest computational requirements. This algorithm has no clearly defined stopping criterion, and various authors run it with a widely differing number of iterations. The iteration count we applied here (50 for Gaussian RBMs and 30 for binary RBMs) is an average value, and follows the work of Seide et al. [5]. Mohamed applies many more iterations [4], while Jaitly et al. use far fewer iterations [6]. However,

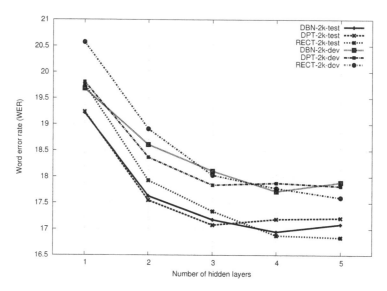

Fig. 2. Word error rates for the broadcast news corpus as a function of the number of hidden layers

no configuration could beat the training time of deep rectifier networks. Discriminative pre-training is also much faster than the DBN-based method, but is still slower than rectifier nets.

Table 1. The training times required by the various methods for 5-layer networks

Training method	Pre-training time	Fine-tuning training time
DBN pre-training	48 hours	14 hours
Discr. pre-training	9 hours	11 hours
Rectifier network	0 hours	14.5 hours

Lastly, although the main goal here was to compare the three deep neural network algorithms, let us now compare the large vocabulary recognition scores with those of a conventional HMM. The same HMM model that was used to generate the training labels attained a word error rate of 20.07% (with maximum likelihood training). However, this result is not fully comparable with those obtained with the hybrid recognizer, because the two systems used different pruning beam widths. Tuning the parameters so that the two systems had a similar real-time factor was also out of the question, as the hybrid model was implemented on a GPU, while the HMM used a normal CPU.

6   Conclusions

It is perhaps no exaggeration to say that deep neural nets have led to a breakthrough is speech recognition. However, they are computationally intensive, and the quest for the optimal network architecture and training method is still continuing. Here, we compared two training methods and a new type of activation function for deep neural nets, and evaluated them on a large vocabulary recognition task. The three algorithms yielded quite similar recognition performances, but based on the training times deep rectifier networks seem to be the preferred choice. Still, the concept of rectified linear units is quite new, and their behavior requires more theoretical study and practical evaluation. We hope that our study can provide a valuable contribution to this new area of research.

References

1. Hinton, G., Deng, L., Yu, D., Dahl, G.E., Mohamed, A.R., Jaitly, N., et al.: Deep neural networks for acoustic modeling in speech recognition: the shared views of four research groups. IEEE Signal Processing Magazine 29, 82–97 (2012)
2. Glorot, X., Bengio, Y.: Understanding the difficulty of training deep feedforward neural networks. In: Proc. AISTATS, pp. 249–256 (2010)
3. Hinton, G.E., Osindero, S., Teh, Y.W.: A fast learning algorithm for deep belief nets. Neural Computation 18, 1527–1554 (2006)
4. Mohamed, A.R., Dahl, G.E., Hinton, G.: Acoustic modeling using deep belief networks. IEEE Trans. Audio, Speech, and Language Processing 20, 14–22 (2012)
5. Seide, F., Li, G., Chen, X., Yu, D.: Feature engineering in context-dependent deep neural networks for conversational speech transcription. In: Proc. ASRU, pp. 24–29 (2011)
6. Jaitly, N., Nguyen, P., Senior, A., Vanhoucke, V.: Application of pretrained deep neural networks to large vocabulary conversational speech recognition. Technical report, Dept. Comp. Sci., University of Toronto (2012)
7. Dahl, G.E., Yu, D., Deng, L., Acero, A.: Context-dependent pre-trained deep neural networks for large-vocabulary speech recognition. IEEE Trans. Audio, Speech, and Language Processing 20, 30–42 (2012)
8. Glorot, X., Bordes, A., Bengio, Y.: Deep sparse rectifier networks. In: Proc. AISTATS, pp. 315–323 (2011)
9. Nair, V., Hinton, G.E.: Rectified linear units improve restricted Boltzmann machines. In: Proc. ICML, pp. 807–814 (2010)
10. Tóth, L.: Phone recognition with deep sparse rectifier neural networks. In: Proc. ICASSP (accepted, in print, 2013)
11. Bourlard, H., Morgan, N.: Connectionist speech recognition: a hybrid approach. Kluwer Academic (1994)
12. Young, S., et al.: The HTK book. Cambridge Univ. Engineering Department (2005)
13. Abari, K., Olaszy, G., Zainkó, C., Kiss, G.: Hungarian pronunciation dictionary on Internet. In: Proc. MSZNY, pp. 223–230 (2006) (in Hungarian)

A Comparison of Two Approaches to Bilingual HMM-Based Speech Synthesis

Miran Pobar[1], Tadej Justin[2], Janez Žibert[3], France Mihelič[2], and Ivo Ipšić[1]

[1] Department of Informatics, University of Rijeka
Radmile Matejčić 2, 51000 Rijeka, Croatia
{mpobar,ivoi}@inf.uniri.hr
http://www.inf.uniri.hr
[2] Faculty of Electrical Engineering, University of Ljubljana
Tržaška 25, 1000 Ljubljana, Slovenia
{tadej.justin,france.mihelic}@fe.uni-lj.si
http://www.luks.fe.uni-lj.si
[3] Faculty of Mathematics, Natural Sciences and Information Technologies,
University of Primorska, Glagoljaška 8, 6000 Koper, Slovenia
janez.zibert@upr.si
http://www.famnit.upr.si/en

Abstract. We compare the performance of two approaches when using cross-lingual data from different speakers to build bilingual speech synthesis systems capable of producing speech with the same speaker identity. One approach treats data from both languages as monolingual, by labeling all data with a manually joined phoneme set. Speaker independent voice is trained using the joined data, and adapted to the target speaker using the CMLLR adaptation.

In the second approach, speaker independent voices are trained for each language separately. State mapping between these voices is derived automatically from minimum Kullback–Leibler divergence between state distributions. The mapping is used to apply the adaptation transformations calculated within one language across languages to the other speaker independent voice.

We evaluate the quality of speech on MOS scale and similarity of synthesized speech characteristics to the target speaker using DMOS on the example of Croatian-Slovene language pair.

Keywords: bilingual, HMM, speech synthesis, phoneme mapping, state mapping, speaker adaptation, Kullback–Leibler divergence.

1 Introduction

In the paper we analyze approaches to bilingual speech synthesis based on cross-lingual adaptation within the HMM speech synthesis framework. The bilingual system should be able to synthesize speech in two languages with the same speaker identity. Creating voices with data from truly bilingual human speaker is possible [1], but not always practical due to possible difficulties of finding and recruiting speakers fluent in both languages for data collection. Cross-lingual speaker adaptation methods that allow construction of such voices without true bilingual data are preferable in these cases. Data

I. Habernal and V. Matousek (Eds.): TSD 2013, LNAI 8082, pp. 44–51, 2013.

from one speaker fluent in only one language is used to adapt models trained on data from multiple other speakers fluent in the other language, so that the adapted models sound like the adaptation speaker. There are multiple approaches to using such bilingual data from multiple speakers, and we compare two of the possible methods for the task of Croatian-Slovene bilingual synthesis. Systems for the Croatian and Slovene languages similar to those proposed in [2] are compared to systems developed using state mapping techniques proposed in [3]. According to [3] better results of cross-lingual speech adaptation in the task of creating personalized voices for speech-to-speech translation were achieved with state mapping between two languages' models than with manual mapping on phoneme level. It is of interest to test whether the same advantage occurs in the case of bilingual Croatian-Slovene synthesis.

We compare two approaches of building a bilingual speech synthesis system, i.e. a system that can produce speech in two different languages with the same speaker voice. It is assumed that the speaker whose voice is chosen for synthesis (the target speaker) does not speak both languages, so methods of cross-lingual speaker adaptation should be used to transform models of language foreign to that speaker to his/her identity.

In the first approach, we built a speaker independent average voice using speaker adaptive training [4] and joined data from both languages. The resulting bilingual average voice was then adapted to a selected target speaker using adaptation data labeled with phonemes from the joined phoneme set of both languages.

The second approach uses automatically derived cross-lingual state-mapping to apply the speaker adaptation transforms, calculated intra-lingually for one language, onto models of another language [3]. State level adaptation mapping techniques were applied in [3] to the problem of customizing voices for speech-to-speech translation. In our study, we decided to apply the similar method for creating a bilingual speech synthesis system, and compare the performance with the previously used phoneme mapping approach. State mapping approach to bilingual synthesis has also been used in [5], but the mapping was established using a corpus of bilingual speech from a single speaker who speaks two languages.

The rest of the paper is organized as follows: in Section 2 the training data, the phoneme mapping and the state mapping approaches are described. In Section 3 we describe the set of experiments and present the evaluation results. Finally we give a discussion of the results and a conclusion.

2   Bilingual Speech Synthesis Systems

We performed the experiments on the Croatian and Slovene languages, which are two closely related south Slavic languages. We chose one speaker from the data in each language as the target speaker, and three speakers for building the speaker - independent models. Then we built two bilingual systems for each target speaker using the two considered approaches. Thus, we obtained two bilingual speech synthesis systems with the voice identity of the target Croatian speaker, and two systems with the voice identity of the target Slovene speaker.

2.1 Training Data and Features

We used data from the Croatian VEPRAD [6] and Slovene VNTV [7] databases, both from radio news and weather report domains. Male speakers sm04 and 02m were chosen as target speakers from Croatian and Slovene corpus respectively. Two hundred utterances that provide rich phonetic coverage were chosen from each of the speakers for adaptation data. The duration of the data is 16 minutes and 26 seconds for the Croatian and 17 minutes and 27 seconds for the Slovene speaker.

For estimating the speaker independent models, we picked the data from three male speakers different than the target speakers from both databases in the same way as the target speaker data. Two hundred utterances with rich phonetic coverage were picked from each speaker, resulting in 53 minutes and 15 seconds of data for the Croatian set and 51 minutes and 50 seconds for the Slovene set.

For all systems we used the HTS speech synthesis toolkit [8], with the same acoustic features and model structure. The acoustic features consisted of 25 MFCC coefficients, their first and second order deltas, log F0 and its first order deltas, and energy. The features were extracted using the Hamming window with 5ms shift. The HMMs were composed of five states with no-skip left-to-right transitions, with one Gaussian mixture for each state. The Mel Log Spectrum Approximation (MLSA) filter [9] was used for synthesis from generated speech parameters.

2.2 Phoneme Mapping Approach System

The first considered approach is building a bilingual speaker independent voice using speaker adaptive training [4] with data from multiple speakers of both languages. To train the voice, all the data had to be labeled using a common phoneme set, which was adopted for this experiment from [2]. It was obtained by manual mapping between both languages' phoneme sets using the SAMPA alphabet [10] and joining the common phonemes, while keeping the phoneme unique to a language separate. The joined phoneme set consists of 44 phonemes, of which 33 were shared between both languages, 7 were unique to the Slovene (further referred to as SI) and 4 were unique to the Croatian language (further referred to as HR). The obtained speaker independent bilingual average voice is then used as basis for adaptation to the target speaker's identity. Target speaker's data is also labeled using the joined phoneme set, and speaker adaptation is done in the usual intra-lingual way by using constrained maximum likelihood linear regression (CMLLR)[11]. We performed the CMLLR adaptation of the same average voice to the two target speakers, giving two bilingual voices with identities of Croatian speaker sm04 and Slovene speaker 02m. Each combination of voice and target language was marked according to labels in Table 1. The resulting voice may be considered truly bilingual, as the same models are used to produce speech in both languages involved. Similar experiments with creating multilingual speaker independent models with phoneme mapping were reported in [12,2].

In order to synthesize the speech in either language, language-specific dictionary or grapheme-to-phoneme mapping rules were used to transform the input text into the phonetic transcription in the joined phoneme set, and then synthesis was performed using the bilingual models.

Table 1. System configurations for the phoneme mapping approach (S1). One system has the identity of speaker sm04, and the other of speaker 02m. In system labels, X denotes different languages of adaptation speaker and for synthesis (cross-lingual situation), I denotes same language of adaptation speaker and synthesis (intra-lingual) and HR and SI denote languages for synthesis.

System label	S1-IHR	S1-XSI	S1-ISI	S1-XHR
Adaptation speaker	sm04		02m	
Adaptation speaker language	HR	HR	SI	SI
Synthesis language	HR	SI	SI	HR

2.3 State Mapping Approach

The second approach uses intra-lingual speaker adaptation for synthesis within the target speaker's language, and cross-lingual speaker adaptation for the language foreign to that speaker. Intra-lingual speech adaptation transforms models of one source speaker using data from another (target) speaker so that the transformed models more closely match the speaker characteristics of the target speaker. The source and the target speaker speak the same language. However, in the cross-lingual adaptation, data of the source and target speakers are not in the same language so the adaptation cannot be applied directly.

For both Slovene and Croatian languages, language - dependent, speaker - independent average voices were trained separately using the same data as with the phoneme mapping approach. The data was labeled in each language's own phoneme set and used to train the average voices in the same manner as above. With the CMLLR speaker adaptation we obtain the language dependent synthetic voice with the characteristics of a target speaker for the intra-lingual task.

To obtain the cross-language synthetic voice we used the transform mapping approach [3]. First we calculated the KLD divergence between the states of previously obtained language-dependent average voice models. We found the nearest state in the target language models with a minimum KLD criteria for each state in the source language models. With the use of CMLLR transform matrix for the target voice obtained from intra-lingual speaker adaptation, we adapt the mapped model states in the target language. Such adaptation was conducted only with spectral and F0 parameters, while duration models were left unchanged. This means that at synthesis time, phoneme durations were always determined by the model trained only on data in the synthesis language. This is in contrast with the first approach, where duration models were trained on joined data, and both languages contributed to estimated duration parameters of the joined phonemes. All the voice and target language configurations for the state mapping approach are marked according to labels in Table 2.

3   Experiments and Evaluation

The evaluation focused on two aspects: quality of synthesized speech in general as perceived by listeners and speaker similarity, i.e. how well do speech characteristics of synthetic speech match the characteristics of the target speaker. Two formal listening tests were conducted in the experiment to obtain the scores. Both Croatian and Slovene were used as target languages, and the tests were done separately in Croatia and Slovenia with native speakers of both languages (14 Croatian and 18 Slovene). The quality of the speech was evaluated on five-point mean opinion score (MOS) [13] scale individually for each system configuration. For the speaker similarity we used a degradation mean opinion score (DMOS) [13], with reference samples from a speaker dependent synthetic voice. Using each system we synthesized 50 sentences randomly picked from the VEPRAD and VNTV databases that have not been used in training or as adaptation data. Summary of all tested system configurations is given in Table 3 with respect to language of data used for average voice, language of adaptation speaker and target language for synthesis. For the MOS scores, the listeners only listened samples in their own language. For each system configuration, listeners were presented with speech sample about 6-9 s long, and rated the speech on the Likert scale from 1 (worst) to 5 (best). The listeners could repeat the sample before giving a grade, but could not go back and change the rating of previously rated system.

In the DMOS scores, listeners were presented with two samples, first the reference speaker-dependent voice sample, and then the tested system's sample. The task was to rate how similar the voices sound with respect to speaker identity and not the content or language of the utterance. Synthesized speech was used as the reference sample instead of a natural sample selected from the database, so that both the reference and the test sample are processed in the same way (using the MLSA filter). Since target speakers only speak one of the languages, in the speaker similarity score task the listeners heard the reference sample in their language once per system, and once in the other language. The listeners were instructed in that case to try and disregard the language spoken and concentrate on the voice characteristics independent of language.

The MOS results are shown in Fig. 1, and the speaker similarity DMOS scores in Fig. 2.

4   Discussion and Conclusion

The evaluation results in Figure 1 and 2 show that in the intra lingual situation the phone mapping approach performs similarly to the state mapping approach for both tested languages, with trend showing slightly better performance of phoneme mapping approach. For the cross-lingual situation, when the target language is Slovene, both approaches

Table 2. System configuration for the state-mapping approach (S2). One system has the identity of speaker sm04, and the other of speaker 02m. In system labels, X denotes the cross-lingual scenario (language for synthesis and average voice is different from the language of adaptation speaker), I denotes the intra-lingual scenario (same language for all data and synthesis) and HR and SI denote languages for synthesis.

System label	S2-IHR	S2-XSI	S2-ISI	S2-XHR
Adaptation speaker	sm04		02m	
Average voice language	HR	SI	SI	HR
Adaptation speaker language	HR	HR	SI	SI
Synthesis language	HR	SI	SI	HR

Table 3. Summary of tested systems language configuration

	Cross-lingual				Intra-lingual			
	S1-XHR	S1-XSI	S2-XHR	S2-XSI	S1-ISI	S1-IHR	S2-ISI	S2-IHR
Average voice language	both	both	HR	SI	both	both	SI	HR
Adaptation speaker language	SI	HR	SI	HR	SI	HR	SI	HR
Synthesis language	HR	SI	HR	SI	SI	HR	SI	HR

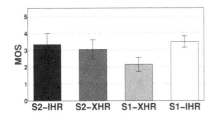 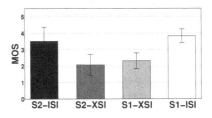

Fig. 1. Mean opinion scores for systems. Left: Croatian synthesis language, right: Slovene synthesis language.

again perform similarly, but the scores were significantly lower than in the intra-lingual situation. For Croatian as the target language, the phoneme mapping approach gives worse results than in intra-lingual case, however the state mapping approach performs almost as good in the cross-lingual scenario as in the intra-lingual.

In both intra-lingual and cross-lingual scenarios, the similarity to the target speaker is as good as or better with phoneme mapping approach. With Croatian as target language, the speaker similarity drops when going from intra-lingual to cross-lingual scenario, as well as when using state mapping instead of phoneme mapping. For Slovene, in all cases the speaker similarity score is about the same, except for state mapping cross-lingual scenario (S2-XSI) which has significantly lower scores.

If we compare systems with the same adaptation speaker, we can see that the system S2-ISI/S2-XHR with speaker identity 02m, which uses the state mapping approach, performs consistently in both langues, with mean scores 3.33 for Slovene S2-ISI and 3 for Croatian S2-XHR. Pair S2-IHR/S2-XSI built using the same approach, with speaker

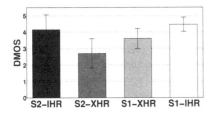 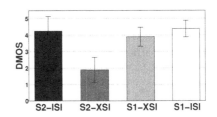

Fig. 2. Similarity to target speaker (degradation mean opinion score). Left: Croatian synthesis language, right: Slovene synthesis language.

identity of speaker sm04, performs worse for Slovene (S2-XSI MOS=1.89) than for Croatian (S2-IHR MOS=3.21).

These results suggest that although there are differences in the two approaches of phoneme mapping, the target speaker itself may also play a role in the performance of the bilingual systems. Therefore additional experiments with different target speakers are needed to verify whether the target language or the target speaker (or both) influence the results.

The relatively good performance of the phoneme mapping approach may be partly explained by the joined data that were used for estimating the average voice. That means that the models were trained using the joined amount of data, which was therefore doubled in the comparison to the data of the state mapping approach, and that possibly resulted in better estimation of acoustic parameters. However, in the phoneme mapping systems S1* the duration model for each phoneme is also shared between the languages and derived from the joined data. Although acoustically similar, these phonemes may have different duration patterns in different languages, and the duration model might not benefit from the joined data. In the systems S2*, the duration models are trained on data from single language and adaptation only affects the spectral parameters of the model. The data from adaptation speaker thus modifies speaker characteristics, but keeps language-dependent duration patterns from a single language.

In the tested case, total amount of data for the average voices in each language was rather small (50 min), and possibly insufficient for good estimation of all models. It should be tested if the same benefit of joined data of the phoneme mapping approach would still hold if more data was used for the average voices. Similarity of the two considered languages may also justify the phoneme mapping approach, where many phonemes can be simply matched together, while this is not the case in languages from different language groups. We believe that in later case the automatically derived data driven phoneme mapping could play an important role in building HMM speech synthesis systems. This is a point in our future work, as well as extending the bilingual approach to multilingual synthesis for similar languages in the Slavic group (Czech, Slovak, Polish).

References

1. Traber, C., Huber, K., Nedir, K., Pfister, B., Keller, E., Zellner, B.: From multilingual to polyglot speech synthesis. In: Proc. of the Eurospeech, vol. 99, pp. 835–838 (1999)
2. Justin, T., Pobar, M., Ipšić, I., Mihelič, F., Žibert, J.: A bilingual HMM-based speech synthesis system for closely related languages. In: Sojka, P., Horák, A., Kopeček, I., Pala, K. (eds.) TSD 2012. LNCS, vol. 7499, pp. 543–550. Springer, Heidelberg (2012)
3. Wu, Y.J., Nankaku, Y., Tokuda, K.: State mapping based method for cross-lingual speaker adaptation in hmm-based speech synthesis. In: Proc. of Interspeech, pp. 528–531 (2009)
4. Yamagishi, J., Masuko, T., Tokuda, K., Kobayashi, T.: A training method for average voice model based on shared decision tree context clustering and speaker adaptive training. In: Proceedings of ICASSP 2003, vol. 1, I–716–I–719 (2003)
5. Liang, H., Qian, Y., Soong, F.K., Liu, G.: A cross-language state mapping approach to bilingual (mandarin-english) tts. In: ICASSP 2008, pp. 4641–4644. IEEE (2008)
6. Martincic-Ipsic, S., Ipsic, I.: Veprad: a croatian speech database of weather forecasts. In: Information Technology Interfaces, ITI 2003, pp. 321–326 (2003)

7. Žibert, J., Mihelič, F.: Slovenian weather forecast speech database. In: Proc, Softcom, vol. 1, pp. 199–206 (October 2000)
8. Zen, H., Nose, T., Yamagishi, J., Sako, S., Masuko, T., Black, A.W., Tokuda, K.: The hmm-based speech synthesis system (hts) version 2.0. In: Proc. of Sixth ISCA Workshop on Speech Synthesis, pp. 294–299 (2007)
9. Imai, S., Sumita, K., Furuichi, C.: Mel log spectrum approximation (MLSA) filter for speech synthesis. Electronics and Communications in Japan (Part I: Communications) 66(2), 10–18 (1983)
10. Wells, J.C.: SAMPA computer readable phonetic alphabet. In: Handbook of Standards and Resources for Spoken Language Systems. Walter de Gruyter, Berlin (1997)
11. Yamagishi, J., Ogata, K., Nakano, Y., Isogai, J., Kobayashi, T.: Hsmm-based model adaptation algorithms for average-voice-based speech synthesis. In: ICASSP 2006 Proceedings, vol. 1, p. 1 (2006)
12. Latorre, J., Iwano, K., Furui, S.: New approach to the polyglot speech generation by means of an hmm-based speaker adaptable synthesizer. Speech Communication 48(10), 1227–1242 (2006)
13. International Telecommunication Union: ITU-T Recommendation P.800.1: Mean Opinion Score (MOS) terminology. Technical report (2006)

A Direct Criterion Minimization Based fMLLR via Gradient Descend

Jan Vaněk and Zbyněk Zajíc

University of West Bohemia in Pilsen, Univerzitní 22, 306 14 Pilsen
Faculty of Applied Sciences, Department of Cybernetics
{vanekyj,zzajic}@kky.zcu.cz

Abstract. Adaptation techniques are necessary in automatic speech recognizers to improve a recognition accuracy. Linear Transformation methods (MLLR or fMLLR) are the most favorite in the case of limited available data. The fMLLR is the feature-space transformation. This is the advantage with contrast to MLLR that transforms the entire acoustic model. The classical fMLLR estimation involves maximization of the likelihood criterion based on individual Gaussian components statistic. We proposed an approach which takes into account the overall likelihood of a HMM state. It estimates the transformation to optimize the ML criterion of HMM directly using gradient descent algorithm.

Keywords: ASR, adaptation, fMLLR, gradient descend, Hessian matrix.

1 Introduction

Nowadays, systems of automatic speech recognition (ASR) are based on Hidden Markov Models (HMMs) with output probabilities described mainly by Gaussian Mixture Models (GMMs) [1]. To recognize the speech from a recording, one could train a Speaker Dependent (SD) model for each of the speaker present in the recording. However, this is in praxis often intractable because of the need of a large database of utterances coming from one speaker. Instead, so called Speaker Independent (SI) model is trained from large amount of data collected from many speakers, and subsequently, the SI model is adapted to better capture the voice of the talking person. Thus, the SD model is acquired.

Well known adaptation methods are Maximum A-posteriori Probability (MAP) technique [2] and Linear Transformations based on Maximum Likelihood (LTML), as model adaptation Maximum Likelihood Linear Regression (MLLR) [3] or feature Maximum Likelihood Linear Regression (fMLLR) [4]. In this paper we have chosen out of LTML based adaptations preferably the feature transformations which are well suited for on-line adaptation, see [5].

The classical fMLLR approach using the row-by-row estimation of the adaptation matrix requires data accumulated with respect to the individual Gaussians [6]. In our proposed method a direct minimization of a criterion function is applied. Our criterion is based on likelihood of whole HMM states. The adaptation parameters are estimated via gradient descend method [7]. We used Newton's method with diagonal Hessian

I. Habernal and V. Matousek (Eds.): TSD 2013, LNAI 8082, pp. 52–59, 2013.

matrix to speed-up a convergence of the estimation process. Moreover, we modified the ML criterion to be less sensitive to phones length.

This paper is organized as follows. In Section 2, an idea of speaker adaptation is described. Particular techniques for feature adaptation, classical fMLLR approach, is presented in Section 3. The proposed approach for finding the fMLLR adaptation matrices using gradient techniques is discussed in Section 4. Experimental results are presented in Section 5.

2   Adaptation Techniques

The difference between the adaptation and ordinary training methods stands in the prior knowledge about the distribution of model parameters, usually derived from the SI model [8]. The adaptation adjusts the model in order to maximize the probability of adaptation data. Hence, the new, adapted parameters can be chosen as

$$\boldsymbol{\lambda}^* = \arg\max_{\boldsymbol{\lambda}} p(\boldsymbol{O}|\boldsymbol{\lambda})p(\boldsymbol{\lambda}), \tag{1}$$

where $p(\boldsymbol{\lambda})$ stands for the prior information about the distribution of the vector $\boldsymbol{\lambda}$ containing model parameters, $\boldsymbol{O} = \{\boldsymbol{o}_1, \boldsymbol{o}_2, \ldots, \boldsymbol{o}_T\}$ is the sequence of T feature vectors related to one speaker, $\boldsymbol{\lambda}^*$ is the best estimate of the SD model parameters. We will focus on HMMs with output probabilities of states represented by GMMs. GMM of the j-th state is characterized by a set $\boldsymbol{\lambda}_j = \{\omega_{jm}, \boldsymbol{\mu}_{jm}, \boldsymbol{C}_{jm}\}_{m=1}^{M_j}$, where M_j is the number of mixtures, ω_{jm}, $\boldsymbol{\mu}_{jm}$ and $\boldsymbol{C}_{jm}$ are weight, mean and variance of the m-th mixture, respectively.

The advantage of LTML techniques over the MAP technique is that the number of available model parameters is reduced via clustering of similar model components [9]. The transformation is the same for all the parameters from the same cluster $K_n, n = 1, \ldots, N$. Hence, less amount of adaptation data is needed. In the extreme case, so-called global adaptation, only one adaptation matrix for all model components is computed from all the adaptation data.

The main difference between these MLLR and fMLLR approaches stands in the area of their interest. MLLR transforms means and covariances of the model, whereas fMLLR transforms directly the acoustic feature vectors. The MLLR method is out of our interest and the adaptation formulas can be found in [3].

3   Feature Maximum Likelihood Linear Regression (fMLLR)

The classical approach to this method is based on the minimization of the auxiliary function [4]:

$$Q(\boldsymbol{\lambda}, \bar{\boldsymbol{\lambda}}) = const - \frac{1}{2} \sum_{jm} \sum_t \gamma_{jm}(t)(const_{jm} + \log|\boldsymbol{C}_{jm}| + \\ + (\bar{\boldsymbol{o}}(t) - \boldsymbol{\mu}_{jm})^{\mathrm{T}} \boldsymbol{C}_{jm}^{-1}(\bar{\boldsymbol{o}}(t) - \boldsymbol{\mu}_{jm})), \tag{2}$$

where $\bar{o}(t)$ represents the feature vector transformed according to the formula:

$$\bar{o}_t = A_{(n)}o_t + b_{(n)} = W_{(n)}\xi(t), \quad (3)$$

where $W_{(n)} = [A_{(n)}, b_{(n)}]$ stands for the transformation matrix corresponding to the $n - th$ cluster K_n and $\xi(t) = [o_t^T, 1]^T$ represents the extended feature vector.

The classical implementation of fMLLR (or other adaptation based on linear transformation) requires four steps [10]:

1. **Alignment**
   of the adaptation utterance to HMM states. This can be done by *forced-alignment* or more time demanding but more accurate *forward-backward algorithm* [8]. Both approaches need transcription of adaptation utterance. This transcription can by done as reference transcription (supervised adaptation) or can by required from the first pass of the ASR (unsupervised adaptation). The result of alignment is probability $p(o(t)|jm)$ that feature $o(t)$ is generated by m-th mixture of the j-th state of the HMM. Posterior probability $\gamma_{jm}(t)$ of feature $o(t)$ is given as

$$\gamma_{jm}(t) = \frac{\omega_{jm}p(o(t)|jm)}{\sum_{m=1}^{M}\omega_{jm}p(o(t)|jm)} \quad (4)$$

2. **Computation**
   of the soft count c_{jm} of mixture m and the first and the second statistics moment, $\varepsilon_{jm}(o)$ and $\varepsilon_{jm}(oo^T)$, of features which align to mixture m in the j-th state of the HMM

$$c_{jm} = \sum_{t=1}^{T}\gamma_{jm}(t), \quad (5)$$

$$\varepsilon_{jm}(o) = \frac{\sum_{t=1}^{T}\gamma_{jm}(t)o(t)}{\sum_{t=1}^{T}\gamma_{jm}(t)}, \quad \varepsilon_{jm}(oo^T) = \frac{\sum_{t=1}^{T}\gamma_{jm}(t)o(t)o(t)^T}{\sum_{t=1}^{T}\gamma_{jm}(t)} \quad (6)$$

Note that $\sigma_{jm}^2 = \text{diag}(C_{jm})$ is the diagonal of the covariance matrix C_{jm}.

3. **Accumulation**
   of the statistics matrices $G_{(n)i}$ and $k_{(n)i}$ for each cluster n of similar model components [9] and for i-row of the adaptation matrix $W_{(n)}$

$$k_{(n)i} = \sum_{m \in K_n}\frac{c_m \mu_{mi}\varepsilon_m(\xi)}{\sigma_{mi}^2}, \quad G_{(n)i} = \sum_{m \in K_n}\frac{c_m\varepsilon_m(\xi\xi^T)}{\sigma_{mi}^2} \quad (7)$$

where

$$\varepsilon_m(\xi) = \left[\varepsilon_m^T(o), 1\right]^T, \quad \varepsilon_m(\xi\xi^T) = \begin{bmatrix} \varepsilon_m(oo^T) & \varepsilon_m(o) \\ \varepsilon_m^T(o) & 1 \end{bmatrix}. \quad (8)$$

4. **Iterative Update**
   of estimated matrix $W_{(n)}$. The auxiliary function (2) can be rearranged into the form [6]

$$Q_{W_{(n)}}(\lambda, \bar{\lambda}) = \log|A_{(n)}| - \sum_{i=1}^{I}w_{(n)i}^T k_i - 0.5w_{(n)i}^T G_{(n)i}w_{(n)i}, \quad (9)$$

To find the solution of equation (9) we have to express $A_{(n)}$ in terms of $W_{(n)}$, e.g. use the equivalency $\log |A_{(n)}| = \log |w_{(n)i}^T v_{(n)i}|$, where $v_{(n)i}$ stands for transpose of the i-th row of cofactors of the matrix $A_{(n)}$ extended with a zero in the last dimension. After the maximization of the auxiliary function (9) we receive

$$\frac{\partial Q(\lambda, \bar{\lambda})}{\partial W_{(n)}} = 0 \Rightarrow w_{(n)i} = G_{(n)i}^{-1} \left(\frac{v_{(n)i}}{\alpha_{(n)}} + k_{(n)i} \right), \tag{10}$$

where $\alpha_{(n)} = w_{(n)i}^T v_{(n)i}$ can be found as the solution of the quadratic function

$$\beta_{(n)} \alpha_{(n)}^2 - \alpha_{(n)} v_{(n)i}^T G_{(n)i}^{-1} k_{(n)i} - v_{(n)i}^T G_{(n)i}^{-1} v_{(n)i} = 0, \tag{11}$$

where

$$\beta_{(n)} = \sum_{m \in K_n} \sum_t \gamma_m(t). \tag{12}$$

Two different solutions $w_{(n)i}^{1,2}$ are obtained, because of the quadratic function (11). The one that maximizes the auxiliary function (9) is chosen. Note that an additional term appears in the log likelihood for fMLLR because of the feature transforms, hence:

$$\log \mathcal{L} \left(o_t | \mu_m, C_m, A_{(n)}, b_{(n)} \right) = $$
$$= \log \mathcal{N} \left(A_{(n)} o_t + b_{(n)}; \mu_m, C_m \right) + 0.5 \log |A_{(n)}|^2. \tag{13}$$

The estimation of $W_{(n)}$ is an iterative procedure. Matrices $A_{(n)}$ and $b_{(n)}$ have to be correctly initialized first, e.g. $A_{(n)}$ can be chosen as a diagonal matrix with ones on the diagonal and $b_{(n)}$ can be initialized as a zero vector. The estimation ends when the change in parameters of transformation matrices is small enough (about 20 iterations are sufficient) [6].

4   Gradient Descent fMLLR

Classical fMLLR is based on a row-by-row estimation of the adaptation matrix $W_{(n)}$ with respect to data accumulated for each Gaussian. The main difference in our gradient descend fMLLR technique is a direct minimization of a criterion function [10]. From classical fMLLR described above, only the first step of the estimation - *alignment* - is identical. The rest of the estimation is modified to direct minimization of the criterion function.

We do not consider individual Gaussians only. We consider negative Maximal Likelihood (ML) criterion that is based on likelihood of whole HMM states (see Figure 1). In contrast with classical fMLLR approach, adapted data are transformed into the center of the HMM state instead of the center of the Gaussian only.

The same approach can be used for various alternative differentiable criteria (e.g. Maximal Mutual Information or other discriminative ones). The minimization of the criteria formally written is similar to the equation (1)

$$\lambda^* = \arg \min_\lambda \mathcal{F}(O, \lambda), \tag{14}$$

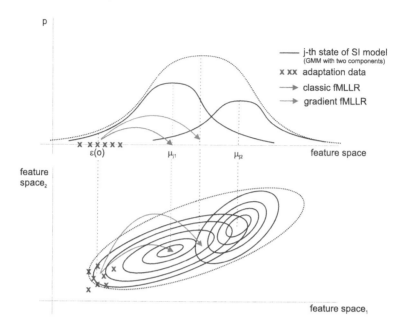

Fig. 1. Visualization of the fMLLR adaptation base on classical estimation and on our proposed estimation using gradient approach

where $\mathcal{F}(O, \lambda)$ is the criterion function which is the negative ML criterion in our case

$$\mathcal{F}(O, \lambda) = -p(O|\lambda)p(\lambda). \tag{15}$$

We choose the gradient descend method to optimize parameters λ because it is the most general optimization technique. Therefore, it can be used with various criteria and it may optimize even other parameters, not only the fMLLR transformation matrix [7]. So, the same framework can be developed further. In our case of ML criterion, even second derivatives - diagonal Hessian - can be easily calculated and the second order Newton optimization method can by employed to reduce a number of the optimization iterations.

For single Gaussian case, the partial derivation of one element a_{ij} of the transformation matrix $A_{(n)}$ is

$$\frac{\partial \mathcal{F}}{\partial a_{ij}} = \frac{\mu_i - \bar{o}_i(t)}{\sigma_i^2} o_j(t), \tag{16}$$

and the diagonal Hessian element - the second partial derivation is

$$\frac{\partial^2 \mathcal{F}}{\partial a_{ij}^2} = -\frac{o_j^2(t)}{\sigma_i^2}. \tag{17}$$

The partial derivations for the fMLLR vector $b_{(n)}$ are

$$\frac{\partial \mathcal{F}}{\partial b_i} = \frac{\mu_i - \bar{o}_i(t)}{\sigma_i^2} \tag{18}$$

and

$$\frac{\partial^2 \mathcal{F}}{\partial b_i^2} = -\frac{1}{\sigma_i^2}. \tag{19}$$

Besides the sum of partial derivatives over all data, the $\log(\det(\boldsymbol{A}_{(n)}))$ derivation needs to be added. The derivative is equal to $inv(\boldsymbol{A}_{(n)})^T$. The second derivative of $\log(\det(\boldsymbol{A}_{(n)}))$ is computed numerically.

The total partial derivatives for entire HMM is a sum of all the individual Gaussians with using the same γ_{jm} as in the equations (5) and (6) obtained during the *alignment* stage.

Then, the new estimate of $\boldsymbol{A}_{(n)}$ in iteration k is

$$\boldsymbol{A}_{(n)}^k = \boldsymbol{A}_{(n)}^{k-1} - \alpha \frac{1}{2} \frac{\frac{\partial \mathcal{F}}{\partial \boldsymbol{A}_{(n)}^{k-1}}}{\frac{\partial^2 \mathcal{F}}{\partial \boldsymbol{A}_{(n)}^{2(k-1)}}}, \tag{20}$$

where α is a stabilization constant from interval $\langle 0, 1 \rangle$. The stabilization together with an iterative approach must be used because we use only the diagonal Hessian which is inaccurate. The used γ_{jm} are also dependent on the derived parameters, but it makes the derivatives too complicated. Therefore, we ignore theirs influence and the gammas are treated as fixed constants. It brings an additional inaccuracy which involves a need of the iterative stabilized approach.

4.1 Modified ML Criterion

A classic ML criterion has uniform influence over all processed feature-vectors. It means that long phones or non-speech models have a higher total influence than shorter phones. Therefore, we modified the criterion to compute per-state means of the ML criterion and than the total sum is calculated from the means. But, some states with a few accumulated feature-vectors may disturb the final estimates. Therefore, we proposed a smooth fade-out of the low-occupied states via soft threshold τ. The per-state means are summed with using a state weight w_j

$$w_j = \frac{\sum_{m=1}^{M} c_{jm}}{\tau + \sum_{m=1}^{M} c_{jm}}. \tag{21}$$

The same weights are used to compute first and second order of the partial derivatives.

5 Experiments

5.1 SpeechDat-East (SD-E) Corpus

For experiment purposes we used the Czech part of SpeechDat-East corpus [11]. In order to extract the features Mel-frequency cepstral coefficients (MFCCs) were utilized, 11 dimensional feature vectors were extracted each 10 ms utilizing a 32 ms hamming window, Cepstral Mean Normalization (CMN) was applied, and Δ, Δ^2 coefficients were added. A 3 state HMM based on triphones with 2105 states total and 8-components GMM with diagonal covariances in each of the states was trained on 700 speakers with 50 sentences for each speaker (cca 5 sec. on a sentence). To test the systems performance, different 200 speakers from SD-E were used with 50 sentences for each speaker, however a maximum of 12 sentences was used for the adaptation. A language model based on trigrams used in the recognition [12]. The vocabulary consisted of 7000 words.

5.2 Results

The results of the experiment are shown in Table 1. The table contains Accuracy (Acc) of the baseline system (recognition done utilizing only the SI model), the classical fMLLR and the gradient fMLLR approach. We estimate only one global matrix via 20 iterations. The setting of the gradient fMLLR: $\tau = 14$ and $\alpha = 0.5$.

Table 1. Accuracy (Acc)[%] of transcribed words for each type of the adaptation

	supervised	unsupervised[%]
SI model	74.27	74.27
classical fMLLR	78.67	77.37
gradient fMLLR	78.99	77.66

As can be seen from Table 1, the proposed gradient fMLLR approach performed better than the classical fMLLR. The margin is not large but significant and it is obtained for both cases, supervised as well as unsupervised adaptation.

6 Conclusion

We proposed an approach which takes into account the overall likelihood of a HMM state. It estimates the transformation to optimize the ML criterion of HMM directly using the gradient descent algorithm. The criterion is based on likelihood of whole HMM states. It is better than the classical fMLLR which considers a likelihood of individual Gaussians only. The experiment results show improvement over the classical fMLLR method. Additional advantage of our approach is a compatibility with other differentiable criteria, especially the discriminative ones.

Acknowledgements. This research was supported by the Technology Agency of the Czech Republic, project No. TA01030476.

References

1. Rabiner, L.R.: A Tutorial on Hidden Markov Models and Selected Applications in Speech Recognition. In: Readings in Speech Recognition, pp. 267–296 (1990)
2. Gauvain, L., Lee, C.H.: Maximum A-Posteriori Estimation for Multivariate Gaussian Mixture Observations of Markov Chains. IEEE Transactions SAP, 2:291–2:298 (1994)

3. Leggeter, C.J., Woodland, P.C.: Maximum Likelihood Linear Regression for Speaker Adaption of Continuous Density Hidden Markov Models. Computer Speech and Language, 9:171–9:185 (1995)
4. Gales, M.J.F.: Maximum Likelihood Linear Transformation for HMM-based Speech Recognition. Tech. Report, CUED/FINFENG/TR291, Cambridge Univ. (1997)
5. Machlica, L., Zajíc, Z., Pražák, A.: Methods of Unsupervised Adaptation in Online Speech Recognition. In: Specom. St.Petersburg (2009)
6. Povey, D., Saon, G.: Feature and Model Space Speaker Adaptation with Full Covariance Gaussians. In: Interspeech, paper 2050-Tue2BuP.14 (2006)
7. Visweswariah, K., Gopinath, R.: Adaptation of front end parameters in a speech recognizer. In: Interspeech, pp. 21–24 (2004)
8. Psutka, J., Müller, L., Matoušek, J., Radová, V.: Mluvíme s počítačem česky, Academia, Praha (2007) ISBN:80-200-1309-1
9. Gales, M.J.F.: The Generation and use of Regression class Trees for MLLR Adaptation. Cambridge University Engineering Department (1996)
10. Balakrishnan, S.V.: Fast incremental adaptation using maximum likelihood regression and stochastic gradient descent. In: Eurospeech, pp. 1521–1524 (2003)
11. Pollak, P., et al.: SpeechDat(E) - Eastern European Telephone Speech Databases, XLDB - Very Large Telephone Speech Databases (ELRA), Paris (2000)
12. Pražák, A., Psutka, J.V., Hoidekr, J., Kanis, J., Müller, L., Psutka, J.: Automatic Online Subtitling of the Czech Parliament Meetings. In: Sojka, P., Kopeček, I., Pala, K. (eds.) TSD 2006. LNCS (LNAI), vol. 4188, pp. 501–508. Springer, Heidelberg (2006)

A Machine Learning Based Approach for Vocabulary Selection for Speech Transcription

Denis Jouvet[1,2,3] and David Langlois[1,2,3]

[1] Speech Group, LORIA
Inria, Villers-lès-Nancy, F-54600, France
[2] Université de Lorraine, LORIA, UMR 7503, Villers-lès-Nancy, F-54600, France
[3] CNRS, LORIA, UMR 7503, Villers-lès-Nancy, F-54600, France
denis.jouvet@inria.fr, david.langlois@loria.fr

Abstract. This paper introduces a new approach based on neural networks for selecting the vocabulary to be used in a speech transcription system. Indeed, nowadays, large sets of text data can be collected from web sources, and used in addition to more traditional text sources for building language models for speech transcription systems. However, web data sources lead to large amounts of heterogeneous data, and, as a consequence, standard vocabulary selection procedures based on unigram approaches tend to select unwanted and undesirable items as new words. As an alternative to unigram-based and empirical manual-based selection approaches, this paper proposes a new selection procedure that relies on a machine learning technique, namely neural networks. The paper presents and discusses the results obtained with the various selection procedures. The neural network based selection experiments are promising and they can handle automatically various detailed information in the selection process.

Keywords: vocabulary selection, neural network, language modeling, speech transcription, speech recognition.

1 Introduction

The choice of the recognition vocabulary of a speech recognition system has a strong impact on the recognition performance. Indeed, in average, each out-of-vocabulary (OOV) word produces 1.2 errors [1].

For a new speech recognition task in a new domain, one has to choose carefully the words to be included in the vocabulary. When only one training corpus is available, and when this corpus is homogeneous, the method to select words is straightforward: the vocabulary is composed of the most frequent words in the training corpus. But, nowadays, numerous and heterogeneous training corpora are available. These training corpora differ strongly in terms of source (journalism, web, radio transcriptions, newswire texts, etc.), size (from a few million words up to more than several hundred million words), time period (very recent or many years old), and domain. It is not suitable to simply concatenate all the corpora and to select the most frequent words. For example, a proper name having a low frequency because it has just recently appeared in the news, may be very relevant for a broadcast news transcription system, and thus should be kept. Such a

I. Habernal and V. Matousek (Eds.): TSD 2013, LNAI 8082, pp. 60–67, 2013.
© Springer-Verlag Berlin Heidelberg 2013

word would have a low frequency at the whole corpus scale, and consequently may be excluded by the frequency criterion. Hence the frequency in the overall corpora should not be the only criterion to use.

To deal with this problem, it is possible to rely on selection methods whether automatic or not which take into account the unigram distributions of the words in different sub-corpora. One approach consists in finding the linear combination of the sub-corpora unigram distributions that provides the closest match with the unigram distribution of the validation set [2]; then the words having the largest unigram values (according to the combined unigram distribution) are selected. The Expectation-Maximization (E.M.) algorithm is used to determine the optimal weights of the linear combination. The objective function the E.M. algorithm optimizes is the Kullback-Leibler distance between the unigram distribution corresponding to the linear combination of the unigrams estimated on each sub-corpus, and the unigram distribution estimated on the validation corpus. The authors have shown that their approach is efficient compared to the algorithm consisting in choosing from each training corpus the most frequent words. Moreover, they have shown that the type of segmentation of the training corpus into several sub-corpora (for example segmentation by year, or by source) impacts on the performance. A similar approach was presented in [3] where the weights of the linear combination are also estimated by the E.M. algorithm. In addition, [3] proposes an alternative approach which estimates the weights given to each training sub-corpus directly according to the Kullback-Leibler distance between the unigram distributions of this sub-corpus and of the validation corpus. The authors show that for very large vocabularies (our case), both methods perform equally well in terms of vocabulary coverage.

In those studies, the decision of selecting a word into the vocabulary depends only on its probability as defined by the linear combination of all training sub-corpora unigrams. We argue that the importance of a word depends also on its use in the language. Two words with the same frequency may appear in few or numerous distinct n-grams. We think that such an information is useful for deciding to keep or not a word, because if a word appears in numerous contexts, it will be useful in more different cases during recognition. Such information has not yet been used in previous works.

In this paper, we propose to use a neural network based approach in order to decide whether to select or not a word on the basis of the word frequencies estimated on each training sub-corpus, and also on its presence in various n-grams. Such approach is prone to further extensions with richer word information (such as number of documents in which the word occurs, tf-idf, and so on [4]...) because neural networks allow easily to increase the size of the input vector.

The article is structured as follows. Section 2 describes the different selection procedures evaluated, and, in particular details the neural network approach. In section 3, we describe our experimental set-up. Section 4 presents and discusses the results in terms of coverage, perplexity and speech recognition performance. Finally, a conclusion and some perspectives end the paper.

2   Vocabulary Selection

This section introduces the text corpora used in the experiments, and describes the selection methods that are evaluated and discussed in section 4.

2.1   Text Corpora

We used various text corpora for vocabulary selection and language modeling. This includes more than 500 million words of newspaper data from 1987 to 2007; several million words from transcriptions of various radio broadcast shows; more than 800 million words from the French Gigaword corpus [5] from 1994 to 2008; plus 300 million words of web data collected in 2011 from various web sources, and thus mainly covering recent years.

All this text data has been divided into 20 subsets with respect to the time period of the collection (roughly from 1987 to 1997, from 1998 to 2005, and finally, 2006 and after), and the nature of the data (newspaper, broadcast transcriptions, newswire, etc.).

2.2   Unigram-Based Selection

The unigram-based method is described in [3] (and was summarized in the introduction of this paper). The SRI Toolkit [6] implements it and was used to obtain the selected vocabulary. Words are sorted according to their unigram score, and the top of the list is selected.

2.3   Manual Empirical Selection

The manual empirical selection proceeds in an incremental way. First, words appearing at least one time every year in the newspaper data are selected. Then, the same process is applied for the Gigaword data: words appearing at least once every year are also selected. The web data were split into 16 subsets according to the web site they are coming from. Words appearing at least once in each of these 16 subsets are selected. This way we hope to select recent data (as the data was collected in 2011) and at the same time avoid undesirable data occurring only in a few subsets (i.e. coming from only a few web sites). The selected word list was then extended by adding words corresponding to the largest French cities, and to the countries. Overall we created a list of almost 85,000 different words.

2.4   Machine Learning Based Approach

In this new proposed approach, a multi-layer neural network (multi-layer perceptron) is used to learn the decision function that will then be used to select the words. The set of candidate words corresponds to all the words occurring in the various text corpora (cf. section 2.1). The idea is to have the neural network module producing a score close to 1 for words that are relevant to the task, and a score close to 0 for words that are irrelevant. The scores produced by the neural network module will then be used to sort the list of candidate words, and the top of the list will be selected to define the recognition vocabulary.

For training the neural network parameters we have to associate a target value to each word. This is done by splitting the candidate word list (of about 2 million items) into 3 parts as represented in Table 1: words we would like to select (target value equal to 1.0), words that may be acceptable, and doubtful words (target value equal to 0.0).

Table 1. Splitting of candidate words list for training

Training phase	Candidate words	Selection phase
Target = 1.0	1/ Words observed in validation set	Apply the NN to every candidate word (including words from block 2/), sort them according to the NN output value, and select top of the list.
Ignored in training	2/ Known words (in bdlex or in in-house lexicons)	
Target = 0.0	3/ Other words (doubtful words)	

The first part, corresponding to the acceptable words, consists of the words present in the Etape [7] train set (used here as a validation set, and described in section 3.1). Ideally, these words should end in the selected lexicon. Hence, those words will have a target value set to 1.0 for training the neural network.

The second part corresponds to words that are present in different lists of known words (BdLex [8], in house proper names and acronyms). These known words are correct words, but we don't know yet if they should be selected or not for the task. Hence, as we have no way of deciding on the optimal target value for those words, so we excluded them from the neural network training data.

The last part corresponds to the other words which are mainly doubtful words which should not be selected. A large part of them come from useless acronyms or pseudos or spelling errors, which are quite frequent especially in web collected data. These words will have a target set to 0.0 for training the neural network.

In the reported experiment, the input feature vector associated to a word contains the 20 counts of occurrences (one count per data subset, cf. section 2.1), plus 16 extra parameters corresponding to the number of different n-grams (n=1,2,3,4) which contain that word for four subsets (newspapers, radio broadcast shows, web data, and Gigaword corpus). Feature vectors and associated target values are used for training the neural network: the input-feature vector is the input layer (36 input neurons); the target value is the output of the unique output layer neuron. There is one hidden layer containing 18 neurons. The experiments were conducted with the FANN toolkit [9].

After training, the neural network is applied to all the words of the candidate list (right part of Table 1), including the ones from the known dictionaries that were not used for training the neural net. We thus obtain a score between 0.0 and 1.0 for each word of the candidate list. After ranking according to this score, the 85,000 words having the highest score are selected as the recognition vocabulary.

3   Experimental Set-Up

Experiments have been conducted in the context of speech transcription of radio and TV shows.

The transcription task used in this paper is the one of the Etape French evaluation campaign [7]. The radio and TV shows were selected in order to include mostly non planned speech and a reasonable proportion of multiple speaker data.

In the vocabulary selection experiments reported in this paper, the Etape training set, which contains about 300,000 running words, was used as a validation corpus for selecting the vocabulary and optimizing the language models (more precisely for optimizing the linear combination weights).

Data from the Etape development set was used only for evaluation purpose (coverage, perplexity, speech recognition performance). This corresponds to about 80,000 running words. The evaluation performance was carried on using the evaluation tools provided by the Etape evaluation campaign. The reported results correspond to the case-independent evaluation, and the overlap speech segments are ignored.

The speech recognition experiments have been conducted using the Sphinx speech recognition toolkit [10]. After the diarization step which segments the audio data according to speakers, and classifies each audio segment with respect to the environment (studio quality vs. telephone quality), a class-based speech decoding is applied for each segment [11]: the class corresponding to the current segment is determined (according to the highest class GMM likelihood), and the speech decoding is then performed with the phonetic acoustic models corresponding to that class. 16 classes are used, that were obtained through automatic clustering of the training data.

Each acoustic model has 8,500 senones (shared densities) and 64 Gaussian components per mixture. Generic phonetic models (one for studio quality and one for telephone quality) are first trained using all the available training data (280 hours of speech from the ESTER2 evaluation campaign [12] and EPAC corpus [13]). Then, the context-dependent acoustic models are adapted to each class using the associated data [11].

For each experiment, a pronunciation lexicon is derived from the selected vocabulary. For the words that are present in available pronunciation dictionaries (BdLex and in house proper names and acronyms dictionaries), the pronunciation variants are directly copied from these available pronunciation dictionaries. The pronunciation variants for the remaining new words are obtained automatically using grapheme-to-phoneme converters [14]: both the Joint Multigram Model and the Conditional Random Field approaches are used for generating the pronunciation variants, and their results combined.

For each selected lexicon, a trigram language model is estimated using the text corpora described in section 2.1. More precisely, a trigram is first estimated on each of the 4 main subsets: newspaper, broadcast, web & gigaword. Then the linear combination weights are determined in order to minimize the perplexity of the Etape training set.

4   Experiments

This section presents and discusses the results obtained with the selection procedures described above. The evaluations are performed on the Etape development set. Three selection methods are evaluated: unigram-based, manual based, and the proposed neural network based approach. The three selected vocabularies have about the same size (85,000 items). In [2] and [3] the results are evaluated only in terms of OOV rates. Although this is true that the OOV rate has a direct negative impact on the word error rate, this should not be the only criterion; undesirable words similar to important vocabulary words can also hurt the recognizer. Consequently, besides the traditional OOV rate, we propose to evaluate our method also in terms of perplexity and word error rates.

4.1 Analysis of the Associated Language Models

Table 2 presents the results in terms of language model size and perplexity (computed with SRILM tools). The perplexity is given for the Etape training corpus and for the Etape development corpus. The results show that manual selection and Neural Network lead to rather similar results.

Table 2. Size of the language models and corresponding perplexity on the Etape train and development sets

Selection method	LM size		Perplexity	
	bigram	trigram	Train	Dev
Unigram	39.9 M	77.4 M	168.6	208.9
Manual	41.3 M	78.2 M	186.3	224.9
Neural Network	41.5 M	77.9 M	175.7	225.3

Table 3 presents the coverage of the vocabularies on the Etape development corpus. When considering the percentage of different words that are out of vocabulary, the Neural Network selection does better than both the unigram and manual selection approaches. The last column, which takes into account the frequency of occurrences of the OOV words, shows that manual and Neural Network selections lead to the best coverage on the development set, better than the unigram approach. Overall, the Neural Network selection is the most performing approach, and leads to better coverage than the unigram selection.

Table 3. Percentage of different words and of occurrences that are out of vocabulary

Selection method	% different words OOV	% occurrences OOV
Unigram	5.38%	1.29%
Manual	5.41%	0.79%
Neural Network	4.98%	0.76%

4.2 Speech Recognition Performance

Table 4 presents the speech recognition performance obtained for each selected vocabulary. The worst recognition results are obtained with the unigram-based selection method, and the best recognition performance is achieved by both the manually selected vocabulary and the neural network selection. A detailed analysis of the errors was conducted to determine the amount of errors specific to each system or common to two systems, in order to apply the McNemar test to compare the systems two by two [15]. The McNemar tests showed that the performance resulting from the Neural Network selection process is significantly better than the performance achieved with the unigram-based selection processes (685 errors are specific to the unigram method and 548 are specific to the NN approach, leading to a χ^2 value of 15 and a p value equal to 10^{-4}). This shows that this automatic approach is able to select relevant vocabularies for a speech transcription task.

Table 4. Word error rates results according to the vocabulary selection method

Selection method	Sub	Del	Ins	Total
Unigram	17.21%	9.08%	4.06%	30.34%
Manual	17.06%	9.11%	4.05%	30.22%
Neural Network	17.10%	9.07%	4.01%	30.18%

A final recognition experiment was conducted by combining the lexicon obtained with manual selection and with Neural Network selection. As many words are common between those two 85K-word lexicons, the resulting lexicon has 97,000 entries. This leads to a word error rate of 30.10%. According to the McNemar tests, this result is significantly better than all the results reported in Table 4.

5   Conclusion

In this paper we have proposed to use a machine learning based approach for selecting the recognition vocabulary to be used in a speech transcription system.

The proposed machine learning based approach, which relies on neural networks, reveals to be an interesting procedure for selecting the recognition vocabulary of a speech transcription system. It allows handling large sets of input information. Good results were achieved by considering a neural network input feature vector corresponding to both the frequency counts in various subsets of text data (here 20 subsets corresponding to various time period and text sources) and the number of different n-grams in which the words occur (in 4 subsets corresponding to various types of data). According to the McNemar test, the resulting recognition performance is significantly better than the performance achieved with the unigram-based selection process.

Further experiments have also shown that this neural-network based selection can be efficiently combined with the manual based selection. Combining both vocabularies led to significant speech recognition performance improvement, compared to the results achieved with any of the previous selection procedures.

References

1. Rosenfeld, R.: Optimizing lexical and ngram coverage via judicious use of linguistic data. In: Proc. EUROSPEECH 1995, 4th European Conf. on Speech Communication and Technology, Madrid, Spain, pp. 1763–1766 (1995)
2. Allauzen, A., Gauvain, J.-L.: Automatic building of the vocabulary of a speech transcription system (in French) "Construction automatique du vocabulaire d'un système de transcription". In: Proc. JEP 2004, Journées d'Etudes sur la Parole, Fès, Maroc (2004)
3. Venkataraman, A., Wang, W.: Techniques for effective vocabulary selection. In: Proc. INTERSPEECH 2003, 8th European Conf. on Speech Communication and Technology, Geneva, Switzerland, pp. 245–248 (2003)
4. Maergner, P., Waibel, A., Lane, I.: Unsupervised Vocabulary Selection for Real-Time Speech Recognition of Lectures. In: Proc. ICASSP 2012, IEEE Int. Conf. on Acoustics, Speech and Signal Processing, Kyoto, Japan (2012)

5. Mendona, A., Graff, D., DiPersio, D.: French Gigaword, 2nd edn. Linguistic Data Consortium, Philadelphia (2009)
6. Stolcke, A.: SRILM - An Extensible Language Modeling Toolkit. In: Proc. ICSLP 2002, Int. Conf. on Spoken Language Processing, Denver, Colorado (2002)
7. Gravier, G., Adda, G.: Evaluations en traitement automatique de la parole (ETAPE). Evaluation Plan, Etape 2011, version 2.0 (2011)
8. de Calmès, M., Pérennou, G.: BDLEX: A Lexicon for Spoken and Written French. In: Proc. LREC 1998, 1st Int. Conf. on Language Resources & Evaluation, Grenade, pp. 1129–1136 (1998)
9. FANN toolkit, `http://leenissen.dk/fann/wp/`
10. Sphinx (2011), `http://cmusphinx.sourceforge.net`
11. Jouvet, D., Vinuesa, N.: Classification margin for improved class-based speech recognition performance. In: ICASSP 2012, IEEE Int. Conf. on Acoustics, Speech and Signal Processing, Kyoto, Japan (2012)
12. Galliano, S., Gravier, G., Chaubard, L.: The Ester 2 evaluation campaign for rich transcription of French broadcasts. In: Proc. INTERSPEECH 2009, Brighton, UK, pp. 2583–2586 (2009)
13. Corpus EPAC: Transcriptions orthographiques. Catalogue ELRA, reference ELRA-S0305, `http://catalog.elra.info`
14. Illina, I., Fohr, D., Jouvet, D.: Grapheme-to-Phoneme Conversion using Conditional Random Fields. In: Proc. INTERSPEECH 2011, Florence, Italy (2011)
15. Gillick, L., Cox, S.J.: Some statistical issues in the comparison of speech recognition algorithms. In: Proc. ICASSP 1989, Int. Conf. on Acoustics, Speech and Signal Processing, pp. 532–535 (1989)

A New State-of-The-Art
Czech Named Entity Recognizer

Jana Straková, Milan Straka, and Jan Hajič

Charles University in Prague, Faculty of Mathematics and Physics
Institute of Formal and Applied Linguistics,
Malostranské náměstí 25,
118 00 Prague, Czech Republic
{strakova,straka,hajic}@ufal.mff.cuni.cz

Abstract. We present a new named entity recognizer for the Czech language. It reaches 82.82 F-measure on the Czech Named Entity Corpus 1.0 and significantly outperforms previously published Czech named entity recognizers. On the English CoNLL-2003 shared task, we achieved 89.16 F-measure, reaching comparable results to the English state of the art. The recognizer is based on Maximum Entropy Markov Model and a Viterbi algorithm decodes an optimal sequence labeling using probabilities estimated by a maximum entropy classifier. The classification features utilize morphological analysis, two-stage prediction, word clustering and gazetteers.

Keywords: named entities, named entity recognition, Czech.

1 Introduction

Named entity recognition is one of the most important tasks in natural language processing. Not only is named entity identification an important component of large applications, such as machine translation, it also belongs to one of the most useful natural language processing applications itself. Therefore it has received a great deal of attention from computational linguists. Multiple shared tasks have been organized (CoNLL-2003 [1], MUC7 [2]) for the English language and the existing state of the art systems reach remarkable results, with almost human annotator performance. Other languages, such as Czech, are with some delay also receiving attention. In this paper, we present a new state of the art named entity recognition system for Czech and English. We significantly outperform the three known Czech named entity recognizers ([3], [4], [5]) and achieve results comparable to English state of the art ([6]). The organization of this work is as follows: We describe the datasets and their evaluation methodology in Chapter 2 and present the related work in Chapter 3. Our methodology is described in Chapter 4 and results in Chapter 5. Chapter 6 concludes the paper.

2 Datasets and Task Description

2.1 English CoNLL-2003 Shared Task

For English, many datasets and shared tasks exist (e.g. CoNLL-2003 [1], MUC7 [2]). In this paper, we used one of the most widely recognized shared task dataset, the

I. Habernal and V. Matousek (Eds.): TSD 2013, LNAI 8082, pp. 68–75, 2013.
© Springer-Verlag Berlin Heidelberg 2013

CoNLL-2003 ([1]). In this task, four classes are predicted: PER (person), LOC (location), ORG (organization) and MISC (miscellaneous). It is assumed that the entities are non-embedded, non-overlapping and annotated with exactly one label. The publicly available evaluation script `conlleval`[1] evaluates the standard measures – precision, recall and F-measure.

2.2 Czech Named Entity Corpus 1.0

In 2007, the Czech Named Entity Corpus 1.0 was annotated ([7], [3]). In this corpus, Czech NEs are classified into a set of 42 classes with very detailed characterization of the predicted entities. For example, instead of English LOC, the Czech local entities are further divided into gc (states), gl (nature areas / objects), gq (urban parts), gs (streets), gu (cities / towns), gh (hydronyms), gp (planets / cosmic objects), gr (territorial names), gt continents and g_ (unspecified) in Czech.

The 42 fine-grained classes are merged into 7 super-classes (called "supertypes" in [4]), which are a (numbers in addresses), g (geographic items), i (institutions), m (media names), o (artifact names), p (personal names) and t (time expressions).

Furthermore, one entity may be labelled with one or more classes, e.g. <oa<gu Santa Barbara>>, where the location (gu) "Santa Barbara" appears in a TV document (oa) "Santa Barbara". Embedded entities are allowed, and frequently appearing patterns of named entities are also embedded in so called "containers", e.g. <P<pf Jan> <ps Stráský>>, where the first name "Jan" and last name "Stráský" are embedded in a name container P.

The fine-grained and possibly embedded classification makes the Czech named entity recognition task more complicated. Our system recognizes named entities described in [7], p. 32, Table 4.1, as well as most related work ([3], [4]).[2] An evaluation script which evaluates precision, recall and F-measure for all entities, one-word entities and two-word entities is available.[3]

3   Related Work

3.1   Systems for English

For the CoNLL-2003 shared task, the winning two systems were [8] and [9].[4] For English, Stanford Named Entity Recognizer ([10]) is available online.[5] The systems which published high scores on the CoNLL-2003 task include [11], [12], and to our knowledge, the best currently known results on this dataset were published in 2009 by [6] and reached 90.80 F-measure on the test portion of the data.

---

[1] http://www.cnts.ua.ac.be/conll2000/chunking/output.html

[2] We do not recognize number usages annotated in the second annotation round.

[3] http://ufal.mff.cuni.cz/tectomt/releases/czech_named_entity_corpus_10/

[4] The difference between the two systems was statistically insignificant.

[5] http://nlp.stanford.edu/software/CRF-NER.shtml

3.2 Systems for Czech

Together with the Czech Named Entity Corpus 1.0 ([7], [3]), a decision tree classifier was published. It achieved 62 F-measure on the embedded fine-grained classification and 68 F-measure on the embedded supertypes classification.

Another Czech named entity recognizer is [4] from 2009. On the Czech task, the authors achieved 68 F-measure for the embedded fine-grained classification and 71 F-measure on the embedded supertypes. The system used a combination of simple n-gram SVM-based recognizers.

In 2011, Konkol and Konopík ([5]) published a maximum-entropy based recognizer. They achieved 72.94 F-measure on the supertypes. The results for the fine-grained classification were not published.

4 Methods

4.1 System Overview

A simple overview of our named entity recognizer is described in Figure 4.1. The system is based on Maximum Entropy Markov Model (MEMM) and a Viterbi decoder decodes probabilities estimated by a maximum entropy classifier.

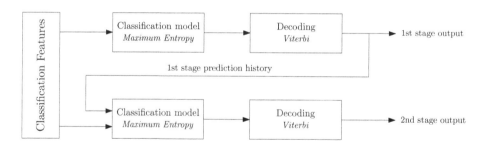

Fig. 1. System overview

First, maximum entropy model predicts for each word in a sentence the full probability distribution of its classes and positions with respect to an entity. Consequently, a global optimization via dynamic programming determines the optimal combination of classes and named entities chunks (lengths). This procedure deals with the most inner embedded entities and the system outputs one label per entity. Finally, the Czech system output is post-edited with four rules to add containers.[6]

The whole pipeline runs two times, utilizing the output from the first stage as additional classification features in the second stage.

---

[6] We automatically selected a subset of embedding patterns appearing in the training data by sequential adding the rule that increased F-measure the most. There are no such rules for English because the dataset does not contain embedded entities.

4.2 Maximum Entropy Classifier

In the first step, the maximum entropy classifier task is to predict for each word the named entity type and position within the entity. The positions are described with a BILOU scheme ([6]): B for multiword entity Beginning, I for Inside multiword entity, L for Last word of multiword entity, U for unit word entity and O for outside any entity. This scheme results in a large combination of predicted classes ($4 \times |C| + 1$, where $|C|$ is the number of classes, 4 is for B-X, I-X, L-X, U-X and $+1$ for O).

For the classifier training, we implemented our own gradient ascent parameter estimation.

4.3 Decoding

For decoding, global sequence decoders are often used, such as HMM ([13]) or CRF ([14]).

We decode the probabilities estimated by the maximum entropy model via dynamic programming. In our implementation of the Viterbi algorithm, we prune the impossible trellis transitions (e.g., once B-X starts, it can be followed either by I-X or L-X). Using this observation we can decode a whole sentence using dynamic programming with $\mathcal{O}(N \cdot C)$ complexity, where N is the number of words in the sentence.

Also, we were concerned with large growth of classes predicted by the maximum entropy classifier in the first step. With the full BILOU scheme, there are 17 classes for English and 169 classes for Czech. With the previous observation, we simplified the BILOU scheme from full B-X, I-X, L-X, U-X and O, to B-X, I, L, U-X and O. With this simplified scheme, the number of predicted classes is halved.

4.4 Classification Features

For maximum entropy classifier, we use a standard set of classification features: form, lemma, tag, chunk (only English) of current word and surrounding words in window ± 2, orthographic features (capitalization, punctuation, lowercase and uppercase form of the word), suffixes and prefixes of length 4 and regular expressions identifying possible year, date and time (in Czech). Feature selection was done by sequentially (manually) adding new classification features to the feature set; we retrained those features that have improved the classification based on development data. In English, we used forms in most of the classification features, while in Czech, we had to use lemmas because of data sparsity due to the Czech being a morphologically rich language.

Apart from the features based on the current word and its immediate vicinity, we tried to incorporate also global features. We use two-stage prediction, that is, we run our system two times in a row and in the second run, we use the predictions made in the first run. We used the information about the prediction of the previous and following five words and about the previous predictions of the candidate word in the preceding window of 500 words.

Named entity recognizers rely substantially on external knowledge. For English, we used 24 gazetteers of $1.8M$ items and for the Czech language, we used 17 gazetteers of $148K$ items. We collected both manually maintained gazetteers and automatically retrieved gazetteers from the English and Czech Wikipedia ([15]). We did not parse the whole Wikipedia article content, we only listed the title in gazetteer when it was filed under an appropriate category (e.g. "people", "births", "cities", etc.)

In the Czech morphology, lemmas are manually annotated with labels marking proper names, such as Y for given names, S for surnames and G for geographical names ([16], p. 121). These labels act as gazetteers built inside morphology.

Furthermore, we utilized Brown mutual information bigram clusters ([17], [18]), which we trained on Czech Wikipedia and downloaded for English.[7] We added these clusters respective to forms (English) and lemmas (Czech) and cluster prefixes of length 4, 6, 10 and 20 (see [6]) as new classification features.

4.5   Preprocessing and other Experiments

We did not use the original morphological analysis annotation in the data and instead, we retagged both the English and the Czech data with Featurama tagger[8], based on average perceptron sequence labeling. The Czech data was lemmatized by Featurama and the English data with an algorithm by [19]. We chunked the English data with TagChunk ([20]).[9]

We also experimented with classifier combination. In English, we used the publicly available Stanford NER ([10]) and interpolated its output with our maximum entropy classifier predicted probability distribution just before the dynamic programming step. The probability distributions were interpolated using a linear combination in which the weight was discovered via grid search.[10] Our future work involves experiments with more English named entity recognizers. Unfortunately in Czech, the previously published named entity recognizers ([3], [4], [5]) are not available for such a combination approach.

5   Results and Discussion

We call "baseline" the simplest model where we used the common set of classification features in maximum entropy model, then decoded the probability distribution given by the classifier with dynamic programming and in Czech, post-edited the result with three automatically discovered rules.

Table 1 shows the effect of more sophisticated classification features or processing: (A) new tagging, lemmatization and chunking, (B) two stage prediction, (C) gazetteers, (D) Brown clusters, (E) linear combination with the Stanford NER. The experiments (A), (B), (C), (D) and (E) show the system improvement after adding the respective feature to the baseline. The last line of the table shows results after combining all features. All new features and preprocessing steps improved the system performance over the baseline and the gains were similar in both languages. In the Czech language, most of the impact of adding gazetteers (C) is formed by the manually annotated proper name

---

[7] http://people.csail.mit.edu/maestro/papers/bllip-clusters.gz

[8] http://sourceforge.net/projects/featurama/

[9] http://www.umiacs.umd.edu/~hal/TagChunk/

[10] Our maximum entropy classifier weight = 10, Stanford NER weight = 3.

Table 1. System development. The experiments (A), (B), (C), (D), (E) show F-measure gains over the baseline on the test portion of the English and Czech data.

	English	Czech
baseline	83.80	74.87
(A) new tags, lemmas and chunks	84.20	75.47
(B) two stage prediction	84.93	76.14
(C) gazetteers	86.20	76.15
(D) Brown clusters	85.88	76.67
(E) linear combination with Stanford NER	84.21	NA
all	89.16	79.23

labels in the morphology and the manually collected and Wikipedia extracted gazetteers did not yield substantial improvement.

Table 2 shows detailed results with precision, recall and F-measure for Czech one-word, two-word and all named entities for comparison with similar tables published in [3] and [4]. Table 3 compares the related work for Czech and English on the respective datasets.

Table 2. Detailed results for Czech language. The table shows results for one-word, two-word and all named entities. The three measures evaluated are precision (P), recall (R) and F-measure (F).

	All NEs			One-word NEs			Two-word NEs		
	P	R	F	P	R	F	P	R	F
Type:	84.46	74.61	79.23	87.70	79.97	83.66	81.85	77.10	79.40
Suptype:	88.27	78.00	82.82	92.07	84.00	87.85	84.12	79.24	81.60
Span:	91.56	82.56	86.83	94.00	87.90	90.85	90.28	86.09	88.13

Table 3. System comparison for English and Czech language (F-measure on test data)

Czech	Types	Supertypes
this work	**79.23**	**82.82**
Konkol and Konopík, 2011 ([5])	NA	72.94
Kravalová and Žabokrtský, 2009 ([4])	68.00	71.00
Ševčíková et al., 2007 ([3])	62.00	68.00

English	Test F-measure
Ratinov and Roth, 2009 ([6])	90.80
Suzuki and Isozaki, 2008 ([11])	89.92
Ando and Zhang, 2005 ([12])	89.31
this work	**89.16**
Florian et al. 2003 ([8])	88.76
Chieu and Ng, 2003 ([9])	88.31
Finkel et al. 2005 ([10], Stanford parser)	86.86

6 Conclusions

We have presented a new named entity recognizer and evaluated it for Czech and English. We have reached 82.82 F-measure for the Czech language and significantly outperformed the existing Czech state of the art. For English, we achieved 89.16 F-measure. Our future work includes publicly releasing the recognizer and experimenting with named entity recognizer combination.

Acknowledgements. This work has been partially supported and has been using language resources developed and/or stored and/or distributed by the LINDAT-Clarin project of the Ministry of Education of the Czech Republic (project LM2010013). This work was also partially supported by SVV project number 267 314. We are grateful to the reviewers of this paper for comments which helped us to improve the paper.

References

1. Tjong Kim Sang, E.F., De Meulder, F.: Introduction to the CoNLL-2003 Shared Task: Language-Independent Named Entity Recognition. In: Proceedings of CoNLL 2003, Edmonton, Canada, pp. 142–147 (2003)
2. Chinchor, N.A.: Proceedings of the Seventh Message Understanding Conference (MUC-7) Named Entity Task Definition. In: Proceedings of the Seventh Message Understanding Conference (MUC-7), 21 pages (April 1998)
3. Ševčíková, M., Žabokrtský, Z., Krůza, O.: Named entities in czech: Annotating data and developing NE tagger. In: Matoušek, V., Mautner, P. (eds.) TSD 2007. LNCS (LNAI), vol. 4629, pp. 188–195. Springer, Heidelberg (2007)
4. Kravalová, J., Žabokrtský, Z.: Czech named entity corpus and SVM-based recognizer. In: Proceedings of the 2009 Named Entities Workshop: Shared Task on Transliteration. NEWS 2009, pp. 194–201. Association for Computational Linguistics (2009)
5. Konkol, M., Konopík, M.: Maximum Entropy Named Entity Recognition for Czech Language. In: Habernal, I., Matoušek, V. (eds.) TSD 2011. LNCS, vol. 6836, pp. 203–210. Springer, Heidelberg (2011)
6. Ratinov, L., Roth, D.: Design challenges and misconceptions in named entity recognition. In: CoNLL 2009: Proceedings of the Thirteenth Conference on Computational Natural Language Learning, pp. 147–155. Association for Computational Linguistics (2009)
7. Ševčíková, M., Žabokrtský, Z., Krůza, O.: Zpracování pojmenovaných entit v českých textech. Technical Report TR-2007-36 (2007)
8. Florian, R., Ittycheriah, A., Jing, H., Zhang, T.: Named Entity Recognition through Classifier Combination. In: Proceedings of CoNLL 2003, Edmonton, Canada, pp. 168–171 (2003)
9. Chieu, H.L., Ng, H.T.: Named entity recognition with a maximum entropy approach. In: Proceedings of the Seventh Conference on Natural Language Learning at HLT-NAACL 2003, CONLL 2003, vol. 4, pp. 160–163. Association for Computational Linguistics (2003)
10. Finkel, J.R., Grenager, T., Manning, C.: Incorporating non-local information into information extraction systems by Gibbs sampling. In: Proceedings of the 43rd Annual Meeting on Association for Computational Linguistics, ACL 2005, pp. 363–370. Association for Computational Linguistics (2005)
11. Suzuki, J., Isozaki, H.: Semi-Supervised Sequential Labeling and Segmentation using Gigaword Scale Unlabeled Data. Computational Linguistics, 665–673 (June 2008)

12. Ando, R.K., Zhang, T.: A high-performance semi-supervised learning method for text chunking. In: Proceedings of the 43rd Annual Meeting on Association for Computational Linguistics, ACL 2005, pp. 1–9. Association for Computational Linguistics (2005)

13. Rabiner, L.R.: A tutorial on hidden Markov models and selected applications in speech recognition. Proceedings of the IEEE 77(2), 257–286 (1989)

14. Lafferty, J.D., McCallum, A., Pereira, F.C.N.: Conditional Random Fields: Probabilistic Models for Segmenting and Labeling Sequence Data. In: Proceedings of the Eighteenth International Conference on Machine Learning, ICML 2001, pp. 282–289. Morgan Kaufmann Publishers Inc. (2001)

15. Kazama, J., Torisawa, K.: Exploiting Wikipedia as External Knowledge for Named Entity Recognition. In: Proceedings of the 2007 Joint Conference on Empirical Methods in Natural Language Processing and Computational Natural Language Learning (EMNLP-CoNLL), pp. 698–707. Association for Computational Linguistics (2007)

16. Hajič, J.: Disambiguation of Rich Inflection: Computational Morphology of Czech. Karolinum Press (2004)

17. Brown, P.F., deSouza, P.V., Mercer, R.L., Pietra, V.J.D., Lai, J.C.: Class-based n-gram models of natural language. Computational Linguistics 18(4), 467–479 (1992)

18. Liang, P.: Semi-Supervised Learning for Natural Language. Master's thesis, Massachusetts Institute of Technology (2005)

19. Popel, M.: Ways to Improve the Quality of English-Czech Machine Translation. Master's thesis, ÚFAL, MFF UK, Prague, Czech Republic (2009)

20. Daumé III, H., Marcu, D.: Learning as search optimization: approximate large margin methods for structured prediction. In: Proceedings of the 22nd International Conference on Machine Learning, ICML 2005, pp. 169–176. ACM (2005)

Algorithms for Dysfluency Detection in Symbolic Sequences Using Suffix Arrays

Juraj Pálfy[1,2] and Jiří Pospíchal[1]

[1] Slovak University of Technology, Faculty of Informatics and Information Technologies,
Bratislava, Slovakia
{palfy,pospichal}@fiit.stuba.sk
[2] Slovak Academy of Sciences, Institute of Informatics, Bratislava, Slovakia

Abstract. Dysfluencies are common in spontaneous speech, but these types of events are laborious to recognize by methods used in speech recognition technologies. Speech recognition systems work well with fluent speech, but their accuracy is degraded by dysfluent events. If dysfluent events can be detected from description of their representative features before speech recognition task, statistical models could be augmented with dysfluency detector module. This work introduces our algorithm developed to extract novelty features of complex dysfluencies and derived functions for detecting pure dysfluent events. It uses statistical apparatus to analyze proposed features of complex dysfluencies in spectral domain and in symbolic sequences. With the help of Support vector machines, it performs objective assessment of MFCC features, MFCC based derived features and symbolic sequence based derived features of complex dysfluencies, where our symbolic sequence based approach increased recognition accuracy from 50.2 to 97.6 % compared to MFCC.

Keywords: dysfluency detection, data mining, bioinformatics, speech.

1 Introduction

Dysfluent speech recognition gets a lot of attention in field of Speech Language Pathology (SLP), where objective evaluation of stuttered speech is still under development [8], [15], [13], [14]. Another aim of dysfluent speech recognition is to improve the accuracy in current Automatic Speech Recognition (ASR) systems, by introducing information about the unknown phenomenon [7], [9].

Dysfluencies are divided into subcategories, where each dysfluent event has its annotation label (e.g. *R1* denotes syllable repetition *re re research* or *P* denotes prolongation *rrrun*). These 'simple' dysfluent events were already studied in many works, which only rarely consider fusion of diverse dysfluent events (like *RS* repetition of words *make make peace* or *RSZ* repetition of verbal phrases *I do my, I do my work.*). However, 'complex' dysfluencies specified as a chaotic mixture of dysfluent events (e.g. prolongation combined with hesitation and repetition) are frequent in stutterers' speech. The statistical distribution of atomic parts of speech (phonemes) is essential to build an ASR, but the sparse regularity of dysfluencies causes a problem, the amount of available data containing dysfluent events is not sufficient to design robust speech recognition system

I. Habernal and V. Matousek (Eds.): TSD 2013, LNAI 8082, pp. 76–83, 2013.
© Springer-Verlag Berlin Heidelberg 2013

(e.g. on the basis of Hidden Markov Models). Another problem is the complexity of such ASR, to define every transition between states which can occur in case of dysfluent events would produce a very complex HMM. In dysfluent speech recognition, the common methodology is to fix a window (e.g. 200 ms, 500 ms, 800 ms) and build a dysfluency recognition system (e.g. Artificial Neural Networks, Support Vector Machines) which can recognize the ('simple') dysfluent events in a fixed interval of speech. Annotations of dysfluent speech show us that dysfluent events frequently do not fit the fixed window, but are dynamically distributed throughout much longer 2-4 s intervals. The above mentioned problems are addressed by our following approach, inspired by DNA analysis.

Part of bioinformatics deals with sequence analysis to discover interesting frequent patterns in a DNA sequences. Before applying sequence analysis tools to dysfluent speech, it was necessary to transform speech signal to symbolic sequences.

Algorithm searching dysfluencies in symbolic sequences was developed in [11] and compared to Naive String Matching Algorithm (NSMA) for stuttered speech containing 'complex' dysfluencies. The new algorithm was faster and able to produce 90.9% accuracy in detection of dysfluent intervals compared to NSMA, which yielded 45.5%. This contribution contains substantial changes to [11]. We enhanced the algorithms accuracy (by 6.7%) by changing the principle of computing matching statistics and shortened the run time by 10%. Instead of direct transformation to symbolic sequences, firstly a short time energy was computed and then transformed to symbolic representation of speech (with alphabet size 10 and word size 500). The analyzed speech data was increased 10 times. Apart from the further developed algorithm, this paper presents a wider statistical analysis of new dysfluency detection functions and a comparison of standard feature extraction method with method based on symbolic sequences.

2   Methodology

In work [4] the authors used 12 selected audio recordings *'working set'* from University College London Archive of Stuttered Speech (UCLASS). We used subset of this working set with 22 050 Hz sampling rate and a total of 19:32 min playing time for our experiments[1].

2.1   Speech Representation and Feature Extraction

Many works dealing with dysfluency detection use Fourier transformation to analyze spectrum and compute derived homomorphic features, for example MFCC, LPCC, PLP [12].

Symbolic Aggregate Approximation (SAX) discretization allows a time series of arbitrary length n to be reduced to a string of arbitrary length w, ($w < n$, typically $w << n$). The alphabet size is an integer a, which satisfies $a > 2$. As an intermediate step between the original time series and its SAX representation, we must create a dimensionality reduced version of the data [6]. We utilize the Piecewise Aggregate Approximation

---

[1] Further information about data: http://sites.google.com/site/georgepalfy/.

(PAA) [5], [16]. Let time series X of length n is represented by a reduced vector $X = x_1, \ldots x_N$. The ith element of X is calculated by: $\overline{x}_i = \frac{N}{n} \sum_{j=\frac{n}{N}(i-1)+1}^{\frac{n}{N}i} x_j$. In order to reduce the data from n dimensions to N dimensions, the data is divided into N equal windows. The mean value of the data falling within a window is calculated and a vector of these values becomes the data reduced representation [5]. SAX produces symbols with equiprobability, since normalized time series have a Gaussian distribution. With Gaussian distribution, we can determine the breakpoints that will produce equivalent areas under Gaussian curve. Breakpoints are a sorted list of numbers $B = \beta_1, \ldots, \beta_{a-1}$ such that the area under an $N(0,1)$ Gaussian curve from β_i to β_{i+1} equals $\frac{1}{a}$ [6]. Symbol concatenation, that represents a subsequence, announces a *word*. A subsequence S of length n can be represented as a word $W = \widehat{w}_1, \ldots \widehat{w}_m$. Let a_i be the ith symbol of the alphabet, i.e. $a_3 = c$, $a_4 = d$. The mapping from a PAA approximation X to a word W is obtained by: $\widehat{w}_i = a_i$, iff $\beta_{j-1} < w_j <= \beta_j$. SAX enables to optimally reduce the size of the speech signal and represent a signal as a set of ordered symbolic sequences.

Feature vectors were preprocessed by standardization. To calculate the speech features (MFCC), we maintain standard method used in Hidden Markov Model Toolkit: Hamming window length (0.025 s), overlapping adjacent frames (0.01 s) and number of bandpass filters (20). Each frame was processed with such attributes to conserve MFCC vector with 13 coefficients. Prior to SAX discrete transformation of speech, we calculated the short time energy over 0.01 s (10 ms) frames (Figure 1). The MFCC co-

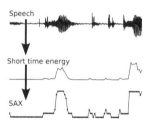

Fig. 1. Converting dysfluent speech (contains:"c can c c can") to symbolic sequences

efficients are in the form of matrix. In the case of SAX, we get for the same interval only vector of symbols. Algorithm's outputs are always moved by one block (100 ms). For every 5 s long speech interval the following three features of 100ms blocks were computed: (1) *patterns average redundancy*, (2) *patterns relative frequency*, and (3) *patterns redundancies sum*.

2.2   Data Structure

Data structure of our algorithm is inspired by bioinformatics, where it is used in DNA analysis. A primary motivation for suffix arrays was to enable efficient execution of online string queries for very long genetic sequences (for example an order of one million or greater symbols long) [10]. Consider a large sequence $C = c_0 c_1 \ldots c_{N-1}$ of length

i	Pos[i]	C[Pos[i] ... n]
1	11	$
2	4	cessing$
3	5	essing$
4	10	g$
5	8	ing$
6	9	ng$
7	3	ocessing$
8	1	processing$
9	2	rocessing$
10	7	sing$
11	6	ssing$

i =	1	2	3	4	5	6	7	8	9	10	11
C =	p	r	o	c	e	s	s	i	n	g	$

Fig. 2. Suffix array construction for string C = *'processing$'*. $Pos[i]$ denotes a constructed suffix array, $C[Pos[i]...n]$ lists all prefixes in an array of suffixes.

N. The suffix of C that begins at position i, shall be expressed by $C_i = c_i c_{i+1} \ldots c_{N-1}$. The basis of this data structure is a lexicographically sorted array, Pos, of the suffixes of C. $Pos[k]$ is the onset position of the kth smallest suffix in the set $C_0, C_1, \ldots C_{N-1}$. We assume that Pos is given. $C_{Pos[0]} < C_{Pos[1]} < \cdots < C_{Pos[N-1]}$, where '<' denotes the lexicographical order [10]. Figure 2 provides an example for suffix array construction.

2.3 Functions for Detecting Minimal Signal Change

We assume that prolongations are characterized by minimal difference between n neighboring frames of prolonged speech. We hope to capture these frames using special functions from 2D equations for video segmentation [2], detecting repeated signal interval. We derived functions $D(x,y)$, $D_b(x,l)$, $D_h(H_x,l)$, which were specially adapted for speech. Let vectors $x = x_1, \ldots x_N$, $y = y_1, \ldots y_N$, be the vectors of length N, then function $D(x,y) = \frac{1}{N} \sum_{i=1}^{N} |x_i - y_i|$, computes vectors difference. The function $D_b(x,l) = \sum_{i=1}^{b} D(x_i, x_{(i+l)})$, measures the difference between vectors x and $x+l$ in the speech, where b is the number of blocks, the parameter l denotes an offset. The value of l is a tradeoff between the detection of abrupt changes ($l = 1$) or smooth transitions ($l > 1$). Let h be the number of H histograms with identical classes. The last function $D_h(H_x,l) = \sum_{i=1}^{h} D(H_x(i), H_x(i + l))$, with two input parameters: H_x histograms of vectors x and offset l computes the difference between distribution of vectors without taking into account their position. Articulation organs in vowel realization emit periodic signal. Our derived functions give us an acceptable score (Figure 3), characterizing a minimal change in these periodic signal changes along analysis windows. Problem in this method occurs in case of consonant (plosives) prolongation and in any type of repetition. In speech these events raise a problem of automatic analysis window adaptation. We see automatic analysis window adaptation as a useful technique for discovery of repeated speech patterns with unpredictable length and chaotic occurrence.

2.4 Symbolic Sequence Searching

We created two algorithms. One for speech pattern searching (*patSearch*) and second for searching repeated patterns in speech (*repSearch*). Last algorithm with help of first pattern searching algorithm, automatically discovers all repeated sequences in a reduced

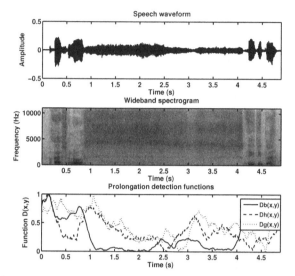

Fig. 3. Derived functions for prolongation detection. The speech contains lexical information: 'personal s:eedee player'. $D_b(x, y)$ denotes block based function, $D_h(x, y)$ computes histogram differences, $D_g(x, y)$ compares local changes in histograms to speech global histogram.

time domain (thanks to SAX representation). In large relational databases, short query set is executed in a large set of data. Our key idea in *patSearch* is the opposite of the search technique used in case of relational database. We query in a short sequence P a "moving window" of a long sequence C. This opposite approach allows execution of our algorithm (*patSearch* - parameters: s is a shift, l is C length) in a reduced memory space.

```
1: while i < n do                                    ▷ Begin: repSearch().
2:     In i-th window 1st block set to P, remaining blocks put to C.
3:     Compute Pos for P.                            ▷ Pos is a suffix array.
4:     With Pos construe Tab for P.                  ▷ Tab is a look up table.
5:     while s < l do                                ▷ Begin: patSearch().
6:         Use Tab to query C in P.
7:         Save patterns position and patterns length.
8:     end while                                     ▷ End: patSearch().
9: end while                                         ▷ End: repSearch().
```

Key feature of *repSearch* algorithm is its adaptation capability to unknown repeated speech pattern length. According to its input parameters (S symbolic speech sequence, w window length, b block length, n number of blocks), it computes proximity measures, which allows to find all occurrences of pattern P in a long symbolic speech sequence C independently of matching length P in C.

3   Results

In order to compare performance of studied features (algorithms output *Specrep* and *Symrep*) to commonly used MFCC coefficients, objective assessment was provided by using feature vectors as inputs for SVM. SVM and their variants and extensions, often called kernel-based methods (or simply kernel methods), have been studied extensively and applied to various pattern classification and function approximation problems [1]. In our experiment, we used sigmoid kernel function, $k(x, y) = \tanh\left(\gamma x^T y + r\right)$. During the process of classifier design one of intermediate step is the measurement of data class separability, which we prove with correlation between two classes (fluent speech and dysfluent speech). In the next stage we studied the data characteristics, for two classes with Mann-Whitney U-test. Nonparametric Mann-Whitney U-tests examine the equality of class medians of random variables X, Y. We use confusion matrix to measure SVM models' performance. In addition to *accuracy*, we compute *sensitivity* and *specificity* of confusion matrix [3]. Table 1 and Table 2 compare features computed from spectral domain (*Specrep* - Dynamic Time Warping on basis of MFCC features) and features computed from symbolic sequences (*Symrep* - our developed algorithms on basis of SAX). Low correlation coefficients r in Table 1 refer to a fact, that between features computed from fluent intervals and from dysfluent intervals of speech there is a low linear dependence. It is evident, that between our proposed features *Specrep average* and *Symrep sum* is the lowest linear dependency.

Table 1. Correlations between group of fluent and dysfluent events in proposed features

Feature	r	p-value
Specrep average	0.0356	0.3654
Specrep frequency	0.3434	0.0
Specrep sum	-0.0971	0.1032
Symrep average	-0.0668	0.0
Symrep frequency	0.1120	0.0
Symrep sum	-0.0296	0.6205

Table 2. Mann-Whitney U-test between group of fluent and dysfluent events along examined features

Feature	h	p-value
Specrep average	1	0.0
Specrep frequency	1	0.0
Specrep sum	0	0.1659
Symrep average	1	0.0
Symrep frequency	0	0.2385
Symrep sum	1	0.0

In case of class separability it is considerable that *p-values* between *Specrep average* and *Symrep sum* are above 0.05. Therefore correlation is not significant, which implies that features are not linearly dependent. According to correlation values, features clearly separate group of fluent and group of dysfluent events.

Results of data characteristic study of proposed features are shown in Table 2. Nonparametric Mann-Whitney U-tests in Table 2 are significant, where their *p-values* are below 0.05 level. h values specify accepted hypotheses. We fail to reject hypotheses $h = 0$ only for features *Specrep sum* and *Symrep frequency*. Rejected hypotheses $h = 0$ in other features depict that features with $h = 1$ do not have equal medians. According to test results, features with $h = 1$ have unequal data distribution for group of fluent and group of dysfluent features.

We divided data to training (80 %) and testing (20 %) sets. In next step, we trained 6 individual SVM with sigmoidal kernel function. In Table 3 are evaluated the MFCC,

Table 3. Testing results of Support vector machines

Feature	Sensitivity	Specificity	Accuracy (%)
MFCC	0.508	0.496	50.2
Specrep avg. / Symrep avg.	0.739 / 0.924	1 / 0.676	85.4 / 80
Specrep freq. / Symrep freq.	0.826 / 0.456	0 / 0.662	46.3 / 55.9
Specrep sum / Symrep sum	0.391 / 1	0 / 0.944	22 / 97.6

Specrep and *Symrep* features. For MFCC feature we get 50.2 % accuracy. *Specrep average* achieved 85.4 % accuracy, which is 35.2 % better than base MFCC feature. SVM model for *Symrep sum sensitivity* do not commit any *false negative* prediction and produced only 6 % *false positive* predictions. According to the represented classification results in Table 3, the *Symrep sum* maintains the upper limit with 97.6 % *accuracy*.

4   Conclusion

The used statistical apparatus shows that to recognize complex dysfluent phenomena, speech features on basis of symbolic sequences are competitive with speech features on basis of spectral domain.

We developed algorithms designed especially for phrase repetition detection in symbolic sequences of speech. Paper shows that speech transformation to discrete symbolic sequences offers an alternative way of speech processing. SAX allows to apply DNA sequence analysis to speech, this potential advantage was enabled by highly efficient data structure, suffix array.

New designated features, capturing phrase repetitions, were statistically analyzed. Objective assessment of new features and MFCC were compared by SVM. SVM trained by MFCC features to recognize repetitions have 50.2% accuracy. SVM in case of our features based on symbolic sequences accomplished 97.6% accuracy on identical groups of speech data.

We presented derived functions for prolongation detection, inspired by domain of video analysis. Presented algorithms in the future may be extended by computing Hamming distance between observed symbolic sequences to obtain higher accuracy. Beside short time energy, we plan to seek speech features in spectral domain to use its symbolic representation to detect other types of dysfluent events. The algorithms running time (30 times faster than DTW) could be further improved with suffix tree, which is a similar data structure to suffix array.

Acknowledgments. This contribution was supported by Grant Agency VEGA SR 1/0553/12, 1/0458/13 and supported by Research & Development Operational Programme funded by the ERDF project RPKOM, ITMS 26240220064. We thank Milan Rusko and Sakhia Darjaa for their insightful recommendations about speech processing. Last but not least we thank Erika Pálfy for her help concerning communication disorders.

References

1. Abe, S.: Support Vector Machines for Pattern Classification. In: Advances in Pattern Recognition. Springer, London (2010)
2. Camastra, F., Vinciarelli, A.: Machine Learning for Audio, Image and Video Analysis: Theory and Applications. Springer-Verlag London Limited (2008)
3. Hamel, L.: Knowledge Discovery with Support Vector Machines. John Wiley & Sons, Inc., Hoboken (2009)
4. Howell, P., Davis, S., Bartrip, J.: The UCLASS archive of stuttered speech. Journal of Speech, Language, and Hearing Research 52, 556–569 (2009)
5. Keogh, E., Chakrabarti, K., Pazzani, M., Mehrotra, S.: Dimensionality Reduction for Fast Similarity Search in Large Time Series Databases. Knowledge and Information Systems 3, 263–286 (2001)
6. Lin, J., Keogh, E., Lonardi, S., Patel, P.: Finding motifs in time series. ACM Special Interest Group on Knowledge Discovery and Data Mining (2002)
7. Liu, Y., Shriberg, E., Stolcke, A., Harper, M.: Comparing HMM, Maximum Entropy, and Conditional Random Fields for Disfluency Detection. In: Proceedings of the European Conference on Speech Communication and Technology (2005)
8. Lustyk, T., Bergl, P., Čmejla, R., Vokřál, J.: Change evaluation of Bayesian detector for dysfluent speech assessment. In: International Conference on Applied Electronics 2011, Pilsen, Czech Republic, pp. 231–234 (2011)
9. Maskey, S., Zhou, B., Gao, Y.: A Phrase-Level Machine Translation Approach for Disfluency Detection Using Weighted Finite State Transducers. In: Interspeech (2006)
10. Manber, U., Myers, G.: Suffix arrays: A new method for on-line string searches. SIAM J. Comput. 22(5), 935–948 (1993)
11. Pálfy, J., Pospíchal, J.: Pattern Search in Dysfluent Speech.. In: Proceedings of the IEEE International Workshop on Machine Learning for Signal Processing, Santander, Spain (2012)
12. Rabiner, L.R., Schafer, R.W.: Introduction to Digital Speech Processing. In: Foundations and Trends in Signal Processing, pp. 1–194 (2007)
13. Ravi Kumar, K.M., Ganesan, S.: Comparison of Multidimensional MFCC Feature Vectors for Objective Assessment of Stuttered Disfluencies. International Journal of Advanced Networking and Applications 2, 854–860 (2011)
14. Świetlicka, I., Kuniszyk-Jóźkowiak, W., Smołka, E.: Hierarchical ANN system for stuttering identification.. Computer Speech and Language 27, 228–242 (2013)
15. Wiśniewski, M., Kuniszyk-Jóźkowiak, W.: Automatic detection and classification of phoneme repetitions using HTK toolkit. Journal of MIT 17 (2011)
16. Yi, B.-K., Faloutsos, C.: Fast Time Sequence Indexing for Arbitrary Lp Norms. In: Proceedings of the 26th International Conference on Very Large Databases, pp. 385–394 (2000)

Analysis and Combination of Forward and Backward Based Decoders for Improved Speech Transcription

Denis Jouvet[1,2,3] and Dominique Fohr[1,2,3]

[1] Speech Group, LORIA
Inria, Villers-lès-Nancy, F-54600, France
[2] Université de Lorraine, LORIA, UMR 7503, Villers-lès-Nancy, F-54600, France
[3] CNRS, LORIA, UMR 7503, Villers-lès-Nancy, F-54600, France
denis.jouvet@inria.fr, dominique.fohr@loria.fr

Abstract. This paper analysis the behavior of forward and backward-based decoders used for speech transcription. Experiments have showed that backward-based decoding leads to similar recognition performance as forward-based decoding, which is consistent with the fact that both systems handle similar information through the acoustic, lexical and language models. However, because of heuristics, search algorithms used in decoding explore only a limited portion of the search space. As forward-based and backward-based approaches do not process the speech signal in the same temporal way, they explore different portions of the search space; leading to complementary systems that can be efficiently combined using the ROVER approach. The speech transcription results achieved by combining forward-based and backward-based systems are significantly better than the results obtained by combining the same amount of forward-only or backward-only systems. This confirms the complementary of the forward and backward approaches and thus the usefulness of their combination.

Keywords: speech transcription, speech recognition, forward-based and backward-based decoding, combining speech recognizer outputs, ROVER.

1 Introduction

Combining several recognition systems is a way of improving speech transcription performance. A well known combination approach consists in combining the outputs of different speech recognition systems through the ROVER procedure [1]. This procedure aligns the different hypotheses at the word level, and relies on a voting procedure for determining the best candidate word results. Good performances are obtained with such an approach, and the procedure has been enriched through the handling of a language model for helping the decision process [2] or through the introduction of a classifier for deciding on the best answer for each word using more detailed information than just the frequency of occurrences of the words and their confidence measures [3]. Other extensions involve dealing with the combination of confusion networks [4] instead of combining just the best hypothesis provided by each system. Tighter combinations of systems are also investigated to take benefit of several systems. This includes for example the exploitation of n-gram generated from the decoding with auxiliary systems for adjusting dynamically the language model used by the main decoder [5].

I. Habernal and V. Matousek (Eds.): TSD 2013, LNAI 8082, pp. 84–91, 2013.

This paper focuses on the combination of forward-based and backward-based speech decoders. The goal of the reported experiments was to get a better understanding on the reasons why combining Sphinx-based and Julius-based decoders lead to much better performance than the combination of the same amount of Sphinx-based decoders only, which, individually had better performance. In other words, why are Sphinx-based and Julius-based decoders complementary? One possible explanation was linked to the way they process the speech signal during decoding. Standard Sphinx systems [6] process the speech frames in a standard time forward way. On the opposite, the Julius decoder [7] do a two-pass decoding. The forward first pass generates a word graph, which is then rescored during the A*-based backward second pass. In order to refine this hypothesis, specific experiments have been conducted. A variant of the speech transcription system was developed in which the speech frames (feature vectors) were given to the Sphinx decoder a time reverse order. Language models and pronunciation dictionaries were also reverse, and training of reverse acoustic models was performed. A similar approach was applied to the Julius-based systems by reversing the speech signal processed. Overall, several speech transcription systems have been developed using different sets of basic units (standard set of phonemes vs. a reduced set) and different types of acoustic features. Results obtained from the combination of forward and backward systems are compared to the results obtained from the combination of forward-only or backward-only systems. These results are discussed in this paper, along with some detailed analysis of the systems aiming at getting a better understanding of their behavior.

The paper is organized as follows. Section 2 describes the experimental set-up and presents the speech corpora that are used as well as the individual systems. Section 3 presents and compares various combinations of speech transcription system outputs. Section 4 analysis and discusses the behavior of the systems, in order to get a better understanding of their complementarity. Finally a conclusion ends the paper.

2   Experimental Set-Up

The experimental set-up chosen is the transcription of speech from radio broadcasts and TV shows, using a set of systems that rely on the Sphinx decoder [6], and another set of systems that rely on the Julius decoder [7].

2.1   Speech Corpora, Lexicons and Language Models

The speech corpora comes from the ESTER2 [8] and the ETAPE [9] evaluation campaigns, and the EPAC [10],[11] project. The ESTER2 and EPAC data are French broadcast news collected from various radio channels, they contain prepared speech, plus some interviews. The ETAPE data corresponds to debates collected from radio and TV channels, and is mainly spontaneous speech.

The acoustic models were trained from the speech data of the ESTER2 and ETAPE train sets, and the transcribed data from the EPAC corpus. This training data amounts to almost 300 hours of signal and 4 million running words.

The speech decoders relies on lexicons of about 95,000 words and n-gram language models. The pronunciation variants were extracted from the BDLEX [12] and other in-house pronunciation lexicons whenever possible; and, for the remaining words, they were automatically obtained using Grapheme-to-Phoneme converters [13]. The language models were trained using the SRILM tools [14] and various text corpora amounting for a total of more than 1,600 million words from various sources: newspapers, transcriptions of radio broadcast shows, French Gigaword corpus [15], and web data collected in 2011.

The speech transcription word error rates are given for the non-African radios of the ESTER2 development and test sets (respectively named Dev-na - about 42,000 running words - and Test-na - about 63,000 running words) and for the whole ESTER2 test set (about 79,000 running words). Results are also provided for the whole ETAPE development and test sets (about 82,000 running words each). Performance on the ESTER2 data is computed using the sclite tool [16] according to the ESTER2 campaign protocol. Performance on the ETAPE data is computed using the new LNE tools according to the ETAPE evaluation protocol [17] (ETAPE results are reported for non-overlapping speech data and for case independent mode).

2.2   Sphinx-Based Systems

The Sphinx-based transcription systems rely on a trigram language model, and on acoustic models specific to gender (male vs. female) and speech quality (studio vs. telephone). Context-dependent phoneme units are used, for a total of 7,500 shared densities (senones), each of them having 64 Gaussian components. The first decoding pass does a decoding of each audio segment using the most adequate acoustic model, and the second decoding pass benefits from VTLN adaptation of the features and MLLR adaptation of the acoustic models.

Several systems have been developed. They differ with respect to basic units and acoustic features. Three sets of acoustic features were considered: Sphinx MFCC, HTK [18] MFCC and HTK PLP features. In every case, 39 coefficients are used (static coefficients plus their first and second temporal derivatives). Two sets of basic units have been considered. One set uses all the phonemes defined in the BDLEX pronunciation lexicon, whereas in the second set the aperture of the vowels is not considered; hence we merge the open and the close /o/, the open and the close /e/, as well as the open and the close /ø/. The six corresponding Sphinx standard (forward-based) systems have rather similar performance (less than 1% absolute difference between them). The results for the best system (HTK MFCC features and set of all phonemes) are reported in Table 1.

A similar set of systems, but based on a reverse processing approach, have also been developed: frames of each audio segment were given to the training tool and to the decoder in a reverse time order (i.e. last frame of each audio segment was given first). The pronunciation of each word in the lexicon was also reversed, and language models were re-estimated after reversing all the text sentences. The corresponding reverse (backward-based) systems achieved similar performance as the standard (forward-based) systems.

2.3   Julius-Based Systems

The second set of transcription systems rely on the Julius decoder. They also use acoustic models dependent on the quality of the signal (studio vs. telephone). Context-dependent phoneme units are used, they are modeled with 6,000 shared states/densities, and each mixture density has 62 Gaussian components. An HLDA transformation is applied on the acoustic features before modeling. These transcription systems run also in two passes; and the second transcription pass benefits from SAT adapted models.

The Julius decoder runs in a forward-backward mode. The forward pass uses a bigram and generates a word graph; then, the backward A* pass explores this graph guided by a reverse 4-gram language model.

Two systems have been developed, one using HTK MFCC features, and the other HTK PLP features. An HLDA transformation is applied on a window of 9 acoustic feature vectors to provide the 40 input modeling coefficients. The phoneme units chosen ignore the aperture of the vowels. The two systems have performance in the same range, and the performance of the best one is reported in Table 1. As for Sphinx, an extra Julius-based transcription system has been developed that processes the speech signal in a reverse time order.

3   Combining Forward and Backward Decoders

Combinations of the outputs of several recognizers using the ROVER [1] approach are investigated here. Confidence measures are not used in the ROVER combinations. Results from the combination of Sphinx-based and Julius-based recognition systems are presented first, then a focus is set on the combination of forward and backward Sphinx-based decoders.

3.1   Combining Sphinx-Based and Julius-Based Systems

Here, five system outputs are combined: three standard Sphinx-based systems (with standard phone units), plus two additional systems (involving the reduced set of phone units), on the one side, two other standard Sphinx-based systems, and on the other side, two Julius-based systems. Results are reported in middle of Table 1.

Combining two Julius-based systems with three Sphinx-based systems leads to word error rates that are about 1.5% absolute lower than those resulting from the combination of 5 Sphinx-based systems. The difference between the two combinations is limited to the two last recognizers involved in each combination. The corresponding four systems (2 Sphinx and 2 Julius) involve the same set of phonetic units (after merging of vowel apertures) and the same input features (one system using HTK MFCC features and the other HTK PLP features in each combination). Although Julius-based recognizers alone provide significantly worse results than the corresponding Sphinx-based recognizers (cf. top of Table 1), their combination with other Sphinx-based recognizers leads to a much larger reduction in the word error rates. This rather large reduction in word error rate may result from the different operating modes of the systems used. The standard Sphinx decoder processes the speech signal in a forward pass, whereas the Julius

Table 1. Word error rates on the ESTER2 and ETAPE data

Speech recognition system	ESTER2			ETAPE	
	Dev-na	Test-na	Test	Dev	Test
Sphinx-based (standard)	20.7%	21.2%	22.9%	27.7%	28.6%
Sphinx-based (reverse)	20.9%	21.1%	22.7%	28.0%	28.7%
Julius-based (standard)	23.5%	23.5%	26.6%	30.1%	30.6%
ROVER: 3 Sphinx + 2 Sphinx	19.0%	19.0%	20.5%	25.8%	26.7%
ROVER: 3 Sphinx + 2 Julius	**17.6%**	**17.7%**	**19.2%**	**24.1%**	**24.9%**
ROVER: 6 Sphinx forward	18.9%	19.0%	20.5%	25.7%	26.6%
ROVER: 6 Sphinx backward	19.1%	19.0%	20.3%	25.7%	26.5%
ROVER: 3 forw. + 3 backw.	**18.0%**	**18.2%**	**19.5%**	**24.7%**	**25.4%**
ROVER: 3 backw. +3 forw.	**18.1%**	**18.3%**	**19.6%**	**24.8%**	**25.5%**

decoder deals with the speech signal in a forward plus a backward process. This should normally lead to different search spaces, and one might expect that one system is likely to recover errors made by the other system.

3.2 Combining Forward and Backward-Based Sphinx Decoders

In order to get the best possible insight on the combination of forward-based and backward-based recognizers, four combinations of six sphinx systems each are compared. Results reported in lower part of Table 1 show that the combination of six forward-based Sphinx systems and the combination of six backward-based Sphinx systems lead to very similar recognition performance. The two other combinations of three forward-based and three backward-based systems also lead to very similar recognition performance, although much better than that of the two previous combinations. Each of these four combinations involves two systems with Sphinx MFCC features, two with HTK MFCC features and two with HTK PLP features. With respect to the phone units, each combination involve three systems using the standard phoneme units and three systems using the reduced set after merging of vowels apertures. Hence, the only difference between the various combinations is the fact that some systems are forward-based whereas some others are backward-based. Comparing the last two lines to the first two lines shows that the combination of forward-based and backward-based systems provides much better results than the combination of forward only or backward only systems. On average, there is a 0.7% to 1.0% absolute error rate reduction due to the simple fact of combining forward-based and backward-based systems.

4 Comparing Forward and Backward-Based Approaches

Speech decoding algorithms search the best path in a huge graph resulting from language, pronunciation and acoustic models. Heuristic search algorithms explore only a small part of the graph. For example, the forward pass of the Julius decoder relies on a beam search for creating a word graph that is explored with more detailed models in the subsequent backward pass. Beam search heuristics limits the number of active paths (and associated words) that end at any given frame. Backtracking the active paths define the starting frame associated to each word. An analysis of word-graphs created in the forward pass of the Julius decoder, shows that the number of words starting and ending in each time frame follow completely different distributions, as shown in Figure 1. The number of words ending in each frame is limited and rather smooth compared to the number of words starting in each frame. Of course the average values of both distributions are equal, as each word in the graph as one starting and one ending frame. The positions of the largest spikes frequently correspond to word boundaries.

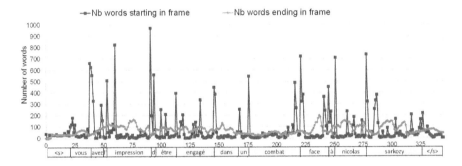

Fig. 1. Analysis of a word graph created in the Julius decoding forward pass

Figure 2 displays the number of words in the word-graphs that originate in a given frame for the standard (blue curve) and for the reversed (red dotted curve) decodings. The recognized words resulting from the standard and reverse decoding are displayed under the graph. Both curves have a similar behavior, although the spikes are not exactly the same. The word-graphs are of similar sizes; each one contains about 1500 different word spellings among which less than 600 (i.e. 40%) are common to both word-graphs; this emphasizes the complementarity of the forward-based and backward-based decoders.

Although forward and backward language models lead to similar perplexity values, the probability values of individual words can largely differ, as for example the word "à" in the sentence "... un combat face à nicolas sarkozy <s >" which has a forward probability of 0.51 after the sequence "un combat face", and a backward probability of only 0.07 given the following words "nicolas sarkozy <s >". These different values of the word probabilities also contribute to the complementarity of the approaches.

Although forward-based and backward-based decoders lead to similar performance, they do not make the same errors: about one fourth of the errors made by the forward-based system correspond to words correctly recognized by the backward-based system, and conversely about one fourth of the errors made by the backward-based system correspond to words correctly recognized by the forward-based system. The fact they do not make similar errors explains why it is beneficial to combine both systems.

In 80% of the cases, the forward and backward decoders provide the same answer (same word hypothesis). And, when both decoders provide the same answer, it was observed that this common answer is a correct answer in more than 91% of the cases.

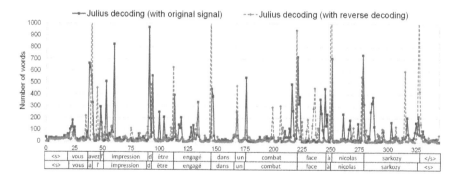

Fig. 2. Number of words originating from a given frame in word graphs created by running the Julius decoder in a standard way and on the reversed signal

Although further detailed analysis is still needed on the few errors observed for common answers, detecting common answers could be useful for improving confidence measures, and they could also be the basis for anchoring island-based refined decoding or rescoring.

5   Conclusion

Several transcription systems were developed and their outputs combined through the ROVER procedure. The systems differ one from the other with respect to the acoustic features, the set of phonetic units, and also the decoding engine. The experiments clearly show that it is much more efficient to combine outputs of recognizers that behave complementary, even if they do not have the best performance, rather than combining only the best performing systems. Among the complementarity aspects of the decoders, the benefit of combining forward and backward processing has been further investigated. To do that, specific experiments were conducted, which showed that the combination of forward-based and backward-based decoders leads to much better results (0.7% to 1.0% absolute error rate reduction) than the combination of forward only or backward only decoders. Finally the search spaces resulting from forward and backward decoders were analyzed, and associated decoding results were compared and discussed.

Overall, results in Table 1 show that the combination of several forward-based and backward-based transcription systems lead to state of the art performance on the ES-TER2 and on the ETAPE transcription tasks.

References

1. Fiscus, J.G.: A post-processing system to yield reduced word error rates: Recognizer Output Voting Error Reduction (ROVER). In: Proc. ASRU 1997, IEEE Workshop on Automatic Speech Recognition and Understanding, pp. 347–354 (1997)
2. Schwenk, H., Gauvain, J.-L.: Combining multiple speech recognizers using voting and language model information. In: Proc. INTERSPEECH 2000, pp. 915–918 (2000)

3. Hillard, D., Hoffmeister, B., Ostendorf, M., Schlüter, R., Ney, H.: iROVER: improving system combination with classification. In: Conf. of the North American Chapter of the Association for Computational Linguistics, Rochester, New-York, pp. 65–68 (2007)
4. Evermann, G., Woodland, P.C.: Posterior probability decoding, confidence estimation and system combination. In: Proc. NIST Speech Transcription Workshop (2000)
5. Bougares, F., Estéve, Y., Deléglise, P., Linares, G.: Bag of n-gram driven decoding for LVCSR system harnessing. In: Proc. ASRU 2011, IEEE Workshop on Automatic Speech Recognition and Understanding, Hawaï, USA (2011)
6. Sphinx (2011), `http://cmusphinx.sourceforge.net/`
7. Julius, `http://julius.sourceforge.jp/en_index.php`
8. Galliano, S., Gravier, G., Chaubard, L.: The Ester 2 evaluation campaign for rich transcription of French broadcasts. In: Proc. INTERSPEECH 2009, 10th Annual Conf. of the Int. Speech Communication Association, Brighton, UK, pp. 2583–2586 (2009)
9. Gravier, G., Adda, G., Paulsson, N., Carré, M., Giraudel, A., Galibert, O.: The ETAPE corpus for the evaluation of speech-based TV content processing in the French language. In: Proc. LREC 2012, Int. Conf. on Language Resources, Evaluation and Corpora, Istanbul, Turkey (2012)
10. Estéve, Y., Bazillon, T., Antoine, J.-Y., Béchet, F., Farinas, J.: The EPAC corpus: Manual and automatic annotations of conversational speech in French broadcast news. In: Proc. LREC 2010, European Conf. on Language Resources and Evaluation, Valetta, Malta (2010)
11. Corpus EPAC: Transcriptions orthographiques. Catalogue ELRA, reference ELRA-S0305, `http://catalog.elra.info`
12. de Calmés, M., Pérennou, G.: BDLEX: A Lexicon for Spoken and Written French. In: Proc. LREC 1998, 1st Int. Conf. on Language Resources & Evaluation, Grenade, pp. 1129–1136 (1998)
13. Illina, I., Fohr, D., Jouvet, D.: Grapheme-to-Phoneme Conversion using Conditional Random Fields. In: Proc. INTERSPEECH 2011, Florence, Italy (2011)
14. Stolcke, A.: SRILM - An Extensible Language Modeling Toolkit. In: Proc. ICSLP 2002, Int. Conf. on Spoken Language Processing, Denver, Colorado (2002)
15. Mendonça, A., Graff, D., DiPersio, D.: French Gigaword Second Edition. Linguistic Data Consortium, Philadelphia (2009)
16. NIST evaluation tools, `http://www.itl.nist.gov/iad/mig//tools/`
17. Gravier, G., Adda, G.: Evaluations en traitement automatique de la parole (ETAPE). Evaluation Plan, Etape 2011, version 2.0 (2011)
18. HTK, `http://htk.eng.cam.ac.uk/`

Annotating Signs of Syntactic Complexity to Support Sentence Simplification

Richard Evans and Constantin Orăsan

Research Group in Computational Linguistics
University of Wolverhampton, United Kingdom
r.j.evans@wlv.ac.uk

Abstract. This article presents a new annotation scheme for syntactic complexity in text which has the advantage over other existing syntactic annotation schemes that it is easy to apply, is reliable and it is able to encode a wide range of phenomena. It is based on the notion that the syntactic complexity of sentences is explicitly indicated by signs such as conjunctions, complementisers and punctuation marks. The article describes the annotation scheme developed to annotate these signs and evaluates three corpora containing texts from three genres that were annotated using it. Inter-annotator agreement calculated on the three corpora shows that there is at least "substantial agreement" and motivates directions for future work.

1 Introduction

Syntactic simplification, framed as the conversion of long syntactically complex sentences into shorter simple sentences, can improve the reliability of numerous tasks in natural language processing (NLP), including information extraction [1–3], machine translation [4], and syntactic parsing [5, 6]. In this context, coordination and subordination are considered key elements of syntactic complexity. Quirk et al. [7] define coordination as a paratactic relationship that holds between constituents at the same level of syntactic structure. The linking function occurs between conjoins[1] that match, to a greater or lesser extent, in terms of form, function, and meaning (1). By contrast, subordination is defined as a hypotactic relationship holding between constituents at different levels of syntactic structure, referred to as superordinate and subordinate constituents (2). In this paper we address the challenges that arise when annotating indicative signs of syntactic complexity involving these phenomena.

(1) She <u>knew the risks</u> and <u>still insisted the operation should go ahead</u>, Dr Addicott said.

(2) McKay, <u>of Wark, Northumberland</u>, denies five charges of contaminating food.

The annotation undertaken in research described in this paper is part of a larger project to implement a system that improves the accessibility of written documents for

---

[1] This paper employs the terminology used by Quirk et al. [7]. In related work, the term *conjunct* has been used rather than *conjoin*, but Quirk et al. use the former term to denote "linking adverbials".

I. Habernal and V. Matousek (Eds.): TSD 2013, LNAI 8082, pp. 92–104, 2013.

people with autistic spectrum disorders (ASD). The system under development does not rely on a parser and simplifies complex sentences in two steps. First it identifies and classifies trigger words, referred to in this paper as *signs of syntactic complexity*. This is done on the basis of a machine learning method trained on the corpus presented in Section 3 [8]. In the second step, manually crafted rules are applied to rewrite sentences. Complex sentences containing coordination can be simplified by generating copies of the sentences in which each coordinated constituent is replaced by a different conjoin of that constituent. In sentences containing subordination, simplification can be performed by deleting the subordinate constituent from the original sentence and generating an additional sentence that expresses the relation between the subordinated constituent and the head that it modifies. Previous work by Evans [3] demonstrated that a similar approach to sentence simplification yields improved performance in information extraction from clinical assessment items.

In this paper, we present an annotation scheme to encode the linking functions of coordinators (conjunctions, punctuation marks, and pairs consisting of a punctuation mark followed by a conjunction) and the bounding functions of subordination boundaries (complementisers, wh-words, punctuation marks, and pairs consisting of a punctuation mark followed by a lexical sign).[2] The annotation scheme also encodes information about false signs that do not have a coordinating or bounding function (e.g. use of the word *that* as an anaphor or specifier).

In this paper, Sect. 2 provides an overview of related work. Section 3 presents the annotated resources developed in the current research and details the annotation scheme applied (Sect. 3.1). Section 4 describes an assessment of the consistency and reliability of the annotation. In Sect. 5, conclusions are drawn and directions for future work considered.

2   Related Work

The main aim of the research described in this paper is to facilitate the development of annotated resources to support research in syntactic simplification. For this reason, the most relevant topics in previous work include the development of syntactically annotated resources; proposals to improve the quality of these resources; and the development of syntactic parsers that can automate the process.

There are currently a wide range of Treebanks available, providing access to syntactically annotated resources in many languages [10–12]. In English, one of the most widely-used is the Penn Treebank [13] which has been widely used in the development of supervised syntactic parsers [14, 15]. Despite several criticisms of this resource, the Penn Treebank continues to be widely exploited in the field of supervised parsing because syntactically annotated data is scarce and expensive to produce. In addition, it has been enhanced with other types of annotation.

---

[2] With regard to punctuation, this research is concerned with the annotation of what Nunberg et al. [9] refer to as *secondary boundary marks*. The annotation of other types of punctuation such as primary terminals, parentheses, dashes, punctuation involved in quotation, citation, and naming, capitalisation, and word-level punctuation is considered to be beyond the scope of the current work.

Maier et al. [16] observed that one shortcoming of the Penn Treebank is that punctuation symbols (commas and semicolons) are not tagged with information on their syntactic function. This lack of information makes it especially difficult to train parsers capable of identifying the conjoins of coordinated constituents when asyndetic coordination [7] is involved. To address this problem, they propose the addition of a second layer of annotation to disambiguate the role of punctuation in the Penn Treebank. They present a detailed scheme to ensure the consistent and reliable manual annotation of commas and semicolons with information to indicate their coordinating function.

An advantage of the approach described in [16] is that the addition of an annotation layer is more cost-effective than the development of new annotated resources from scratch. By leveraging the original layer of annotation, minimal human effort and expertise is required. However, this methodology can be criticised in two ways. First, the scheme encodes only coarse-grained information, with no discrimination between subclasses of coordinating and non-coordinating functions. Second, although production of the second annotation layer is inexpensive, application of the proposed scheme is costly as it depends on the availability of the original syntactic annotation layer. This limits the portability of the approach.

The annotation scheme presented in the current paper tags coordinators with more detailed information about their conjoins. Resources produced using this scheme and the scheme proposed in [16] can thus be regarded as complementary.

As noted earlier in this section, the Penn Treebank has been exploited in the development of supervised approaches to syntactic parsing. Given that this type of processing, if done with sufficient accuracy, could serve as the basis of any syntactic processing or syntactic simplification system, there has been considerable research activity towards improving the performance of syntactic parsing. Many of these involve techniques specifically designed to improve the parsing of coordinated structures [14, 17–23]. However, it must be noted that supervised methods trained on the Penn Treebank are likely to generate syntactic analyses subject to the criticisms levelled at that training data. A better prospect is to exploit such traditional resources in combination with others, such as the second annotation layer proposed in [16].

The scheme presented in this paper is derived from the one proposed in [3], which aimed to improve performance in information extraction by simplifying input documents syntactically. In that scheme, signs of syntactic complexity were considered to belong to one of two broad classes: *coordinators* and *subordinators*. These groups were annotated with information on the syntactic projection level and grammatical category of conjoins linked and subordinated constituents bounded by those signs. The annotation of a limited set of signs was exploited to develop an automatic classifier used in combination with a part-of-speech tagger and a set of rules to rewrite complex sentences as sequences of simpler sentences. Extrinsic evaluation showed that the simplification process evoked improvements in information extraction from clinical documents.

One weakness of the approach presented in [3] is that the classification scheme was derived by empirical analysis of rather homogeneous documents from a specialised source. Their consistency, together with the restricted range of linguistic phenomena manifested, imposes limits on the potential utility of the resources annotated. The scheme is incapable of encoding the full range of syntactic complexity encountered in documents of other genres.

3   The Annotated Corpus

This section presents the annotation scheme used to encode information on the syntactic function of a range of indicative signs of syntactic complexity (Sect. 3.1). The characteristics of three text collections annotated in accordance with this scheme are presented in Table 1.[3]

Table 1. The three collections of documents from which resources annotated for syntactic complexity were derived

Collection	Genre	#Docs	#Sents	#Words	#Signs	
					Present	Annotated
1. METER corpus	News	674	22,858	295,718	30,459	12,796
2. *www.patient.co.uk*	Healthcare	752	79,684	1,174,460	97,244	8,987
3. Gutenberg	Literature	24	4,468	87,712	11,031	11,031

Of the three sets of annotated resources, one was derived from a collection of news articles, the second from a collection of patient healthcare information leaflets and the third from a collection of literature. The texts were selected for annotation in line with the requirements identified in the larger project.

3.1   Annotation Scheme

Development of the annotation scheme described in this paper was informed by corpus analysis and consultation of sources covering a range of issues related to syntactic complexity, coordination, and subordination [7] and the function and distribution of punctuation marks [9]. Taking the scheme presented in [3] as a starting point, *subordinators* were re-designated and divided into *leftmost subordination boundaries* and *rightmost subordination boundaries*. Both types (coordinators and subordination boundaries) were extended to include a larger number of signs in the current paper. A far wider range of conjoins and subordinated constituents can be distinguished in the scheme presented in the current paper than was possible in that used in [3].

The annotation scheme is intended to encode information on the linking and bounding functions of different signs of syntactic complexity. Figure 1 presents the scheme graphically. It depicts each type of complexity, with the central nodes representing core functions of these signs. Coordinating functions are depicted on the left side of the diagram and subordinating functions on the right.

---

[3] The annotated resources and the annotation guidelines will be made available online.

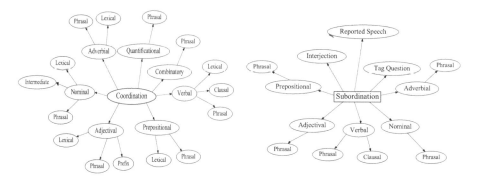

Fig. 1. Typology of coordinators (left hand side) and subordination boundaries (right hand side) annotated in the corpus

As noted by Quirk et al. [7], and supported by corpus evidence (see Sect. 4), coordinators usually link constituents of the same syntactic category. Nodes in the diagram directly connected to the central one representing coordination denote different syntactic categories of coordination. The leaf nodes represent the different levels of syntactic projection that coordinated conjoins may take. In terms of syntactic complexity, conjoins were observed in a range of projection levels: zero (morphemic and lexical), intermediate, maximal (phrasal), and extended (clausal).

In Fig. 1 (right), nodes directly connected to the central one representing subordination denote different types of subordinated constituent. These include five syntactic categories, reported speech, and two other units typical of spontaneous or colloquial language (tag questions and interjections). The leaf nodes in the diagram represent the different syntactic projection levels that these subordinated constituents may take [24]. Subordinated constituents of the five syntactic categories all occur at the maximal projection level. Subordinated verbal constituents may also be extended projections. By contrast, the other two types of subordinated constituent can occur at numerous levels of projection, from zero to extended.

To summarise, the annotation scheme described in this paper is used to develop resources in which coordinators are annotated with information about the specific type of coordination that they embody and therefore the specific types of conjoins that they link. Subordination boundaries are annotated with information about the specific type of subordinated constituent that they bound. As discussed in the next two sections, there is a limited set of signs which potentially indicate syntactic complexity. For this reason, they can be automatically detected with great reliability by a specially developed annotation tool and presented to human annotators for classification. One advantage of this approach is that the annotation task does not require the level of expertise in syntax that would be required in the development of a treebank. The annotation presented in the current paper involves a much less detailed analysis of the syntactic structure of each sentence and does not depend on strict adherence to a specfiic linguistic theory. As a result, the task is less onerous and can be performed more rapidly.

3.2 Coordinators

There are three major types of coordinator. *Conjunctions* ([*and*], [*but*] and [*or*]) have an exclusively coordinating function; *punctuation* marks ([,] and [;]), may occur as coordinators in some contexts and as subordination boundaries in others. It follows from this that other signs of syntactic complexity involving punctuation-conjunction pairs ([, *and*], [; *and*], [, *but*], [; *but*], [, *or*] and [; *or*]) are similarly ambiguous.

Signs that have been identified as coordinators are classified by assigning them to one of the subclasses (leaf nodes connected to coordination) displayed in Fig. 1. The class labels used in the annotation scheme are acronyms that indicate the function of each annotated coordinator.[4]

- The first part of the acronym indicates the coordinating function (C).
- The second part indicates the syntactic projection level of the conjoins linked by the coordinator. These include constituents at the morphemic (P), lexical (L), intermediate (I), maximal (M), and extended (E) levels of syntactic projection.
- The third part of the acronym indicates the syntactic category of the coordinated conjoins. These may be verbal (V), nominal (N), adjectival (A), adverbial (Adv), prepositional (P), or quantificational (Q).
- The fourth part of the acronym is optional and takes a numerical value. It is used to distinguish between the coordination of different types of nominal and verbal maximal projection. These sub-types are:
  1. default maximal projections (CMA1, CMN1, CMV1);
  2. maximal projections in which the head of the second conjoin has been elided (CMN2, CMV2);
  3. maximal projections in which the complement of the head of the first conjoin has been elided (CMV3);
  4. maximal projections in which the head of the first conjoin has been elided (CMN4).

To illustrate, the class label CLA indicates that the sign is a coordinator of two lexical projections of an adjective (3) whereas CMP indicates that the sign is a coordinator of two maximal projections of a preposition (4).

(3) "He had a <u>stable</u> [and] <u>loving</u> family."

(4) "But the melancholy experience <u>of the courtrooms</u> [and] <u>of life</u> is that people have a good character in some respects and not in others."

Several additional class labels may be assigned to coordinators. COMBINATORY is used to tag markables indicating combinatory coordination, typically used in fixed phrases, proverbs, and aphoristic sentences. These coordinations are usually atomic and require a different type of processing in tasks such as syntactic simplification. CXE is

---

[4] Guidelines accessible from
http://clg.wlv.ac.uk/resources/SignsOfSyntacticComplexity/
provide examples of the use of all the classes specified in this scheme. Due to space restrictions, they are not provided in the current paper.

used to denote coordinators linking conjoins with unusual patterns of ellipsis (5) while CMX is assigned to coordinators linking syntactically ill-assorted conjoins[5] (6).

(5) The 38-page judgment stated that Mrs Coughlan, a tetraplegic, was entitled to free nursing care because her primary need for accommodation was$_i$ a health need [and] her nursing needs ϕ_i not 'incidental'.[6]

(6) "Name something that is currently on BBC1 that gets people excited [and] talking about it.

3.3 Subordination Boundaries

There are six major types of subordination boundary. Of these, two involve lexical signs (complementisers ([*that*]) and wh-words ([*what*], [*when*], [*where*], [*which*], [*while*] and [*who*])) while four involve the use of punctuation either in isolation ([,], [;] and [:]) or in a pair, followed by any other sign of syntactic complexity. *Complementisers* and *wh-words* exclusively serve to bound subordinated constituents. As noted in Sect. 3.2, signs of syntactic complexity involving commas and semicolons may serve as coordinators in some contexts and as subordination boundaries in others.

Signs that have been identified as subordination boundaries are classified by assigning them to one of the subclasses (leaf nodes connected to subordination) displayed in Fig. 1. The class labels used in the annotation scheme are acronyms that indicate the function of each annotated subordination boundary.

– The first part of the acronym can indicate the rightmost boundary of a subordinated constituent (ES) or the leftmost boundary of a subordinated constituent (SS).
– The second part indicates the syntactic projection level of the bounded constituent. These include constituents at the maximal (M) and extended (E) levels of syntactic projection.
– The third part of the acronym indicates the syntactic category of the bounded constituent. These may be verbal (V), nominal (N), adjectival (A), adverbial (Adv), or prepositional (P).

The scheme includes class labels for annotation of the following additional types of subordinated constituent:

– Interjections (SSMI/ESMI),
– direct quotes (SSCM/ESCM),
– tag questions (STQ),[7]
– constituents of ambiguous syntactic category (SSMX/ESMX).[8]

---

[5] This term denotes pairs of conjoins that do not match in terms of grammatical category [7].

[6] The position of the elided element is indicated via the symbol ϕ.

[7] It should be noted that the rightmost boundary of a tag question is usually a sentence boundary (question mark). In the research supporting the current paper, sentence boundaries were not considered markable. As a result, no signs of syntactic complexity serving as the rightmost boundaries of tag questions were encountered in the corpora presented in Sect. 4.

[8] There are just 25 instances of this class in the 32,814 signs annotated so far.

To illustrate, the class label ESMV indicates that the sign is a rightmost subordination boundary of the subclass bounding the maximal projection of a verb (7).

(7) "Being put into a psychiatric ward with people with long-term mental illnesses who are shaking with the drugs they are taking[,] there's no way you can feel normal and be OK with yourself," she told BBC TV's *That's Esther* programme with Esther Rantzen.

3.4 False Signs

Finally, the annotation scheme includes the class label SPECIAL to denote false signs of syntactic complexity such as use of the word *that* with a specifying (8) or referential (9) function.

(8) "I'm quite happy to abandon [that] specific point" he said.

(9) 'Because of your involvement in the past with trying to stop all [that] in your work, you more than anybody else should have known the misery of people who had become addicted.'

4 Corpus Analysis

This section provides an evaluation of the annotated resources developed in this research. The descriptions include analysis of the distribution of different signs of syntactic complexity and the classes to which they belong. Assessments of inter-annotator agreement are presented to provide insight into the reliability and consistency of the annotation.

A subset of the signs of syntactic complexity occurring in Collections 1 and 2 (Table 1) were manually classified in accordance with the annotation scheme presented in Sect. 3.1. All of the signs occurring in Collection 3 were annotated. During the annotation process, annotators were provided with access to annotation guidelines and to a grammar of English [7].

4.1 Sign and Class Distribution

In documents of all genres, the comma, the conjunction [*and*], and the complementiser [*that*] were the most frequently occurring signs of syntactic complexity. Use of the sign [*, and*] was more characteristic of the genre/domain of literature (15.95%) than news (2.49%) or patient healthcare (3.30%). In comparison with documents of the other genres, the conjunction [*or*] was most frequent in those of patient healthcare. Use of the semicolon was relatively frequent in nineteenth/twentieth century literature.

In the three genres, the classes to which different signs of syntactic complexity most frequently belonged were leftmost boundaries of subordinated clauses (SSEV), coordinators of verb phrases (CMV1), and coordinators of noun phrases (CMN1). The comma is used with a wide range of coordinating and bounding functions in all three domains/genres, with its function as a subordination boundary being considerably more

Table 2. Relative frequency of indicative signs and classes in the three collections of annotated documents

Relative frequency	Collection 1 (News)	Collection 2 (Healthcare)	Collection 3 (Literature)
High	Sign [:] of class SSCM Sign [*who*] of class SSEV	Sign [*or*] of classes CLN, CMN1, and CLA Signs of class ESMAdv Signs of class ESMN	Sign [;] of classes CEV, SSEV, and ESEV Signs of class SSCM
Low		Signs of class CEV	Signs of classes SSMN and ESMN

frequent than its function as a coordinator. In the genre of literature, the comma is used slightly more often as the rightmost boundary of a subordinated constituent (in 56% of its occurrences) while in patient healthcare documents, it is used this way considerably more often (81% of its occurrences). In news articles, it is used slightly more frequently as the leftmost boundary of a subordinated constituent (52% of its occurrences).

Due to the large number of classes and signs used in the annotation, it is difficult to visualise their distribution in an practical way.[9] Table 2 presents information on the relative frequency of signs and classes characteristic of all three document collections. These features are indicative of various linguistic properties of each one:

– Collection 1 (News)

1. Frequent use of the colon [:] to bound reported speech (SSCM).
2. Frequent provision of additional explanatory information about the people mentioned in news articles.

– Collection 2 (Healthcare)

1. Frequent presentation of lists of alternative possibilities for treatment options, symptoms, anatomical locations, and medical procedures.
2. Frequent occurrence of sentence-initial adverbials to contextualise symptoms or treatment options.
3. Frequent use of appositions to provide explanatory definitions of terms.
4. Sentences are of limited overall complexity, making the documents accessible by a wide a range of readers.

– Collection 3 (Literature)

1. Terms used in literary documents are rarely defined via explanatory subordinated noun phrases (SSMN and ESMN).
2. There is greater variation in the occurrence of direct speech and reporting clauses than is the case for news articles.

---

[9] The raw data is accessible at
http://clg.wlv.ac.uk/resources/SignsOfSyntacticComplexity/

4.2 Consistency/Reliability of Annotation

A subset of 1000 annotations of each of the three collections was cross-annotated and a confusion matrix plotted. The values of Kappa obtained for the annotations were 0.80 for signs annotated in news articles, 0.74 for those in documents conveying patient healthcare information, and 0.76 for those in literary documents. These levels imply a minimum of "substantial agreement" between annotators [25].

In the annotation of news articles, the most common disagreement (8.67% of the cases) occurred for signs that one annotator considered to indicate clause coordination (CEV) and the other considered to indicate verb phrase coordination (CMV1), especially in cases involving imperative clauses and complex VP conjoins with clausal arguments and modifiers.

Of the most frequent types of disagreement in the genres of patient healthcare information and literature, many were of the same type as those in news articles, though some genre-specific types were also in evidence.

In the genre of patient healthcare information, hyphenated items were a cause of disagreement, with annotators disagreeing on the projection level of nominal conjoins. Disagreement on the annotation of signs of syntactic complexity in bibliographic references was also evident in this genre.

In the genre of literature, many disagreements involved the coordination of verb phrases. It was observed that in cases where the second conjoin of a coordinator has an adverbial pre-modifier consisting of a *wh*-complementiser and its clausal complement, the coordinator is misclassified as linking two clauses (10).

(10)   The king comforted her and said: 'Leave your bedroom door open this night, and my servants shall stand outside[, and] when he has fallen asleep shall go in, bind him, and take him on board a ship which shall carry him into the wide world.

Another type of disagreement arising in the annotation of signs in literary text involves signs used in colloquial constructions such as interjections. At times, interjections are nested and there is disagreement about whether a sign is serving as the right boundary of an interjection or the left boundary of one that is subordinated (11).

(11)   "Ah[,] well! We did our best, the dear knows."

While it is difficult to motivate any particular choice of class label (between SSMI and ESMI) for such signs, it is expected that consistency of annotation can be improved by adding to the guidelines an instruction for annotators to select SSMI (for example), when encountering signs in such contexts.

Approximately 12 months after the initial annotation, a set of 500 signs of syntactic complexity were re-annotated by one annotator. Of the signs involved, 99.97% were assigned the same class label ($\kappa = 0.9997$). This result indicates that the annotation task is well defined. Despite the protracted period between the annotation sessions, in nearly all cases, the second session led to the same syntactic functions being assigned to signs of syntactic complexity. This implies that fidelity of annotation is based on annotators' comprehension of the task rather than guesswork or chance.

5   Conclusions and Directions for Future Work

This paper has presented a new annotation scheme to enable the encoding of information on syntactic complexity (coordination and subordination) in English texts. The annotation process is inexpensive for two main reasons. First, only a limited number of signs of syntactic complexity are considered markable, and these can be automatically detected with great reliability by the annotation tool developed in this research. Annotators are not required to encode a complete syntactic analysis of the text. Instead, markables are tagged to indicate their syntactic function as either coordinators or subordination boundaries and to indicate the syntactic category and projection level of the conjoins linked by the former or bounded by the latter. Second, the annotation is not dependent on other, potentially expensive layers of annotation, as is the case for that presented in [16]. The scheme is thus portable and can be used to derive new language resources from unrestricted English text.

It has been shown that resources produced using this annotation scheme are of high levels of reliability and consistency (Sect. 4). Assessments of inter-annotator consistency in three genres show that, with appropriate guidelines, non-experts can annotate text from specialised domains with little degradation in reliability.

The paper points to two main directions for future work. The first concerns expansion of the annotation scheme. Currently, the class labels denoting signs that bound clauses (SSEV and ESEV) subsume several potential subclasses. This is due to the variety of types of clause that may be subordinated, which include nominal *that*-clauses, adverbial *when*-clauses, nonfinite clauses,[10] and some verbless clauses. As a result, SSEV is by far the most frequent class label occurring in the annotated corpora. It is possible that the inclusion of such a wide-ranging class will cause confusion in automatic classification of signs of syntactic complexity. In future work, it will be interesting to investigate the effect of extending the set of class labels to include different ones for different types of clause.

While encoding a wide range of phenomena related to syntactic complexity, it may be argued that additional signs should be included in the annotation scheme presented in the current paper. Though not reported here, the annotation of parentheses has been included in the most recent version of the scheme. They frequently bound subordinated constituents, especially in documents providing patient healthcare information, where they bound subordinated noun phrases in the great majority of cases, as well as verb phrases and clauses. Previous work has described a wide range of functions of hyphens or dashes [9], which is another candidate for inclusion in the set of signs of syntactic complexity to be annotated.

The second possible direction of future work involves extrinsic evaluation of the resources developed via exploitation by NLP applications. Following Maier et al. [16], the automatic classification of signs of syntactic complexity would enable the addition of a second annotation layer to resources such as the Penn Treebank, which

---

[10] Including infinitive clauses (bare infinitive clauses and *to*-infinitive clauses whose leftmost boundary is a *wh*-word) but excluding *to*-infinitive clauses whose leftmost boundary is a punctuation mark and *ed*-participle clauses, both of which are considered subordinated verb phrases in the current paper.

could then be exploited by supervised parsing methods. Applications such as machine translation and automatic summarisation provide further scope for indirect evaluation of the resources presented in the current paper.

To conclude, the development of syntactic processing systems is expected to benefit from the availability of the annotated resources presented in the current paper. Insights gained from the empirical analysis of these resources will enable developers to prioritise the accurate processing of the most prevalent types of syntactic complexity occurring in texts of different genres.

Acknowledgments. The research described in this paper was partially funded by the European Commission under the Seventh (FP7 - 2007- 2013) Framework Programme for Research and Technological Development (FP7-ICT-2011.5.5 FIRST 287607). This publication reflects the views only of the Authors, and the Commission cannot be held responsible for any use which may be made of the information contained therein. The Authors gratefully acknowledge the contributions of the annotators, Emma Franklin and Zoe Harrison, who identified problematic cases and prompted useful revisions of the annotation scheme.

References

1. Agarwal, R., Boggess, L.: A simple but useful approach to conjunct identification. In: Proceedings of the 30th Annual Meeting for Computational Linguistics, Newark, Delaware, pp. 15–21. Association for Computational Linguistics (1992)
2. Rindflesch, T.C., Rajan, J.V., Hunter, L.: Extracting molecular binding relationships from biomedical text. In: Proceedings of the Sixth Conference on Applied Natural Language Processing, Seattle, Washington, pp. 188–195. Association of Computational Linguistics (2000)
3. Evans, R.: Comparing methods for the syntactic simplification of sentences in information extraction. Literary and Linguistic Computing 26 (4), 371–388 (2011)
4. Gerber, L., Hovy, E.: Improving translation quality by manipulating sentence length. In: Farwell, D., Gerber, L., Hovy, E. (eds.) AMTA 1998. LNCS (LNAI), vol. 1529, pp. 448–460. Springer, Heidelberg (1998)
5. Tomita, M.: Efficient Parsing for Natural Language: A Fast Algorithm for Practical Systems. Kluwer Academic Publishers, Norwell (1985)
6. McDonald, R.T., Nivre, J.: Analyzing and integrating dependency parsers. Computational Linguistics 37, 197–230 (2011)
7. Quirk, R., Greenbaum, S., Leech, G., Svartvik, J.: A comprehensive grammar of the English language. Longman (1985)
8. Orăsan, C., Evans, R., Dornescu, I.: Towards multilingual Europe 2020: A Romanian perspective, pp. 287–312. Romanian Academy Publishing House (2013)
9. Nunberg, G., Briscoe, T., Huddleston, R.: Punctuation, pp. 1724–1764. Cambridge University Press (2002)
10. Brants, S., Dipper, S., Hansen, S., Lezius, W., Smith, G.: The TIGER treebank. In: Proceedings of the Workshop on Treebanks and Linguistic Theories, Sozopol (2002)
11. Simov, K., Popova, G., Osenova, P.: HPSG-based syntactic treebank of Bulgarian (BulTreeBank), pp. 135–142. Lincom-Europa, Munich (2002)

12. Hajič, J., Zemánek, P.: Prague arabic dependency treebank: Development in data and tools. In: Proceedings of the NEMLAR International Conference on Arabic Language Resources and Tools, pp. 110–117 (2004)
13. Marcus, M.P., Santorini, B., Marcinkiewicz, M.A.: Building a large annotated corpus of english: The penn treebank. Computational Linguistics 19, 313–330 (1993)
14. Charniak, E., Johnson, M.: Coarse-to-fine n-best parsing and MaxEnt discriminative reranking. In: Proceedings of the 43rd Annual Meeting of the ACL, Ann Arbor, pp. 173–180 (2005)
15. Collins, M., Koo, T.: Discriminative reranking for natural language parsing. Computational Linguistics 31, 25–69 (2005)
16. Maier, W., Kübler, S., Hinrichs, E., Kriwanek, J.: Annotating coordination in the penn treebank. In: Proceedings of the Sixth Linguistic Annotation Workshop, Jeju, Republic of Korea, pp. 166–174. Association for Computational Linguistics (2012)
17. Ratnaparkhi, A., Roukos, S., Ward, R.T.: A maximum entropy model for parsing. In: Proceedings of the International Conference on Spoken Language Processing (ICSLP), Yokohama, Japan, pp. 803–806 (1994)
18. Rus, V., Moldovan, D., Bolohan, O.: FLAIRS Conference. AAAI Press (2002)
19. Kim, M.Y., Lee, J.H.: S-clause segmentation for efficient syntactic analysis using decision trees. In: Proceedings of the Australasian Language Technology Workshop, Melbourne, Australia (2003)
20. Nakov, P., Hearst, M.: Using the web as an implicit training set: Application to structural ambiguity resolution. In: Proceedings of Human Language Technology Conference and Conference on Empirical Methods in Natural Language Processing (HLT/EMNLP), Vancouver, Association for Computational Linguistics, pp. 835–842 (2005)
21. Hogan, D.: Coordinate noun phrase disambiguation in a generative parsing model. In: Proceedings of the 45th Annual Meeting of the Association of Computational Linguistics, Prague, Czech Republic, pp. 680–687. Association for Computational Linguistics (2007)
22. Kawahara, D., Kurohashi, S.: Coordination disambiguation without any similarities. In: Proceedings of the 22nd International Conference on Computational Linguistics (Coling 2008), Manchester, England, pp. 425–432 (2008)
23. Kübler, S., Hinrichs, E., Maier, W., Klett, E.: Parsing coordinations. In: Proceedings of the 12th Conference of the European Chapter of the ACL, Athens, Greece, pp. 406–414. Association for Computational Linguistics (2009)
24. Chomsky, N.: Knowledge of language: its nature, origin, and use. Greenwood Publishing Group, Santa Barbara (1986)
25. Viera, A.J., Garrett, J.M.: Understanding interobserver agreement: The kappa statistic. Family Medicine 37, 360–363 (2005)

Application of LSTM Neural Networks in Language Modelling

Daniel Soutner and Luděk Müller

University of West Bohemia, Faculty of Applied Sciences, Department of Cybernetics,
Univerzitní 22, Plzeň, Czech Rep.
{dsoutner,muller}@kky.zcu.cz
www.kky.zcu.cz

Abstract. Artificial neural networks have become state-of-the-art in the task of language modelling on a small corpora. While feed-forward networks are able to take into account only a fixed context length to predict the next word, recurrent neural networks (RNN) can take advantage of all previous words. Due the difficulties in training of RNN, the way could be in using Long Short Term Memory (LSTM) neural network architecture.

In this work, we show an application of LSTM network with extensions on a language modelling task with Czech spontaneous phone calls. Experiments show considerable improvements in perplexity and WER on recognition system over n-gram baseline.

Keywords: language modelling, recurrent neural networks, LSTM neural networks.

1 Introduction

Statistical language models (LM) play an important role in the state-of-art large vocabulary continuous speech recognition (LVCSR) systems. Statistically computed n-gram models and class-based LMs are the main models used in LVCSR systems, however, subsequent models are becoming more important supplement to existing techniques.

In recent years feed forward neural networks (FFNN) [12] attracted attention due their ability to overcome biggest disadvantage of n-gram models: even when the n-gram is not observed in training, FFNN estimates probabilities of the word based on the full history [15]. That is in contrast to n-gram, where back-off model estimates unseen n-grams with $(n-1)$-gram.

To avoid handling with the parameter n (number of words in n-gram and in FNN LM) we can use the recurrent neural network (RNN) architecture [2]. The RNN is going further in model generalization: instead of considering only the several previous words (parameter n) the recursive weights are assumed to represent short term memory. More in general we could say that RNN sees text as a signal consisting of words.

Long Short-Term Memory (LSTM) neural network [8] is different type of RNN structure. As was shown, this structure allows to discover both long and short patterns in data and eliminates the problem of vanishing gradient by training RNN. LSTM approved themselves in various applications [8][1] and it seems to be very promising course also for the field of language modelling [3].

I. Habernal and V. Matousek (Eds.): TSD 2013, LNAI 8082, pp. 105–112, 2013.

In this work we present an application of LSTM language model as an extension to the basic n-gram model and the influence of this modification to the perplexity and word error rate analysed on English and Czech corpora.

2   LSTM Neural Networks

The vanishing gradient seems to be problematic during the training of RNN as shown in [8]. This led authors to re-design of the network unit, in LSTM called as a cell. Fig. 1 shows that every LSTM cell contains *gates* that determine when the input is significant enough to remember, when it should continue to remember or forget the value, and when it should output the value. So designed cells may be interpreted as a differentiable memory.

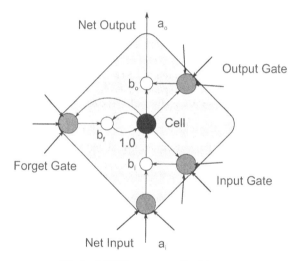

Fig. 1. LSTM memory cell with gates

2.1   LSTM Topology

Typical NN unit consists of the input activation which is transformed to output activation with activation function (usually sigmoidal).

The LSTM cell provides this more comprehensively: The three cell inputs called *gates* determine when values are allowed to flow into or out of the block's memory. Firstly, the activation function is applied to all gates. When *input gate* outputs a value close to zero, it zeros out the value from the net input, effectively blocking that value from entering into the next layer. When *forget gate* outputs a value close to zero, the block will effectively forget whatever value it was remembering. The *output gate* determines when the unit should output the value in its memory.

Depends on type of LSTM, consecutions may slightly differ (some modifications and enhancements were introduced), but the main principals are the same. The training algorithm and complete equations of LSTM neural network could be found e.g. in [8] [9].

Due this specific topology of LSTM, especially because of a constant error flow, regular back-propagation could be effective at training an LSTM cell to remember values for very long durations. LSTM can be also trained by evolution strategies or genetic algorithms in reinforcement learning applications [13].

2.2 LSTM Language Model

The LSTM NN was successfully introduced to the field of language modelling [3]. The topology of our model (shown in Figure 2) is similar to common RNN language models ([2] [3]) and is based on these principals:

- The input vector is word encoded as 1-of-N coding.
- There is a softmax function used in output layer to produce normalized probabilities.
- The cross entropy is used as training criterion.

Normalization of input vector which is generally advised for neural networks is not needed due the 1-of-N input coding.

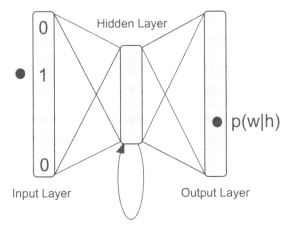

Fig. 2. Neural network LM architecture

3 Input Vector Modifications

The standard input vector in various neural net language models is mostly 1-of-N. The words on input are encoded by 1-of-N coding, where N is number of words in vocabulary.

We also intended and applied two extensions of the basic model - added Latent Dirichlet Allocation [7] (LDA) for better modelling of longer context [16] and the class extension for dealing with similar words in the same context. Both of these extensions are described in sections bellow.

3.1 Latent Dirichlet Allocation Extension

Language models with cache brings improvement in perplexity but not when measured on WER in speech recognition. Thus, to exploit more information from the long span context we decided to use the Latent Dirichlet Allocation (LDA) [7] in our experiments as proposed by T. Mikolov and G. Zweig in [16], where this model is closely described. The LDA process converts word representation of document to low-dimensional vector which represents probability of to topic.

LDA represents documents as mixtures of topics that split out words with certain probabilities. It assumes that documents are produced in the following fashion:

– Deciding on the number of words N the document will have by sampling from a Poisson distribution.
– Choosing a topic mixture for the document (according to a Dirichlet distribution over a fixed set of K topics).
– Generating each word w_i in the document by:
  • Picking a topic (according to the multinomial distribution that you sampled above).
  • Using the topic to generate the word itself (according to the topics multinomial distribution).

Assuming this generative model for a collection of documents, LDA tries to backtrack from the documents to find a set of topics that are most likely to generate the collection.

In our experiments we fixed the length of the word cache for computing topic distribution. For Czech training corpus every phone call is equal to one document, for English Penn Treebank we divided text to documents of 10 non-overlapping sentences; the input vector of NN is modified as original 1-of-N coding and proposed additional LDA feature. The models were created with *gensim* tool [6]. We explored several configurations of trained models with a different number of topics (from 20 to 70) and cache length (50 and 100).

3.2 Class Extension

In both - written and spoken language - we use different words for expressing a similar topic or the same fact. There are many approaches in language modelling that are trying to deal with this aspect of language i.e. the class-based models [10]. Assuming this, we tried to investigate whether word classes are able to help us in LSTM LM.

The similar words could be split up to the classes with a lot of different ways, we decided to use one based on inducing word classes from n-gram statistics. Word classes induced from distributional statistics are produced so as to minimize perplexity of a class-based n-gram model given the provided word n-gram counts. This means that words occurring in the similar context should be found in the same class. The classes were prepared using the SRI toolkit [4].

We have modified input vector for our purposes analogously as in LDA extension: we added to the standard input vector (1-of-N) a vector 1-of-C, where C is number of classes. We trained models with various number of classes, from 100 to 2300.

4   Experimental Results

4.1   Perplexity Results

To maintain comparability with the other experiments, we chose well-known Penn Tree-bank (PTB) [11] portion of the Wall Street Journal corpus for testing our models. Following preprocessing was applied to the corpora: the vocabulary was short-listed to 10k most frequent words, all numbers were unified into $\langle N \rangle$ tag and punctuation was removed. The corpus was divided into 3 parts (training, development and test) with 42k, 3.3k and 3.7k tokens.

The second part of experiments was performed with Czech spontaneous phone calls (BH). This corpus is further described in Section 4.2.

First, we trained LDA model, word classes and both extensions together on training data and with these models we trained our LSTM neural networks on the same data. All models were trained with 20 cells in hidden layer. Afterwards, we chose models with the best parameters measured on development data. The final results for PTB achieved on test data are shown in Table 1, models were combined with baseline model by linear combination. The proposed extensions seem to be promising, they improve the perplexity of the baseline model by $\approx 2 - 12\%$.

The influence of word cache and number of topics (as parameters of LDA extension) was tested on BH, the Table 2 shows the result, which suggests a cache of length 100 and 50 topics. The influence of the number of classes is shown in Fig. 3, for 500 classes we obtained the best perplexity values.

Table 1. Perplexity results with PTB

model	PPL
KN5	140
LSTM +KN5	120
LSTM LDA +KN5	117
LSTM CLASS +KN5	113
LSTM LDA & CLASS +KN5	105

Table 2. Perplexity results of LSTM with LDA extension on phone calls (BH), different number of topics (20-100) and word cache (50 and 100)

#topics	cache	
	50	100
20	154.1	163.9
30	164.4	117.2
40	128.5	109.5
50	110.4	**103.7**
70	110.2	107.6
100	**109.6**	125.6

4.2   Model Evaluation

As a training and test data for models evaluation we used Czech spontaneous speech which was recorded from phone calls. These calls were acquired as "Free calls" where people could phone for free while giving the permission to use anonymously their calls for the speech recognition experiments. We had to deal with the task where recorded data were very different from a common written Czech language. This is not a trivial task, as shown in previous work [5], where the records with spontaneous speech were also processed.

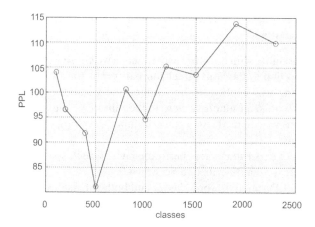

Fig. 3. Perplexity results of LSTM with class extension combined with KN5 model depending on number of classes; measured on phone calls (BH) corpora

The data are specified by:

- a high inflection of Czech language (cases, various verb forms,...)
- word inflection is partially different from written Czech language
- unusual words used by speakers (slang, diminutives,...)
- only a small set of data available (about 2.8M words)
- the records contain a lot of non-speech events
- the sentences are relatively short
- the vocabulary is relatively small (about 120k words)

The statistics of the used corpus are shown in Table 3; the corpora was divided into tree parts: training, development and test set. The characteristics of the test phone records for models evaluation on speech recognition are shown in Table 4.

Table 3. BH text data

	Sentences	Words	OOVs
Train	400k	2.2M	
Dev	3k	13k	350
Test	3k	14k	385

Table 4. BH test records

Records	
Length h:mm	2:16
Sentences	3582
Speakers	50

We took our state-of-the-art LVCSR system, as a language model we used 3-gram Knesser-Ney back-off model and finally n-best hypothesis with $n = 1000$ from the lattices were extracted. Hereafter, this n-best list will be the base for our experiments with language models.

As the baseline model we used 5-gram Knesser-Ney statistical model (KN5) [14] trained from the same corpus with a full vocabulary. The LSTM language models were trained from data with a limited vocabulary, where only 10k most frequent words were used, and all were combined with KN5 model with linear interpolation. The width of hidden layer was again fixed to 20 cells for all models. As described above, we chose the parameters of models on development part of data, in terms of perplexity.

We advanced the KN5 baseline by $\approx 3.7\%$ in relative, models with extended features slightly overcome the basic LSTM model, according to perplexity results; the improvement is statistically significant with $p = 0.05$. The complete results are shown in Table 5. The theoretical maximum that we eventually could obtain while we are rescoring this n-best list is 73.7% in accuracy.

Table 5. Evaluating on speech recognition (1000-best list rescore)

model	Acc in %
LSTM LDA&CLASS +KN5	52.25
LSTM CLASS +KN5	52.05
LSTM LDA +KN5	51.95
LSTM +KN5	51.54
KN5 baseline model	50.41

5   Conclusions

We have applied the LSTM neural network language model to spontaneous Czech speech. We explored several extensions to this approach: with LDA to explore long span context in dialogue and with classes to find similarities in topics.

We gained some not breakthrough but significant improvements in comparison to the basic model while applying these models in terms of perplexity and speech recognition. For future work it seems interesting to further discover the influence of input feature vectors and realize more experiments with another corpora.

Acknowledgements. This research was supported by the Ministry of Culture Czech Republic, project No. DF12P01OVV022.

References

1. Frinken, V., Zamora-Martinez, F., Espana-Boquera, S., Castro-Bleda, M.J., Fischer, A., Bunke, H.: Long-short term memory neural networks language modeling for handwriting recognition. In: 21st International Conference on Pattern Recognition (ICPR), November 11-15, pp. 701–704 (2012)
2. Mikolov, T., Kombrink, S., Burget, L., Cernocky, J.H.: Extensions of recurrent neural network language model. In: 2011 IEEE International Conference on Acoustics, Speech and Signal Processing (ICASSP), May 22-27, pp. 5528–5531 (2011)
3. Sundermeyer, M., Schlüter, R., Ney, H.: LSTM Neural Networks for Language Modeling. In: INTERSPEECH 2012 (2012)

4. Stolcke, A.: SRILM – An Extensible Language Modeling Toolkit. In: Proc. Intl. Conf. on Spoken Language Processing, Denver, vol. 2, pp. 901–904 (2002)
5. Soutner, D., Loose, Z., Müller, L., Pražák, A.: Neural Network Language Model with Cache. TSD 2012:528-534
6. Řehůřek, R., Sojka, P.: Software Framework for Topic Modelling with Large Corpora. In: Proceedings of LREC 2010 Workshop New Challenges for NLP Frameworks, p. 5. University of Malta, Valletta (2010) ISBN 2-9517408-6-7
7. Blei, D.M., Ng, A.Y., Jordan, M.I., Lafferty, J.: Latent dirichlet allocation. Journal of Machine Learning Research 3 (2003)
8. Hochreiter, S., Schmidhuber, J.: Long Short-term Memory. Neural Computation 9(8), 1735–1780 (1997)
9. Gers, F.: Long Short-Term Memory in Recurrent Neural Networks, Ph.D. Thesis. École Polytechnique Fédérale de Lausanne, Switzerland (2001)
10. Brown, P.F., Della Pietra, V.J., de Souza, P.V., Lai, J.C., Mercer, R.L.: Class-Based n-gram Models of Natural Language. Computational Linguistics 18(4), 467–479 (1992)
11. Charniak, E.: BLLIP 1987-89 WSJ Corpus Release 1, Linguistic Data Consortium, Philadelphia (2000)
12. Bengio, Y., Ducharme, R., Vincent, P., Janvin, C.: A neural probabilistic language model. J. Mach. Learn. Res. 3, 1137–1155 (2003)
13. Schmidhuber, J., Wierstra, D., Gagliolo, M., Gomez, F.: Training Recurrent Networks by Evolino. Neural Computation 19(3), 757–779 (2007) PDF (preprint)
14. Kneser, R., Ney, H.: Improved backing-off for M-gram language modeling. In: 1995 International Conference on Acoustics, Speech, and Signal Processing, ICASSP 1995, May 9-12, vol. 1, pp. 181–184 (1995)
15. Oparin, I., Sundermeyer, M., Ney, H., Gauvain, J.: Performance analysis of Neural Networks in combination with n-gram language models. In: ICASSP, pp. 5005–5008 (2012)
16. Mikolov, T., Zweig, G.: Context Dependent Recurrent Neural Network Language Model. Microsoft Research Technical Report MSR-TR-2012-92 (2012)

Automatic Laughter Detection
in Spontaneous Speech Using GMM–SVM Method

Tilda Neuberger and András Beke

Departement of Phonetics, Research Institute for Linguistics: Hungarian Academy of Sciences,
Benczúr u. 33, Budapest, Hungary
{neuberger.tilda,beke.andras}@nytud.mta.hu
www.nytud.hu

Abstract. Spontaneous conversations frequently contain various non-verbal vocalizations (such as laughter). The accuracy of a speech recognizer may decrease in the case of spontaneous speech because of these non-verbal vocalization phenomena. The aim of the present research is to develop an accurate and efficient method in order to recognize laughter in spontaneous utterances. We used GMM in modeling the data and SVM for differentiating laughter from other speech events. The training and testing of the laughter detector were carried out using the BEA Hungarian spoken language database. The results show that the GMM–SVM system seems to be a particularly good method for solving this problem.

Keywords: laughter, classification, GMM–SVM, spontaneous speech.

1   Introduction

The non-verbal communication plays an important role in human speech comprehension. Beside the visual cues (gestures, facial expressions, eye contacts, etc.), messages of some type can be transferred by non-verbal vocalizations (laughter, throat clearing, breathing) as well. Spontaneous speech frequently contains such non-verbal vocalizations of various functions where laughter is one of the most frequent phenomena. From the perspective of a human being, laughter is an inborn, species-specific indicator of affection that provides information about the emotional state of the speaker. It can also be a social signal, a socially constituted and easily decodable phenomenon. It has various functions in everyday conversations. Laughter can be part of social interactions in early infancy [19]-[20], part of aggressive behavior (laughing at someone), or a part of appeasement in situations of dominance/subordination. In meetings, it serves to regulate the flow of the interactions (backchannel sound sequences), mitigates the semantic context of the preceding utterance, or serves as a stress-reducing strategy [21]. Laughter was investigated by researchers coming from various fields. It was observed in relation to the psychology of humor [2], [11], [13], [17] on the one hand, and its acoustic properties were studied [1], [3], [21] on the other hand. The perceivable sound sequence(s) of typical laughter is usually like those of breathy CV syllables (e.g., /hV/ syllable).

I. Habernal and V. Matousek (Eds.): TSD 2013, LNAI 8082, pp. 113–120, 2013.

Speech and laughter were found to be quite similar to each other in their syllable durations and in the number of syllables uttered per second. In addition, fundamental frequency, formant structure, and RMS (Root-Mean-Square) amplitude of laughter seem to be also rather speech-like [3]. Previous studies found that the average duration of laughter appeared between 395 ms and 915 ms [21], and the mean fundamental frequency of laughter was between 160 Hz and 502 Hz in women and between 126 Hz and 424 Hz in men [1]. Previous acoustic measurements [3], [22] showed that there were measurable differences between laughter and speech depending on mean pitch values and the pattern of voiced/unvoiced portions. Speech recognition research has focused on non-lexical sounds (such as laughter), because these events (besides disfluencies) may cause an important decrease in the accuracy of a speech recognizer in spontaneous speech compared to read speech [5], [10]. Various types of features (spectral, cepstral, prosodic, perceptual ones) were investigated for laughter detection using diverse classification techniques. Gaussian Mixture Models (GMMs) were trained with Perceptual Linear Prediction (PLP), pitch and energy, pitch and voicing, and modulation spectrum features to model laughter and speech by Troung and van Leeuwen [22]. Their results showed equal error rates ranging from 7.1% to 20.0% of the cases. For detection of overlapping laughter, Kennedy and Ellis [14] used Support Vector Machines (SVMs) trained with MFCCs + Δ, modulation spectrum, and spatial cues. The result showed a true positive rate of 87% of the cases. In [23] Troung and van Leeuwen developed a gender-independent laugh detector using different classification techniques and also their fusion (GMM, SVM, MLP) with various types of features. They observed that SVM performs better than GMM in most of the cases, but the fusion of the classifiers improved the performance of the classification (lower equal error rate of around 3% were obtained). Presegmented laughter and speech segments were classified appropriately in 88% of the test segments by Lockerd and Mueller [16] using Hidden Markov Models (HMMs). Cai et al. [6] modeled laughter using HMMs and Mel-Frequency Cepstral Coefficients (MFCCs) together with perceptual features (short-time energy, zero crossing rate). These methods achieved average recall and precision percentages of 92.95% and 86.88%, respectively. Campbell [7] measured pitch, power, duration, and spectral shape in the analysis of laughter and laughing speech. Neural networks (ANN) were successfully trained to identify the nature of the interlocutor (social or intercultural relationships). In [15] Knox's and Mirghafori's method for non-presegmented frame-by-frame laughter recognition produced an equal error rate of 7.9% of the cases. They used Neural Networks (ANN) trained on MFCC, AC PEAK, and F0 features. The present study investigates the acoustic characteristics of laughter and compares their characteristics to those of speech. Our aim is to develop an accurate method in order to recognize laughter events in spontaneous speech. The task of our laughter detector is to differentiate laughter and speech. This detector is supposed to classify a given acoustic signal as either laughter or speech. All laughter and speech segments selected were presegmented (however, the definition of onset and offset of the laughter segments was not the task of the classifier). The segment boundaries were identified by human transcribers.

2   Subjects, Material and Method

2.1   Subjects and Material

We used the BEA Hungarian Spoken Language Database [12] to train and test the detector. It is the largest speech database in Hungarian, which contains material of 260 hours produced by 280 speakers (aged between 20 and 90 years). For the present study, we used conversational speech material, a total of 75 meetings, whose average duration was 16 minutes. The recordings were made in the same sound-proof room, using three microphones (AT4040), GoldWave sound editing software, at an audio sampling rate of 44.1 kHz (storage: 16 bits). Our presegmented data contains 332 laughter and 321 speech segments (in this case words), we used 1/3 of the data in the testing set and the 2/3 of them in the training set.

2.2   Method

Feature. The analyzed features in spontaneous speech were extracted using Praat voice analysis software [4]. All features were extracted in 25 ms Hamming window shifted with 10 ms steps using MATLAB 7.12 software. The statistical analysis was carried out by SPSS 13.0 software. The short-term features were MFCC and PLP (12 coefficients $(1–12)$ + log energy + Δ + $\Delta^{'}\Delta$). Acoustic Parameters (APs):

- F0: mean, standard deviation of fundamental frequency,

- Jitter (mean), Shimmer (mean): a measure of the cycle-to-cycle variations of F0 and waveform amplitude

- RMS: local root mean squared energy,

- HNR (mean): harmonic-to-noise ratio,

- ZCR (mean): zero crossing rate,

- Spectral slope(mean): amount of decreasing of the spectral amplitude (computed by linear regression)

- LPC-CoG (mean): center of gravity

Feature Selection. Feature selection is used to identify features of large discrimination power and also those which can be ignored being irrelevant in the given decision or classification task. For the present research, selection of features was based on Receiver Operating Characteristic (ROC) curves and the Area Under Curve (AUC) measure calculated for the ROC. AUC is a strong predictor of performance, especially in imbalanced data classification problems, where it can be used for feature ranking. The measurement of the ROC AUC is very simple and fast compared to other feature selection algorithms [24]. We applied this feature selection method to all features used (APs, MFCC, PLP). We used this feature selection technique only on the train set supplied by validaion set from the train set.

Modeling Method. GMM–SVM hybrid method was used for laughter and speech segment classification. Gaussian Mixture Model (GMM) is an approximation of the originally observed feature probability density functions by a mixture of weighted Gaussians. The mixture coefficients were computed using the Expectation Maximization (EM) algorithm. Support Vector Machine (SVM) is a statistical algorithm with a great potential to generalize, that can be successfully used in pattern recognition and information retrieval tasks. The basic task during the training of SVM is to build a hyperplane as a decision boundary between two categories. This study combines the generalized GMM–UBM classifier and the discriminative SVM classifier. The laughter and speech are modeled by GMM (the number of components was 4 and we used diagonal covariance matrices of GMM) and adapted Universal Background Model (UBM) to laughter and speech, respectively. UBM is trained on large set of elements from the database. In this experiment only means were adapted to UBM based on MAP (Maximum A-Posteriori Probability) method. After GMM–UBM modeling the supervector can be obtained by concatenating each of the mixture component mean vectors. This GMM-supervector was the input feature for the SVM [9]. In the case of testing procedure the same processing was applied as in the training method (Fig. 1).

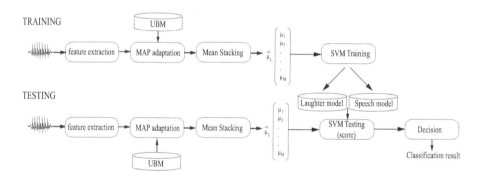

Fig. 1. Training and testing method

We used SVMs with an RBF (Radial Basic Function) kernel. The RBF has two parameters which have to be optimized. These parameters are the kernels parameters, and the soft margin parameter C. 3-fold cross-validation and grid search method were used on the training set to choose the optimal combination of these parameters. We calculated the performance of the classifier using the test set in various ways. We computed the accuracy, the precision, the recall and the F-measure. In addition, we calculated the ROC AUC value and the EER (equal error rate), and represent the results by means of DET (detection error tradeoff) curve [18].

2.3   Result

The average duration of laughter was 911 ms and the standard deviation was 605 ms (Fig. 2). For comparison, the average laughter duration was 1615 ms with a standard deviation of 1241 ms in the Bmr subset of the ICSI Meeting Recorder Corpus [15].

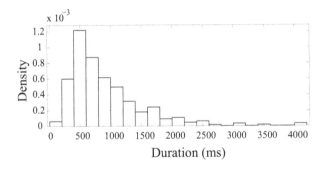

Fig. 2. Distribution of the laughter duration

We measured various features to characterize the laughter segments of spontaneous speech and analyzed them statistically (using ANOVA) to identify those features that provide significant differences between laughter and speech segment. The results showed that there were significant differences in mean values between laughter and speech segment in most of the cases (Jitter: $F(1, 651)=51.15$; p=0.001; Shimmer: $F(1, 651)=27.78$; p=0.001; HNR: $F(1, 651)=38.93$; p=0.001; Spectral slope: $F(1, 651)=0.02$; p=0.8944; LPC COG: $F(1, 651)=10.37$; p=0.0013; F0 means: $F(1, 651)=635.88$; p=0.001; ZCR: $F(1, 651)=46.14$; p=0.001; RMS: $F(1, 651)=79.08$; p=0.001). There is only one feature, the Spectral slope, in which case no statistically confirmed difference was found between laughter and speech segments. The separation ability of the APs was tested using ROC method. The F0 had the largest discriminate power out of all tested APs (Fig. 3).

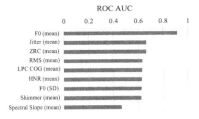

Fig. 3. ROC AUC values of features

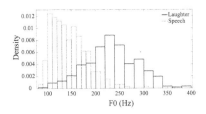

Fig. 4. Probability distribution of the F0 mean

F0 values were found to be 207 (SD: 49) Hz in males' laughter, and 247 (SD: 40) Hz in females' laughter, while 165 (SD: 45) Hz in males' and 198 (SD: 39) Hz in females' speech segments (Fig. 4). Previous studies showed that laughter was highly variable in fundamental frequency. The mean F0 values of laughter are different across studies (138 Hz in men and 266 Hz in women [3], 284 in man and 421 Hz in women [1], 424 in man and 472 Hz in women [21]). GMM–SVM system was trained with various sets of features. We wanted to know which of the features provides the best performance of the classifier. We focused on the features in the classification, as opposed to classification algorithms. The result shows that the classifier based on MFCC provided the best result (Table 1). The EER value of the MFCCs system is equal to that of the PLPs system, while the recall, F-value and ROC AUC value are higher in case of MFCCs system. The poorest result was yielded by the classifier based on APs (Fig. 5 and Fig. 7).

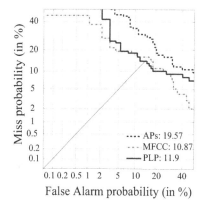

Fig. 5. DET plot of classifier trained with various features

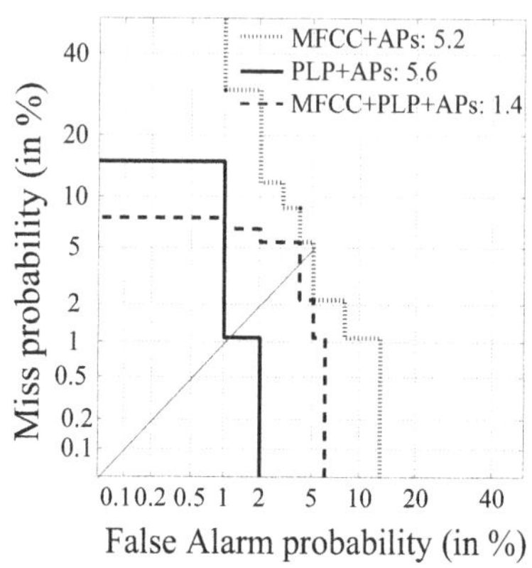

Fig. 6. DET plot of classifier trained with combined features

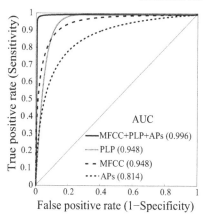

Fig. 7. ROC plot of classifier trained with combined features

We trained and tested our classifier system with different combinations of features. Previous researches showed that the classifier method trained with short- and long-term combined features can improve the performance [15]. We combined the short-term features with the long-term features (MFCC+APs; PLP+APs; MFCC+PLP+APs). Generally, the result shows that classifier trained with combined features gives better

Table 1. The performance of the laughter classifier trained with different features

Value	MFCC	PLP	APs	MFCC+APs	PLP+APs	MFCC+PLP+APs
Accuracy	90.53%	90.00%	77.89%	96.84%	95.26%	95.79%
Precision	87.04%	87.62%	72.58%	100.0%	98.90%	100.0%
Recall	95.92%	93.88%	91.84%	96.84%	95.24%	95.74%
F-measure	91.26%	90.64%	81.08%	98.39%	97.03%	97.82%

result than single features (Table 1). The classifier using MFCC+PLP+APs combined features yielded the best results. Using this combined feature the EER value decreased from 10.87 to 1.4 (Fig. 6).

2.4 Conclusion and Future Plans

The present study aimed at classifying the laughter and speech segments in spontaneous speech. We could determine the most efficient parameters for laughter detection using ROC AUC and EER analyses. We investigated several feature sets whereby Mel-Frequency Cepstral Coefficients (MFCCs) gave the best results. The reason of the poor classification performance of the APs might be that only the F0 had a relatively high discriminating power. The result shows that the classifier based on short-term acoustic features is better than that based on long-term acoustic features. The high AUC value and the low EER value confirmed that GMM–SVM supervector system based on short-term features is suitable for automatic classification of the laughter segments and speech segments in Hungarian spontaneous speech. We tested whether a classifier trained with combined features reduces or not the EER values. We can observe that the classifier using combined features gives lower EER values. It means that both short- and long-term features play an important role in laughter recognition. The development of new features, and further analysis of acoustic similarities and differences between laughter and speech will be used in our planned future study. We plan to test our model using corpora of other languages. Culture specific laughs are supposed to exist; however, no significant differences were found in laughter between Italian and German students [21], Japanese laughter may be somewhat different from that is commonly produced in Western cultures [8].

References

1. Bachorowski, J.-A., Smoski, M.J., Owren, M.J.: The acoustic features of human laughter. Journal of the Acoustical Society of America 110(3), 1581–1597 (2001)
2. Berlyne, D.E.: Laughter, humor, and play. In: Lindzey, G., Aronson, E. (eds.) Handbook of Social Psychology, pp. 223–240. Addison-Wesley, Reading (1969)
3. Bickley, C., Hunnicutt, S.: Acoustic analysis of laughter. In: Proceedings of the International Conference on Spoken Language Processing, Banff, Canada, pp. 927–930 (1992)
4. Boersma, P., Weenink, D.: Praat: doing phonetics by computer (Version 5.3.02) [Computer program] (2011), http://www.praat.org (retrieved October 10, 2011)

5. Butzberger, J., Murveit, H., Shriberg, E., Price, P.: Spontaneous speech effects in large vocabulary speech recognition applications. In: Proceedings of the 1992 DARPA Speech and Natural Language Workshop, pp. 339–343. Morgan Kaufmann, New York (1992)

6. Cai, R., Lu, L., Zhang, H.-J., Cai, L.-H.: Highlight Sound Effects Detection in Audio Stream. In: Proceedings of the IEEE International Conference on Multimedia and Expo, pp. 37–40 (2003)

7. Campbell, N.: Who we laugh with affects how we laugh. In: Proceedings of the Interdisciplinary Workshop on the Phonetics of Laughter, Saarbrücken, Germany, pp. 61–65 (2007)

8. Campbell, N., Kashioka, H., Ohara, R.: No laughing matter. In: Proceedings of the Interspeech 2005, Lisbon, Portugal, pp. 465–468 (2005)

9. Campbell, W.D., Sturim, D.E., Reynolds, D.A., Solomonoff, A.: SVM Based Speaker Verification Using a GMM Supervector Kernel and NAP Variability Compensation. In: Proc. of ICASSP, pp. 97–100 (2006)

10. Furui, S.: Recent progress in corpus-based spontaneous speech recognition. IEICE-Transactions on Information and Systems E88-D(3), 366–375 (2005)

11. Goldstein, J.H., McGhee, P.E. (eds.): The Psychology of Humor. Academic Press, New York (1972)

12. Gósy, M.: BEA A multifunctional Hungarian spoken language database. The Phonetician 105-106, 51–62 (2012)

13. Holland, N.N.: Laughing, a psychology of humor. Cornell University Press, Ithaca (1982)

14. Kennedy, L.S., Ellis, D.P.W.: Laughter detection in meetings. In: Proceedings of the NIST Meeting Recognition Workshop at the IEEE Conference on Acoustics, Speech and Signal Processing, pp. 118–121 (2004)

15. Knox, M.T., Mirghafori, N.: Automatic laughter detection using neural net-works. In: Proceedings of Interspeech 2007, pp. 2973–2976 (2007)

16. Lockerd, A., Mueller, F.: LAFCam leveraging affective feedback camcorder. In: Proc. CHI 2002, pp. 574–575 (2002)

17. Martin, R.A.: The psychology of humor. Elsevier Academic Press, Burlington (2007)

18. Martin, A., Doddington, G., Kamm, T., Ordowski, M., Przybocki, M.: The DET curve in assessment of detection task performance. In: Proc. Eurospeech 1997, Rhodes, Greece, pp. 1899–1903 (1997)

19. Nwokah, E., Fogel, A.: Laughter in mother-infant emotional communication. Humor: International Journal of Humor Research 6/2, 137–161 (1993)

20. Papoušek, M.: Vom ersten Schrei zum ersten Wort, Anfänge der Sprachentwicklung in der vorsprachlichen Kommunikation. Huber, Bern (1994)

21. Rothgänger, H., Hauser, G., Cappellini, A.C., Guidotti, A.: Analysis of laughter and speech sounds in Italian and German students. Naturwissenschaften 85, 394–402 (1998)

22. Troung, K.P., van Leeuwen, D.A.: Automatic detection of laughter. In: Proceedings of Interspeech 2005, pp. 485–488 (2005)

23. Troung, K.P., van Leeuwen, D.A.: Automatic discrimination between laughter and speech. Speech Communication 49, 144–158 (2007)

24. Wang, R., Tang, K.: Feature Selection for Maximizing the Area Under the ROC Curve. In: Proceedings of the IEEE International Conference on Data Mining Workshops, pp. 400–405 (2009)

Automatic Machine Translation Evaluation with Part-of-Speech Information

Aaron L.-F. Han, Derek F. Wong, Lidia S. Chao, and Liangye He

University of Macau, Department of Computer and Information Science
Av. Padre Toms Pereira Taipa, Macau, China
{hanlifengaaron,wutianshui0515}@gmail.com,
{derekfw,lidiasc}@umac.mo

Abstract. One problem of automatic translation is the evaluation of the result. The result should be as close to a human reference translation as possible, but varying word order or synonyms have to be taken into account for the evaluation of the similarity of both. In the conventional methods, researchers tend to employ many resources such as the synonyms vocabulary, paraphrasing, and text entailment data, etc. To make the evaluation model both accurate and concise, this paper explores the evaluation only using Part-of-Speech information of the words, which means the method is based only on the consilience of the POS strings of the hypothesis translation and reference. In this developed method, the POS also acts as the similar function with the synonyms in addition to its syntactic or morphological behaviour of the lexical item in question. Measures for the similarity between machine translation and human reference are dependent on the language pair since the word order or the number of synonyms may vary, for instance. This new measure solves this problem to a certain extent by introducing weights to different sources of information. The experiment results on English, German and French languages correlate on average better with the human reference than some existing measures, such as BLEU, AMBER and MP4IBM1.

Keywords: Natural language processing, Machine translation evaluation, Part-of-Speech, Reference translation.

1 Introduction

With the rapid development of Machine Translation systems, how to evaluate each MT system's quality and what should be the criteria have become the new challenges in front of MT researchers. The commonly used automatic evaluation metrics include the word error rate WER [2], BLEU [3] (the geometric mean of n-gram precision by the system output with respect to reference translations), and NIST [4]. Recently, many other methods were proposed to revise or improve the previous works.

METEOR [5] metric conducts a flexible matching, considering stems, synonyms and paraphrases, which method and formula for computing a score is much more complicated than BLEU's [1]. The matching process involves computationally expensive word alignment. There are some parameters such as the relative weight of recall to precision, the weight for stemming or synonym that should be tuned. Snover [6] discussed that one

I. Habernal and V. Matousek (Eds.): TSD 2013, LNAI 8082, pp. 121–128, 2013.

disadvantage of the Levenshtein distance was that mismatches in word order required the deletion and re-insertion of the misplaced words. They proposed TER by adding an editing step that allows the movement of word sequences from one part of the output to another. AMBER [7] including AMBER-TI and AMBER-NL declare a modified version of BLEU and attaches more kinds of penalty coefficients, combining the n-gram precision and recall with the arithmetic average of F-measure. F15 [8] and F15G3 perform evaluation with the F1 measure (assigning the same weight on precision and recall) over target features as a metric for evaluating translation quality. The target features they defined include TP (be the true positive), TN (the true negative), FP (the false positive), and FN (the false negative rates), etc. To consider the surrounding phrases for a missing token in the translation they employed the gapped word sequence kernels [9] approach to evaluate translations. Other related works include [10], [11] and [12] about the discussion of word order, ROSE [13], MPF and WMPF [14] about the employing of POS information, MP4IBM1 [15] without relying on reference translations, etc.

The evaluation methods proposed previously tend to rely on too many linguistic features (difficult in replicability) or no linguistic information (leading the metrics result in low correlation with human judgments). To address this problem, this paper explores the performance of a novel method only using the consilience of the POS strings of the hypothesis translation and reference translation. This ensures that the linguistic information is considered in the evaluation but it is a very concise model.

2   Linguistic Features

As discussed above, language variability results in no single correct translation and different languages do not always express the same content in the same way. To address the variability phenomenon, researchers used to employ the synonyms, paraphrasing or text entailment as auxiliary information. All of these approaches have their advantages and weaknesses, e.g. the synonyms are difficult to cover all the acceptable expressions. Instead, in the designed metric, we use the part-of-speech (POS) information (also applied by ROSE [13], MPF and WMPF [14]). If the translation sentence of system outputs is a good translation then there is a potential that the output sentence has a similar semantic information with the reference sentence (the two sentences may not contain exactly the same words but with the words that have similar semantic meaning). For example, "there is a big bag" and "there is a large bag" could be the same expression since "big" and "large" has the similar meaning (with POS as adjective). To try this approach, we conduct the evaluation on the POS of the words instead of the words themselves and we do not use other external resources such as synonym dictionaries. We also test the approach by calculating the correlation score of this method with human judgments in the experiment. Assume that we have two sentences: one reference and one system output translation. Firstly, we extract the POS of each word. Then, we calculate the similarity of these two sentences through the alignment of their POS information.

3   Calculation Methods

3.1   Design of hLEPOR Metric

First, we introduce the mathematical harmonic mean for multi-variables (n variables $(X_1, X_2, \ldots, X_n)$).

$$Harmonic(X_1, X_2, ..., X_n) = \frac{n}{\sum_{i=1}^{n} \frac{1}{X_i}} \qquad (1)$$

where n means the number of variables (also named as factors). Then, the weighted harmonic mean for multi-variables is:

$$Harmonic(w_{X_1} X_1, w_{X_2} X_2, ..., w_{X_n} X_n) = \frac{\sum_{i=1}^{n} w_{X_i}}{\sum_{i=1}^{n} \frac{w_{X_i}}{X_i}} \qquad (2)$$

where w_{X_i} presents the weight assigned to the corresponding variable X_i. Finally, the proposed evaluation metric $hLEPOR$ (tunable Harmonic mean of Length Penalty, Precision, n-gram Position difference Penalty and Recall) is designed as:

$$hLEPOR = Harmonic(w_{LP} LP, w_{NPosPenal} NPosPenal, w_{HPR} HPR) \quad (3)$$

$$= \frac{\sum_{i=1}^{n} w_i}{\sum_{i=1}^{n} \frac{w_i}{Factor_i}} = \frac{w_{LP} + w_{NPosPenal} + w_{HPR}}{\frac{w_{LP}}{LP} + \frac{w_{NPosPenal}}{NPosPenal} + \frac{w_{HPR}}{HPR}}$$

where LP, $NPosPenal$ and HPR are three factors in $hLEPOR$ and will be introduced in the following. Three tunable weights parameters w_{LP}, $w_{NPosPenal}$ and w_{HPR} are assigned to the three factors respectively.

3.2   Design of Internal Factors

Length Penalty. In the Eq. (3), LP means Length penalty to embrace the penalty for both longer and shorter system outputs compared with the reference translations:

$$LP = \begin{cases} e^{1-\frac{r}{c}} & : & c < r \\ 1 & : & c = r \\ e^{1-\frac{c}{r}} & : & c > r \end{cases} \qquad (4)$$

where c and r mean the sentence length of candidate translation and reference translation respectively.

N-gram Position Difference Penalty. In the Eq.(3), the $NPosPenal$ is defined as:

$$NPosPenal = e^{-NPD} \qquad (5)$$

where NPD means n-gram position difference penalty. The $NPosPenal$ value is designed to compare the POS order in the sentences between reference translation and output translation. The NPD is defined as:

$$NPD = \frac{1}{Length_{output}} \sum_{i=1}^{Length_{output}} |PD_i| \qquad (6)$$

where $Length_{output}$ represents the length of system output sentence and PD_i means the n-gram position difference value of aligned POS between output and reference sentences. Every POS from both output translation and reference should be aligned only once (one-to-one alignment). When there is no match, the value of PD_i will be zero as default for this output POS.

To calculate the NPD value, there are two steps: aligning and calculating. To begin with, the context-dependent n-gram alignment task: we use the n-gram method and assign higher priority on it, which means we take into account the surrounding context (surrounding POS) of the potential POS to select a better matching pairs between the output and the reference. If there are both nearby matching or there is no matched POS around the potential pairs, then we consider the nearest matching to align as a backup choice. The alignment direction is from output sentence to the references.

See example in Figure 1. In the second step (calculating step), we label each POS with its position number divided by the corresponding sentence length for normalization, and then using the Eq. (6) to finish the calculation.

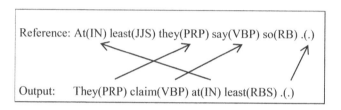

Fig. 1. Example of n-gram POS alignment

We also use the example in Figure 1 for the NPD introduction (Figure 2). In the example, when we label the position number of output sentence we divide the numerical position (from 1 to 5) of the current POS by the sentence length 5. For the reference sentence it is the similar step. After we get the NPD value, using the Eq. (5), the values of $NPosPenal$ are calculated.

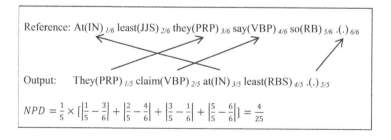

Fig. 2. Example of NPD calculation

Precision and Recall. Precision is designed to reflect the accurate rate of outputs while recall means the loyalty to the references. In the Eq. (3), HPR means the weighted Harmonic mean of precision and recall i.e. $Harmonic(\alpha R, \beta P)$, with parameters α and β as the tunable weights for recall and precision respectively.

$$Harmonic(\alpha R, \beta P) = \frac{\alpha + \beta}{\frac{\alpha}{R} + \frac{\beta}{P}} \tag{7}$$

$$P = \frac{aligned_{num}}{system_{length}} \tag{8}$$

$$R = \frac{aligned_{num}}{reference_{length}} \tag{9}$$

where $aligned_{num}$ represents the number of successfully aligned (matched) POS appearing both in translation and reference, $system_{length}$ and $reference_{length}$ specify the sentence length of system output and reference respectively.

System-level hLEPOR. We have introduced the calculation of $hLEPOR$ on single output sentence, and we should consider a proper way to calculate the value when the cases turn into document (or system) level. We perform the system-level $hLEPOR$ as below.

$$hLEPOR_{sys} =$$
$$Harmonic(w_{LP}LP_{sys}, w_{NPosPenal}PosPenalty_{sys}, w_{HPR}HPR_{sys}) \tag{10}$$

As shown in the formula, to calculate the system-level score $hLEPOR_{sys}$, we should firstly calculate the system-level scores of its factors LP_{sys}, $PosPenalty_{sys}$ and HPR_{sys}. The system level factor scores are calculated by the arithmetic means of the corresponding sentence-level factor scores.

4   Experiments

We trained $hLEPOR$ and tuned the parameters on the public ACL WMT 2008[1] data. There are five languages in the WMT 2008 data including English, Spanish, German, French and Czech; however, we currently did not find proper parser tools for the Spanish and Czech languages. So we tested on the English, German and French languages using the Berkeley parsers [16] to extract the POS information of the tested sentences. Thus, there are four language pairs in our tested corpora: from German and French to English, and the inverse. The parameter values on all language pairs are shown in Table 1.

The tested corpora we used are from ACL WMT 2011[2]. There are more than one hundred MT systems offering their output translation results, with most MT systems

---

[1] http://www.statmt.org/wmt08/
[2] http://www.statmt.org/wmt11/

Table 1. Values of tuned parameters

Parameters		
(α, β)	n-gram POS Alignment	Weights(HPR:LP:NPosPenal)
(9,1)	2-gram	3:2:1

statistical-based except for five rule-based ones. The gold standard reference data for those corpora consists of 3003 sentences. For each language pairs, there are different numbers of participated MT systems. Automatic MT evaluation systems are differed by calculating their Spearman rank correlation coefficient with the human judgment results [17].

Table 2. Correlation coefficients with human judgments

	Correlation Score with Human Judgment				
	Other-to-English		English-to-Other		
Metrics	DE-EN	FR-EN	EN-DE	EN-FR	Mean score
$hLEPOR$	0.83	0.74	0.84	0.82	**0.81**
MPF	0.69	0.87	0.63	0.89	0.77
WMPF	0.66	0.87	0.61	0.89	0.76
AMBER-TI	0.63	0.94	0.54	0.84	0.74
AMBER	0.59	0.95	0.53	0.84	0.73
AMBER-NL	0.58	0.94	0.45	0.83	0.7
METEOR-1.3	0.71	0.93	0.3	0.85	0.70
ROSE	0.59	0.86	0.41	0.86	0.68
BLEU	0.48	0.85	0.44	0.86	0.66
F15G3	0.48	0.88	0.3	0.84	0.63
F15	0.45	0.87	0.19	0.85	0.59
MP4IBM1	0.56	0.08	0.91	0.61	0.54
TER	0.33	0.77	0.12	0.84	0.52

We compare the experiments results with several classic metrics including BLEU, METEOR, TER and some latest ones (e.g. MPF, ROSE, F15, AMBER, MP4IBM1). The system level correlation coefficients of these metrics with human judgments are shown in the Table 2 which is ranked by the mean correlation scores of the metrics on four language pairs. Several conclusions from the results could be drawn: first, many evaluation metrics performed well in certain language pairs but weak on others, e.g. WMPF results in 0.89 correlation with human judgments on English-to-French corpus but down to 0.61 score on English-to-German, F15 gets 0.87 score on French-to-English but 0.45 on German-to-English, ROSE performs well on both French-to-English and English-to-French but worse on Germen-to-English and English-to-German. Second, recently proposed evaluation metrics (e.g. MPF and AMBER) generally perform better than the traditional ones (e.g. BLEU and TER), showing an improvement of the research work.

5   Conclusion and Perspectives

To make the evaluation model both accurate and concise, instead of using the synonyms vocabularies, paraphrasing and text entailment that are commonly used by other researchers, this paper explores the use of the POS information of the words sequences. What is noticing is that some researchers have used the n-gram method on the words alignment (e.g. BLEU using 1-gram to 4-gram), other researchers used the POS information in the similarity calculation by counting the number of corresponding POS. However, this paper employs the n-gram method on the POS alignment. Since this developed metric only relies on the consilience of the POS strings of the evaluated sentence even without using the surface words, it has a potential to be further developed as a reference independent metric. The main difference of this paper and our previous work LEPOR (the product of factors, perform on words) [18] is that this method groups the factors based on mathematical weighted harmonic mean, instead of the simple product of factors, and this method is performed on the POS instead of words. The overall weighted harmonic mean allows to tune the model neatly according to different circumstances, and the POS information can act as part of synonyms in addition to the syntactic or morphological behaviour of the lexical item.

Even though the designed metric has shown promising performances on the tested language pairs (EN to DE and FR, and the inverse direction). There are several weaknesses of this metric. Firstly, the POS codes are not language independent, so this method may not work well on the distant languages e.g. English and Japanese. Secondly, the parsing accuracy will effect the performance of the evaluation. Thirdly, to make the evaluation model concise, this work only uses the POS information without considering the surface words. To address these weaknesses, in the future work, more language pairs will be tested, other POS generation tools will be explored, and the combination of surface words and POS will be employed.

Acknowledgments. The authors wish to thank the anonymous reviewers for many helpful comments.

References

1. Koehn, P.: Statistical Machine Translation (University of Edinburgh). Cambridge University Press (2010)
2. Su, K.-Y., Wu, M.-W., Chang, J.-S.: A New Quantitative Quality Measure for Machine Translation Systems. In: Proceedings of the 14th International Conference on Computational Linguistics, Nantes, France, pp. 433–439 (July 1992)
3. Papineni, K., Roukos, S., Ward, T., Zhu, W.J.: BLEU: a method for automatic evaluation of machine translation. In: Proceedings of the ACL 2002, Philadelphia, PA, USA, pp. 311–318 (2002)
4. Doddington, G.: Automatic evaluation of machine translation quality using n-gram co-occurrence statistics. In: Proceedings of the Second International Conference on Human Language Technology Research, San Diego, California, USA, pp. 138–145 (2002)
5. Banerjee, S., Lavie, A.: Meteor: an automatic metric for MT evaluation with high levels of correlation with human judgments. In: Proceedings of ACL-WMT, Prague, Czech Republic, pp. 65–72 (2005)

6. Snover, M., Dorr, B., Schwartz, R., Micciulla, L., Makhoul, J.: A study of translation edit rate with targeted human annotation. In: Proceedings of the Conference of the Association for Machine Translation in the Americas, Boston, USA, pp. 223–231 (2006)

7. Chen, B., Kuhn, R.: Amber: A modified bleu, enhanced ranking metric. In: Proceedings of ACL-WMT, Edinburgh, Scotland, UK, pp. 71–77 (2011)

8. Bicici, E., Yuret, D.: RegMT system for machine translation, system combination, and evaluation. In: Proceedings ACL-WMT, Edinburgh, Scotland, UK, pp. 323–329 (2011)

9. Taylor, J.S., Cristianini, N.: Kernel Methods for Pattern Analysis. Cambridge University Press (2004)

10. Wong, B.T.-M., Kit, C.: Word choice and word position for automatic MT evaluation. In: Workshop: MetricsMATR of the Association for Machine Translation in the Americas, Waikiki, Hawai, USA, 3 pages (2008)

11. Isozaki, H., Hirao, T., Duh, K., Sudoh, K., Tsukada, H.: Automatic evaluation of translation quality for distant language pairs. In: Proceedings of the 2010 Conference on EMNLP, Cambridge, MA, pp. 944–952 (2010)

12. Talbot, D., Kazawa, H., Ichikawa, H., Katz-Brown, J., Seno, M., Och, F.: A Lightweight Evaluation Framework for Machine Translation Reordering. In: Proceedings of the Sixth ACL-WMT, Edinburgh, Scotland, UK, pp. 12–21 (2011)

13. Song, X., Cohn, T.: Regression and ranking based optimisation for sentence level MT evaluation. In: Proceedings of the ACL-WMT, Edinburgh, Scotland, UK, pp. 123–129 (2011)

14. Popovic, M.: Morphemes and POS tags for n-gram based evaluation metrics. In: Proceedings of ACL-WMT, Edinburgh, Scotland, UK, pp. 104–107 (2011)

15. Popovic, M., Vilar, D., Avramidis, E., Burchardt, A.: Evaluation without references: IBM1 scores as evaluation metrics. In: Proceedings of the ACL-WMT, Edinburgh, Scotland, UK, pp. 99–103 (2011)

16. Petrov, S., Barrett, L., Thibaux, R., Klein, D.: Learning accurate, compact, and interpretable tree annotation. In: Proceedings of the 21st ACL, Sydney, pp. 433–440 (July 2006)

17. Callison-Bruch, C., Koehn, P., Monz, C., Zaidan, O.F.: Findings of the 2011 Workshop on Statistical Machine Translation. In: Proceedings of ACL-WMT, Edinburgh, Scotland, UK, pp. 22–64 (2011)

18. Han, A.L.-F., Wong, D.F., Chao, L.S.: LEPOR: A Robust Evaluation Metric for Machine Translation with Augmented Factors. In: Proceedings of the 24th International Conference on Computational Linguistics (COLING 2012): Posters, Mumbai, India, pp. 441–450 (2012)

Automatic Extraction of Polish Language Errors from Text Edition History

Roman Grundkiewicz

Adam Mickiewicz University,
Faculty of Mathematics and Computer Science,
ul. Umultowska 87, 61-614 Poznan, Poland
romang@amu.edu.pl

Abstract. There are no large error corpora for a number of languages, despite the fact that they have multiple applications in natural language processing. The main reason underlying this situation is a high cost of manual corpora creation. In this paper we present the methods of automatic extraction of various kinds of errors such as spelling, typographical, grammatical, syntactic, semantic, and stylistic ones from text edition histories. By applying of these methods to the Wikipedia's article revision history, we created the large and publicly available corpus of naturally-occurring language errors for Polish, called PlEWi. Finally, we analyse and evaluate the detected error categories in our corpus.

Keywords: error corpora, language errors detection, mining Wikipedia.

1   Introduction

Error corpora are widely applied in the natural language processing, especially in the course of developing proofreading tools. Gathering the corpus containing annotated naturally-occurring errors in the traditional way is very costly, because it usually entails the manual annotation of text. Consequently, there are no large digital error corpora for a number of languages, as is the case with the Polish language. Admittedly, there exist error corpora of foreign language learners (mainly of English language learners), but non-native errors are quite different [1] and tools developed based on such data may not be sufficiently robust to detect errors made by native-speakers.

To reduce the time and cost of manual work required for collecting language mistakes, corrections made by teachers in written assignments or the history of text editions are subject to analysis. The acquisition of such documents, especially in the electronic form, poses a major challenge as edition history is usually not stored. The exceptions are Wikipedia and other Wiki family members (e.g. WikiNews and other smaller wiki-like websites), services such as Google Docs or even files inside the control version systems.

In this paper we will present the automatic method used for building the corpus of naturally-occurring language errors from the Polish Wikipedia, called PlEWi (*Polish Language Errors from Wikipedia*). In Sect. 2 we will describe the technical aspects of Wikipedia mining and we will present our solution for the detection and extraction of language errors from edition history. Finally, we will analyse and evaluate the collected data in Sections 3 and 4.

I. Habernal and V. Matousek (Eds.): TSD 2013, LNAI 8082, pp. 129–136, 2013.

1.1 Wikipedia as Source of Language Errors

The advantages of Wikipedia are its size, availability and the contribution made by a diversified community. Moreover, its content is generally considered reliable [2]. But as has been pointed out by Miłkowski [3], Wikipedia probably cannot represent the average language due to its digital form, rather formal and restricted style, uncommon scope of topics and a higher education level of its users. The mere encyclopaedic style of Wikipedia's can be viewed as inconvenient, because some changes are imposed only by the style unification requirements.

Nevertheless, Wikipedia can be perceived as an accurate source of language error corrections as the aim of the majority of its editors is to improve the quality of the content of articles[1]. A high average education level of Wikipedia users may confirm this statement. Please note, that Wikipedia with its community pages is also an up-to-date record of the living language.

1.2 Related Works

The idea of using edition histories of documents for the purpose of collecting certain types of language errors abounds in literature. Miłkowski [3] proposed the building of error corpora using Wikipedia revisions based on the hypothesis that the majority of frequent minor edits are the corrections of spelling, grammar, style and usage mistakes. This hypothesis, although very accurate, does not yield the expected result in the form of a wide range of error types, e.g. inflectional errors, because they are rarely repeated.

The work of Max and Wisniewski's [4] has led to the creation of WiCoPaCo — a corpus of naturally-occurring corrections and paraphrases[2]. One of the applications of the WiCoPaCo was the construction of a set of spelling error corrections and its application in the evaluation of the spell checker. However, it did not contain certain types of errors, such as repetitions or omissions of words, and corrections that refers only to punctuation or case modification.

Zesch [5] extracted the samples of real-word spelling errors and their contexts from Wikipedia's revision histories. Collected data were used to evaluate statistical and knowledge-based measures applied in contextual fitness in the task of real-word spell checking. He confirms the opinion that such natural errors are better suited for evaluation purposes than artificially created ones.

2 Extracting Language Errors

We have accessed the Wikipedia data with script iterates over each two adjacent revision in every article on Wikipedia's dump file in XML format[3]. Edited text fragments from these revisions were extracted using the longest common subsequence (LCS) algorithm and cleaned from the Wikipedia format markups. Next PSI-Toolkit toolbox [6] has been used on all fragments for sentence segmentation and lemmatization in the further stage

---

[1] The issue of vandalism will be discussed in Sect. 2.3.

[2] http://wicopaco.limsi.fr/

[3] http://dumps.wikimedia.org/plwiki/

as well. All editions involving only the addition or deletion of the article content were disregarded.

After that, if two edited sentences met certain surface conditions, such as (1) the sentence length is between 4 and 80 tokens, (2) the difference in length is less than 4 tokens, (3) a ratio of words to non-word tokens is higher than 0.75, and (4) a number of non-letter characters is less than a quarter of all the characters, the LCS algorithm was run again. For the time being it worked on tokens instead of lines, so we obtained all edition instances per each sentence.

2.1 Language Errors Recognition

As a result of the initial stage, for each sentence pair it is obtained the sequence of editions $((u_0, v_0), (u_1, v_1), \dots)$, where each edition (u, v) is a pair of the older and the newer word(s).

Because too many editions in a sentence imply a rewording or an extension of the sentence rather than error corrections, we rejected the sentences containing more than four editions. But there is no restriction to the mere single word and non-empty editions (i.e. u, v may consist of two words or be empty)[4]. Next, each edition is classified into a defined error category (modeled on Bušta's work [7]) through hand-crafted heuristics, or rejected.

Simple Errors. First, the word u and its edition v are tested using surface conditions that do not imply the use of any natural language processing tool. In particular, the edition can be easily discarded if (1) v occurs in the list of vulgarisms, (2) u and v differ only in more than one punctuation mark (e.g. *what→what???*), (3) the change involves abnormal case modifications like *word→WoRd*.

The following types of errors are detected: (1) misused punctuation marks (e.g. missing of a comma or a full stop), (2) misspellings connected with separable and inseparable writing when u and v differ only in the space character or the hyphen, (3) wrong letter case (e.g. *polska→Polska [Poland]*).

Spelling Errors. If the edition has not been recognised as error correction by surface conditions, it is labeled using the spell checker[5] with a dictionary D as (1) a non-word spelling correction if $u \notin D$ and $v \in D$, as (2) a real-word error correction if $u, v \in D$, as (3) an act of vandalism if $u \in D$ and $v \notin D$, and as (4) "out of dictionary" if $u, v \notin D$.

For the non-word spelling corrections, there is made a distinction between the misspellings involving in the omission of diacritical signs and the other misspellings. The real-word editions are further classified into one of the grammatical error types, whereas the editions appearing to be examples of vandalism are discarded. As Kukich's studies showed [8], most of language errors are in the short edit distance. Hence, in the case

---

[4] Further in this work we will use a term *word* even if it is a sequence of words.
[5] We used Hunspell spell checker: http://hunspell.sourceforge.net/.

when both words are out of dictionary, and if the edit distance[6] for u and v is smaller than 4, the edition is categorised as "probable misspelling" and the process is stopped.

Grammatical Errors. The use of a lemmatiser enables the analysis of real-word editions and their classification into one of the more specific grammatical error types: inflection, syntactic or semantic.

The examples of inflectional errors are of prime relevance as they are very frequent in languages with rich morphology, e.g. a grammatical gender disagreement:

– *System of a Down jest pierwszą grupą, która dwa razy w ciągu jednego roku (miał→miała) dwa albumy na szczycie.* [*System of a Down is the first group which ({he→she} has) two albums at the top of the chart in one year.*]

Even if their correction is an area of interest of current research [9], there is no tool that would handle the problem effectively. This kind of error indicates that u and v have equal lemmas. But it cannot be ascertained for sure whether the correction is a grammar or only style-related. The greatest confusion is about verbs differing only in tense or aspect, or nouns with the only change in the number, so we labelled all of them separately, e.g.:

– *Każdy odcinek (trwa→trwał) około pół godziny.* [*Each episode (takes→took) about half an hour.*]
– *Energie mają wyznaczone (miejsce→miejsca) w widmie elektromagnetycznym.* [*Energies have designated (place→places) in electromagnetic spectrum.*]

Editions in which u and v have different lemma are likely to be (1) a syntactic error correction if u and v belong to different grammatical classes and (2) semantic ones in the other case. Like in the case of inflectional errors, also the semantic errors which differ only in degree or aspect are classified separately. What is more, some editions recognised as semantic can be structural or pragmatic error corrections (according to Kukich classification [8]) and their automatic detection is probably impossible, e.g.:

– *Armia straciła ok. 1000 czołgów i (samolotów→samochodów) pancernych.* [*The army had lost about 1 000 tanks and armoured (planes→cars).*]

Other types of errors that are detected at this stage are insertions, deletions or substitutions of prepositions, pronouns and conjunctions.

Style Errors. Editions within abbreviations and acronyms are captured with the additional information provided by a lemmatiser. Other style adjustment editions are recognised by a thesaurus[7]. Using it before the grammatical errors detection prevents classifying style errors in the short edit distance into wrong category.

---

[6] As an edit distance we chose Damerau-Levenshtein distance.
[7] http://synonimy.ux.pl/

2.2 Conditions of Acceptance

We allow at most one discarded edition in a sentence with the exception that the remaining editions are recognised as any type of the grammatical error. This is based on the observation that some editions (u_i, v_i) may be dictated by other editions (u_j, v_j) in the same sentence as in the case of inflection changes dictated by rewriting parts of a sentence, e.g.

- *Arytmetyka (jest→—) (najstarszą→najstarsza) i najbardziej (podstawową→ podstawowa) (gałęzią→gałąź) matematyki.* [*Arithmetic (is→—) the oldest and most elementary branch of mathematics.*]

The above sentence is rejected because the edition (*jest,* —) is not recognised with our heuristics and the rest of editions are grammatical error corrections. There is no such assumption in the case of spelling errors.

In order to avoid the situation when an edition is reverted (once or more times), for a given sentence with editions (u_i, v_i) we also check backward if the previously collected sentences from the same article are equal up to reversed editions (v_i, u_i). These sentences get cancelled altogether. It may imply an act of vandalism, or only a hesitation or differences in opinion of the editors, nevertheless, we do not take into account the editions of controversial words.

Finally, we perform post-processing included the entire sentence, as during the error recognition process we did the local analysis without consideration for a wider context. This stage has an impact on editions mainly concerning punctuation and latter case modification. For instance, we reject sentences for which the only change is: (1) deletion of a full stop from the end, (2) addition of a colon at the end, or (3) the conversion the first letter to lower case. We also discard text fragments addressing Wikipedia-specific content.

2.3 Difficulties

One of the main difficulties we had to deal with were the changes made by vandals [10], which was due to the edition of Wikipedia content being freely available to everybody. Vandalism usually involves minor changes that can be classified according to our heuristics as language error corrections.

The problem is solved on three levels:

1. Some acts of vandalism are reverted in the revision process and a relevant comment is added. So the revision with such comment (like *cancelled edits, revert after vandalism,* etc.) and the first previous editing of non-logged user are omitted.
2. Editions that include a vulgarisms or popular internet acronyms are rejected.
3. The backward checking is done as described in the previous section.

Another issue is an automatic edition done by Wikipedia's robots[8] mainly concerning data format, common abbreviations and the replacement of some HTML entities. From our point of view it is not, however, that relevant who made the correction, but that the error had occurred.

---

[8] Editions done by robot are easily detectable through *username* XML attribute.

3   Error Corpus

We have applied proposed method to the Polish Wikipedia revision history creating the PlEWi corpus of naturally-occurring language errors and their corrections[9]. The dump of Wikipedia[10] contains 1,747,083 pages with about 910,000 articles.

The number of extracted text fragments with at least one potential language error is 1,532,275, including 1,303,806 (85.1%) of well-formed sentences. By "well-formed sentence" we mean each text fragment beginning with a capital letter, a number or quotation or hyphen mark and ending with the regular sentence delimiter: .;?!"". The remaining text fragments (228,469) are phrases, texts from tables, picture descriptions etc. About 23.0% of all the editions comes from anonymous users which may confirm a rather low risk of vandalism.

The total number of collected editions is 1,713,835, including 157,043 (10.9%) editions labelled as "probable misspelling", i.e. words not existing in dictionary, but for which edit distance is smaller than 4. 70,828 corrections (0.05%) concern multi-word editions, whereas the number of deletions and insertions is 33,279 and 34,339, respectively (both constitute 0.02%). Detailed distribution of extracted error types is listed in Table 1.

Table 1. Error frequencies in PlEWi corpus

Category	Error type	#
simple	punctuation	308,802
	case modification	220,533
	separable and inseparable writing	13,782
spelling	modification of diacritics (contextual)	241,777 (39,529)
	spelling	356,762
grammar	inflection (tense or number)	164,659 (43,340)
	syntactic	19,443
	semantic (aspect or degree)	64,600 (13,757)
	pron., prep., conj., particle-adverb	94,578
style	synonym	29,596
	abbreviation (year or age)	38,812 (21,431)
	"probable misspelling"	157,043

The number of corrections concerning diacritic modification together with the number of detected grammatical errors (but without the most doubted ones involving only of aspect, tense or number modification) can be considered as the total number of real-word errors. If editions identified as spelling error corrections would be considered as

---

[9] Corpus in YAML format and scripts, including a detailed documentation of presented heuristics, are publicly available at http://www.staff.amu.edu.pl/~romang/wiki_errors.php

[10] The XML file of about 330 GB size from 14th July, 2012.

the non-word errors, then about 29.3% of all of them would be real-word. Including editions labelled as "probable misspellings" to non-word errors the ratio decreased to 24.1%. To the best of our knowledge it is the first estimation of this factor for the Polish language, when for English the range between 25% and 40% is used for the current research [8]. The lower relative number of real-word errors may be due to the fact that the average length of words in Polish is larger than in English.

4   Evaluation

To evaluate the effectiveness of the presented method for the language errors extraction and the quality of PlEWi corpus itself, we have manually checked 200 random text fragments from each category. For each first edition in each example we verified whether it is the right mark of the error type or not — it means that we calculated the precision value. The results are presented in Table 2.

Table 2. The evaluation of the selected error categories in PlEWi corpus

Category	#	Overall precision	#	Well-formed sentences
simple	200	0.86	146	0.90
spelling	200	0.98	173	0.99
grammar	200	0.73	170	0.71
style	200	0.99	169	0.99
probable misspelling	200	0.86	164	0.87

In the "simple" category (i.e. editions which recognition as error correction did not require any NLP tool) there were 14% of not valuable error corrections. Most of them were a faulty letter case modification and wrong insertion of comma, and next entire text fragments were syntactically incorrect as they probably came from a paragraph header or a bulleted list.

The application of the spell checker can explain a high precision value (98%) for recognition of spelling error corrections, and the lack of intentionally wrong editions.

On the other hand, 19% of editions among grammatical error corrections were connected only with the style improvement, such as grammatical aspect modification or the updating of the tense of some verbs (for instance, because time reference has been changed). The next 7% of them were pragmatic error corrections or context was not enough to decide if the edition was necessary. But most of these editions are marked by our heuristics separately, and after removing them from evaluation data set, the precision value increases to 0.80. In general, only 4 (2%) of all editions in this category result in an error form of word.

23% of editions which have not been recognised directly as any error correction were neither spelling nor grammatical error correction. But only in 10 (5%) cases an incorrect word was replaced by another incorrect form. As many as 112 (56%) editions concern the correction of named entities, 35 (18%) of them are inflectional error corrections, 11 (6%) stylistic changes, and 7 (4%) corrections of English or German word. Finally,

31 (16%) of them were proper spelling error corrections but concern less common or technical words.

The average precision value is 0.88. We do not present the recall value because the calculation of it would require a manually annotation of a quite large part of Wikipedia history.

5 Summary

In this paper, we presented automatic methods for the collection of a wide range of language errors and their corrections from histories of document editions. By applying them to the Polish Wikipedia revision history, we created the PlEWi corpus containing about 1.7 million naturally-occurring errors, including above 160 thousands of inflectional errors. As evaluation shows, the corpus is characterised by a high reliability of spelling error annotations (98.0%) and quite high for grammatical errors (72.5%).

We hope that the PlEWi corpus will become an important resource for developing and evaluating proofreading techniques for Polish.

References

1. Leacock, C., Chodorow, M., Gamon, M., Tetreault, J.: Automated Grammatical Error Detection for Language Learners. Morgan and Claypool Publishers (2010)
2. Zeng, H., Alhossaini, M.A., Ding, L., Fikes, R., McGuinness, D.L.: Computing trust from revision history. In: Proceedings of the 2006 International Conference on Privacy, Security and Trust (2006)
3. Miłkowski, M.: Automated building of error corpora of polish. In: Corpus Linguistics, Computer Tools, and Applications State of the Art, pp. 631–639. Peter Lang (2008)
4. Max, A., Wisniewski, G.: Mining naturally-occurring corrections and paraphrases from wikipedia's revision history. In: Proceedings of the Seventh International Conference on Language Resources and Evaluation (2010)
5. Zesch, T.: Measuring contextual fitness using error contexts extracted from the wikipedia revision history. In: Proceedings of the 13th Conference of the European Chapter of the Association for Computational Linguistics, pp. 529–538 (2012)
6. Graliński, F., Jassem, K., Junczys-Dowmunt, M.: PSI-Toolkit: Natural language processing pipeline. Computational Linguistics - Applications, 27–39 (2012)
7. Bušta, J., Hlaváčková, D., Jakubíček, M., Pala, K.: Classification of errors in text. In: RASLAN 2009: Recent Advances in Slavonic Natural Language Processing, pp. 109–119 (2009)
8. Kukich, K.: Techniques for automatically correcting words in text. ACM Comput. Surv., 377–439 (1992)
9. Kapłon, T., Mazurkiewicz, J.: The method of inflection errors correction in texts composed in polish language – A concept. In: Duch, W., Kacprzyk, J., Oja, E., Zadrożny, S. (eds.) ICANN 2005. LNCS, vol. 3697, pp. 853–858. Springer, Heidelberg (2005)
10. Chin, S.C., Street, W.N., Srinivasan, P., Eichmann, D.: Detecting wikipedia vandalism with active learning and statistical language models. In: Proceedings of the 4th Workshop on Information Credibility, pp. 3–10 (2010)

Bilingual Voice Conversion by Weighted Frequency Warping Based on Formant Space

Young-Sun Yun[1,*] and Richard E. Ladner[2]

[1] Dept. of Information and Communication Engineering,
Hannam University, Daejeon, Republic of Korea 306-791
ysyun@hannam.kr
[2] Dept. of Computer Science and Engineering,
University of Washington, Seattle, Washington, USA 98195
ladner@cs.washington.edu

Abstract. Voice conversion is a technique that transforms the source speaker's individuality to that of the target speaker. In this paper, we propose a simple and intuitive voice conversion algorithm that does not use training data between different languages, but uses text-to-speech generated speech rather than real recorded voices. The suggested method finds the transformed frequency by formant space warping. The formant space comprises four representative monophthongs for each language. The warping functions are represented by piecewise linear equations using pairs of four formants at matched monophthongs. Experimental results show the potential of the proposed method.

Keywords: voice conversion, weighted frequency warping, formant space.

1 Introduction

Recently, social interests regarding old, weak, and disabled people have increased, and discussions in this area are actively progressing. Many studies on accessibility have also been conducted. According to the general definition [1], accessibility is a general term used to describe the degree of which a product, device, service, or environment is available to as many people as possible. Accessibility is often used to focus on people with disabilities or special needs and their right of access to entities, through the use of assistive technology. Accessibility has become a significantly greater issue, as computers can now handle various types of information, including interaction with audio-visual interfaces on the Internet or actual life.

There are some assistive devices and programs to assist computer operation, but they have not been available to all because of cost and the operating platforms. A popular assistive tool is the screen reader, using which the visual interfaces and information on the computer screen are converted to speech for the disabled, particularly low-vision and blind people. The screen reader plays a very important role for the disabled using computers, but is limited in that it depends on platforms and has a very high cost. To overcome these difficulties, server-based assistive tools are preferred. Server-based

---

* He was a visiting scholar at the University of Washington from Aug. 2012 to Jul. 2013.

I. Habernal and V. Matousek (Eds.): TSD 2013, LNAI 8082, pp. 137–144, 2013.

assistive tools are installed on the server and usually serviced free of charge. They can be easily used on public computers because of no need for installation of specific software.

WebAnywhere, which is a typical example of a server-based tool, was proposed in 2008 [2] and is a web-based reader that enables blind users to access the web by generating speech from a given text. We deployed *WebAnywhere* to support Korean low-vision and blind people who access the Internet [1]. If the system supports only one language, the user should switch the web reader to another when he/she wants to access other pages in different languages. Therefore, to navigate the different language sites, a text-to-speech (TTS) system must support different language outputs based on one or multiple speakers. In this paper, we propose a voice conversion algorithm to make the serviced speech to sound as if it were generated by the same person, even though the TTS system is developed based on multiple speakers. The proposed system uses a simple formant space transformation that is independent of text/context information rather than other methods.

2 Related Works

Voice conversion is a technique that modifies a speaker's individuality. That is, speech uttered by one speaker is transformed to different speech as if another speaker had generated it [3].

It is known that any single specific acoustic parameter alone does not carry all of the individuality information, and various parameters affect the characteristics of speech. In these parameters, the formant frequency is considered to be one of the most important parameters characterizing speech and a speaker's individuality [3,4]. There are many approaches to manipulate the formant frequency by subspace codebook mapping [3], transformation using artificial neural networks [5], and vocal tract length normalization (VTLN) techniques [8,11]. VTLN attempts to normalize the speaker-dependent vocal tract lengths by warping the frequency axis of the phase and magnitude spectrum. In speech recognition, VTLN removes the speaker's individuality and improves the recognition performance [6]. The same techniques are introduced in voice conversion and modify source speech as it is uttered by a target speaker [7]. The frequency warping approaches based on VTLN are implemented in various techniques, such as bilinear transformation [12] and piecewise linear transformation [8,11]. Some of these methods are performed under text-dependent conditions for the same language [3,5,8,11,12], whereas others are performed text-independent or in different language environments [7,10].

While these approaches address the manipulation of various parameters for voice conversion, most algorithms are implemented with a speech synthesis system. In our considered web reader, we are limited to using only the generated speech waveforms from a TTS system. Therefore, we have not used many control factors to adjust the individuality. From this limitation, we consider a simple voice conversion algorithm using the formant space transformation.

---

[1] A current version of *Korean WebAnywhere* without the voice conversion feature can be found at http://phoenix.hnu.kr/wa_beta/

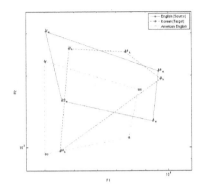

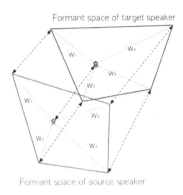

Fig. 1. F1-F2 diagrams for typical American English [13], English, and Korean speech for four representative monophthongs (*/iy, oo, a, ael*)

Fig. 2. Basic idea of frequency warping based on formant spaces between source and target speakers

3 Weighted Frequency Warping

In this section, we describe the idea of weighted frequency warping based on formant space and its algorithm.

3.1 Formant Space

The vocal tract can be modeled as a linear filter with resonances. The formants correspond to the harmonics of the fundamental frequency at natural resonances of the vocal tract cavity position for the vowels. We can easily find the formants by showing dark horizontal bands in the spectrogram of the given speech. The major resonances of the oral and pharyngeal cavities for vowels are called F1 and F2, the first and second formants, respectively. They are determined by tongue placement and oral tract shape in vowels and determine the characteristic timbre or quality of the vowel [13]. Figure 3.1 compares the F1-F2 diagram of typical American English [13], English, and Korean speech for the four representative monophthongs (*/iy, oo, a, ael*). English and Korean monophthongs are obtained from TTS outputs.

In general, F1 and F2 are considered sufficient to differentiate vowels, F3 is important in determining the phonemic quality, and F4 and higher formants are significant for voice qualities. Therefore, we used four formants in this work to construct the *formant space* of each speaker. These formants are used to build piecewise linear warping functions [7] of the frequency from the source speaker to the target speaker.

3.2 Outline of Frequency Warping

As shown in Fig. 3.1, the formant spaces of the source and target speakers are different because of the speaker's individuality and language environment. Unlike previous studies [7,9,11], our system does not depend on the text and does not use phonemic clusters by segmentation or speech units obtained by speech recognition. Therefore, it requires reference features to transform the frequency of the source speaker to that of the target speaker. We selected four monophthongs that are commonly found in both the English and Korean languages.

The frequency warping processes are briefly described as follows. First, the source speaker's speech, which is generated from a TTS system, is divided into frames, which are transformed to the frequency domain. Next, for voiced frames, we compute the relative position of the input frequency from the formant space of the source speaker. The relative position is represented as weights to the formants of each monophthong in the formant space. The target frequency position is estimated by applying weights to frequency warping functions from the source formant space to the target formant space (Fig. 3.1). As in the previous studies [5,11], it is known that the transformation of the voiced sounds is significantly more important than that of unvoiced sounds. Thus, our system transforms voiced frames, whereas unvoiced frames are not modified. Lastly, the estimated frequency spectra are reconstructed to the speech waveforms.

3.3 Frequency Warping Function

In our proposed method, to determine the voiced frame, we use the zero crossing rate, log energy, and autocorrelation ratio, as many approaches have. If the given frame is identified as a voiced frame, the frame is transformed to the frequency domain, and the formants are extracted. The formants are compared to those of reference monophthongs, and the relative location is calculated in the formant space.

To compare formants of the given frame with those of reference monophthongs, we modify the distance measures to avoid endpoint restrictions of the general dynamic programming (DP) method in formant comparison. One of weakness of DP is the fixed endpoints problem, in which the initial and terminal points are fixed. Because it is difficult for reliable formants to be extracted from the speech waveform, the first and last formants can be inaccurate. To consider the comparison of unreliable formants, we adjust the distance measure as follows:

$$D_k(i) = \begin{cases} \min_k d(1, k) & i = 1, 1 \le k \le n \\ \min_k \{D_{k'}(i - 1) + d(i, k)\} & 1 < i \le m, 1 \le k' \le k \le n \end{cases} \tag{1}$$

$$D(\mathbf{f}_s, \mathbf{f}_t) = \left\{ D_p(m) + \tilde{D}_q(n) \right\} / 2, \quad 1 \le p \le n, 1 \le q \le m \tag{2}$$

where m and n are the number of formants in matched formant vectors $\mathbf{f}_s$ and $\mathbf{f}_t$ (usually $m = n = 4$), $d(.)$ is the Euclidean distance between two formants, and $D(.)$ and $\tilde{D}(.)$ are the forward and backward comparisons of two formant vectors, respectively. Equation (2) is used to make the distance symmetric.

To find the location of a given frame in the formant space, the weights of the input frame are calculated by comparison with the monophthong's formants organizing the formant space. The weights are easily obtained by (3), and the formant location can be represented in a source speaker's formant space.

$$W(\mathbf{f}, \mathbf{S}_j) = \begin{cases} \alpha \cdot \min D(\mathbf{f}, \mathbf{S}_k) + (1 - \alpha) \cdot \mathbf{I} & \mathbf{f} \notin \mathcal{S}_{fs} \\ \alpha \cdot \min_1 D(\mathbf{f}, \mathbf{S}_k) + \beta \cdot \min_2 D(\mathbf{f}, \mathbf{S}_k) + (1 - \alpha - \beta) \cdot \mathbf{I} \\ \qquad \text{one of } \mathbf{f} \in \text{one of axis ranges of } \mathcal{S}_{fs} \\ D(\mathbf{f}, \mathbf{S}_j) / \sum_{k=1}^{4} D(\mathbf{f}, \mathbf{S}_k) & \mathbf{f} \in \mathcal{S}_{fs} \end{cases} \tag{3}$$

where $\mathbf{f}$ is a formant vector of the input frame, and $\mathbf{S}_j$ denotes the jth reference formant vector of the source speaker. α and β are the weights contributing to each monophthong, $\mathcal{S}_{fs}$ is the formant space comprising four monophthongs, $\mathbf{I}$ is the identity transform function, and $\min_k$ denotes the kth minimum value. In (3), if formant $\mathbf{f}$ is far from the formant space of the source speaker, the weights are calculated by interpolation of the nearest monophthong formant and the identity transform function. If formant is partially overlapped in the range of each formant axis, the new formant is estimated by a combination of the nearest two monophthong formants and the identity warping function.

If weights $W(\mathbf{f}, \mathbf{S}_k), k = 1, ..., 4$ of the input formant $\mathbf{f}$ are obtained, we can estimate the frequency warping function, $\mathcal{T}(\mathbf{f})$, to the target speech.

$$\mathcal{T}(\mathbf{f}) = \sum_{k=1}^{4} W(\mathbf{f}, \mathbf{S}_k) \cdot \mathbf{T}(\mathbf{S}_k, \mathbf{T}_k), \tag{4}$$

where $\mathbf{T}(.)$ is the transform function from the source formant vector $\mathbf{S}_k$ to the target formant vector $\mathbf{T}_k$. The transform function is a piecewise linear warping function, which has line segments corresponding to the frequency pairs of four formants.

4 Preliminary Experiments

Some experiments have been performed to evaluate the validity and potential of the proposed algorithm.

4.1 Speech Samples

The speech samples utilized in this work are generated by the Voiceware TTS system [14]. English and Korean voices were recorded in 2008 and 2003, respectively. The English woman was born in Pennsylvania in 1974. The Korean speaker was born in 1971 and speaks the standard Korean language. The English and Korean speech samples are generated by a pitch synchronous overlap and add (PSOLA) TTS system at a sampling frequency of 16 kHz.

4.2 Results and Discussion

To obtain the frequency warping function, formants based on linear prediction coefficients (LPCs) are calculated for English monophthongs and the corresponding Korean monophthongs. Because phonemes are not exactly matched between two languages,

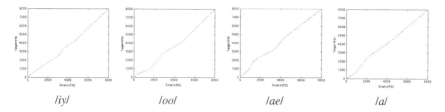

/iy/   /oo/   /ae/   /a/

Fig. 3. Piecewise warping functions for four representative monophthongs between the English and Korean languages

similar vowels are used. The piecewise linear warping functions for matched monophthongs between two languages are displayed in Fig. 3.

To transform the source speaker's individuality (English) to the target speaker's individuality (Korean), each frame of the English speech is transformed to the frequency domain, and its formants are calculated. The relative formant location of the given frame is obtained, and the frequency warping function is estimated by the weighted summation of each transform function of the four representative monophthongs /iy, oo, ae, a/.

The voice conversion experiments are performed in three cases: 1) both magnitude and phase information are used, 2) only magnitude information is used, and 3) transform the satisfied frame with the condition that it is placed within the source formant space. Figure 4 shows that many formants (marked as "x") are placed outside of the source formant space (boxed area). This phenomenon causes a degradation in speech quality in voice conversion. Therefore, for considering qualities, if the formants are placed in the formant space, the frame is converted by the estimated frequency warping function. For three sentences, six Korean participants evaluated the preferences and quality of the transformed speech.

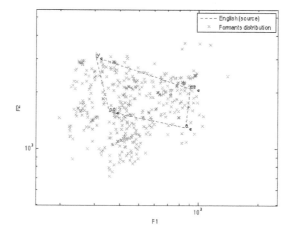

Fig. 4. English formant space and distribution of formants from source speaker

Table 1. Results of the ABX test (target speech preference) for speaker individuality and MOS test for voice quality for 3 different approaches

	ABX (target speech)	MOS (voice quality)
1) phase and magnitude	50%	2.00
2) magnitude only	50%	3.14
3) in formant space	33%	4.29

After each participant listened to the source speech (English), target speech (Korean), and three converted recordings, he/she determined whether the transformed speech was close to Korean, English, or "none of these" (ABX evaluation for speaker individuality) on the whole. Furthermore, they were asked to assess the overall voice quality of the converted speech on a mean opinion score (MOS) scale between 1 (very poor) and 5 (excellent, very good). Table 1 shows the results of the ABX test and MOS rating for each case [2]. It was interesting that none of the participants selected "none of these" ("x"). Because this system converts from English to Korean speech, and participants evaluated the speaker's overall individuality, they could prefer the converted speech to one of two languages as a result.

From the experimental results, we found that the formant in the formant space (monophthong trapezoid) has good voice quality (case 3), and when all information is converted together, the participants thought that the converted speech is close to the target speech (cases 1 and 2). However, it is necessary to use other parameters, such as line spectrum frequencies (LSF) [11], rather than direct frequency warping to improve voice quality, including both phase and magnitude information. In cases 1) and 2), the degraded voice quality caused many frames to be placed in the out-of-formant space (out of a trapezoid consisting of representative monophthongs) as in Fig. 4 and caused phase information to not be correctly converted.

5 Conclusion

A simple and intuitive voice conversion algorithm has been proposed. The presented algorithm uses weighted piecewise warping functions based on formant space information. The formant space consisted of matched vowels between the source and target speech. Our method estimated the warping functions by the weighted summation of each transform function between corresponding monophthongs based on the interpolation method. Therefore, the method can be used in case short voice samples are provided. From the experimental results, we found the potential of the proposed system. If we adopt the parameters, such as LSF, for improved conversion of phase information, performance will improve.

Acknowledgment. This work was supported by the research fund of Hannam University in 2013. We would like to thank the reviewers for their valuable remarks and suggestions.

---

[2] The experimental description and sample waveforms are found at `http://phoenix.hnu.kr/ysyun/voice_conv/eval_voice.html`

References

1. Accessibility, http://en.wikipedia.org/wiki/Accessibility
2. Bigham, J.P., Prince, C.M., Ladner, R.E.: WebAnywhere: A Screen Reader On-the-Go. In: Proc. of the Int. Cross-Disciplinary Conference on Web Accessibility (W4A), pp. 73–82 (2008)
3. Mizuno, H., Abe, M.: Voice conversion algorithm based on piecewise linear conversion rules of formant frequency and spectrum tilt. Speech Communication 16, 153–164 (1995)
4. Kuwabara, H., Sagisaka, Y.: Acoustic characteristics of speaker individuality: Control and conversion. Speech Communication 16, 165–173 (1995)
5. Narendranath, M., Murthy, H.A., Rajendran, S., Yegnanarayna, B.: Transformation of formants for voice conversion using artificial neural networks. Speech Communication 16, 207–216 (1995)
6. Pye, D., Woodland, P.C.: Experiments In Speaker Normalisation And Adaptation For Large Vocabulary Speech Recognition. In: Proc. of IEEE Int. Conference on Acoustics, Speech and Signal Processing, pp. 1047–1050 (1997)
7. Sundermann, D., Ney, H., Hoge, H.: VTLN-Based Cross-Language Voice Conversion. In: Proc. of IEEE Automatic Speech Recognition and Understanding Workshop, pp. 676–681 (2003)
8. Sundermann, D., Bonafonte, A., Ney, H.: Time Domain Vocal Tract Length Normalization. In: Proc. of IEEE Int. Symposium on Signal Processing and Information Technology, pp. 191–194 (2004)
9. Sundermann, D., Strecha, G., Bonafonte, A., Hoge, H., Ney, H.: Evaluation of VTLN-Based Voice Conversion for Embedded Speech Synthesis. Int. In: Proc. of Int. Conference on Spoken Language Processing, pp. 3–6 (2005)
10. Sundermann, D., Hoge, H., Bonafonte, A., Ney, H., Black, A., Narayanan, S.: Text-independent Voice Conversion Based on Unit Selection. In: Proc. of IEEE Int. Conference on Acoustics, Speech and Signal Processing, pp. I-81–84 (2006)
11. Erro, D., Moreno, A., Bonafonte, A.: Voice Conversion Based on Weighted Frequency Warping. IEEE Tr. on Audio, Speech, and Language Processing 18(5), 922–1931 (2010)
12. Saheer, L., Dines, J., Garner, P.N.: Vocal Tract Length Normalization for Statistical Parametric Speech Synthesis. IEEE Tr. on Audio, Speech, and Language Processing 20(7), 2134–2148 (2012)
13. Huang, X., Acero, A., Hon, H.-W.: Spoken Language Processing - A guide to Theory, Algorithm, and System Development. Prentice Hall (2001)
14. Voiceware Co., http://www.voiceware.co.kr/english/index.html

Building a Hybrid: Chatterbot – Dialog System

Alexiei Dingli and Darren Scerri

Department of Intelligent Computer Systems, University of Malta
{alexiei.dingli,dsce0006}@um.edu.mt

Abstract. Generic conversational agents often use hard-coded stimulus-response data to generate responses, for which little to no effort is attributed to effectively understand and comprehend the input. The limitation of these types of systems is obvious: the general and linguistic knowledge of the system is limited to what the developer of the system explicitly defined. Therefore, a system which analyses user input at a deeper level of abstraction which backs its knowledge with common sense information will essentially result in a system that is capable of providing more adequate responses which in turn result in a better overall user experience.

From this premise, a framework was proposed, and a working prototype was implemented upon this framework. The prototype makes use of various natural language processing tools, online and offline knowledge bases, and other information sources, to enable it to comprehend and construct relevant responses.

Keywords: Dialog Systems, External Knowledge Acquisition, RDF, Knowledge Bases, Chatterbots.

1 Introduction and Background

Conversational agents are deployed in various forms and designed to cater for different domains and goals, ranging from automated hotel booking agents, to personal assistants, companions and entertainment purposes. Moreover, one can categorise conversational agents into two main types based on how these process the input and generate their output. These two types can be realised as being *chatterbots* and *dialog systems* [5].

The main difference between these two types of conversational agents lies in what these systems are designed to model. Chatterbots model, or rather simulate a conversation in its basic sense, and intend to fool the user that he is communicating with an intelligent entity that does in fact understand what is being said. On the other hand, dialog systems attempt to model the actual dialog process which also incorporates the task of analyzing and understanding the input, which in turn aids in the generation of an adequate dynamic response.

Due to the nature of conversational agents, and by implication, the notion of natural language, the user is given the opportunity to provide less restricted input compared to other types of systems using other more conventional types of user interfaces. Therefore, conversational agents must provide a certain tolerance with regards to unexpected input, in other words, *robustness*. This issue is normally tackled by the use of stimulus-response methodology, or rather, pattern-matching techniques in chatterbots, since this

I. Habernal and V. Matousek (Eds.): TSD 2013, LNAI 8082, pp. 145–152, 2013.

provides a certain level of control over the system for which other approaches are not currently able to provide. Dialog systems carry out more complex input analysis to achieve a more meaningful representation of the input than simple pattern-matching techniques [5].

Conversational agents used mainly for generic conversation, *chatterbots*, depend on large amounts of inflexible language data while dialog systems make use of more refined technologies and approaches, including the integration of knowledge and the use of methods originating from Computational Linguistics [5]. It has also been shown that dialog systems which make use of certain knowledge bases and ontologies benefit in terms of *recognition, interpretation*, and *generation* [8]. Even though this study has been carried out on strict domains, the same approach could be used for broader and generic domains, similar to those which are typically associated with chatterbots.

Studies have shown that dialog systems which make use of an open world model produce a "more realistic conversation than a system without the open world model" [2]. This notion refers to the ability of a dialog system to use external knowledge bases, and in certain cases unstructured text, to gather knowledge about topics, concepts and entities which it does not have any information about. By the presence of this feature, dialog systems are able to more deeply understand the input provided by the user, and by implication, respond in a more intelligent and dynamic way, unlike simpler stimulus-response agents, which respond in a pre-defined verbatim manner.

2 Aims and Objectives

> "Building a system that could understand open-ended natural language utterances would require **common sense reasoning**, the huge open-ended mass of sensory-motor competencies, **knowledge and reasoning skills** which human beings make use of in their everyday dealings with the world" [6].

The aim of this project is to provide a proof-of-concept generic-conversation (unrestricted domain) framework for conversational agents, and a working prototype which can be categorised as being a hybrid between a chatterbot and a dialog system.

With regards to this hypothesis, the system makes use of modern natural language processing technologies and tools to analyze user dialog input while simultaneously using information that is obtainable from external sources to attempt to further understand the input and ultimately generate appropriate responses to the user. Another goal is to merge these various sources to create a single, local, knowledge base which enables the system to keep track of the world of the user, i.e. the relationship and interaction of the user with various entities. These sources include content and information on different concepts, including common sense knowledge and knowledge about specific people and world entities.

The aims and objectives of this project can be summarized as follows:

1. To build an expandable proof-of-concept system that provides a syntactic, semantic and pragmatic understanding of input.

2. To simulate intelligence by providing adequate output and logical conclusions derived from dialogue input and local and external knowledge bases.

Moreover, the proposed system will be as customizable, flexible, and modular as possible, so that it would require minimum effort to upgrade and adapt the system to handle input of varying complexity and topics.

3   Design

The developed prototype consists mainly of three phases: Natural Language Understanding (NLU) phase, intermediary processing phase, and finally the output generation phase.

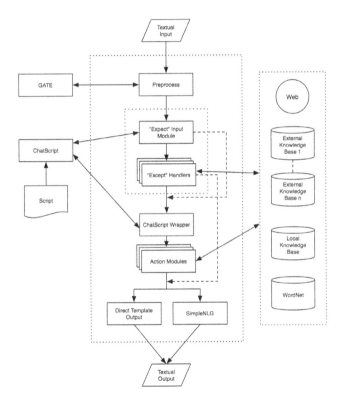

Fig. 1. System Block Diagram

The NLU module primarily makes use of ChatScript [9], an award-winning chatterbot engine. ChatScript employs various linguistic technologies to provide support for more flexible linguistic input in terms of syntax and semantics, and more expressive and semantics-oriented pattern matching rules, where matching patterns of meaning (semantics and pragmatics) is considered more important than matching patterns of words (syntax). These technologies and processes include the use of WordNet as a semantic network, Part-of-Speech tagging, pronoun resolution, conceptual relations, and preprocessing abilities.

ChatScript is used by the system mainly as a "normalisation phase". This phase is essentially a mapping process that maps natural language input into a more formal representation in XML. This is essential in order to allow the creation of rules that allow matching of a number of input utterances that are effectively semantically equivalent. This phase is analogous to how Faade maps text to discourse acts.

Input is preprocessed using two approaches. Using GATE and ANNIE [3,1], the system attempts to resolve pronouns into their respective named entities. Moreover, ChatScript includes a preprocessing phase in itself, performing actions such as spell checking and term substitution.

The normalisation phase allows the system to perform further intermediary processing on the input, such as querying local and external knowledge bases to allow better input comprehension. The conversational agent forwards the input to ChatScript after some surface pre-processing for which in turn it expects a normalised input in order to allow the system to further process the input. This is achieved by a number of processing modules (called "Action Modules") that can be developed to handle specific types of input.

For example, an Action Module can be designed and developed to handle any form of *"who"* questions. Upon an utterance which can be viewed as being a *"who"* question is detected by ChatScript, the input is then normalised and forwarded to the conversational agent for processing. The agent will then appoint an action module (in this case, the *"who" Action Module* to process that input and ultimately generate a response which can be dynamically generated from external knowledge bases (Figure 2).

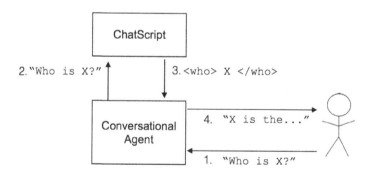

Fig. 2. Data flow between conversational agent and ChatScript

The system's dialog manager employs a similar technique used in CONVERSE [4] with regards to these processing modules. The system incorporates a polling technique for which the input is passed through all implemented processing modules, or Action Modules, for which each module advertises a score indicating how much it is confident to process the input.

Each action module is able to perform various tasks, include querying and managing local and external knowledge bases through global wrappers, triggering other action modules, querying ChatScript for further normalisation, adding and retrieving topics from the dialog manager, generating output, and alerting the dialog manager that the subsequent user utterance is expected to contain certain type of information (such as names, feelings, numbers, etc.).

The knowledge bases used in the system can be grouped into two types: the local RDF knowledge base, and other external knowledge bases (can be either offline or online), which are all accessible throughout the system. The local RDF knowledge base is managed using Apache Jena [7] and backed by its TDB component.

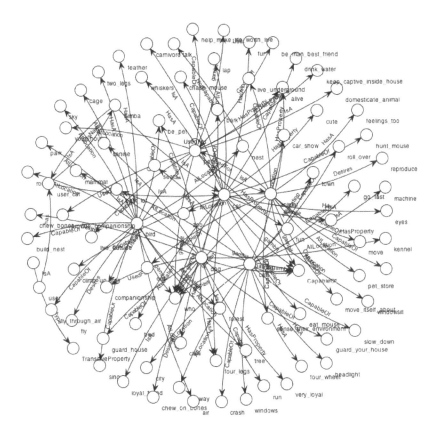

Fig. 3. An Example of a Generated RDF Graph

YAGO and ConceptNet are used by the system as external knowledge bases allowing the system to gather and make use of both entity specific and common sense knowledge. The retrieved data from these knowledge bases can be merged with the local RDF knowledge base to create an amalgamated network of knowledge, combining the world of the user with external knowledge.

4 Implementation and Evaluation

A prototype system was developed upon the described framework. This prototype included the development of a number of action modules, in conjunction with their respective ChatScript scripts which attempt to handle a specific set of user utterances, ranging from possession and sentiment statements, to questions regarding general and specific world entities.

The evaluation was carried out two-fold: hands-on conversation with the system (prototype), and a questionnaire filled by a number of users who engaged in a dialogue with the system following a pre-defined context and a set of example tasks. These users tested this prototype and answered questions on a five-point Likert scale which were aimed to be conclusive in whether the system actually achieved its aims and objectives. All results were biased towards the higher end (mean values for answers: 1 - 2.05%, 2 - 6.77%, 3 - 25.84%, 4 - 49.73%, 5 - 15.54%), which imply that the overall result is positive.

The following example shows the system's ability to keep track of user-defined information and depicts a level of understandability:

User: I have a red car.
System: OK, noted!
User: What is the color of my car?
System: Red.
User: I also have a cat named Simba.
System: Glad to know that you have a cat named Simba.
User: Do you know who is Simba?
System: Simba is your cat.
User: Do you know who is Barack Obama?
System: Barack Hussein Obama II is the 44th and current President of the United
        States. Also, I have heard a lot of positive things about Barack Obama. What
        's your view?
User: I think he is a great president .
System: Why do you think that Barack Obama is great?
User: Because he is an exemplary politician .
System: Ok then, so you think that Barack Obama is great, because he is an
        exemplary politician .
User: Are cats capable of chasing mice?
System: Yes
User: Can cats fly?
System: No, I don't think so
User: Is a cat an animal?
System: Yes
User: Is it a machine?
System: No
User: Do you know anything about Bill Gates?
System: Bill Gates is born on date 1955-10-28.
User: Anything else?
System: Bill Gates is married to Melinda Gates.
User: Tell me about cats
System: Cats are capable of eating meats, killing birds, drinking milks, eating
        mice, sleeping, climbing up trees and chasing mice.

5   Conclusions and Future Work

It is concluded that the results obtained are positive and show that the proposed framework has potential in supporting the implementation of dynamic conversational agents where generic knowledge need not be explicitly defined in the system beforehand as hard-coded data. Satisfying the project's aims and objectives for open-ended conversation would require a system which surpasses the ability to pass the Turing Test since it would also require responding in an informed and knowledgeable manner, and not just fooling a human that the system is a human and not a machine.

Naturally, there are still possible areas for improvement since the evaluation was done on a prototype with strict boundaries. Possible future work can be carried out in the integration of more refined dialog management techniques which will enable the system to be used in practical environments especially in complicated task-based environments. Another area for which the system falls short is its NLG capabilities.

Moreover, information extraction techniques can be used to gather information about generic and specific entities for which data is not found in the structured external knowledge bases.

An agent similar to Siri, i.e. a personal assistant, can be developed using this framework and the implemented modules, since such agent would not require complex dialog management techniques, but simply answering to the user's requests, in a task-oriented environment. If such system is to be developed, one would need to adapt the system to not only be able to reply, but also is able to talk to the user when necessary, e.g. as a reminder.

In addition, the system can be further improved by the implementation of new modules to achieve higher quality results in terms of input comprehension, mainly with regards to semantics and therefore, the actual meaning of the user utterances.

The system's ability to make use of external sources to support and enhance its knowledge of the real world can be considered as being a step forward towards the implementation of more natural and human-like conversational systems. The system is able to exploit the vast amounts of data found in structured knowledge databases, and are consulted to both understand and answer in a natural and informed manner.

References

1. Bontcheva, K., Cunningham, H., Maynard, D., Tablan, V., Saggion, H.: Developing reusable and robust language processing components for information systems using gate. In: 3rd International Workshop on Natural Language and Information Systems (NLIS 2002). Aix-en-Provence, pp. 223–227. Society Press (2002)
2. Catizone, R., Wilks, Y.: A companionable agent. In: Conversational Agents and Natural Language Interaction Techniques and Effective Practices, p. 302 (2011)
3. Cunningham, H., Humphreys, K., Gaizauskas, R.: Gate - a tipster-based general architecture for text engineering. In: Proceedings of the TIPSTER Text Program (Phase III) 6 Month Workshop. DARPA. Morgan Kaufmann (1997)
4. David Levy, B.L.C., Batacharia, B., Catizone, R., Krotov, A., Wilks, Y.: Converse: a conversational companion. In: Proc. of the 1st International Workshop on Human-Computer Conversation (1997)

5. Klüwer, T.: From chatbots to dialog systems. International Journal, 1–22 (2011)
6. Mateas, M., Stern, A.: Natural language understanding in façade: Surface-text processing (2004)
7. McBride, B.: Jena: A semantic web toolkit. IEEE Internet Computing 6(6), 55–59 (2002)
8. Milward, D., et al.: Ontology-based dialogue systems (2003)
9. Wilcox, B.: Chatscript documentation (2012),
   http://chatscript.sourceforge.net/Documentation/

CRF-Based Czech Named Entity Recognizer and Consolidation of Czech NER Research

Michal Konkol and Miloslav Konopík

Department of Computer Science and Engineering
Faculty of Applied Sciences
University of West Bohemia
Univerzitní 8, 306 14 Plzeň, Czech Republic
{konkol,konopik}@kiv.zcu.cz
nlp.kiv.zcu.cz

Abstract. In this paper, we present our effort to consolidate and push further the named entity recognition (NER) research for the Czech language. The research in Czech is based upon a non-standard basis. Some systems are constructed to provide hierarchical outputs whereas the rests give flat entities. Direct comparison among these system is therefore impossible. Our first goal is to tackle this issue. We build our own NER system based upon conditional random fields (CRF) model. It is constructed to output either flat or hierarchical named entities thus enabling an evaluation with all the known systems for Czech language. We show a 3.5 – 11% absolute performance increase when compared to previously published results. As a last step we put our system in the context of the research for other languages. We show results for English, Spanish and Dutch corpora. We can conclude that our system provides solid results when compared to the foreign state of the art.

Keywords: named entity recognition, conditional random fields, Czech Named Entity Corpus.

1 Introduction

Named entity recognition (NER) has proven to be an important preprocessing step for many natural language processing tasks including question answering [1], machine translation [2] or summarization [3]. The purpose of NER is to identify text phrases which carry a particular predefined meaning. The phrases are classified to a given set of classes, e.g. person, organization or location.

In this paper, we will mainly focus on NER used for texts in the Czech language. Czech is highly inflectional Slavic language with free word order. All published results [4,5,6,7] were evaluated on the Czech Named Entity Corpus [4], which is the only publicly available corpus for Czech.

Four systems were presented for Czech so far. The first two systems are similar in the architecture, but they use different methods (decision trees [4] and support vector machines [5]) and slightly different feature set. Both systems use non-standard architecture compared to conventional NER systems. They use different classifiers for one-word

I. Habernal and V. Matousek (Eds.): TSD 2013, LNAI 8082, pp. 153–160, 2013.

named entities (NEs), two-word NEs and NEs with more words. The results of these classifiers are combined to a final result that contains structured tags. The third system [6] can be viewed as a traditional approach to NER using a maximum entropy classifier. The fourth system [7] is based on conditional random fields. The output of the latter two is not structured. This difference in the output is the reason for incompatible evaluation metrics making the direct comparison of these systems impossible.

We will present our new NER system based on conditional random fields. Our system was directly compared with all previously published systems for Czech using the original evaluation scripts and outperformed them.

We have also transformed the Czech Named Entity Corpus into CoNLL format and evaluated our system using standard CoNLL evaluation. Then we have evaluated our system on the English, Spanish and Dutch CoNLL corpora and compared the results between languages.

The paper is structured as follows. The second section describes basic properties of the Czech Named Entity Corpus. Section 3 is focused on the evaluation metrics. Section 4 presents our CRF-based NER system. Section 5 provides results of our experiments and compares them with other systems. The last section summarizes our results and provides some ideas for future research.

2   Czech Named Entity Corpus

The Czech Named Entity Corpus (CNEC) [4] is created from the Czech National Corpus. Sentences were chosen by heuristic to contain more entities than completely random sentences. The NE annotation scheme uses hierarchical types with two levels. The first level is denoted as *supertypes* and the second level as *types*. The entities in the corpus can be embedded and form structured tags. We will use the term *top level entities* for entities that are not embedded. The corpus was created in three rounds. In the first round, 7 supertypes and 42 types were used. The subsequent rounds used an extended set with 10 supertypes and 62 types. Unfortunately, this extension was not used retrospectively for the first part. The evaluation in [4,5] is done only on the not-extended set, which is consistent across the whole corpus, but no information about this problem was given in the papers. The following example demonstrates the corpus. The NE classes are encoded using two letters, first for supertypes, second for types.

```
Generální ředitel <if Škody <gu Plzeň>> <P<pf Lubomír>
      <ps Soudek>> je stále největším akcionářem
            plzeňského strojírenského gigantu .
```

The CNEC consists of approximately 6000 sentences with 150.000 tokens. The corpus can be considered small when compared to CoNLL corpora [8,9] which have over 300.000 tokens. While the CoNLL corpora use 4 NE types, CNEC uses 7 (10) resp. 42 (62) types, which makes the data much sparser and therefore the NER task harder.

We have transformed the corpus into the standard CoNLL format enriched with lemmas and POS tags.[1] Only supertypes and top level entities are used. This transformed

---

[1] http://nlp.kiv.zcu.cz

corpus is used for our CoNLL evaluation and can be easily used by multilingual NER systems for their evaluation on Czech, because these systems usually work with the CoNLL corpus format.

3   Evaluation Metrics

NER is a typical example of multiclass classification problem, where the NE classes are highly skewed compared to the non-entity class. For such tasks precision, recall and f-measure are the metrics of choice. Unfortunately, it is possible to define correct answer in various ways, especially in corpora with hierarchical types and structured annotations.

3.1   Structured Metric

The structured metric is used in [4,5]. All entities (including the embedded) from the gold data set are used for the evaluation. Conventional NER systems, however, do not allow embedding. Under such conditions, their performance may seem worse then it actually is, because the embedded entities (which form an indispensable portion) are unreachable for them.

In the structured metric, an entity marked by a system is considered correct if the span and type are the same as in the gold data. All the levels in the structured annotations have the same weights. In [4,5], the results are given for one-word NEs, two-word NEs and all NEs. We believe, that it would be better to provide results for the top level NEs and then gradually for lower levels, because it is hard to interpret usefulness of entities based on number of words and the results can be hardly compared with other systems and other languages.

3.2   Word by Word Metric

The word by word metric is used in [6]. It does not allow embedding, so only the top types from the corpus are used for the evaluation. In this metric, each word is a self-standing unit with only one type. For the NE `<if Škody <gu Plzeň>>` spanning over two words we would have two independent objects with one type `if`. The NER system output is transformed in individual objects as well. For each word, the gold data type is compared with the type of the system and is considered correct if both are the same. The problem of this metric is that it ignores the presence of multi-word entities.

3.3   Standard CoNLL Metric

The standard CoNLL metric is used on the CoNLL-2002 and CoNLL-2003 conferences [8,9] and it is used in the majority of papers as well. It does not allow embedding, so the embedded entities must be ignored. The output NE is considered correct, only if its span and type is exactly the same as the span and type in the gold data. It is equivalent to the structured metric if only the top level of structured annotations is used.

It was also used in [7], but they removed structure of the tags by adding higher priority to embedded entities, e.g. `<if Škody <gu Plzeň>>` became `<if Škody> <gu Plzeň>`. This choice is questionable, because the top level is significantly more useful. It can also produce weird entity sequences, e.g. `<ic <ps Smith> & <ps Delvin> Ltd.>` produces `<ps Smith> <ic &> <ps Delvin> <ic Ltd.>`.

3.4   Lenient Metric Extension

The lenient metric extension is based on the GATE evaluator.[2] It is used as a supplement to the CoNLL evaluation metric. It adds the information about cases, where the system correctly guessed the type, but not the span.

4   NER System

4.1   Conditional Random Fields

Conditional random fields (CRF) are undirected graphical models [10]. Simple chain CRF are currently the state-of-the-art method for NER [11]. They model probability distribution $p(\mathbf{y}|\mathbf{x})$, where $\mathbf{x}$ is sequence of words and $\mathbf{y}$ is sequence of labels. The modeled probability distribution $p(\mathbf{y}|\mathbf{x})$ has the following form.

$$p(\mathbf{y}|\mathbf{x}) \propto \prod_{j=1}^{N} \exp \sum_{i=0}^{M} \lambda_i f_i(y_{j-1}, y_j, x_j) \tag{1}$$

N is number of tokens and M number of features. The parameters λ_i are estimated using L-BFGS method [12]. Gaussian prior is used for regularization [13]. Typically, the feature functions f_i are binary and are similar to the following example.

$$f_i(y_{j-1}, y_j, x_j) = \begin{cases} 1 & \text{if } y_{j-1} \text{ is CITY, } y_j \text{ is CITY and } x_j \text{ is 'York'} \\ 0 & \text{otherwise} \end{cases} \tag{2}$$

4.2   Feature Set

Our CRF system works with the feature set described in this section. All the features use a window of two previous and two next words.

The feature set does not include morphological tags, which are usually used in language dependent systems. We have tried some variations of their usage, but none of them improved the results.

Lemma – The PDT [14] lemmatizer is used to get the lemmas. The lemma has to appear at least twice to be used as a feature.
Affixes – Prefixes and suffixes of length from 2 to 4 were used. Both affixes are taken from lemmas and have to appear at least 5 times.

---

[2] http://gate.ac.uk/sale/tao/splitch10.html#x14-26900010.2

Bag of words – Bag of words is similar to lemma feature, but ignores order of the lemmas in the window.

Bi-grams – Bi-grams of lemmas have to appear at least twice to be used. Higher level n-grams did not improve the results, probably due to the size of the corpora.

Orthographic features – Standard orthographic features. Specifically: *firstLetterUpper; allUpper; mixedCaps; contains ., ', –, &; upperWithDot; acronym*

Orthographic patterns – Orthographic pattern [15] is a lemma, where every lower case letter is rewritten to a, upper case letter to A, number to 1 and symbol to -. We are using these patterns directly and also compressed, where every sequence of the same type is represented by only one character. The uncompressed pattern has to appear at least 5 times, compressed 20 times.

Orthographic word pattern – Combines compressed orthographic patterns (by joining them) of all the words in the window. This pattern has to appear at least 5 times.

Gazetteers – We use various gazetteers. The majority of them are from publicly available sources (e.g. list of cities provided by the Czech Ministry of Regional Development).

Table 1. Comparison of Czech NER systems

	structured	word by word	CoNLL(inner)	CoNLL	Lenient
DecTree [4]	68%	—	—	—	—
SVM [5]	71%	—	—	—	—
MaxEnt [6]	—	72.94%	—	—	—
CRF [7]	—	—	58.00%	—	—
Our system	79%	75.61%	62.07%	74.08%	79.50%

5 Experiments

5.1 Czech Systems Comparison

Our priority for the Czech language is to compare our system with all the other Czech systems using their evaluation scripts. All the systems use Czech lemmatiaztion and morphological analysis on a similar level, but use different gazetteers.

For comparison with [4,5], we have altered the system slightly to get structured output. We have added one more CRF model with the same features, but trained only on the embedded entities. For all the multiword entities found by the original CRF model, the second model tries to find entities inside.

We have also evaluated our system using the CoNLL evaluation. Comparison between the systems is given in table 1. Detailed results for individual classes are given in table 2.

Our system outperforms the other systems by a large margin. An interesting fact is, that marking the embedded entities (using the original structural evaluation) is actually easier task than finding only the not-embedded entities.

Table 2. Detailed CoNLL evaluation on the Czech Named Entity Corpus of our system

	Lenient			Strict		
	Precission	Recall	F-measure	Precission	Recall	F-measure
Time	92.11%	88.62%	90.33%	88.17%	85.24%	86.68%
Geography	80.13%	80.65%	80.39%	77.10%	77.61%	77.37%
Person	83.62%	88.41%	85.95%	79.82%	84.91%	82.29%
Address	50.00%	70.00%	58.33%	50.00%	70.00%	58.33%
Media	76.92%	30.30%	43.48%	69.23%	27.27%	39.13%
Other	67.70%	66.55%	67.13%	56.25%	55.67%	55.96%
Institution	79.95%	70.06%	74.68%	71.52%	63.05%	67.02%
Total	80.46%	78.56%	79.50%	74.81%	73.38%	74.08%

5.2   Comparison among Languages

We have evaluated our system on the CoNLL corpora [8,9] for other languages. The CoNLL corpora use four types of entities – organizations (ORG), persons (PER), locations (LOC) and miscellaneous (MISC). The results for English, Spanish and Dutch are shown in table 3. Our system is not tweaked for these languages. Lemmas are replaced with words and gazetteers are removed, because they are unavailable for some languages. For Spanish and Dutch we compare the results with the best system [16] from the original CoNLL conference. For English we report the best results from the conference [17] along with the overall best result so far [11]. For the latter we present two results – a language independent baseline (LIB) and a complete system.

The results of our system are obviously worse then from systems tweaked for given languages, because we have not tweaked our feature set (features and their parameters) for given languages at all. We still believe that this comparison shows the state-of-the-art quality of our system. It is also possible to roughly compare Czech with the other languages. We can see, that Czech has the lowest f-measure, even though our system uses some Czech language specific features for the Czech corpus. This can be partly caused by the smaller size of the Czech corpus and more entity types, but we believe that the major reason is the difficulty of Czech NER task.

6   Conclusion and Future Work

We have developed new CRF-based Czech NER system which outperforms all other published systems by a large margin. Using the standard CoNLL evaluation metric we achieved f-measure 74.08%. We have also evaluated the system on English, Spanish and Dutch and discussed the differences in performance. The whole paper has set the new state of the art for Czech language.

Table 3. Results for the (a) Spanish, (b) Dutch and (c) English CoNLL corpus. The comparison with other systems is shown at the bottom.

	(a)					(b)		
	Precission	Recall	F-measure			Precission	Recall	F-measure
ORG	74.54%	64.40%	69.10%		ORG	79.48%	80.79%	80.13%
PER	80.97%	82.51%	81.73%		PER	86.35%	89.52%	87.91%
LOC	81.54%	76.49%	78.93%		LOC	80.21%	77.77%	78.97%
MISC	75.71%	71.19%	73.38%		MISC	62.58%	54.12%	58.04%
Total	78.18%	73.86%	75.97%		Total	79.77%	79.12%	79.44%
Best [16]	77.83%	76.29%	77.05%		Best [16]	81.38%	81.40%	81.39%

	(c)		
	Precission	Recall	F-measure
ORG	81.15%	75.68%	78.32%
PER	87.28%	86.58%	86.93%
LOC	87.81%	86.81%	87.31%
MISC	78.55%	74.07%	76.24%
Total	84.64%	81.89%	83.24%
Best CoNLL [17]	88.99%	88.54%	88.76%
Best LIB [11]	—	—	83.78%
Best [11]	—	—	90.90%

We believe that the results can be further improved by feature extraction. Another possibility is to combine various machine learning methods into one model.

Acknowledgements. This work was supported by grant no. SGS-2010-028, by grant no. SGS-2013-029 Advanced computing and information systems, by the European Regional Development Fund (ERDF). Access to the MetaCentrum computing facilities provided under the program "Projects of Large Infrastructure for Research, Development, and Innovations" LM2010005, funded by the Ministry of Education, Youth, and Sports of the Czech Republic, is highly appreciated.

References

1. Mollá, D., Van Zaanen, M., Smith, D.: Named entity recognition for question answering (2006)
2. Babych, B., Hartley, A.: Improving machine translation quality with automatic named entity recognition. In: Proceedings of the 7th International EAMT Workshop on MT and Other Language Technology Tools, Improving MT Through Other Language Technology Tools: Resources and Tools for Building MT, EAMT 2003, pp. 1–8. Association for Computational Linguistics, Stroudsburg (2003)

3. Nobata, C., Sekine, S., Isahara, H., Grishman, R.: Summarization System Integrated with Named Entity Tagging and IE pattern Discovery. In: LREC (2002)
4. Ševčíková, M., Žabokrtský, Z., Krůza, O.: Named entities in Czech: annotating data and developing NE tagger. In: Matoušek, V., Mautner, P. (eds.) TSD 2007. LNCS (LNAI), vol. 4629, pp. 188–195. Springer, Heidelberg (2007)
5. Kravalová, J., Žabokrtský, Z.: Czech named entity corpus and SVM-based recognizer. In: Proceedings of the 2009 Named Entities Workshop: Shared Task on Transliteration, NEWS 2009, pp. 194–201. Association for Computational Linguistics, Stroudsburg (2009)
6. Konkol, M., Konopík, M.: Maximum entropy named entity recognition for Czech language. In: Habernal, I., Matoušek, V. (eds.) TSD 2011. LNCS, vol. 6836, pp. 203–210. Springer, Heidelberg (2011)
7. Král, P.: Features for Named Entity Recognition in Czech Language. In: KEOD, pp. 437–441 (2011)
8. Tjong Kim Sang, E.F.: Introduction to the CoNLL-2002 shared task: language-independent named entity recognition. In: Proceedings of the 6th Conference on Natural Language Learning, COLING 2002, vol. 20, pp. 1–4. Association for Computational Linguistics, Stroudsburg (2002)
9. Tjong Kim Sang, E.F., De Meulder, F.: Introduction to the CoNLL-2003 shared task: language-independent named entity recognition. In: Proceedings of the Seventh Conference on Natural Language Learning at HLT-NAACL 2003, CONLL 2003, vol. 4, pp. 142–147. Association for Computational Linguistics, Stroudsburg (2003)
10. Lafferty, J.D., McCallum, A., Pereira, F.C.N.: Conditional Random Fields: Probabilistic Models for Segmenting and Labeling Sequence Data. In: Proceedings of the Eighteenth International Conference on Machine Learning, ICML 2001, pp. 282–289. Morgan Kaufmann Publishers Inc., San Francisco (2001)
11. Lin, D., Wu, X.: Phrase clustering for discriminative learning. In: Proceedings of the Joint Conference of the 47th Annual Meeting of the ACL and the 4th International Joint Conference on Natural Language Processing of the AFNLP, ACL 2009, vol. 2, pp. 1030–1038. Association for Computational Linguistics, Stroudsburg (2009)
12. Nocedal, J.: Updating Quasi-Newton Matrices with Limited Storage. Mathematics of Computation 35, 773–782 (1980)
13. Chen, S.F., Rosenfeld, R.: A gaussian prior for smoothing maximum entropy models (1999)
14. Hajič, J.: Disambiguation of Rich Inflection (Computational Morphology of Czech). Karolinum, Charles University Press, Prague, Czech Republic (2004)
15. Ciaramita, M., Altun, Y.: Named-Entity Recognition in Novel Domains with External Lexical Knowledge (2005)
16. Carreras, X., Màrques, L., Padró, L.: (named entity extraction using adaboost). In: Proceedings of CoNLL 2002, Taipei, Taiwan, pp. 167–170 (2002)
17. Florian, R., Ittycheriah, A., Jing, H., Zhang, T.: (named entity recognition through classifier combination). In: Daelemans, W., Osborne, M. (eds.) Proceedings of CoNLL 2003, Edmonton, Canada, pp. 168–171 (2003)

Comparison and Analysis of Several Phonetic Decoding Approaches

Luiza Orosanu[1,2,3] and Denis Jouvet[1,2,3]

[1] Speech Group, LORIA
Inria, Villers-lès-Nancy, F-54600, France
[2] Université de Lorraine, LORIA, UMR 7503, Villers-lès-Nancy, F-54600, France
[3] CNRS, LORIA, UMR 7503, Villers-lès-Nancy, F-54600, France
{luiza.orosanu,denis.jouvet}@loria.fr

Abstract. This article analyzes the phonetic decoding performance obtained with different choices of linguistic units. The context is to later use such an approach as a support for helping communication with deaf people, and to run it on an embedded decoder on a portable terminal, which introduces constrains on the model size. As a first step, this paper compares the performance of various approaches on the ESTER2 and ETAPE speech corpora. Two baseline systems are considered, one relying on a large vocabulary speech recognizer, and another one relying on a phonetic n-gram language model. The third model which relies on a syllable-based lexicon and a trigram language model, provides a good tradeoff between model size and phonetic decoding performance. The phone error rate is only 4% worse (absolute) than the phone error rate obtained with the large vocabulary recognizer, and much better than the phone error rate obtained with the phone n-gram language model. Phone error rates are then analyzed with respect to SNR and speaking rate.

Keywords: syllables, deaf, speech recognition, embedded system.

1 Introduction

Support for deaf people or for people with hearing impairment is an application area of automatic speech processing technologies [1]. Their objective is to become a communication aid for disabled persons. Over the past decades, scientists have tried to offer a better speech understanding, by displaying phonetic features to help lipreading [2], by displaying signs in sign language through an avatar [3], and of course by displaying subtitles, generated in a semi-automatic or fully automatic manner. The ergonomic aspects and the conditions for using speech recognition to help deaf people were analyzed in [4]. One of the main drawbacks of speech recognition systems is their incapacity of recognizing the words that do not belong to their vocabulary. Given the limited amount of speech training data, it is impossible to conceive a system that covers all the words, let alone the proper names or abbreviations. Furthermore, recognition systems are not perfect, it happens quite frequently that a word is confused with another one which is pronounced the same (homophone) or almost the same. The performance is very far

I. Habernal and V. Matousek (Eds.): TSD 2013, LNAI 8082, pp. 161–168, 2013.
© Springer-Verlag Berlin Heidelberg 2013

from human performance [5] and even degrades rapidly in the presence of noise. Therefore, in the context of communication aids for deaf people, displaying the orthographic form of the recognized words may not be an ideal solution.

IBM has thus tested subtitling the phonetic speech of a speaker, with the system called LIPCOM [6]. The application was based on a phonetic decoding (with no prior defined vocabulary) and the result was displayed as phonemes coded on one or two letters. More recent studies have measured the contribution of confidence measures [7] within the use of automatic transcription for deaf people [8]. Subjective tests have shown a preference for displaying the phonetic form of the words with a low confidence score.

An alternative solution is to use multi-phone sub-word units, like the syllable. Its appeal lies in its close connection to human speech perception and articulation, since it's more intuitive for representing speech sounds. The use of syllable-size acoustic units in speech recognition has been investigated in the past [9,10], for large vocabulary continuous speech recognition (usually in combination with context dependent phones) [11,12] or for phonetic decoding only [13]. In this last case [13], because of the structure of the acoustic units, coarticulation was modeled between phonemes inside the syllable unit, but no context-dependent modeling was taken into account between syllable units, moreover the language model applied at the syllable level was a bigram. Besides, to overcome the limited size of any speech recognizer lexicon, studies have been conducted in extending the word-based lexicon with fragments, typically sequences of phonemes determined in a data driven way; this extension helped providing better acoustic matches on out-of-vocabulary portions of the speech signal, which globally led to a smaller phonetic error rate [14].

In this paper we shall investigate the use of syllables at the lexical level. The syllables are described in terms of phonemes, which are modeled with context-dependent 3-states HMM. The language model applied on the syllables is a trigram. We have followed the rules proposed in a recent study for detecting syllables boundaries within a sequence of phonemes [15]. These rules are used to derive the syllables from the phonetic forced-aligned training data, and some criteria are applied to reduce the list of syllables constituting the lexicon. Performance is reported in terms of phoneme error rate, and evaluations are conducted on two large French speech corpus.

The work presented in this paper is part of the RAPSODIE project, which aims at studying, deepening and enriching the extraction of relevant speech information, in order to support communication with deaf or hard of hearing people. Therefore, the optimal solution should determine the best compromise for the recognition model and the best way of presenting the recognized information (words, syllables, phonemes or combinations), within the constraints of limited available resources (the memory size and computational power of an embedded system).

The paper is organized as follows. The first section provides a description of the various linguistic units used in our analysis, that is phonemes, syllables and words. The second part of the paper is devoted to the description of experiments and the discussion of results. The different approaches, based on phoneme, syllable and word units, are compared on the ESTER and ETAPE data. Then, a detailed analysis of the performance is carried out with respect to signal-to-noise ratio (SNR) and speaking rate.

2 Linguistic Units

This section describes the linguistic units used in our analysis: the phonemes, as the basic and smallest linguistic unit, the syllables, as the phonological "building blocks" of words, and the words, as the largest linguistic unit, but at the same time the smallest linguistic element which caries a real meaning. Note that the choice of linguistic units impacts on the language model. In the experiments reported later, the acoustic unit is always the phoneme and the language models are always trigram statistical models.

2.1 Phonemes

Regarding the pronunciation lexicon, the pronunciation of a phoneme is the phoneme itself. Using this type of linguistic unit, we minimize the size of our vocabulary (less than 40 phonemes for the French language) and therefore the size of our language model. But unfortunately, with less modeling power usually comes worse performance.

2.2 Words

The word lexicon contains the mappings from words to their pronunciations in the given phoneme set. Given that French is a non-phonetic language, some letters can be pronounced in different ways or sometimes not at all, and a normally silent consonant at the end of a word can be pronounced at the beginning of the word that follows it ("liaison"). So, in order to make the automatic phonetic transcription as fluid as the real speech, the dictionary usually contains several pronunciation variants for each word. Using words as linguistic units leads to a large vocabulary (about 97,000 words in our dictionary) and therefore also to a large language model. This kind of model usually gives the best performance, but with the cost of large memory use and slow computational time (hence, not ideal for embedded systems).

2.3 Syllables

Regarding the vocabulary, the pronunciation of a "phonetic" syllable is its decomposition into the phonemic components. In order to account for the "liaison" events, the words are not processed individually. The training corpora is entirely phonetized and the resulting continuous list of phonemes is processed by the syllabification tool. The phonetization process is realized by force-aligning the manual transcriptions. Note that a word can have several pronunciation variants, and that one or more phonemes might be missing in some of them. Our syllabification tool is based on the rules described in [15], which follow two main principles: a syllable contains a single vowel and a pause designates a syllable's boundary. Therefore, the syllabification algorithm will give out a list of syllables and pseudo-syllables. The pseudo-syllables are the units where one vowel is surrounded by a large number of consonants, which normally should not belong to a single syllable. In order to filter some of the pseudo-syllable models, we have chosen to create different lists corresponding to two criteria : a minimum number of occurrences within the training corpora, and a maximum number of phonemes per syllable. The number of linguistic units of each list varies between 4,000 (maximum 3 phonemes, minimum 10 occurrences) and 16,000 (minimum 1 occurrence). Using syllables as linguistic units leads to a compromise between the memory use and computational time (ideal for embedded systems).

3   Experiments and Results

This section describes the data sets and tools used in our experiments, along with the corresponding results.

3.1   Data

The speech corpora used in our experiments come from the ESTER2 [16] and the ETAPE [17] evaluation campaigns, and the EPAC [18] project. The ESTER2 and EPAC data are French broadcast news collected from various radio channels, thus they contain prepared speech, plus interviews. A large part of the speech data is of studio quality, and some parts are of telephone quality. On the opposite, the ETAPE data correspond to debates collected from various radio and TV channels. Thus this is mainly spontaneous speech. The speech data of the ESTER2 and ETAPE train sets, as well as the transcribed data from the EPAC corpus, were used to train the acoustic models. The training data amounts to almost 300 hours of signal and almost 4 million running words. The phoneme-based language model and the syllable-based language models were also trained on the results of the forced-alignments of ESTER2, ETAPE and EPAC corpora, on about 12 million running phonemes and on about 6 million running syllables.

For the creation of the word-based language model, various text corpora were used: more than 500 million words of newspaper data from 1987 to 2007; several million words from transcriptions of various radio broadcast shows; more than 800 million words from the French Gigaword corpus [19] from 1994 to 2008; plus 300 million words of web data collected in 2011 from various web sources, and thus mainly covering recent years. For the word-based lexicon, the vocabulary of about 97,000 words, was developed for the ETAPE evaluation campaign. The pronunciation variants were extracted from the BDLEX lexicon [20] and from in-house pronunciation lexicons, when available. For the missing words, the pronunciation variants were automatically obtained using JMM-based and CRF-based Grapheme-to-Phoneme converters [21].

3.2   Configuration

The SRILM tools [22] were used to create the statistical language models. The Sphinx3 tools [23] were used for training the acoustic models and for decoding the audio signals. The MFCC (Mel Frequency Cepstral Coefficients) acoustic analysis computes 13 coefficients (MFCC and energy) every 10 ms. The acoustic HMM models were modeled with a 64 Gaussian mixture, and adapted to male and female data.

3.3   Results

The development sets of the ESTER2 (non-African radios, about 42,000 running words and 142,000 running phonemes) and ETAPE (entire set, about 82,000 running words and 263,000 running phonemes) data are used in the experiments reported below.

The COALT (Comparing Automatic Labelling Tools) software [24] was used for the analysis of results (phoneme error rates). The compared files are the hypothesis .ctm file (resulting from the decoding process) along with the reference .stm file. The CTM file consists of a concatenation of time-marked phonemes. The STM (segment time marked) file describes the reference transcript and consists of the results of the forced-alignment (sequences of phonemes).

Table 1. Characteristics of language models

LM	# of n-grams			Size [MB]
	n=1	n=2	n=3	
phonemes	40	1347	30898	0.21
syl_min4occ	8.3K	0.38M	1.73M	9.97
words	97.3K	43.35M	79.30M	1269.81

Table 1 describes some of the language models (LM) used in our experiments. With phoneme-based language model, the number of 3-grams is around 30,000 which leads to a minimum disk usage. With syllable-based language model, the number of 3-grams is around 1.7 M (for the list of syllables seen at least 4 times in the training data set) which leads to an average disk usage. Using a large vocabulary, the number of 3-grams is around 79.3 M which leads to the largest disk usage.

Table 2. Performance analysis on ETAPE and ESTER2 corpora [%]

LM	Results on ETAPE				Results on ESTER2			
	PER	Ins	Del	Sub	PER	Ins	Del	Sub
phonemes	38.19	2.82	15.40	19.97	34.09	3.53	11.64	18.92
syl_min4occ	22.05	3.34	8.50	10.21	16.13	3.94	4.88	7.31
words	18.21	3.11	8.01	7.09	12.44	3.41	4.62	4.40

Table 2 presents some of the results obtained on the ETAPE and ESTER2 development sets, described in terms of phoneme error rates (PER), along with their corresponding percentages of insertions (Ins), deletions (Del) and substitutions (Sub). As expected, the best results were obtained with the large vocabulary recognizer. By using only the syllables seen at least 4 times within the training data set, we limit the size of the lexicon (about 8,000) and the size of the language model (only about 10MB), and we achieve nevertheless good phonetic decoding performance. The phone error rate is only 4% worse (absolute) than the phone error rate obtained with the large vocabulary recognizer, and much better than the phone error rate obtained with the phone n-gram language model. All the other syllable lists give more or less the same results. Which means that starting with a minimum number of 7,000 linguistic units we can achieve similar results as with the total number of ~16,000 units. Given that ESTER2 contains mainly prepared speech and that ETAPE contains mainly spontaneous speech, the results obtained on ESTER2 are, as expected, better than the ones obtained on ETAPE.

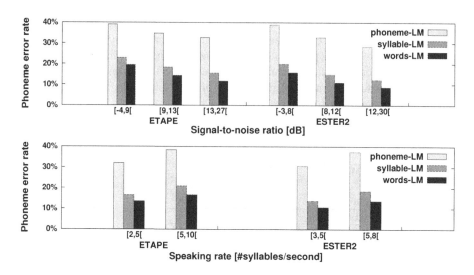

Fig. 1. Analysis of the phoneme error rates on the ETAPE *(left)* and ESTER2 *(right)* corpora, with respect to signal-to-noise ratio *(top)* and speaking-rate *(bottom)*

$$SNR_{dB} = 10log_{10}\frac{P_{signal}}{P_{noise}} \approx 10 \left(\frac{ln\left(\overline{C_0}_{vowel}\right)}{ln\left(10\right)} - \frac{ln\left(\overline{C_0}_{noise}\right)}{ln\left(10\right)} \right) \qquad (1)$$

Figure 1 reports an analysis of the performance with respect to the SNR ratios and the speaking rates of both speech corpora, limited to speech segments longer than 5 seconds. We have observed that the performance improves when the SNR ratio increases and that the performance degrades when the speaking rate increases. The SNR values were obtained from the average values of the C_0 MFCC coefficients (as computed by *sphinx_fe* tool) of vowels relative to noise/silence, converted afterwards to dB (cf. eq. 1). The speaking rates were computed as the number of vowels per second.

Table 3 presents the results obtained with the syllable language model on both corpora (speech segments longer than 5 seconds), with respect to both the SNR and the speaking rates criteria. We can naturally observe that the best results are obtained with the highest SNR ratio and the lowest speaking rate (less than five syllables per second).

Table 3. Analysis of the phone error rate(%) on ETAPE *(left)* and on ESTER2 *(right)*, with respect to signal-to-noise ratios and speaking rates criteria

SNR	Speaking rate			SNR	Speaking rate	
	[2,5[	[5,10[			[3,5[	[5,8[
[-4,9[	19.70	24.98		[-3,8[	17.66	22.38
[9,13[	16.54	19.71		[8,12[	12.17	17.76
[13,27[	13.74	17.28		[12,30[	10.93	14.01

The results then degrade in both directions: when the SNR ratio decreases and when the speaking rate increases.

4   Conclusions

This paper presented a detailed study on the phonetic decoding performance on two French speech corpora (ETAPE and ESTER2). We were interested in finding the best compromise between computational cost and usability of results, constrains that must be met in order to be able to create an embedded speech recognition decoder on a portable terminal. The context is to later use such an approach as a support for helping communication with deaf people. Two baseline systems were considered. The first one relies on a large vocabulary speech recognizer; it gives the best results ($\sim$18% phoneme error rate (PER) on ETAPE and $\sim$12% PER on ESTER2), but it uses a lot of memory and computational power. The second one relies on a phonetic n-gram language model; it does not use much memory, nor computational power, but it does not give good results neither ($\sim$38% PER on ETAPE and $\sim$34% PER on ESTER2). Then syllable language models were investigated. Keeping only the most frequent syllables leads to a limited-size lexicon and language model, which nevertheless provides good phonetic decoding performance. The phone error rate is only 4% worse (absolute) than the phone error rate obtained with the large vocabulary recognizer, and much better than the phone error rate obtained with the phone n-gram language model. Finally, a detailed analysis of the phoneme error rate was conducted with respect to SNR and speaking rate.

Future work will focus on the best, suitable way of presenting the recognized information (phonemes, syllables, words or combinations), based on relevant confidence measures, so that it maximizes communication efficiency with deaf people.

Acknowledgements. The work presented in this article is part of the RAPSODIE project, and has received support from the "Conseil Régional de Lorraine" and from the "Région Lorraine" (FEDER) (http://erocca.com/rapsodie).

References

1. Schönbächler, J.: Le traitement de la parole pour les personnes handicapées. Travail de séminaire (2003)
2. Sokol, R.: Réseaux neuro-flous et reconnaissance de traits phonétiques pour l'aide à la lecture labiale. Thèse Université de Rennes (1996)
3. Cox, S., Lincoln, M., Tryggvason, J., Nakisa, M., Wells, M., Tutt, M., Abbott, S.: Tessa, a system to aid communication with deaf people. In: Proceedings of the 5th International ACM Conference on Assistive Technologies, pp. 205–212 (2002)
4. Woodcock, K.: Ergonomics and automatic speech recognition applications for deaf and hard-of-hearing users. Technology and Disability 7, 147–164 (1997)
5. Lippmann, R.: Speech recognition by machines and humans. Speech Communication 22, 1–15 (1997)
6. Coursant-Moreau, A., Destombes, F.: LIPCOM, prototype d'aide automatique à la réception de la parole par les personnes sourdes. Glossa 68, 36–40 (1999)

7. Jiang, H.: Confidence measures for speech recognition: A survey. Speech Communication 45(4), 455–470 (2005)
8. Razik, J., Mella, O., Fohr, D., Haton, J.-P.: Transcription automatique pour malentendants: amélioration à l'aide de mesures de confiance locales. Journées d'Etude de la parole (2008)
9. Zhang, L., Edmondson, W.H.: Speech recognition using syllable patterns. In: 7th International Conference on Spoken Language Processing (2002)
10. Tachbelie, M., Besacier, L., Rossato, S.: Comparison of syllable and triphone based speech recognition for Amharic. In: Proceedings of the LTC, pp. 207–211 (2011)
11. Ganapathiraju, A., Hamaker, J., Ordowski, M., Doddington, G., Picone, J.: Syllable-based large vocabulary continuous speech recognition. IEEE Transactions on Speech and Audio Processing 9(4), 358–366 (2001)
12. Hämäläinen, A., Boves, L., de Veth, J.: Syllable-Length Acoustic Units in Large-Vocabulary Continuous Speech Recognition. In: Proceedings of SPECOM (2005)
13. Blouch, O., Collen, P.: Reconnaissance automatique de phonemes guide par les syllables. Journées d'Etude de la parole (2006)
14. Rastrow, A., Sethy, A., Ramabhadran, B., Jelinek, F.: Towards using hybrid, word, and fragment units for vocabulary independent LVCSR systems. In: Proceedings Interspeech (2009)
15. Bigi, B., Meunier, C., Bertrand, R., Nesterenko, I.: Annotation automatique en syllabes d'un dialogue oral spontané. Journées d'Etude de la parole (2010)
16. Galliano, S., Gravier, G., Chaubard, L.: The ESTER 2 evaluation campaign for rich transcription of French broadcasts. In: Proceedings INTERSPEECH (2009)
17. Gravier, G., Adda, G., Paulson, N., Carre, M., Giraudel, A., Galibert, O.: The ETAPE corpus for the evaluation of speech-based TV content processing in the French language. In: Proceedings LREC (2012)
18. Est'eve, Y., Bazillon, T., Antoine, J., Béchet, F., Farinas, J.: The EPAC corpus: Manual and automatic annotations of conversational speech in French broadcast news. In: Proceedings LREC (2010)
19. Mendonça, Â, Graff, D., DiPersio, D.: French gigaword third edition. Linguistic Data Consortium (2011)
20. de Calmès, M., Pérennou, G.: BDLEX: a Lexicon for Spoken and Written French. In: Language Resources and Evaluation, pp. 1129–1136 (1998)
21. Illina, I., Fohr, D., Jouvet, D.: Grapheme-to-Phoneme Conversion using Conditional Random Fields. In: Proceedings INTERSPEECH (2011)
22. Stolcke, A.: SRILM an Extensible Language Modeling Toolkit. In: 7th International Conference on Spoken Language Processing (2002)
23. Placeway, P., Chen, S., Eskenazi, M., Jain, U., Parikh, V., Raj, B., Ravishankar, M., Rosenfeld, R., Seymore, K., Siegler, M., Stern, R., Thayer, E.: The 1996 Hub-4 Sphinx-3 System. Carnegie Mellon University (1996)
24. Fohr, D., Mella, O.: CoALT: A Software for Comparing Automatic Labelling Tools. In: Language Resources and Evaluation (2012)

Concatenation Artifact Detection Trained from Listeners Evaluations⋆

Jakub Vít and Jindřich Matoušek

University of West Bohemia, Faculty of Applied Sciences, Dept. of Cybernetics,
Univerzitní 8, 306 14 Plzeň, Czech Republic
{vit89,jmatouse}@kky.zcu.cz

Abstract. Unit selection is known for its ability to produce high-quality synthetic speech. In contrast with HMM-based synthesis, it produces more natural speech but it may suffer from sudden quality drops at concatenation points. The danger of quality deterioration can be reduced (but, unfortunately, not eliminated) by using very large speech corpora. In this paper, our first experiment with automatic artifact detection is presented. Firstly, a brief description of artifacts is given. Then, a listening test experiment, in which listeners evaluated speech synthesis artifacts, is described. The data gathered during the listening test were then used to train an SVM classifier. Finally, results of the SVM-based artifact detection in synthetic speech are discussed.

Keywords: speech synthesis, unit selection, error detection.

1 Introduction

This paper presents our first experiments on automatic detection of artifacts in concatenation speech synthesis. Following the principle of unit-selection speech synthesis, any concatenation point can be a source of an audible artifact. Although speech segments are selected and concatenated to meet various criteria (commonly known as target and join costs) that should prevent artifacts, the artifacts do occur in synthesized speech. This could indicate that the cause of artifacts is not adequately covered by the criteria. So, the analysis and detection of the artifacts can help to design the criterion function better.

Taking into account the fact that artifacts are perceived differently and very subjectively among listeners, a listening test was carried out, in which the listeners marked certain segments in an utterance to contain artifacts. Based on these data, a classifier was trained using many features which were extracted during the unit selection process, computed from synthetic speech signal or extracted from the source speech recording in the speech corpus. The trained classifier is then employed to detect artifacts in synthetic speech.

Problems of listening experiment based artifact detection in diphone synthesis were also presented in [1,2] where the origin of artifacts was limited to spectral discontinuities. An application of a classifier for synthesis error detection was described e.g. in

⋆ The work has been supported by the Technology Agency of the Czech Republic, project No. TA01011264, and by the grant of the University of West Bohemia, project No. SGS-2013-032.

I. Habernal and V. Matousek (Eds.): TSD 2013, LNAI 8082, pp. 169–176, 2013.

[3]. In this paper, listening test inspired by [4] was used to reveal artifacts occurring at concatenation points in synthetic speech, and an SVM classifier was then trained from artifacts manifested by a wide variety of discontinuities (not only from spectral discontinuities).

2   Artifact Evaluation

Since the perception of presence/absence of artifacts in synthetic speech (and synthetic speech perception as such) is very subjective to the listener, there is no definite way to decide whether a concatenation point would be generally accepted by listeners or whether it would be generally judged as producing an artifact. To cope with this problem, a listening test was carried out with the aim to identify concatenation points in which majority of listeners perceived some artifacts (some sort of disruptive effects like various discontinuities, vocal glitches etc.). In this way, a set of reference artifacts was collected.

Stimuli for the listening test were synthetic utterances generated by a Czech unit-selection speech-synthesis system ARTIC [5] employing a female voice [6] and using diphones as the basic speech units. Hence, the concatenation points were at diphone boundaries.

2.1   Listening Test Design

The listening test was realized as a simple website. Listeners could work from their homes at any time and could evaluate any amount of available synthetic utterances. They were instructed to use headphones. Every listener was given a set of synthesized utterances. Utterances from recent news articles were selected for this purpose. To keep listeners' attention throughout the whole utterance, only utterances containing 5 to 8 words (none of them being of foreign origin) were chosen.

In each utterance, the listener should mark any phoneme-like segment that was found disturbing. He or she could even mark a sequence of phonemes if a precise position of an artifact was hard to locate. The maximum allowed length of such a sequence was 5 phonemes. All not marked phoneme segments were further considered not to contain any artifact. Listeners did not know which concatenation points corresponded to a concatenation of diphones which neighbored in the source speech corpus and which points corresponded to a concatenation of non-neighboring diphones.

Every listener got randomly ordered set of utterances from the global sentence database. In order for us to compare and evaluate listeners, the same utterances (10% of all utterances) were played to all listeners. Other utterances were randomly distributed, so that every utterance was played approximately to 50% of listeners. This strategy was chosen to get a large number of different artifacts and, simultaneously, to ensure (from verification reasons) that each potential artifact could be evaluated by more listeners.

For the listening test, listeners both with and without experience with synthesized speech were chosen. 1910 utterances were used as stimuli. Listeners submitted 7200 evaluations and marked more than 4700 suspicious segments. To cope with the fact that some listeners participated very actively while others submitted only few evaluations, listeners were weighted upon the number of completed evaluations. Throughout the whole experiment, 21 listeners participated in the listening test and 10 listeners participated very actively.

2.2   Cheat Prevention and Intra-listener Confidence Score

To prevent cheating, approximately every 20th utterance was played twice to the listener. To prevent the listener from remembering his/her evaluation from the first listening, a long time between re-playing was set. Listeners did not know about this feature during the listening test. Other attributes like date, number of re-plays and time spent on the web page with each particular utterance were also submitted together with the evaluation itself. These attributes were used to compute statistics about the listeners' behavior and included the cheat prevention system.

By evaluating results from multiply evaluated utterances, cheat prevention and listener's consistency was computed. For each listener, this was done by computing Cohen's kappa on utterances which were evaluated twice by the listener. Based on the Cohen's kappa, each listener l was assigned a confidence score cs_l which was used when completing the reference set of representative artifacts and also when training the artifact detection classifier.

Looking more closely at the actual numbers, the average intra-listener kappa was 0.58 but there were listeners who reached 0.8. Average time spent on one utterance was 26 seconds. Average number of utterance re-plays was 7. Since no listener was found cheating, nobody was excluded from the listening test.

2.3   Inter-listener Consistency Evaluation

In addition to the intra-listener evaluation, an inter-listener similarity/variability evaluation was also carried out. To compare two users, Cohen's kappa was computed using the evaluations from utterances evaluated by both listeners. For every concatenation point (located on a diphone boundary), evaluations on the phoneme which encompasses the concatenation point were compared between the listeners. The average kappa computed in this way was 0.27 ± 0.07 with maximum and minimum being 0.43, or 0.08, respectively.

As the determination of the precise location of an artifact is very hard and subjective, another method, more tolerant to slight differences in the location, was used as well. Similarly as before, Cohen's kappa was computed on concatenation points but evaluations were now taken as denoting the same artifact even if the listeners marked any of the three phonemes both preceding and succeeding the concatenation point. With that method of comparing, the average kappa increased to 0.53 ± 0.1 with maximum and minimum being 0.69, or 0.26, respectively.

To measure reliability of agreement among all listeners, Fleiss' kappa was computed on the utterances evaluated by all listeners. The number of such utterances was 137. The value of Fleiss' kappa among all active listeners who participated in the listening test was 0.29.

3   Artifact Selection Procedure

Upon the results of the listening test, a reference set of representative artifacts was collected. The set consisted of both "positive" and "negative" samples. The positive samples refer to concatenation points in which an artifact is perceived by listeners. On the other hand, the negative samples refer to concatenation points which are not accompanied by audible artifacts.

The artifact selection procedure can be summarized in the following steps (with the schematic view being shown in Fig. 1):

1. For every synthesized utterance i, each concatenation point j corresponding to diphones which did not neighbor in the source speech corpus (i.e., most often diphones from different source utterances) was taken as a candidate $a_{i,j}$ for an artifact and was assigned a score $s_{i,j} = 0$.
2. For each listener l who evaluated the i-th utterance the candidate score $s_{i,j}$ was incremented with the listener confidence score cs_l, if the listener marked a phoneme (or a sequence of phonemes) which encompasses the concatenation point j (see Sec. 2.2).
3. The candidate scores $s_{i,j}$ were normalized so that $s_{i,j} = 1$ meant the concatenation point j was evaluated as an artifact by all listeners who evaluated the utterance i.
4. In order to prefer artifact candidates $a_{i,j}$ from utterances i evaluated by more listeners, scores $s_{i,j}$ were multiplied by an utterance confidence score

$$n_i = \sum_{l \in L_i} cs_l$$

$$ucs_i = 1 - x^{-n_i}$$

   where L_i is a set of listeners who evaluated the synthetic utterance i and n_i is total sum of confidence scores from L_i for utterance i. The exponential function is used to favor utterances evaluated by more than few people but to saturate this score for utterances which were evaluated by many listeners. In our experiments, we chose $x = 1.3$ to achieve $ucs_i = 0.9$ provided that utterance i was evaluated approximately by the average number of listeners (10 in our case).
5. All artifact candidates $a_{i,j}$ with score $s_{i,j} > 0$ were taken as positive artifact samples and stored to the reference set of artifacts.
6. The artifact candidates $a_{i,j}$ with score $s_{i,j} = 0$ were taken as negative artifact samples. These candidates were never marked by any listener to contain an artifact.

The reference set of artifacts was used for classifier learning. The positive artifact samples were used to train the classifier with examples of what was perceived as disturbing by listeners. The negative artifact samples were then used to train the classifier with examples in which listeners did not perceive any distortion.

4   Artifact Detection

For the detection of artifacts a standard two-class classification scheme was used with one class being represented by positive samples (i.e., the concatenation points in which artifacts occurred) and the other one represented by negative samples (i.e., the concatenation points in which no listener perceived an artifact).

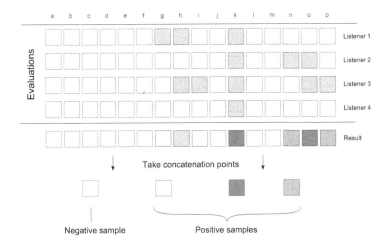

Fig. 1. Scheme of artifact selection procedure

4.1 Classifier Training

For the purposes of our work we utilized support vector machine (SVM) classifier with radial basis function (RBF). Due to its favorable properties (e.g., it is not prone to over-fitting), such a classifier has been successfully used in a wide variety of classification tasks. SVM classifier was trained using LIBSVM toolkit [7] and its parameters were optimized by grid search using 10-fold cross-validation.

Alternatively, to take artifact scores $s_{i,j}$ described in Sec. 3 into account, the training samples were weighted according to their scores during the training of the classifier.

Features used for classification were extracted at concatenation points. The following groups of features were used:

- **Static acoustic parameters:**
  - *Static F0 difference* – Absolute difference between $F0$ values at the concatenation point.
  - *Static energy difference* – Absolute difference between energy values at the concatenation point.
  - *Static spectral difference* – Absolute difference between MFCC vectors at the concatenation point.
  - *Static duration ratio* – Ratio r between phoneme segments a diphone is composed of (with $r \approx 1$ denoting a very similar duration and $r \gg 1$ or $r \ll 1$ denoting duration much different from what was expected).
- **Contextual parameters:**
  - *V/U characteristic* – Three binary values that indicate whether the three phonemes around the concatenation point are voiced (V) or unvoiced (U).
  - *Word boundary* – This binary value indicates whether the concatenation point represents a word boundary.
  - *Vowel* – This binary value indicates whether the phoneme around the concatenation point is a vowel.

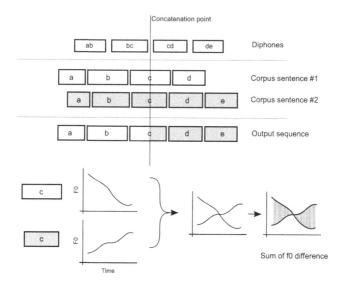

Fig. 2. Illustration of dynamic F0 difference between to-be-concatenated phoneme instances

– **Dynamic parameters.** The following features compare dynamic differences of tracked attributes between the two phoneme instances from source speech corpus which correspond to the phoneme before/after the concatenation point (see Fig. 2 for an example of dynamic F0 differences):

- *Total F0 differences* – sum of differences between F0 values
- *Total energy differences* – sum of differences between energy values
- *Total spectral differences* – sum of differences between first four formant values.
- *Total V/U differences* – sum of differences between V/U values.

4.2   Experiments

Artifact detection algorithm was designed to work on single concatenation points. This assumes that an artifact is caused by a single concatenation point, and the surrounding concatenation points do not contribute to the perception of the artifact. To verify this hypothesis, experiments were carried out both with all artifacts (EXP3, EXP4) and only with "isolated artifacts", i.e., such artifacts which did not immediately neighbor with other artifacts (EXP1, EXP2).

As described in Sec. 3, each artifact was assigned a score $s_{i,j}$. The higher the score, the more listeners perceived this artifact. Hence, artifacts with high scores should represent more relevant positive samples that should be more discriminative against the negative samples, and vice versa. A classifier was then trained with all artifacts (EXP2, EXP4) and also with only n-best artifacts in terms of $s_{i,j}$ (EXP1, EXP3).

In all experiments the training data was balanced so that each class contained the same number of both positive (N_p) and negative (N_n) samples.

4.3 Detection Evaluation

To evaluate the experiments, standard classification measures, precision (P), recall (R), $F1$ measure, and accuracy (A), were computed.

For the case when the classifier was trained with the weighted samples, number of true positive detections (tp), number of false positive detections (fp), number of true negative detections (tn), and number of false negative detections (tf) used to compute the classification measures (e.g. $P = \frac{tp}{tp+fp}$) were reformulated to take the artifact scores $s_{i,j}$ into account. The reformulation of tp and fp is shown here (tn and fn are reformulated in a similar way):

normal version

weighted version

$$tp = \sum_{a_{i,j}} \begin{cases} 1 & y_{i,j}^{(t)} = 1 \wedge y_{i,j}^{(p)} = 1 \\ 0 & \text{otherwise} \end{cases} \qquad tp = \sum_{a_{i,j}} \begin{cases} s_{i,j} & y_{i,j}^{(t)} = 1 \wedge y_{i,j}^{(p)} = 1 \\ 0 & \text{otherwise} \end{cases}$$

$$fp = \sum_{a_{i,j}} \begin{cases} 1 & y_{i,j}^{(t)} = 0 \wedge y_{i,j}^{(p)} = 1 \\ 0 & \text{otherwise} \end{cases} \qquad fp = \sum_{a_{i,j}} \begin{cases} s_{i,j} & y_{i,j}^{(t)} = 0 \wedge y_{i,j}^{(p)} = 1 \\ 0 & \text{otherwise} \end{cases}$$

The terms $y_{i,j}^{(p)}$ and $y_{i,j}^{(t)}$ stand for the predicted and true value of the artifact candidate $a_{i,j}$ (with $y(.) = 1$ being a positive sample corresponding to an artifact and $y(.) = 0$ being a negative sample corresponding to a non-disturbing concatenation point).

Table 1. Artifact detection results

	No. samples				Non-weighted SVM				Weighted SVM			
	N_p	N_n	$N_p^{(all)}$	$N_n^{(all)}$	R	P	$F1$	A	R	P	$F1$	A
EXP1	500	500	1574	3605	0.72	0.80	0.76	0.74	0.79	0.88	0.83	0.78
EXP2	1574	1574	1574	3605	0.63	0.80	0.71	0.67	0.80	0.95	0.87	0.79
EXP3	1000	1000	2458	4025	0.68	0.81	0.74	0.71	0.79	0.91	0.85	0.78
EXP4	2458	2458	2458	4025	0.61	0.76	0.68	0.64	0.79	0.95	0.86	0.79

5 Discussion and Conclusion

In this paper, our first experiments with automatic artifact detection were presented. The results of "standard" non-weighted classification may show that it is advantageous to train the classifier with the isolated artifacts (EXP1 vs EXP3, EXP2 vs EXP4). This could indicate that a single artifact could be hardly detectable when immediately surrounded by adjacent artifacts and that the perceived discontinuity could be caused by more artifacts. Another improvement can be seen when only artifacts with the highest scores are chosen (EXP1 vs EXP2, EXP3 vs EXP4). This could indicate that the artifact candidate scores $s_{i,j}$ are well correlated with artifact prominence and that it is better not to use less prominent artifacts during the training process.

The results for the weighted version of SVM show that the classifier tends to be more stable with respect to training data size. In this case, the preselection of artifacts (either of the isolated artifacts or the artifacts with the highest scores) does not seem to have such an influence on results. This could indicate that the weights were defined correctly as they reflect the actual contribution of each particular artifact. Consequently, all artifacts can be used for training and yet to achieve good detection results.

In our future work, other ways of describing artifacts and of optimizing their features will be searched for. A contextual classifier which takes neighboring concatenation points into account will be experimented with as well. Most importantly, artifact detection will be incorporated into our speech synthesis system [5] where it can be utilized implicitly as a part of the criterion function when selecting to-be-concatenated diphones, or explicitly to mark concatenation points which should be handled, for instance by using a signal modification technique.

References

1. Klabbers, E., Veldhuis, R.: On the reduction of concatenation artefacts in diphone synthesis. In: Proc. ICSLP, Sidney, Australia, pp. 1983–1986 (1998)
2. Pantazis, Y., Stylianou, Y., Klabbers, E.: Discontinuity detection in concatenated speech synthesis based on nonlinear speech analysis. In: Proc. INTERSPEECH, Lisbon, Portugal, pp. 2817–2820 (2005)
3. Lu, H., Wei, S., Dai, L., Wang, R.H.: Automatic error detection for unit selection speech synthesis using log likelihood ratio based SVM classifier. In: Proc. INTERSPEECH, Makuhari, Japan, pp. 162–165 (2010)
4. Legát, M., Matoušek, J.: Analysis of data collected in listening tests for the purpose of evaluation of concatenation cost functions. In: Habernal, I., Matoušek, V. (eds.) TSD 2011. LNCS, vol. 6836, pp. 33–40. Springer, Heidelberg (2011)
5. Tihelka, D., Kala, J., Matoušek, J.: Enhancements of Viterbi search for fast unit selection synthesis. In: Proc. INTERSPEECH, Makuhari, Japan, pp. 174–177 (2010)
6. Matoušek, J., Romportl, J.: Recording and annotation of speech corpus for Czech unit selection speech synthesis. In: Matoušek, V., Mautner, P. (eds.) TSD 2007. LNCS (LNAI), vol. 4629, pp. 326–333. Springer, Heidelberg (2007)
7. Chang, C.C., Lin, C.J.: LIBSVM: A library for support vector machines. ACM Trans. Intell. Syst. Technolog. 2, 27:1–27:27 (2011)

Configuring TTS Evaluation Method
Based on Unit Cost Outlier Detection

Milan Legát, Daniel Tihelka, and Jindřich Matoušek*

University of West Bohemia, Faculty of Applied Sciences,
New Technologies for the Information Society,
Univerzitní 8, 306 14, Plzeň, Czech Republic
{legatm,dtihelka,jmatouse}@kky.zcu.cz

Abstract. This paper presents a new analytic method that can be used for analyzing perceptual relevance of unit selection costs and/or their sub-components as well as for automated tuning of the unit selection weights. In particular, configuration options of the method are discussed in detail. A simple guidance on how to leverage the proposed method for the evaluation of a newly designed unit selection cost is also given in the paper. The advantage of using the proposed method is that different unit selection system configurations and tunings can automatically be evaluated without a need to conduct listening tests for each of them.

Keywords: TTS evaluation, unit selection costs, unit selection tuning.

1   Introduction

The unit selection method has seemed to be getting abandoned as a research topic over the last few years. There is no question that a huge amount of efforts have already been invested in improving the quality of synthetic speech delivered by unit selection based TTS systems since the introduction of the approach [1]. The method has been analyzed from almost all possible angles. Many works have dealt with experiments introducing different speech parameterizations and distances, which could be used for measuring the quality of concatenations [2], [3]; the target cost components; pruning of large unit databases; tuning weights of the costs [4]; and last but not least optimizing the search routine to lower the computational costs [5], [6], to name some. Still, we believe that the most important problem related to the unit selection—the "haphazard" presence of audible artifacts—has not been investigated thoroughly enough. Recently, some papers addressing this particular problem have been published (e.g. [7]).

Generally speaking, there are three main sources of these quality drops. First, any unit database, no matter how thoroughly it is verified, contains mislabelings at different levels. Second, the costs that are used while searching for the optimal sequences of units are not always well correlated with human perception. Third, the traditional implementation of the search algorithm allows, as long as the cost of the whole sequence

---

* Support for this work was provided by the TA CR, project No. TA01011264 and by the European Regional Development Fund (ERDF), project "New Technologies for the Information Society" (NTIS), European Centre of Excellence, ED1.1.00/02.0090.

I. Habernal and V. Matousek (Eds.): TSD 2013, LNAI 8082, pp. 177–184, 2013.

of units is minimum, for selecting units that should locally be avoided according to their assigned costs. This can especially be observed when the unit database is small. In theory, the same behavior can be observed in large footprint systems as well.

Little has also been invested in analyzing the audible artifacts and real understanding of the latent constructs that influence human perception of them. This is predominantly a consequence of not having reliable objective methods for TTS quality evaluation as well as large costs and labor intensiveness of the subjective methods. In this paper, we present a method, which provides insight mainly into the second source of audible artifacts mentioned above. The proposed method represents in its nature an analytic complement to the traditional TTS quality evaluation techniques such as MOS or ABX preference tests.

The rest of this paper is organized as follows. The next section presents the analytic method proposed in this paper. In Section 3, the method description is extended by considering configurable method parameters. In Section 4, a basic procedure for leveraging the proposed method is given. Finally, in Section 5, we draw conclusions and outline the intension for our future work.

2   Analytic Method Description

2.1   Outlier Detection

As already mentioned in the introduction, this work is aiming at the audible artifacts haphazardly appearing in the output of unit selection systems. Let us assume that the unit selection costs correlate reasonably well with human perception. If this was true, most of the selected units of extreme costs should lead to audible artifacts in the TTS output.

In order to see whether or not such units are being selected at all, the box-and-whisker diagrams (boxplots) can be used. The boxplots of values of all concatenation cost sub-components and also of the costs as such of the units forming the optimal sequences of units in a test set of utterances generated by our TTS system [8] are shown in Fig. 1. The plots indeed show that some units of rather outlying costs tend to appear in the selected sequences of units.

2.2   Annotation of Synthesized Sentences

Having confirmed that some outlying units exist in the optimal unit sequences generated by the system, the next step is to investigate whether or not these units coincide with audible artifacts. For this purpose, an annotation experiment using a set of 50 randomly selected synthesized sentences was conducted. At least one unit of an outlying cost or sub-cost was found in approximately 80% of the selected sentences.

The task of listeners was to mark segments of these sentences, which they found unnatural or containing any sort of distortion. The shortest segment that could be annotated was a phoneme. As can be seen in Fig. 2, most of the participants were typically marking segments of an approximate length of 3–5 phonemes.

The test was conducted using a web interface allowing the listeners to work from home. It was, however, stressed in the test instructions that the annotations shall be

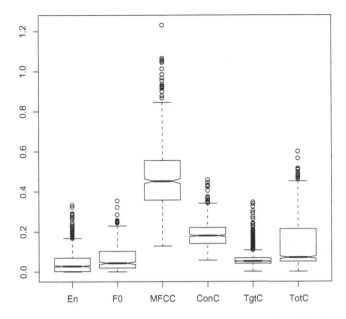

Fig. 1. Boxplots of costs of the units forming the optimal sequences found by the unit selection search algorithm

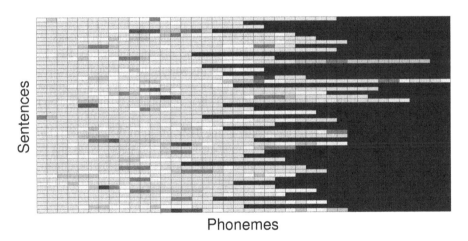

Fig. 2. Annotation of artifacts. Each row in the checkerboard plot represents a sentence, each column a phoneme. The darker a particular cell in the plot is the more agreement listeners found regarding a presence of an artifact at the particular phoneme. Note that the test sentences have different lengths, navy blue is used as a figure background.

done in a silent environment and using headphones. Since the annotation of audible artifacts is not a simple task, only experienced listeners were invited to participate. In total, 8 listeners finished the listening test, 5 of them being TTS researchers.

2.3 Alignment of Outliers and Audible Artifacts

Generally speaking, it is not a simple matter, to evaluate an annotation listening test. One of the typical concerns is how to identify non-reliable listeners. This particular issue was not a problem in our study as all participants were highly motivated to provide good quality annotations.

Another issue is different sensitivity of each participant to various kinds of artifacts. In order to evaluate the perceptual relevance of the outliers, keeping the sensitivity issue in mind, the H_L score (1) was introduced:

$$H_L(i) = \frac{\sum_{n=i-L}^{i+L} D_n}{(2L+1)N}. \tag{1}$$

L stands for a tolerance interval length in phonemes, i is the index of a phoneme and N is a number of listeners. The number of annotations of the particular phoneme, D_n, is defined as follows:

$$D_n = \sum_{j=1}^{N} h_n(j). \tag{2}$$

$h_n(j)$ is the annotation of the phoneme n defined as:

$$h_n(j) = \begin{cases} 1 & n \in A_j \\ 0 & n \notin A_j \end{cases}, \tag{3}$$

where the set A_j is the list of indeces of phonemes annotated by the j-th listener.

Having the H_L score defined, each outlier can be assigned its value. Since outliers have been defined as units of extreme costs present in the selected sequences of units, i.e. diphones in our case, and the annotations obtained from the listening test are phoneme based, an alignment had to be done. This is reasonable as the concatenation points are located in the middle of phonemes.

Fig. 3 shows an illustration of the results of the assignment of the values of the H_L score to all outliers found for the concatenation cost sub-components. Note that based on the above mentioned observation that most of the listeners were using maximum 3–5 phoneme long segments for annotating, the appropriate setting of the maximum length of the tolerance interval can be $L = 2$.

To further quantify the perceptual relevance of the outliers, a *perceptual relevance threshold* (hereafter referred to as thr) is defined for the sum of $H_L(i)$ scores $S_2(i)$ given by:

$$S_2(i) = \sum_{L=0}^{2} H_L(i). \tag{4}$$

Summing the H_L scores up to the length L allows for normalizing the relevance of artifacts annotated exactly at a particular phoneme with those annotated less precisely. Let us further, using the *perceptual relevance threshold*, define the *Hit Rate* as follows:

$$Hit\ Rate = \frac{N_{hit}}{N_{outl}} \times 100\ [\%], \tag{5}$$

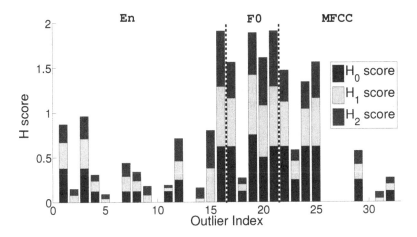

Fig. 3. H_L *scores* of outliers of the concatenation cost sub-components (illustrative example). Each detected outlier can be assigned a H_L *score* using annotations obtained by listeners. The higher the value of the H_L *score* is the more perceptually relevant the given outlier is.

where N_{hit} is a number of outliers of a given cost or a cost sub-component for which the condition $S_2(i) \geq thr$ is fulfilled, and N_{outl} stands for a number of all outliers found for a given cost or cost sub-component.

It is equally important to know how many annotated artifacts, i.e. phonemes fulfilling the condition $S_2(i) \geq thr$, are not identified by the outliers. For that purpose, the *Missed Rate* was defined as:

$$Missed\ Rate = \frac{N_{mis}}{N_{annot}} \times 100\ [\%], \qquad (6)$$

where N_{mis} is a number of annotated artifacts that do not match any outlier position, and N_{annot} is the total number of annotated artifacts.

3   Configuring the Analytic Method

3.1   Configuration Parameters

There are three parameters of the proposed analytic method that can be tuned—the summation length for the listeners ratings (set as $L = 2$ for the H_L *score* (1)), the *perceptual relevance threshold*, i.e. the threshold for deciding whether or not an outlier is perceptually relevant, and the span of the whiskers—*outlier detection threshold*—for the identification of the cost outliers (set as $1.5 \times IQR$[1], Fig. 1).

It can also be useful in some applications of the proposed method to plot values of the *Hit Rates* and *Missed Rates* for different parameter settings. We call these different settings operating points of the method. By connecting the operating points, the method operating curves can be obtained. An example of such a plot is shown in Fig. 4.

---

[1] IQR=Inter-Quartile Range.

3.2 Summation Window Length

Setting of the length L of the summation window as $L = 2$ in our default configuration can be justified by the observation that listeners were inclined to use 3–5 phoneme long segments for annotating the audible unit selection artifacts. Shorter window can be beneficial for identifying clear concatenation artifacts. Longer window is, on the other hand, appropriate for analyzing supra-segmental characteristics of a tested system, e.g. quality of intonation.

3.3 Perceptual Relevance Threshold

Many phonemes can theoretically be identified as locations of audible artifacts in the annotation experiment. Obviously, the more annotators are invited to provide annotations, the more likely any phoneme in the synthesized sentences becomes a location of an audible artifact due to being annotated by at least one of the listeners. This is explained by the subjectivity of the TTS quality perception as well as by the uncertainty concerning the exact locations of the artifacts. The H_L *score* (1) was introduced to measure the perceptual relevance of the detected outliers. The score function can actually be used to assign a value to every phoneme in the test data, no matter if it coincides with an outlier location or not. As already explained, it is beneficial for the better robustness to use the summation $S_2(i)$ of the H_L scores.

Different settings of the *perceptual relevance threshold* give a useful indication of how much perceptually relevant different costs and their sub-components are. One possible way of determining the *perceptual relevance threshold* directly from the annotated data is to calculate the $S_2(i)$ value for all phonemes in the test sentences and to sort the phonemes according to the assigned values. The threshold can then be defined by a certain percentile. In our experiments, we use by default 20th, 40th, 60th, 80th and 100th percentiles. By increasing the threshold, only phonemes that are identified as the locations of artifacts by a large number of listeners are taken into account. For instance, the 20th percentile would only take into account 20 % of the most frequently annotated artifacts. The *Hit Rates* and *Missed Rates* for the individual *perceptual relevance thresholds* can be seen as points on the operating curves in Fig. 4.

3.4 Outlier Detection Threshold

Lowering the *outlier detection threshold* can obviously lead to increasing the chance of hitting annotated artifacts. At the same time, more false alarms, i.e. outliers that do not correspond to any annotated artifact, must be expected. In order to get more insight into the sensitivity of the outlier detection, four different thresholds were experimented with—$0 \times IQR$, $0.5 \times IQR$, $1.0 \times IQR$ and $1.5 \times IQR$. Fig. 4 shows the result of the sensitivity analysis of the energy related concatenation cost sub-component of our system. The figure shows that for this particular sub-component, the best performance can be achieved by the most aggressive setting of the *outlier detection threshold* as the operating curves shift to lower *Missed Rates* while keeping the *Hit Rates* unchanged when comparing the three most aggressive settings. The threshold $1.5 \times IQR$ is not sensitive enough as its *Missed Rate* remains almost equal for all operating points representing different *perceptual relevance thresholds*.

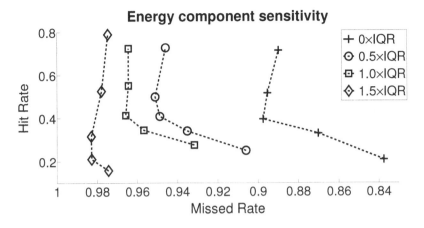

Fig. 4. Outlier sensitivity analysis—Energy related concatenation cost sub-component

4   Guidance on Usage of the Analytic Method

The analytic method introduced in the previous sections can generally be used for analyzing any unit selection based TTS system. Moreover, it can also be used for tuning the unit selection costs and/or testing new cost functions without a need to conduct additional listening tests.

The basic procedure for evaluating a new cost function can be summarized as follows:

1. Synthesize a random set of sentences by a unit selection based TTS system.
2. Annotate audible artifacts present in the synthesized sentences so that they can be related to the system base units.
3. Identify outliers of the original cost function of interest and calculate *Hit Rates* and *Missed Rates*.
4. Introduce a new cost function and resynthesize the set of sentences while keeping the original sequences of units fixed.
5. Calculate *Hit Rate* and *Missed Rate* of the new cost function and compare them to the original ones. From the difference in the rates, performance of the new cost function can be seen.

5   Conclusions and Future Work

A new analytic method for the evaluation of unit selection based TTS systems was proposed in this paper. The method makes use of manually annotated synthesized sentences. The positions of the annotated artifacts are aligned to the positions of units of outlying costs and/or sub-costs forming the synthesized sentences. The perceptual relevance of the system's costs and/or their sub-components is derived from this alignment.

The proposed method can be leveraged for automated tuning of the unit selection weights as well as experimenting with new costs and/or their sub-components without a need to conduct additional listening tests. The basic procedure to validate a newly proposed cost has also been given in this paper.

We plan to utilize the proposed approach for analyzing different unit selection based systems. We encourage the interested readers to contact us and to cooperate on the evaluation of their systems. We also want to look more closely at the audible artifacts that do not correspond to units of any outlying costs or sub-costs with the aim to be able to propose new perceptually relevant metrics that could be used by any of the TTS methods.

References

1. Hunt, A., Black, A.: Unit selection in a concatenative speech synthesis system using a large speech database. In: ICASSP 1996, Atlanta, Georgia, vol. 1, pp. 373–376 (1996)
2. Klabbers, E., Veldhuis, R.: Reducing audible spectral discontinuities. IEEE Transactions on Speech and Audio Processing 9, 39–51 (2001)
3. Vepa, J.: Join cost for unit selection speech synthesis. Ph.D. thesis, University of Edinburgh (2004)
4. Chen, J.D., Campbell, N.: Objective distance measures for assessing concatenative speech synthesis. In: EUROSPEECH 1999, Budapest, Hungary, pp. 611–614 (1999)
5. Tihelka, D., Kala, J., Matoušek, J.: Enhancements of Viterbi search for fast unit selection synthesis. In: INTERSPEECH 2010, Makuhari, Japan, pp. 174–177 (2010)
6. Sakai, S., Kawahara, T., Nakamura, S.: Admissible stopping in Viterbi beam search for unit selection in concatenative speech synthesis. In: ICASSP 2008, Las Vegas, USA, pp. 4613–4616 (2008)
7. Lu, H., et al.: Automatic error detection for unit selection speech synthesis using log likelihood ratio based SVM classifier. In: INTERSPEECH 2010, Makuhari, Japan, pp. 162–165 (2010)
8. Matoušek, J., Tihelka, D., Romportl, J.: Current state of Czech text-to-speech system ARTIC. In: Sojka, P., Kopeček, I., Pala, K. (eds.) TSD 2006. LNCS (LNAI), vol. 4188, pp. 439–446. Springer, Heidelberg (2006)

Development and Evaluation of Spoken Dialog Systems with One or Two Agents through Two Domains

Yuki Todo[1], Ryota Nishimura[2], Kazumasa Yamamoto[3], and Seiichi Nakagawa[3]

[1] Department of Computer Sciences and Engineering,
Toyohashi University of Technology, Japan
ytodo@slp.cs.tut.ac.jp
[2] Nagoya Institute of Technology, Japan
nishimura.ryota@nitech.ac.jp
[3] Department of Computer Sciences and Engineering,
Toyohashi University of Technology, Japan
{nakagawa,kyama}@slp.cs.tut.ac.jp

Abstract. Almost all current spoken dialog systems treat dialog as that where a single user talks to an agent. On the other hand, we set out to investigate a multiparty dialog system that deals with two agents and a single user. We developed a three person (one user and two agents) and a two person (one user and one agent) dialog system to consider the same dialog tasks, that is, "Which do you prefer, *udon* or *ramen* (Japanese noodle or Chinese noodle)?" and 'Which do you want to travel to *Hokkaido* or *Okinawa* (snowy region or tropical region)?" and compared them with respect to user behavior and satisfaction. According to the results of the experiments, the three person dialog system performed better in terms of lively conversation, and user can talk with the agents more like chatting.

Keywords: spoken dialog system, multi-party dialogue, two agents, chat.

1 Introduction

Recently, the demand for speech recognition interfaces has increased and thus spoken dialog systems have been developed. Previously, we developed a spoken dialog system, which has scope for improvement in terms of achieving a more natural dialog [1,2]. Our existing dialog system mimics the interaction between human beings in spontaneous conversation and generates natural responses, including *aizuchi* (back channeling), collaborative completions, and turn-taking, whilst considering response timing. A decision tree, which refers to prosodic information and surface linguistic information as features, was employed to determine the appropriate response timings. The existing system is able to deal with *repetition*, overlap response, and barge-in.

In this study, we aim to develop a more enjoyable dialog system [1]. To achieve this, we have extended our previous system [2], which allowed interaction between a single agent and the user, to handle two agents interacting with a user. In so doing we have formed a new dialog paradigm, and it is expected that the proposed system will achieve a dialog that was impossible in the previous system. Moreover, we deal with agents whose knowledge differs from hierarchical relationships. Thus, there is the possibility

I. Habernal and V. Matousek (Eds.): TSD 2013, LNAI 8082, pp. 185–192, 2013.

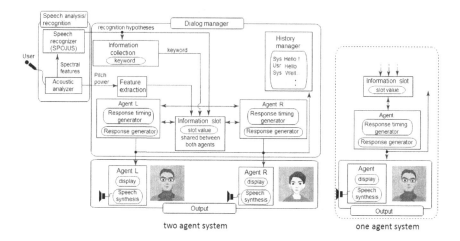

two agent system   one agent system

Fig. 1. Schematic diagram of dialog system with one or two agents

that by conversing with agents with different viewpoints, the user may be prompted with new ideas.

Recently, multi-party dialog has been actively studied. In the multi-party dialog between people, Dielmann [3] learned a model for granting Dialog Act of multi-party dialog automatically. Shriberg et al. [4] investigated overlap/interrupt in the meeting speech data, and showed that interrupts are associated with some events (such as disfluencies) in the foreground speech. Among humans and a conversation agent [5,6] or multi dialog agents [7,8], Fujie et al. conducted a real field experiment; the dialog system with a robot performed a quiz game with elderly people in an adult day-care center, and was able to become a game media which naive users such as elderly people can use and participate easily. In Dohsaka et al. [9], the agent decides the action depending on the situation in a multi-player conversation between humans and the conversation agents. The dialog takes place in a text-based dialog system and two users and two agents participate in the interaction.

Thus, the interaction of multiple agents can lead to an improvement in user satisfaction and activation of the dialog. Based on these considerations, we have developed a chat-like spoken dialog system to handle multiple conversational agents and to increase satisfaction for the user.

2   Dialog System

The spoken dialog system which we previously developed deals with dialog between one user and one agent. The system is now extended to the multi-party conversation, such as interaction between "two agents with different characteristics and one user". A multi-party dialog system has the following advantages:

- The conversation becomes more lively.
- Various interactive controls become possible.

– By using these functions, we can expect the range of new applications of spoken dialog systems to widen.

Figure 1 shows a schematic diagram of the dialog system for multi-party conversation with two agents. This system generates a response sentence using template matching from the result of the automatic speech recognizer (ASR). Moreover, the response type and timing are decided by inputting prosodic features into the decision tree [1]. Details are given in the following paragraphs.

2.1 Domain

It is desirable to choose a conversation domain that everyone can talk about, and is interested in. Therefore, we chose the topic of liking/disliking two things. In the actual experiment, the topic discussed is "Which do you like, *udon* (Japanese noodle) or *ramen* (Chinese noodle)?",'**Food Domain**" and 'Which do you want to travel to *Hokkaido* (snowy region) or *okinawa* (tropical region)?",'**Travel Domain**". 'Food domain" is more chat-like domain than 'Travel domain". The vocabulary sizes are about 270 words and 430 words, respectively.

In our dialog, two agents explain/state good points and bad points, respectively, about "*udon* (*Hokkaido*)" and "*ramen* (*Okinawa*)". In this case, it is possible to draw users into one of the opinions by ensuring that the agents have conflicting opinions. Moreover, we introduce strategies for arranging the different agents' opinions, and for drawing the user into a specific opinion.

2.2 Speech Analysis and Recognition

The speech recognizer SPOJUS [10] was employed to recognize the user input. There are two versions of SPOJUS; an n-gram based large vocabulary continuous speech recognizer, and a CFG (Context Free Grammar) based one, of which we used the latter in our system. The grammar can generate filled pauses and some irregular sentences / fragments for the robustness.

2.3 Dialog Management

Figure 1 gives details of the dialog manager, which consists of five sub-components ("Information collection", "Feature extraction", "Response timing generator", "Response generator", and "History manager"), and which generates response sentences using the hypotheses and prosodic information.

The recognition results and intermediate hypotheses output by SPOJUS are sent to the information collection component, which saves the information in information slots. The slot information is sent to the response generator, which generates responses using the information. The system generates multiple patterns of responses simultaneously and the decision tree selects the most appropriate response in real-time.

Information Collection. The necessary information is extracted from the ASR result and stored in the slot. The slot value is used for response generation which is possible to consider the context. Examples of values stored in the slot are shown in Table 1.

Table 1. Examples of slot and values

Slot name	Food	Travel
the user's favorite one	*udon*	okinawa
the user's favorite kind / location	*miso*	naha
favorite ingredient / local dish	deep-fried *tofu*	bitter melon
reason why he/she likes it	delicious	beautiful ocean
reason why the other is disliked	unhealthy	too cold

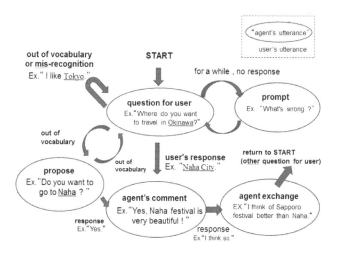

Fig. 2. State transitions in a three person dialog

Response Generator. Template matching is used to generate responses in the proposed system. By comparing the speech recognition result with the response templates, a response sentence is prepared based on the matched one. Furthermore, a response sentence that considers the dialog context can be generated by using slot information. As a response strategy, a conversation that considers the context is possible by defining a subtask (sub-scenario).

Fig. 2 shows the state transition of the three person spoken dialog system with two agents used in this study. Speech production is carried out in the system according to the state transitions. In the figure, encircled utterances denote utterances by agents, while those depicted without circles denote user utterances. In our system, the dialog begins with a question posed to the user in the start state, "question for user". If the system does not receive any response from the user, it prompts the user to respond. If the user's utterance contains unknown words or does not match a rule defined by the system,the agent provides an example that the user can talk about. If the utterance matches a rule, the agent comments on the utterance, and the system then switches between the current agent and the other one. After the change, the dialog state returns to the start state and the dialog is repeated. In a two person dialog system, one agent comments twice on a user's utterance instead of the agent being exchanged in order to convey the same information as in the three person dialog system. Both agents are prevented from uttering the same content continuously through the use of information slots. The slot values determine which agent speaks to user in the three person dialog system.

Response Timing Generator. Previously, we proposed a decision tree-based response timing generator [1], but this was only able to produce a response after detecting the pause (at the end of the user utterance). We have modified this method to enable it to generate overlapping responses by scanning all segments (each segment length is 100 ms) continuously while the user is speaking.

2.4 Output Component

In the output component, each agent is displayed on separate screens by using TVML (TV program Making Language) [11]. The agent's output speech is also output from two separate loud-speakers and we use a text to speech synthesized voice (GalateaTalk [12] or OpenJTalk [13]). The three person dialog system consists of male and female agents, the two person dialog system's agent consists of a male agent(Food) or female agent(Travel) only.

2.5 Construction of a Two Person Dialog System from a Three Person Dialog System

We developed a two person dialog system (one user and one agent) by removing one agent from a three person dialog system (one user and two agents) and having one agent fill the role of two agents. The two person dialog system uses the same speech recognizer, grammar, vocabulary, and templates as the three person system. So, in the three person dialog system, each agent recommends his/her favorite one, *udon*(Hokkaido) or *ramen*(Okinawa), to user. On the other hand, in the two person dialog system, agent recommend both to user.

3 Experimental Results

3.1 Setup

Subjects in the experiment consisted of twenty males (Food) and twelve males (Travel) in their twenties. Each subject evaluated both the three person and two person dialog systems by interacting with them. Subjects first viewed a video about the systems, and then used the dialog systems for a few minutes to become familiar with how to use them. We told the subjects that they had to talk with agents as long as possible until we signaled. Thereafter, each subject interacted with both dialog systems for about 5 minutes, and then stopped talking. After using both systems, subjects completed a survey questionnaire. Half the subjects used the two systems in reverse order. The questionnaire included the following questions:

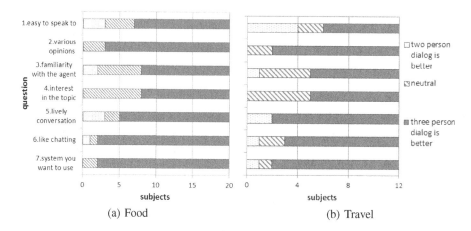

Fig. 3. Relative evaluation: "Two person dialog is better" represents those who gave a 1 or 2 point answer, while "three person dialog is better" represents those who gave a 4 or 5 point answer to the question. Neutral subjects were those who gave a 3 as their answer to a question.

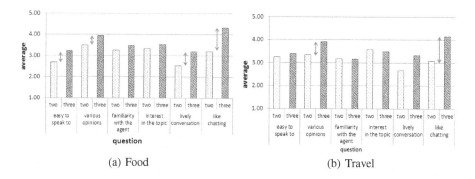

Fig. 4. Absolute evaluation: average

3.2   Subjective Evaluation

Relative Evaluation. Each subject evaluated the two and three person dialog systems using an relative evaluation scale ranging from (two dialog system) 1–5 (three dialog system) for questions such as "Which system is easier to interact with ?" Answers to the survey questions are summarized in Fig.3 . Based on the answers to questions 2, and 5–7, most subjects preferred the three person dialog system(z-test, two-sided, $p < 0.05$) for both domains.

Regarding questions 2 , seventeen of the twenty subjects (Food) and ten of the twelve subjects (Travel), respectively, preferred the three person dialog system. With regard to question 6, eighteen of the twenty subjects (Food) chose the three person dialog system; an example response was: "the conversation with the two person dialog system feels like a question-answering system".

Table 2. Speech recognition performance (words correct) and frequency of dialog phenomena in two and three person systems (Travel Domain)

speaker	Correct [%]		OOV [%]		dialog duration		# user turns		# system turns	
	two	three	two	three	two	three	two	three	two	three
1	57.3	71.3	1.6	0.0	5'00"	4'49"	45	41	61	58
2	65.3	62.6	2.0	2.2	5'04"	5'48"	47	51	65	70
3	72.3	47.1	1.1	3.7	4'55"	4'46"	44	45	55	58
4	61.2	56.8	3.8	2.3	5'11"	4'48"	43	41	58	45
9	42.6	50.9	14.6	8.4	5'14"	5'09"	49	50	72	78
10	43.3	49.5	1.0	2.0	5'19"	5'05"	56	52	80	70
11	42.8	50.0	4.9	7.5	6'18"	4'59"	53	38	79	58
12	26.5	27.2	5.7	3.0	5'46"	5'54"	60	60	74	65
average	53.1	52.9	3.9	3.6	5'14"	5'14"	47.3	47.0	65.3	62.0
correlation with Correct	-0.49	-0.38					-0.75	-0.66		

Absolute Evaluation. In addition to the relative evaluation, each subject evaluated the two and three person dialog systems using an absolute evaluation scale ranging from (disagree) 1–5 (agree) for questions such as " Is it easy to talk to the agent(s)?" Answers to the survey questions are given in Fig. 4 . Responses to all the questions with respect to the three person dialog system were rated more highly than those for the two person dialog system, especially the evaluation of "easy to speak to"(T-test, $p < 0.1$, Food), "various opinions(Food,Travel)", "lively conversation(Food)" and "like chatting"(each $p < 0.05$, Food,Travel) . Thus, the results of the experiments show that the three person dialog system was rated more highly in terms of ease of conversation and users can talk with the agents more like chatting.

3.3 Objective Evaluation

As an objective evaluation, Table 2 shows a part of the automatic speech recognition (ASR) performance (Word Correct), Out Of Vocabulary rate (OOV), and frequency of dialog phenomena, that is, for only typical 8 speakers(users) out of 12 speakers(Travel domain). Speakers 1-4 have the best 4 ASR performances (Correct) and speakers 9-12 have the worst 4 ASR performances (Correct). Included in the system's turn is *aizuchi*. All the dialogs comprised about 100 turns over five minutes. Regarding the correlation between ASR performance and the OOV (two, three) indicates a significant correlation.

Interestingly, in all speakers, regarding the correlation between ASR performance (Correct) and "like chatting" indicates a significant correlation 0.40 (Food) and 0.51 (Travel) in the two person dialog system in absolute evaluation and 0.13(Food) and -0.21(Travel) in the three person dialog system. On the other hand, "like chatting" of absolute evaluation is a higher evalutaion in the three person dialog system than the two person dialog system as shown in Fig. 4 . So, the subjects felt like that the conversation with the three person dialog system is chat, independent of ASR' performance.

4   Conclusion

In this paper, a spoken dialog system consisting of one user and one agent was extended to a three person conversation system with two agents. Both systems were compared in terms of user behavior and satisfaction through two different domains. Based on the results of the experiments, the three person dialog system achieved better results in terms of "familiarity with the agent", "interest in the topic", especially, "easy to speak to" , "various opinion", "lively conversation" and "like chatting". In future work, we intend to compare synthesized speech with recorded voice with regard to the response speech.

References

1. Nishimura, R., Nakagawa, S.: Response timing generation and response type selection for a spontaneous spoken dialog system. In: Proceedings of 2009 IEEE Workshop on ASRU 2009, pp. 462–467 (2009)
2. Itoh, T., Kitaoka, N., Nishimura, R.: Subjective experiments on influence of response timing in spoken dialogues. In: Proceedings of the Interspeech 2009, pp. 1835–1838 (2009)
3. Dielmann: DBN Based Joint Dialogue Act Recognition of Multiparty Meetings. In: Proceedings of ICASSP 2007, pp. 133–136 (2007)
4. Shriberg, E., Stolcke, A., Baron, D.: Observations on Overlap: Findings and Implications for Automatic Processing of Multi-Party Conversation. In: Proceedings of the Interspeech 2009, pp. 1359–1362 (2009)
5. Klotz, D., et al.: Engagement-based Multi-party Dialog with a Humanoid Robot. In: SIGDIAL Conference 2011, pp. 341–343 (2011)
6. Fujie, S., Kobayashi, T., et al.: Conversation Robot Participating in and Activating a Group Communication. In: Proceedings of the Interspeech 2009, pp. 264–267 (2009)
7. Swartout, W., et al.: Ada and Grace: Toward Realistic and Engaging Virtual Museum Guides. In: Allbeck, J., Badler, N., Bickmore, T., Pelachaud, C., Safonova, A. (eds.) IVA 2010. LNCS (LNAI), vol. 6356, pp. 286–300. Springer, Heidelberg (2010)
8. Traum, D., Marsella, S.C., Gratch, J., Lee, J., Hartholt, A.: Multi-party, Multi-issue, Multi-strategy Negotiation for Multi-modal Virtual Agents. In: Prendinger, H., Lester, J.C., Ishizuka, M. (eds.) IVA 2008. LNCS (LNAI), vol. 5208, pp. 117–130. Springer, Heidelberg (2008)
9. Dohsaka, K., Asai, R.: Effects of Conversational Agents on Human Communication in Thought-Evoking Multi-Party Dialogues. In: SIGDIAL, pp. 217–224 (2009)
10. Kai, A., Nakagawa, S.: A frame-synchronous continuous speech recognition algorithm using a top-down parsing of context-free grammar. In: ICSLP, pp. 257–260 (1992)
11. TVML, http://www.nhk.or.jp/strl/tvml/
12. Kawamoto, S., Shimodaira, H., Sagayama, S.: Open-source software for developing anthropomorphic spoken dialog agent. In: Proc. of PRICAI 2002, International Workshop on Life-like Animated Agents, pp. 64–69 (2002)
13. Open JTalk, http://open-jtalk.sourceforge.net/

Distant Supervision Learning of DBPedia Relations

Marcin Zając[1,2] and Adam Przepiórkowski[2]

[1] University of Warsaw
Krakowskie Przedmieście 26/28, 00-927 Warsaw, Poland
marcin.zajac@students.mimuw.edu.pl
[2] Institute of Computer Science, Polish Academy of Sciences
Jana Kazimierza 5, 01-248 Warsaw, Poland
adamp@ipipan.waw.pl

Abstract. This paper presents DBPediaExtender, an information extraction system that aims at extending an existing ontology of geographical entities by extracting information from text. The system uses distant supervision learning – the training data is constructed on the basis of matches between values from infoboxes (taken from the Polish DBPedia) and Wikipedia articles. For every relevant relation, a sentence classifier and a value extractor are trained; the sentence classifier selects sentences expressing a given relation and the value extractor extracts values from selected sentences. The results of manual evaluation for several selected relations are reported.

Keywords: information extraction, distant supervision learning, ontology construction, DBPedia, Wikipedia, Semantic Web.

1 Introduction

1.1 Wikipedia

Wikipedia is a free and multilingual Internet encyclopedia edited by thousands of users. Its Polish version has almost million articles, making it the eighth-largest Wikipedia edition overall.

Due to its broad scope and high quality, Wikipedia is a useful source of information. However, because it supports only keyword-based search, it does not allow the user to ask more sophisticated queries, for example to return all European countries with more than a million inhabitants or to return all geographic entities in a given radius from a specified location. Although all necessary information to answer such queries is present in Wikipedia, it is distributed among separate articles making it hard to compile the answer.

Wikipedia is also a useful resource in information extraction – not only has it broad scope, but it is also relatively homogenous – its articles are written in similar style. Moreover, Wikipedia contains a lot of useful metadata, including links to other articles and links to articles in other languages.

Not all information present in Wikipedia has the form of free text – some of it is contained in infoboxes. An infobox is a list of <attribute, value> pairs describing the most important facts about an entity. For example, an infobox for a country would

I. Habernal and V. Matousek (Eds.): TSD 2013, LNAI 8082, pp. 193–200, 2013.

Fig. 1. An illustration of the aim of the developed system. The article has an infobox, but the value of population is absent from it. However, the fact that the city has 234 inhabitants is expressed in the text. The system is expected to be able to extract this information and create a triple <"Nason, Illinois", population, 234>.

contain, among others, information about its capital, population, area and currency. An example of an infobox is given in Fig. 1.

1.2   DBPedia

DBPedia is a free knowledge base, created by processing Wikipedia infoboxes. DBPedia uses the Resource Description Framework (RDF) to represent the extracted information and the SPARQL query language to enable querying the data. As of March 2013, the Polish version of the ontology contains 9 milion triples describing 800 thousand entities, out of which 2 million describe almost 200 thousand geographic entities.

Building an ontology from infoboxes is a difficult task, because the same concept may be expressed using different names, e.g., *population* and *number of inhabitants*. DBPedia, with help from contributors, developed a mapping from different infobox properties into an ontology, which helps reduce synonyms into single concepts. The extraction algorithm is described in detail in [1].

1.3   Goal

Many Wikipedia articles do not have infoboxes and existing infoboxes are often incomplete, which means that DBPedia contains only a fraction of information contained in Wikipedia.

The practical goal of the current work is to extend the Polish DBPedia ontology of geographic relations by extracting semantic relations from the free text of Polish Wikipedia. Given this task, any extracted information should be reliable, even if it is not complete; in other words, we concentrate on achieving high precision, even if it means reducing recall.

1.4   Related Work

The general distant supervision method of constructing training data adopted in this paper was first presented in [2]. The authors developed Kylin, a system that creates new infoboxes or completes existing ones by extracting information from Wikipedia text. They evaluated their results on a few types of infoboxes. They achieved precision ranging from 74% to 97% and recall ranging from 60% to 96%. In [3] they showed that they managed to improve their results by using dependency parsing.

A later paper, [4], describes a system called iPopulator, which automatically populates infoboxes of Wikipedia articles. Like Kylin, it was tested on whole infoboxes, achieving a precision of 91% and recall of 66%.

In contrast, the system presented in this paper was evaluated on selected relations, enabling us to analyze in detail differences in performance among them. Unfortunately, the different approach in evaluation makes it difficult to compare results achieved by our system with those reported in papers cited above.

Methods used by more recent work ([5], [6]) are more general, performing extraction on corpora unrelated to the knowledge base. However, the approach presented in this paper, where the text corpus used is tighty aligned with the knowledge base, allows for higher precision.

All referenced works processed the English version of Wikipedia. The authors of this paper do not know any use of similar approach for other languages, in particular for Polish.

2   Algorithm

The algorithm used by the system works in three phases, independently for every relation. At first the system uses distant supervision learning to construct training data. Then, using the created data, two models are trained: a sentence classifier and a value extractor. Finally, the trained models are used to extract information from Wikipedia articles. The architecture of the system is presented in Fig. 2.

2.1   Distant Supervision Learning

Supervised learning of semantic relations requires the existence of a labeled training corpus with annotated relations expressed in text. To the authors' best knowledge no such corpus exists for Polish and a creation of such a corpus, even one of moderate size, would be a very time-consuming task.

Therefore, a distant supervision approach was implemented involving an unlabeled corpus (Wikipedia) and a knowledge base (DBPedia), the operating assumption being that, in a Wikipedia article on a certain subject, any sentence mentioning an object which is related to this subject via a relation defined in DBPedia, will – with high probability – also express this relation between them.

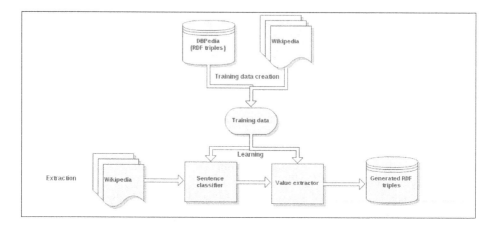

Fig. 2. System's architecture

2.2 Constructing Training Data

Training data is generated separately for every relation. At first all <subject, object> pairs that are in a given relation in DBPedia are retrieved. Then articles about each of the subjects are processed. In the articles, the system looks for sentences that contain the value of the object. If there is a single such sentence in the article, the sentence and the value are simply used as training data. If the value occurs in more than one sentence, only sentences that contain the name of the relation or its synonym are used.

The data constructed this way is very imbalanced because the vast majority of sentences are negative examples. Since the employed classifier is designed to maximize accuracy, it may predict the majority class much more often than the minority class. To avoid this problem, undersampling is performed – randomly selected negative examples are removed, so that in the end there is the same number of positive and negative training examples.

2.3 Wikipedia Articles Processing

The processing of Wikipedia articles consists of a few successive steps. At first the WikipediaExtractor[1] tool is used to convert the articles into plain text. The text is then processed by a part-of-speech tagger for Polish – Pantera ([7]), which also performs tokenization and sentence detection.

In a language with rich morphology, like Polish, proper names inflect for case. Pantera is a state-of-the-art POS tagger, but unfortunately it does not handle proper names very well. It recognizes only a small fraction of geographic names, often performing lemmatization incorrectly,[2] which prevents the extraction of appropriate relations.

---

[1] WikipediaExtractor's website: `medialab.di.unipi.it/wiki/Wikipedia_Extractor`.

[2] This limitation is a reflection of the limited scope of the underlying morphological dictionary ([8,9]).

Table 1. A sentence with the correct lemmatization and a lemmatization suggested by Pantera

tokens	Jeleń – struga na granicy Roztocza Środkowego , prawy dopływ Tanwi
lemmata	Jeleń – struga na granica Roztocze Środkowe , prawy dopływ Tanew
Pantera	jeleń – struga na granica roztocz środkowy , prawy dopływ Tanwi

Table 1. shows a sentence from Wikipedia with the correct lemmatization and an incorrect one suggested by Pantera. It is worth noting that Pantera often treats proper names as though they were common names, even if they start with capital letters in the middle of a sentence (e.g., *Roztocza* is lemmatized to *roztocz*). Also, when Pantera does not know the word, it simply returns the analysed segment as the lemma (see *Tanwi* in the Table).

Because of these deficiencies it was necessary to perform additional processing to correct the lemmatization of proper names. Wikipedia articles, annotated with a markup language, contain other useful information besides plain text, including links to other articles. When a concept or entity is mentioned for the first time and it is the subject of another article, the link to the other article appears in text. The links consist of two parts: the visible one, which is a part of the text, and the invisible one – the name of the concept or entity referenced. Because in Polish Wikipedia the latter is usually a noun group in the nominative case, it is the same as the correct lemma.

Therefore, the following strategy was used to correct the lemmata suggested by Pantera: while extracting plain text from Wikipedia articles, a dictionary of segments and their corresponding lemmata is constructed. For example, upon encountering the segment *Polski* (the genitive of *Polska 'Poland'*) with a link to an article about Poland, the system stores the mapping *Polski ⇒ Polska* and uses this information to correct the possibly erroneous lemmatization proposed by Pantera.

2.4   Sentence Classifier

The sentence classifier tries to predict if a given sentence expresses a given relation. It is a binary, yes/no classifier. The bag-of-words model is used – every sentence is represented as a set of lemmata of words occurring in the sentence. Support Vector Machines (SVMs) are used as the classification algorithm.

2.5   Value Extractor

The value extractor tries to extract a value from a sentence returned by the sentence classifier. This problem can be seen as a tagging problem, where all segments in a sentence should be labeled as positive (indicating that it is a part of the value to extract) or negative. Because some extracted values may consist of multiple continuous segments, the IOB ([10]) tagging method is used. Conditional Random Fields (CRFs, [11]) are a natural choice for this task.

The extractor uses features from a window of 3 tokens to the left and to the right. Experiments showed that increasing the context to 4 or more segments does not improve performance. Following features are used by the extractor:

- the segment
- its base form (lemma)
- some grammatical categories – POS tag, case, number, person and gender
- whether the segment starts with a capital letter
- whether it is a punctuation mark
- whether it is an integer or a numerical value
- whether it is likely a recent year (between 1990 and 2012)

3   Evaluation

3.1   Methodology

For each relation, a separate subcorpus was created by randomly selecting 100 articles, which were read by an annotator whose task was to indicate if, in his opinion, the relation was expressed in the article and, if it was, select its value.

During evaluation, first the system is trained on automatically created training data. In the next step, extraction on articles from the test corpus is performed and its results are compared to the gold standard created by the annotator.

3.2   Results

Table 2. shows evaluation results for all selected relations. For all of them, the system achieved precision exceeding 90%. Recall ranges from over 60% (for just a couple of relations) to almost 100%.

Table 2. Evaluation results

relation name	relation name (in English)	precision	recall	F-measure
populacja	population	98%	94%	96%
stolica	capital	98%	62%	76%
gmina	gmina	98%	86%	92%
powiat	powiat	96%	95%	95%
region	region	99%	82%	89%
prowincja	province	99%	76%	86%
hrabstwo	county	94%	93%	93%
stan	state	100%	95%	97%
gęstość	density	100%	99%	99%
powierzchnia	area	100%	98%	99%
uchodziDo	riverMouth	97%	67%	79%
długośćRzeki	riverLength	100%	95%	96%
powierzchniaDorzecza	drainageBasinArea	98%	88%	93%
średniPrzepływ	averageFlow	96%	96%	96%

3.3   Error Analysis

Two predicates – *capital* and *riverMouth* – achieved relatively low recall, which was caused by the fact that both these relations are expressed in Wikipedia articles using many different phrases, which makes the extraction task more challenging.

Besides that, *capital* is a predicate for which it is most difficult to collect training data, which follows from the fact that names of cities which are capitals appear in many sentences expressing other relations (e.g., being the largest city). To avoid adding such sentences into the training dataset, a heuristic described in section 2.2. was applied – sentences that do not contain the name of the predicate or its synonyms are rejected. Unfortunately this approach lowers recall, because – in the version of the Polish Wordnet used here ([12,13]) – *capital* has no synonyms, whereas in Wikipedia the relation *capital* is often expressed using phrases: *ośrodek administracyjny* (*administrative center*) or *siedziba władz administracyjnych* (*seat of government*).

False positives, which lower precision, occur when system selects a value it should not select. For example from the sentence *W roku 1850 miasto miało 427 mieszkańców* (meaning *In 1850 the city had 427 inhabitants* in English) the system extracted *427* as the current value of population.

4   Conclusions

This paper presents DBPediaExtender – a system that learns new relations about geographic entities. The system uses distant supervision learning – it is trained on automatically generated data based on matches between values from infoboxes (taken from DBPedia) and Wikipedia articles.

Its performance is evaluated on several relations using manually labeled data. For all selected relations the system achieved precision exceeding 90% and precision ranging from 60% to almost 100%.

The final effect of the system is the creation of a database of semantic relations describing geographic entities (mostly populated places and rivers). The database consists of more than 40 thousand RDF triples, which were not present in the Polish DBPedia.

Acknowledgements. This research was funded within CESAR (CEntral and South-east europeAn Resources), a CIP ICT-PSP project (grant agreement 271022).

References

1. Auer, S., Lehmann, J.: What have Innsbruck and Leipzig in common? Extracting semantics from wiki content. In: Franconi, E., Kifer, M., May, W. (eds.) ESWC 2007. LNCS, vol. 4519, pp. 503–517. Springer, Heidelberg (2007)
2. Wu, F., Weld, D.S.: Autonomously semantifying Wikipedia. In: Proceedings of the Sixteenth ACM Conference on Information and Knowledge Management, CIKM 2007, pp. 41–50. ACM, New York (2007)
3. Wu, F., Weld, D.S.: Open information extraction using Wikipedia. In: Proceedings of the 48th Annual Meeting of the Association for Computational Linguistics, ACL 2010, pp. 118–127. Association for Computational Linguistics, Stroudsburg (2010)

4. Lange, D., Böhm, C., Naumann, F.: Extracting structured information from Wikipedia articles to populate infoboxes. In: Proceedings of the 19th ACM International Conference on Information and Knowledge Management, CIKM 2010, pp. 1661–1664. ACM, New York (2010)

5. Mintz, M., Bills, S., Snow, R., Jurafsky, D.: Distant supervision for relation extraction without labeled data. In: Proceedings of the Joint Conference of the 47th Annual Meeting of the ACL and the 4th International Joint Conference on Natural Language Processing of the AFNLP, ACL 2009, vol. 2, pp. 1003–1011. Association for Computational Linguistics, Stroudsburg (2009)

6. Riedel, S., Yao, L., McCallum, A.: Modeling relations and their mentions without labeled text. In: Balcázar, J.L., Bonchi, F., Gionis, A., Sebag, M. (eds.) ECML PKDD 2010, Part III. LNCS, vol. 6323, pp. 148–163. Springer, Heidelberg (2010)

7. Acedański, S.: A morphosyntactic Brill tagger for inflectional languages. In: Loftsson, H., Rögnvaldsson, E., Helgadóttir, S. (eds.) IceTAL 2010. LNCS, vol. 6233, pp. 3–14. Springer, Heidelberg (2010)

8. Woliński, M.: Morfeusz — a practical tool for the morphological analysis of Polish. In: Kłopotek, M.A., Wierzchoń, S.T., Trojanowski, K. (eds.) Intelligent Information Processing and Web Mining. Advances in Soft Computing, pp. 503–512. Springer, Berlin (2006)

9. Saloni, Z., Gruszczyński, W., Woliński, M., Wołosz, R.: Słownik gramatyczny języka polskiego. Wiedza Powszechna, Warsaw (2007)

10. Ramshaw, L.A., Marcus, M.P.: Text chunking using transformation-based learning. In: Proceedings of the ACL Third Workshop on Very Large Corpora, Cambridge, MA, pp. 82–94 (1995)

11. Lafferty, J.D., McCallum, A., Pereira, F.C.N.: Conditional random fields: Probabilistic models for segmenting and labeling sequence data. In: Proceedings of the Eighteenth International Conference on Machine Learning, ICML 2001, pp. 282–289. Morgan Kaufmann Publishers Inc., San Francisco (2001)

12. Piasecki, M., Szpakowicz, S., Broda, B.: A Wordnet from the Ground Up. Oficyna Wydawnicza Politechniki Wrocławskiej, Wrocław (2009)

13. Maziarz, M., Piasecki, M., Szpakowicz, S.: Approaching plWordNet 2.0. In: Proceedings of the 6th Global Wordnet Conference, Matsue, Japan (2012)

Downdating Lexicon and Language Model for Automatic Transcription of Czech Historical Spoken Documents

Josef Chaloupka, Jan Nouza, Petr Červa, and Jiří Málek

SpeechLab, Faculty of Mechatronics, Informatics and Interdisciplinary Studies,
Technical University of Liberec, Studentská 2, 461 17, Liberec, Czech Republic
{josef.chaloupka,jan.nouza,petr.cerva,jiri.malek}@tul.cz

Abstract. This paper deals with the task of adaptation of an existing Czech large-vocabulary speech recognition (LVCSR) system to the language used in previous historical epochs (before 1990). The goal is to fit its lexicon and language model (LM) so that the system could be employed for the automatic transcription of old spoken documents in the Czech Radio archive. The main problem is the lack of texts (in electronic form) from the 1945-1990 period. The only available and large enough source is digitized copies of Rudé Právo, the newspaper of the former Communist party of Czechoslovakia, the actual ruling body in the state. The newspaper has been scanned and converted into text via an OCR software. However, the amount of OCR errors is very high and so we have to apply several text pre-processing techniques to get a corpus suitable for the lexicon and language model 'downdating' (i.e. adaptation to the past). The proposed techniques helped us a) to reduce the number of out-of-vocabulary strings from 8.5 to 6.4 millions, b) to identify 6.7 thousand history-conditioned word candidates to be added to the lexicon and c) to build a more appropriate LM. The adapted LVCSR system was evaluated on broadcast news from 1969-1989 where its word-error-rate decreased from 17.05 to 14.33%.

Keywords: historical speech recognition, oral archives, lexicon.

1 Introduction

A significant progress has been made in the field of automatic speech recognition (ASR) during the last 10 years. Nowadays, large-vocabulary continuous-speech recognition (LVCSR) systems are deployed in various practical applications. One of them is automatic transcription of broadcast (TV and radio) programs. After the transcription, the text, audio and video are indexed, stored in archives, and used for full search [1–3].

A system designed for daily monitoring and transcribing of broadcast news must be regularly updated in order to learn new words, phrases and context. The best, fast and easily available source for these updates are electronic versions of newspapers and other similar text sources on Internet. They can be systematically analyzed to identify out-of-vocabulary (OOV) words and to measure their frequency, which is used for the decision whether they should be included in the updated lexicon. The collected texts serve for upgrading the corresponding language model (LM).

I. Habernal and V. Matousek (Eds.): TSD 2013, LNAI 8082, pp. 201–208, 2013.

When the LVCSR system is to be applied for the transcription of historical broadcast archives, which is the main goal of our recent project [4], the situation becomes more complicated. The main problem is a lack of texts (in electronic form) from previous historical epochs. For example, the first Czech newspapers available both in the printed and electronic format date to early 1990s. There are very few digitized texts from the period before 1990. Another problem is related to the language evolution that happened during the last decades. While the contemporary Czech reflects the plurality of opinions and free use of speech, the official language in former Czechoslovakia (before 1990) was significantly influenced by the communist political regime and its propaganda. The lexicon, in spite of its very large size (500K+ words), and the LM in the current version of the LVCSR are not fitted to that type of language. Therefore we search for another source of text data that could help us in 'downdating' the linguistic part of the transcription system, i.e. in its adaptation to the past. So far, the largest resource found is the scanned copies of Rudé Právo (RP), the official newspaper of the communist party and the government published in the 1945-1989 period. It has been recently digitized via the optical-character recognition (OCR) technology. Unfortunately, the quality of the OCR output is not sufficient for direct use.

In this paper, we describe several methods and techniques applied to the collection of scanned RP newspaper whose main aim is to reduce the amout of OCR errors, identify relevant candidates to be added to the lexicon and to prepare a text corpus applicable for the LM retraining. In the experimental part of the paper, we demonstrate how the adapted linguistic part of the LVCSR system helped to improve the accuracy of transcribed broadcast news from the 1969-1989 era.

2   Text Corpus for Lexicon and LM Downdating

A large part of all the RP newspaper issues has been scanned and optical-character recognized (OCR) for the database of the National library of the Czech Republic, specifically, the issues from 1945 to 1983 years. In total, it is 79293 scanned and OCRed pages. The OCR error ranges from 5 to 20 %.

Since the 1984-1989 issues have not been scanned, yet, we tried to cover this more recent period with another alternative text sources. In our case it was the subtitles made for the TV series Před 25 lety (25-years ago) which replays the official TV news broadcast on the Czechoslovak TV exactly a quarter of century ago. Although, the size of this source is much smaller compared to the RP newspaper, as shown in Table 1.

Table 1. Sources of historical text corpuses from the 1969-1989 era

Text corpus	Epoch	Size [MB]	No. strings	OOVS [%]
RP newspapers	1969-1983	942.16	132,772,115	6.44
25 years ago	1986-1988	4.27	580,940	1.01

3   Post-processing of Newspaper Text Corpus

In order to make the historical RP archive usable, we had to pre-process the digitized data. The texts from the 1969-1983 period had been scanned and OCRed, but only to a limited extent. In the provided text, the OCR was applied only on the character level and no higher level algorithms were employed. The main aim of our pre-processing stage was to reduce the frequency of OCR error in the recognized texts and to identify historically relevant lexical items not yet included in our ever-growing lexicon for automatic broadcast transcription. Many OCR errors in the digitized text corpus were due to poor printing quality of the original material. A possible way to refine the OCRed text was to manually correct each wrong word. The part of the RP archive we are talking about contains 34,700 scanned pages with 132,772,115 strings. It would have been unacceptably time-consuming to review and manually correct the output text from the OCR, therefore, we chose a partly automatic approach.

There exist several dozens of methods and algorithms for automatic OCR text correction. One of them is based on dictionary error correction [5] where the OCRed words are spell-checked and corrected with the use of a lookup dictionary. In the dictionary, pairs of strings are stored, each made of one string for the misspelled word (e.g. hlll) and the other for the corresponding correct (existing) word (e.g. hill). The correction is done by replacing the detected error string with the corresponding correct word. Other methods utilize genetic algorithms [6], Markov Models [7], or most commonly, statistical language models (LM) [8]. The main aim of our project was to identify historical words that were not in our vocabulary and hence that were not covered by our existing LM. That is why we could not use the LM based approach and decided for the dictionary based one. Our scheme for automatic text correction was arranged into five steps. It had been designed with the objective to correct the recognized text without introducing new errors, namely those where a historical word would be replaced by a modern one.

In the first step, we removed some less relevant characters. Hyphenation was deleted completely, while tabulators, brackets and quotation marks were substituted with spaces. Multiple spaces were replaced with single ones.

In the second step, we aimed at removing the blocks of text that often repeated and did not carry historical value. It was mainly advertisements and TV, radio, cinema or theater programs. These were identified by a simple keyword-based classifier. The main reason for this step was to reduce the influence of these specific texts on word and N-gram statistics. It was relatively easy to reliably identify the key words at the beginnings of the deleted parts, because they remained unchanged for many years. The keywords for different volumes of the newspaper were selected manually when randomly chosen issues were analyzed before the automatic text pre-processing.

In the third step, we analyzed the newspaper text with respect to the existing ASR lexicon (containing 551K words). We searched for Out-of-Vocabulary Strings (OOVS), which could be either OCR based error strings [9], misspelled words, real OOV words or historically conditioned terms and names - as shown in Fig. 1. The resulting list of OOVS contained 8.5 million strings (1.8 million different). There were only about 10 % of existing words from the RP newspaper archive on our OOVS list. The rest of the list consisted of error strings from real words and from non-word error strings.

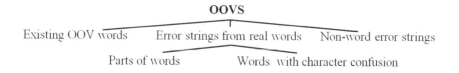

Fig. 1. Taxonomy of Out-Of-Vocabulary strings

The non-word error strings are typically random sequences of characters, e.g. the incorrect string '@lb#k' or alpha-numeric sequence 'B4k12L'. It is often possible to identify existing words in error strings from real words. These error strings can be incomplete words or words with typos where characters have been wrongly recognized. The incomplete words are mainly due to incorrect recognition of space in the OCR process or to malfunction in the pre-processing, if a character has been wrongly identified as a bracket or a quotation mark.

The fourth step was to replace the error strings originated from real words with the correct ones and thus to reduce the number of OCR errors in the texts. To achieve this, the error word strings with high frequency on the OOVS list were selected and saved in the lookup dictionary for OCR error correction. Each error string was paired with an existing word. The lookup dictionary was created as follows: First, we identified the frequent letter confusions in the most frequent error strings, see Table 2.

Table 2. The most frequent letter confusion OCR errors

Confusion err.	Error string	Real word	English meaning
'í' by 'i'	zemědělstvi	zemědělství	agriculture
'í' by 'l'	mlr	mír	Peace
'y' by 'v'	zprávv	zprávy	News
'í' by 'f'	zasedánf	zasedání	Meeting
'i' by 'j'	kapitaljsmus	kapitalismus	Capitalism
'ě' by 'é'	Némecko	Německo	Germany
'ý' by 'y'	bolševicky	bolševický	Bolshevik

Subsequently, an automatic algorithm was developed for the selection of the error word strings. We focused only on the most frequent error word strings with just one letter confusion. It was a relatively safe way to minimize pre-processing errors which could otherwise cause a real OOV word candidate being lost or a word in the text changing its meaning, e.g. the wrongly OCR recognized word 'Madarska' might be 'Mad'arská' (Hungarian - adjective) or 'Mad'arska' (Hungary - noun).

The correction was limited to replacing one letter ('i' by 'í', 'l' by 'í', 'v' by 'y') in a string from the OOVS list. If a new string was found in our 551K lexicon, a new pair of error string and correct word was created and saved in the lookup dictionary. As the result of this step, 13,043 different strings with the total number 1,076,725 occurrences were identified.

Another 505 most frequent error words from the OOVS list (with the total occurrence of 879,729) were found manually and they were added to the lookup dictionary together with their substitute words. They were mainly strings where the diacritics were missing, e.g. 'Strougal' instead of 'Štrougal' (Czechoslovak Prime Minister in 1970-1988), 'CSSR' instead of 'ČSSR' (Czechoslovak Socialist Republic) or where more than one letter has been confused, e.g. 'ostredni' instead of 'ústřední' (central).

Using the final version of the lookup dictionary, we replaced the total number 1,956,454 error strings in the RP archive (1969-1983), as shown in Table 3.

Table 3. Frequency of confusions in corrected text corpus

Confusion	Frequency in OOVS	Frequency in text
'i' by 'í'	8,008	744,341
'l' by 'í'	2,962	137,674
'v' by 'y'	576	94,132
'f' by 'í'	438	60,836
'j' by 'i'	42	16,839
'é' by 'ě'	165	14,676
'y' by 'ý'	347	8,227
Other words	505	879,729
Total number	13,043	1,956,454

In the last fifth step, the OOVS numbers were computed for each page of the text archive. We decided to remove those pages from the text corpus whose OOVS rate exceeded 15 %. This rule was applied to 846 pages 2.4 % of the text corpus.

The above described scheme helped to reduce the list of OOVS from 8.5 million strings to 6.4 million and enabled the creation of historical lexicon and language model.

4   Building Lexicon and LM for 1969-1989 Era

Having the frequency-ordered OOVS list, we could start the selection of word candidates that were to be added to the lexicon. Unfortunately, this task was hard to be automated namely because:

1. Most of the OOVS (about 90 %) were still OCR-based errors that could not be removed in the preprocessing stage.
2. Many OOVS were proper names (mainly names of Czech and foreign politicians from the former Soviet bloc). Most of these names were not commonly known to younger generation and also many included mistakes caused either by wrong spelling or by OCR errors.

We decided to check manually all the items in the OOVS list whose frequency was higher than 20. That part contained about 30 thousands strings. Eventually, we have selected 6,616 word candidates. The majority of them (4,645) were proper names (person names, geographical names and company names). We added also frequent abbreviations, such as 'NDR' (German democratic republic), or 'BKS' (Bulgarian communist party), as well as collocated abbreviations, like 'V KSSS' (Central Bureau of the Communist Party of the Soviet Union), that were often used in official spoken language. The rest of the selected words were mainly regime and propaganda specific words, like 'anti-sovietism', 'comradeship' or 'anti-imperialistic', including words imported from Russian 'druzhba', 'komsomol', 'kolchoz', etc. The lexicon for the 1969-1989 era was

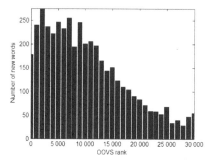 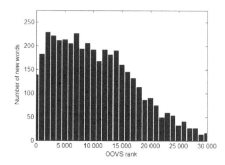

Fig. 2. Numbers of selected words from two OOVS lists split into group of thousands. (The right plot was computed from the corrected RP corpus, the left one from the original.)

made by adding these selected candidates to those words that occurred at least once in the already transcribed broadcast news of the contemporary era (1990-2012) and in the 25-years ago corpus.. Its size was 496K words. The corresponding LM was computed using all the available Czech corpora, i.e. both the broadcast transcripts as well as all newspaper texts, including the RP collection from 1969-1983 years and the 25-years ago texts.

5   Building Lexicon and LM for 1969-1989 Era

The LVCSR system was designed in our lab during the last decade [4]. Recently, it has been used in various applications tasks, the broadcast program transcription being one of them. The system can work with vocabularies larger than 500K words. Its basic structure is shown in Fig. 3.

An input audio signal is preprocessed and parameterized into vectors of 39 Mel-Frequency Cepstral Coefficients (MFCC). Next, the signal is segmented into speech and non-speech parts. The former are further split using a speaker change detector. The decoder employs an acoustic model (AM), a lexicon and a language model (LM). The AM is made of triphone hidden Markov models (HMM) and covers 41 phonemes and 7 types of noise. The Czech AM has been trained on 320 hours of (mainly broadcast) data. The LM is based on bigrams smoothed by the Kneser-Ney algorithm.

6   Evaluation on the Broadcast News from 1969-1989

To evaluate the impact of the adapted lexicon and LM, we prepared a large test set. It consisted in 13 complete radio news randomly picked from the 1969-1989 period.

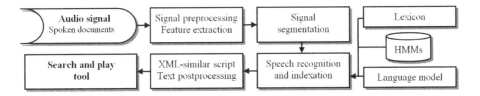

Fig. 3. The principle of broadcast transcription system

These we automatically transcribing and manually corrected. The test set has a total duration of 7.2 hours and contains 47,217 words in total.

In the first experiment, we compared the original lexicon and LM (created on the contemporary text corpora) with that described in section 4. We can see that the downdated lexicon helped to reduce the OOV rate by 0.32 %. The impact of the new LM was even larger as the WER decreased from 17.05 % to 14.72 % when the historic texts were used as they were provided. Another WER reduction to 14.33 % was reached after the application of all the text cleaning and correction techniques described in section 3.

A further significant improvement can be achieved if the LVCSR system runs in the 2-pass mode when speaker and channel adaptation is applied after the first pass. In this case, the WER was reduced to 12.50 %.

Table 4. Comparing different lexicons, language models and LVCSR operation modes on test set made 13 complete broadcast news from 1969-1989 period

Lexicon and language model	WER [%]	OOV[%]
Lexicon and LM of contemporary language	17.05	1.64
Lexicon and LM of adapted on the historic data (RP + 25years ago) - without any text correction	14.72	1.32
Lexicon and LM of adapted on the historic data - with text correction described in section 3.	14.33	1.32
Adapted lexicon and LM, two pass recognition	12.50	1.32

7   Conclusion and Future Work

In this paper, we present a process of adapting the linguistic part of the Czech LVCSR system for transcription of spoken documents in the historical archive of Czech Radio from the 1969-1989 period. The most important part of this adaptation task was to identify the frequent history-conditioned words and add them to the existing lexicon. The second goal was to collect (and correct) a representative corpus of historical texts and utilize it for retraining the statistical language model.

The archive of digitized and optically recognized newspapers Rud Prvo was used as the main source of historical texts. Because they contained many OCR-based errors we had to propose and apply an error correction and OCR pre-processing scheme consisting of several steps. The scheme helped to eliminate a large portion of errors in the text and reduced the size of out-of-vocabulary-string list. The list was used to identify 6,616 word candidates that were added to the existing lexicon.

The impact of the adapted lexicon and language model was evaluated on the test set made of 13 complete radio news from the 1969-1989 period. The WER was reduced from 17.05 to 14.33 %. When the two-pass recognition mode was employed, the WER decreased further to 12.50 %.

Recently, we continue on processing the historical newspapers and preparing the lexicons and language models for the older epochs, namely 1948-1968 and 1923-1948.

Acknowledgments. The research was supported by the Czech Ministry of Culture - project no. DF11P01OVV013 in program NAKI.

References

1. Chen, S.S., Eide, E.M., Gales, M., Gopinath, R.A., Kanevsky, D., Olsen, P.: Recent improvements to IBM's speech recognition system for automatic transcription of broadcast news. In: IEEE International Conference on Acoustics, Speech, and Signal Processing, vol. 1, pp. 37–40 (1999)
2. Gauvain, J.L., Lamel, L., Adda, G.: The LIMSI Broadcast News transcription system. Speech Communication 37(1-2), 89–108 (2002)
3. Chu, S.M., Kuo, H., Liu, Y.Y., Qin, Y., Shi, Q., Zweig, G.: The IBM Mandarin Broadcast Speech Transcription System. In: IEEE International Conference on Acoustics, Speech and Signal Processing, ICASSP 2007, vol. 2, pp. II-345–II-348 (2007)
4. Nouza, J., Blavka, K., Bohac, M., Cerva, P., Zdansky, J., Silovsky, J., Prazak, J.: Voice Technology to Enable Sophisticated Access to Historical Audio Archive of the Czech Radio. In: Grana, C., Cucchiara, R. (eds.) MM4CH 2011. CCIS, vol. 247, pp. 27–38. Springer, Heidelberg (2012)
5. Niwa, H., Kayashima, K., Shimeki, Y.: Postprocessing for Character Recognition Using Keyword Information. In: IAPR Workshop on Machine Vision Applications, Tokyo, pp. 519–522 (1992)
6. Svitak, J.J.: Genetic algorithms for optical character recognition. Doctoral Dissertation, City University of New York, USA (2008) ISBN: 978-0-549-58576-3
7. Guyon, I., Pereira, F.: Design of a Linguistic Postprocessor Using Variable Memory Length Markov Models. In: Proc. 3rd Int. Conf. Document Analysis and Recognition, Montreal, Canada, pp. 454–457 (1995)
8. Smith, R.: Limits on the application of frequency-based language models to ocr. In: IEEE International Conference on Document Analysis and Recognition, pp. 538–542 (2011)
9. Tong, X., Evans, D.A.: A Statistical Approach to Automatic OCR Error Correction in Context. In: Proc. of the Fourth Workshop on Very Large Corpora, pp. 88–100 (1996)

Dynamic Threshold Selection Method
for Multi-label Newspaper Topic Identification

Lucie Skorkovská

University of West Bohemia, Faculty of Applied Sciences, Dept. of Cybernetics
Univerzitní 8, 306 14 Plzeň, Czech Republic
lskorkov@kky.zcu.cz
www.kky.zcu.cz

Abstract. Nowadays, the multi-label classification is increasingly required in modern categorization systems. It is especially essential in the task of newspaper article topics identification. This paper presents a method based on general topic model normalisation for finding a threshold defining the boundary between the "correct" and the "incorrect" topics of a newspaper article. The proposed method is used to improve the topic identification algorithm which is a part of a complex system for acquisition and storing large volumes of text data. The topic identification module uses the Naive Bayes classifier for the multiclass and multi-label classification problem and assigns to each article the topics from a defined quite extensive topic hierarchy - it contains about 450 topics and topic categories. The results of the experiments with the improved topic identification algorithm are presented in this paper.

Keywords: topic identification, multi-label text classification, language modeling, Naive Bayes classification.

1 Introduction

The goal of the text classification (or topic identification) is to categorize a set of documents into predefined set of topic classes or categories. Usually in the field of text classification we are considering only the multiclass classification, where unlike in the binary classification there is more than two possible classes. The simplest task of the text classification is to assign one topic to each document, but real world applications including e-mail routing, web content topical organization or news topic identification require the multi-label classification - each document can belong to more than one topic.

Our topic identification algorithm is a part of a complex system for acquisition and storing large volumes of text data [1]. The system was implemented to gather the training data for the estimation of the parameters of statistical language models for natural language processing (automatic speech recognition, machine translation, etc.). Since it has been shown that not only the size of the training data is important, but also the right scope of the language models training texts is needed [2], the topic identification algorithm is used for large scale language modeling data filtering [3].

Two main approaches to the text classification can be identified - the discriminative techniques like support vector machines(SVMs) [4][5], decision trees [6] and neural

I. Habernal and V. Matousek (Eds.): TSD 2013, LNAI 8082, pp. 209–216, 2013.

networks; and generative techniques like Naive Bayes classifier (NBC) [7][8] and Expectation Maximization based methods.

This paper describes a method based on general topic model normalisation for finding a threshold defining the boundary between the "correct" and the "incorrect" topics of a newspaper article in the generative classification techniques. The generative classifier outputs a distribution of probabilities (or likelihood scores) and a method for processing this distribution into the sets of the "correct" and the "incorrect" topics is needed. The proposed method is used to improve the results of the NBC in the topic identification module.

2   Multi-label Text Classification

The existing methods for multi-label classification can be divided into two main categories - *data transformation (DT) methods* and *algorithm adaptation methods*. The methods of the first group transform the problem into the single-label classification problem and the methods in the second group extend the existing algorithms to handle the multi-label data directly. According to [9] we can divide the existing *data transformation methods*:

First two methods, marked as *DT1* and *DT2*, simply transforms the multi-label data set into single-label [10]. Method DT1 selects only one label from the multiple labels for each data instance and method DT2 discards every multi-label data instance from the set. These methods cannot be really used in a multi-label classification since they remove all the multi-label information from the data set.

The third data transformation method *DT3* considers each set of labels as one label together. The single label classifier then could be used, choosing for each data item one of the predefined sets of labels. The disadvantages of this methods are clear - first, we can end with large number of label sets with only few examples of training data for each set; and second, we cannot assign different combination of labels to the classified data than those previously seen in the training data. This method was used in the works [10][8].

The most common data transformation method *DT4* trains a binary one-vs.-rest classifier for each class. The labels for which the binary classifier yields a positive result are then assigned to the tested data item. The disadvantage of this method is that you have to transform the data set into $|L|$ data sets, where L is the set of possible labels, containing only the positive and negative examples. The second disadvantage is that you have to find the threshold for each binary classifier. This method was used in [4][11][12][5] and also is often used as a baseline for other methods testing [10][8][5].

The *DT5* method decomposes each training data with n labels into n data items each with only one label. One classifier with the distribution of probabilities or likelihoods for all labels is then learned from the transformed data set. The distribution is then processed to find the correct labels of the data item. This approach is used in the work [13] and also in our experiments. The problem of finding the border between correct and incorrect topics is further addressed in Section 2.1.

The last method *DT6* decomposes each training data item into $|L|$ data items each with only one label l and a value $Y(l)$, where $Y(l) = 1$ for the labels which belong to the data item and $Y(l) = -1$ otherwise.

The *algorithm adaptation methods* are methods handling the multi-label data directly or methods that somehow combine one of the DT methods with an existing classification method. For example work [14] uses the adapted C4.5 algorithm; the two extensions of AdaBoost - AdaBoost.MH and AdaBoost.MR were implemented with the combination of DT6 method in the work [6]; in the work [8] the DT3 method is used in combination with the Naive Bayes classifier, the distribution for the sets of label is estimated with the expectation maximization algorithm; the work [5] improves the DT4 method in the combination with SVMs; the adaptation of kNN classifier (ML-kNN) with combination of DT4 method was used in [11].

2.1   Threshold Definition for DT5 Method

As the topic identification module in our system uses a Naive Bayes classification algorithm (the motivation for choosing the NBC is described in Section 3.1) we tried to find out some related work on the problem how to select the set of correct topics from the output distribution of the NBC. A straightforward approach is to select the labels for which the likelihood is greater than a specific threshold (e.g. 0.5) or select a predefined number of topics. In the work [7] the training data with only one label was selected (methods DT1 or DT2) and only the one best label is assigned to each news article, therefore it could not be considered a multi-label classification. In our later work, we selected 3 topics for each article [3]. To our knowledge, the only work concerning the finding of a threshold for choosing the correct topics in the output of a distribution classifier is described in [13]. The classifier used in this work outputs a likelihood distribution of topics for the tested article and the dynamic threshold is set as the mean plus one standard deviation of the topic likelihoods. The assumption is that topics that have a likelihood greater than this threshold are the best choices for the article. The method for finding a threshold proposed in this paper is described in Section 3.1.

3   System for Acquisition and Storing Data

The topic identification module is a part of a system designed for collecting a large text corpus from Internet news servers described in [1]. The system consists of a SQL database and a set of text processing algorithms which use the database as a data storage for the whole system. One of the important features of the system is its modularity - new algorithms can be easily added as modules.

For the topic identification experiments the most important parts of the system are the text preprocessing modules. Each new article is obtained as a HTML page, then the *cleaning* algorithm is applied - it extracts the text and the metadata of the article. Then the *tokenization* and *text normalization* algorithms are applied - text is divided into a sequence of tokens and the non-orthographical symbols (mainly numbers) are substituted with a corresponding full-length form. The tokens of a normalized text are processed with a *vocabulary-based substitution* algorithm. Large vocabularies prepared by experts are used to fix the common typos, replace sequences of tokens with a multi-word or to unify the written form of common terms. *Decapitalization* is also performed - substitutes the capitalized words at the beginning of sentences with the corresponding lower-case variants. The output of each of the preprocessing algorithm is stored as a text record in the database.

Lemmatization has been shown to improve the results when dealing with sparse data in the area of information retrieval [15] and spoken term detection [16] in highly inflected languages, on that account the experiments on the effects of lemmatization in the field of topic identification was performed [17]. As a result of these experiments the automatic *text lemmatization* is also applied in our work. The lemmatization module uses a lemmatizer described in the work [18]. The lemmatizer is automatically created from the data containing the pairs full word form - base word form. A lemmatizer created in this way has been shown to be fully sufficient in the task of information retrieval [18].

3.1   Topic Identification Module

The purpose of the topic identification module in our system is to filter the huge amount of data according to their topics for the future use as the language modeling training data. So far, the topic identification module (which is further described in [3]) used a Naive Bayes based classification algorithm and assigned 3 topics chosen from a hierarchical system - a "topic tree" to each article.

The topic hierarchy built in a form of a topic tree is based on our expert findings in topic distribution in the articles on the Czech favorite news servers like *ČeskéNoviny.cz* or *iDnes.cz*. The topic tree has 32 generic topic categories like `politics` or `sports`, each of this main category has its subcategories, the deepest path in the tree has a length of four nodes. Totally it contains about 450 topics and topic categories, which correspond to the keywords assigned to the articles on the mentioned news servers. The articles with these "originally" assigned topics are used as training texts for the identification algorithm.

Identification Algorithm. Current version of the topic identification module uses a multinomial Naive Bayes classifier (NBC), chosen due to the results of experiments published in [3]. NBC is known to be the fastest learning classifier [5], although having worse accuracy than support vector machines (SVMs), for our task is the best possible choice. As mentioned before, our topic identification runs in a real application. The articles are stored in a database, so the "training" of the identification is done simply by counting the statistics containing the number of occurrences of each word in the whole collection, number of occurrences of each word in each document and the number of occurrences of each word in the documents belonging to a topic.

New articles are downloaded every day and they are instantly processed - the articles which we use as training data since they have the "originally" assigned keywords are used to update the word occurrence statistics tables - as a result, our topic training data update every day. To the rest of the downloaded articles the topic identification module assigns the topics from our topic hierarchy. Every day more than 600 new articles are downloaded to our database and they contain more than 130 new topic training articles, so we had to choose the topic identification algorithm which will be fast and can use the easily updatable statistics stored in the database tables as the trained classifier data. This is why we have chosen to use the NBC over the SVMs.

In the Naive Bayes classifier the probability $P(T|A)$ of an article A belonging to a topic T is computed as

$$P(T|A) \propto P(T) \prod_{t \in A} P(t|T) \tag{1}$$

where $P(T)$ is the prior probability of a topic T and $P(t|T)$ is a conditional probability of a term t given the topic T. The probability is estimated by the maximum likelihood estimate as the relative frequency of the term t in the training articles belonging to the topic T:

$$\hat{P}(t|T) = \frac{tf_{t,T}}{N_T} \tag{2}$$

where $tf_{t,T}$ is the frequency of the term t in T and N_T is the total number of tokens in articles of the topic T. The uniform prior smoothing was used in the estimation of $P(t|T)$.

The goal is to find the most likely or the maximum a posteriori topic (or topics) T of an article A - for each article the topics with the highest probability $P(T|A)$ are chosen:

$$T_{map} = \arg \max_T \hat{P}(T|A) = \arg \max_T \hat{P}(T) \prod_{t \in A} \hat{P}(t|T) . \tag{3}$$

The prior probability of the topic $\hat{P}(T)$ was implemented as the relative frequency of the articles belonging to the topic in the training set, but we found out that it has only small to no effect on the identification results.

General Topic Model Normalisation Method for Finding the Dynamic Threshold.
In our topic identification module we use the combination of the data transformation method DT5 (the article is used as training data for each topic label it has) and the threshold for the selection of the topics to assign to an article. So far we have been selecting the best 3 topics for each article. This is not the best way, because some short articles can concern only one topic, on the other hand some long articles, especially from the politics category often incorporate many other topics. The right way to select the "correct" topics for an article would be setting a dynamic threshold, which should be somehow dependent on the article topic likelihood distribution.

The *General topic model normalisation method (GTMN)* for finding the threshold we propose is inspired by the World model normalisation technique (WMN) used in the speaker recognition task [19][20]. The multinomial NBC is formally equal to the language modeling approach in the information retrieval [21], each topic is described by an unigram language model. In addition to the different topic models, a general topic model is also created as a language model of the whole collection.

First, the NBC classifier is used to output a likelihood topic distribution. Then, the topic likelihood scores $\hat{P}(T|A)$ are normalised with the score of the general model $\hat{P}(G|A)$:

$$\hat{P}(T|A)_{GTMN} = \frac{\hat{P}(T|A)}{\hat{P}(G|A)} \tag{4}$$

Now we have a list of the likelihoods normalised by the general topic model, specifically we have the list of how better the topics describe the article in comparison with the general topic model. We select only the topics which are better scoring than the general topic model and we make the assumption that the topics which have at least 80 percent of the normalised score of the best scoring topic are the "correct" topics to be assigned.

4   Evaluation

In this section the proposed General topic model normalisation method for finding the threshold is compared to the previously used selection of 3 topics for each article and also to selection only one topic as used in [7] and setting the threshold as the mean plus one standard deviation (MpSD) of the topic likelihoods used in the work [13]. For the experiments the smaller collection containing the articles from the news server *ČeskéNoviny.cz* separated from the whole corpus was used [17]. The collection contains 31 419 articles, divided into 27 000 training and 4 419 testing articles.

The evaluation of the result of the multi-label classification requires different metrics than those used in evaluation of single-label classification. We have chosen the metrics somewhat similar to the evaluation used in the field of information retrieval (IR), where each newly downloaded article is considered to be a query in IR and precision and recall is computed for the answer topic set. Similar measures was used in [5] and [10]. For the article set D and the classifier H precision ($P(H, D)$) and recall ($R(H, D)$) is computed:

$$P(H, D) = \frac{1}{|D|} \sum_{i=1}^{|D|} \frac{T_C}{T_A}, \qquad R(H, D) = \frac{1}{|D|} \sum_{i=1}^{|D|} \frac{T_C}{T_R} \tag{5}$$

where T_A is the number of topics assigned to the article, T_C is the number of correctly assigned topics and T_R is the number of relevant reference topics. The $F_1(H, D)$-measure is then computed from the $P(H, D)$ and $R(H, D)$ measures:

$$F_1(H, D) = 2\frac{P(H, D) \cdot R(H, D)}{P(H, D) + R(H, D)}. \tag{6}$$

The results of our experiments are shown in Table 1, from which we can draw following conclusions:

– When choosing only one topic, the precision is quite high, because the first topic is usually correct.

Table 1. Comparison of different threshold finding methods

metric / method(H)	1 topic	3 topics	MpSD	GTMN
$P(H, D)$	0.8123	*0.5859*	0.0554	**0.5916**
$R(H, D)$	0.3191	*0.6155*	0.9611	**0.6992**
$F_1(H, D)$	0.4582	*0.6003*	0.1048	**0.6409**

- The MpSD method achieves high recall, because it selects about 50 topics for each article, on the other hand precision is really low. We believe it is because the method was proposed for the document collection with only 10 topics, unfortunately in our case (450 topics) the method fails.
- The proposed GTMN method achieved the best results and we believe it is more universal than the MpSD method, since, thanks to the general topic model normalisation, the topic set can be of any size.

5   Conclusions and Future Work

The performed experiments with the topic likelihood threshold finding for distribution classifiers suggest that the new proposed General topic model normalisation method for finding the threshold performs better than other previously published tested methods. We have done the same evaluation on a different collection of documents separated from our database and the results were the same. In the future work, we will test the proposed method on other collections with different number of topic categories to confirm the universality of this method.

The advantage of the hierarchical organization of the topics is currently used only for the selection of documents to be used as the training data for the estimation of the parameters of statistical language models for natural language processing. For the future work, we would like to take the advantage of hierarchical topic tree and the relations between the topics also in the topic identification algorithm as described in [5].

Acknowledgments. The work has been supported by the Ministry of Education, Youth and Sports of the Czech Republic project No. LM2010013 and by the University of West Bohemia, project No. SGS-2013-032.

References

1. Švec, J., Hoidekr, J., Soutner, D., Vavruška, J.: Web text data mining for building large scale language modelling corpus. In: Habernal, I., Matoušek, V. (eds.) TSD 2011. LNCS, vol. 6836, pp. 356–363. Springer, Heidelberg (2011)
2. Psutka, J., Ircing, P., Psutka, J.V., Radová, V., Byrne, W., Hajič, J., Mírovský, J., Gustman, S.: Large vocabulary ASR for spontaneous Czech in the MALACH project. In: Proceedings of Eurospeech 2003, Geneva, pp. 1821–1824 (2003)
3. Skorkovská, L., Ircing, P., Pražák, A., Lehečka, J.: Automatic topic identification for large scale language modeling data filtering. In: Habernal, I., Matoušek, V. (eds.) TSD 2011. LNCS, vol. 6836, pp. 64–71. Springer, Heidelberg (2011)

4. Joachims, T.: Text categorization with support vector machines: Learning with many relevant features. In: Nédellec, C., Rouveirol, C. (eds.) ECML 1998. LNCS, vol. 1398, pp. 137–142. Springer, Heidelberg (1998)
5. Godbole, S., Sarawagi, S.: Discriminative methods for multi-labeled classification. In: Dai, H., Srikant, R., Zhang, C. (eds.) PAKDD 2004. LNCS (LNAI), vol. 3056, pp. 22–30. Springer, Heidelberg (2004)
6. Schapire, R.E., Singer, Y.: Boostexter: A boosting-based system for text categorization. In: Machine Learning, pp. 135–168 (2000)
7. Asy'arie, A.D., Pribadi, A.W.: Automatic news articles classification in indonesian language by using naive bayes classifier method. In: Proceedings of the 11th International Conference on Information Integration and Web-based Applications & Services, iiWAS 2009, pp. 658–662. ACM, New York (2009)
8. McCallum, A.K.: Multi-label text classification with a mixture model trained by em. In: AAAI 1999 Workshop on Text Learning (1999)
9. Tsoumakas, G., Katakis, I.: Multi-label classification: An overview. Int. J. Data Warehousing and Mining, 1–13 (2007)
10. Boutell, M.R., Luo, J., Shen, X., Brown, C.M.: Learning multi-label scene classification (2004)
11. Zhang, M.L., Zhou, Z.H.: A k-nearest neighbor based algorithm for multi-label classification. In: 2005 IEEE International Conference on Granular Computing, vol. 2, pp. 718–721 (2005)
12. Yang, Y.: An evaluation of statistical approaches to text categorization. Journal of Information Retrieval 1, 67–88 (1999)
13. Bracewell, D.B., Yan, J., Ren, F., Kuroiwa, S.: Category classification and topic discovery of japanese and english news articles. Electron. Notes Theor. Comput. Sci. 225, 51–65 (2009)
14. Clare, A., King, R.D.: Knowledge discovery in multi-label phenotype data. In: Siebes, A., De Raedt, L. (eds.) PKDD 2001. LNCS (LNAI), vol. 2168, pp. 42–53. Springer, Heidelberg (2001)
15. Ircing, P., Müller, L.: Benefit of Proper Language Processing for Czech Speech Retrieval in the CL-SR Task at CLEF 2006. In: Peters, C., Clough, P., Gey, F.C., Karlgren, J., Magnini, B., Oard, D.W., de Rijke, M., Stempfhuber, M. (eds.) CLEF 2006. LNCS, vol. 4730, pp. 759–765. Springer, Heidelberg (2007)
16. Psutka, J., Švec, J., Psutka, J.V., Vaněk, J., Pražák, A., Šmídl, L., Ircing, P.: System for fast lexical and phonetic spoken term detection in a czech cultural heritage archive. EURASIP J. Audio, Speech and Music Processing (2011)
17. Skorkovská, L.: Application of lemmatization and summarization methods in topic identification module for large scale language modeling data filtering. In: Sojka, P., Horák, A., Kopeček, I., Pala, K. (eds.) TSD 2012. LNCS, vol. 7499, pp. 191–198. Springer, Heidelberg (2012)
18. Kanis, J., Skorkovská, L.: Comparison of different lemmatization approaches through the means of information retrieval performance. In: Sojka, P., Horák, A., Kopeček, I., Pala, K. (eds.) TSD 2010. LNCS, vol. 6231, pp. 93–100. Springer, Heidelberg (2010)
19. Sivakumaran, P., Fortuna, J., Ariyaeeinia, M.A.: Score normalisation applied to open-set, text-independent speaker identification. In: Proceedings of Eurospeech 2003, Geneva, pp. 2669–2672 (2003)
20. Zajíc, Z., Machlica, L., Padrta, A., Vaněk, J., Radová, V.: An expert system in speaker verification task. In: Proceedings of Interspeech, vol. 9, pp. 355–358. International Speech Communication Association, Brisbane (2008)
21. Manning, C.D., Raghavan, P., Schütze, H.: Introduction to Information Retrieval. Cambridge University Press, New York (2008)

Efficiency of Multi-tap Text Entry Method on Interactive Television

Ondřej Poláček, Tomáš Pavlík, and Adam J. Sporka

Faculty of Electrical Engineering, Czech Technical University in Prague, Karlovo nam. 13,
12135 Praha 2, Czech Republic
{polacond,pavlito5,sporkaa}@fel.cvut.cz

Abstract. The paper investigates the effect of response time and keystroke loss ratio on text entry on an Interactive Digital Television system using a remote control. Both parameters were identified based on previous empirical evidence. They are usually caused by poor hardware, software, or the quality of buttons. We conducted an experiment with 15 participants using a Multi-tap text entry method to perform text-copy tasks under different conditions. Entry rate, error rate, and subjective rating were measured in the experiment. The limit values, which have only minimal effect on user performance, are 200 ms response time and 2% keystroke loss ratio. However, these values still cause a significant decrease in subjective user comfort. Based on this observation we determined the maximum response time to 100 ms and the keystroke loss ratio to 1%.

Keywords: Text Entry Methods, Measuring Performance, Text Input, Interactive Digital Television, Remote Control, Multi-tap, User Study.

1 Introduction

Interactive Digital Television (IDTV) introduces new patterns of interaction with television to our every-day lives, such as social TV watching, search content by title, web browsing, accessing social media etc. These patterns require text input to interact with the content of the IDTV. Multi-tap [1] is a frequent method of text entry on the IDTV as it is well-known by the users from mobile phones and requires only a 9-button keypad which is present on most infrared remote controls today. The efficiency of the text input is determined by various attributes of the remote control, known from the empirical evidence: response time to a keystroke, how many keystrokes would not be received by the TV set, type of a keystroke feedback, keypad layout, physical characteristics of individual keys, etc. From these parameters, we selected the response time and keystroke loss ratio as these two parameters received a relatively little interest of the research community. The response time is induced mostly by slow hardware and software used in the IDTV. Keystroke losses are caused by the quality of buttons on the remote control or a poor reception of the infrared signals. Poor performance of these parameters is more a serious problem in text input due to the amount of keystrokes. We present a study describing an effect of the response time and the keystroke loss ratio on the efficiency of the Multi-tap text entry method.

I. Habernal and V. Matousek (Eds.): TSD 2013, LNAI 8082, pp. 217–224, 2013.

Our results show that the maximum response time and the maximum keystroke loss ratio which still do not cause a significant decrease of performance are 200 ms and 2% respectively. Although these values have only minimal effect on the user performance, they significantly decrease the subjective level of user's comfort. Our results indicate that the subjectively acceptable values of the response time and keystroke loss ratio of a remote control do not exceed 100 ms and 1% respectively.

2   Related Work

Probably the most popular text entry method on keyboards with a limited number of keys is the Multi-Tap, commonly available on the mobile telephones. Each key is as-signed three or four characters. To enter a character, the user has to repeatedly press the same key until the desired character appears. The cursor advances upon a prede-fined timeout (user's inactivity). According to MacKenzie et al. [2] the KSPC measure (keystrokes per character) is 2.2. There were several modifications reducing the KSPC: Less-Tap (=1.6) [1], where the sequence of letters on the keyboard was no longer al-phabetical, or LetterWise (=1.15) [3] where the layout of keys is dynamically shifted according to the previously entered text which was on the expense of the clarity of the layout.

The T9 by Tegic Communications [4] is another method intended for use on mobile telephones that can be implemented on a TV remote control. The alphabet is subdivided in the same way as the Multi-tap. The user selects the desired characters by pressing the corresponding keys and after a sequence of keys is entered, the word is disambiguated using a dictionary. Even though T9 is almost optimal in terms of the KSPC (close to 1), many users find it too difficult to use [5].

Using The Numpad Typer (TNT) [6], all characters can be entered by two keystrokes on a 9-button keypad on the remote control. The characters are laid out in two levels of 3×3 grids. The first keystroke selects a group. The second keystroke selects a character in that group. The speed of the keyboard varied from 9.3 up to 17.7 WPM (words per minute). A similar entry method, in which characters are entered by two keystrokes, was patented by Kandogan et al. [7] and later modified by Sporka et al. [8]. This method is designed for a 12-button keypad commonly used on mobile phones. Twist&Tap [9] is designed for text input on a TV using a remote control with accelerometers. Vega-Oliveros et al. [10] explored a multimodal text input on IDTV combining speech input, Multi-tap, and a virtual keyboard.

Iatrino et al. [11] compared Multi-tap and a virtual QWERTY keyboard for text input on IDTV. The virtual QWERTY keyboard was a grid reminiscent of a standard PC keyboard. Arrow keys were used to select a character from this grid. The "OK" button confirmed the selection. In a study with 36 participants, Multi-tap outperformed the virtual QWERTY in terms of speed and user satisfaction. However, a similar study by Geleijnse et al. [12] indicated that the virtual QWERTY keyboard is preferred by the users. The type rate of the virtual QWERTY, Multi-tap, and T9 are similar. Even though a lot of different text entry methods exist on the IDTV, this paper describes a study which aims on the most classic one – the Multi-Tap.

The response time of a general interactive system was first studied by Miller [13]. He estimated the maximum latency of an interactive system to 100 ms. Dabrowski and Munson [14] specified this latency to 150 ms for text entry on a computer keyboard and 195 ms for clicking the mouse. Thus, the perceived latency of an interactive system may depend on the input modality or device.

3   Experiment

The aim of the experiment was to analyze the effect of response time and keystroke loss ratio on the efficiency of Multi-tap text input method on IDTV. The experiment was conducted with 15 unpaid volunteers (3 female, 12 male, mean age = 24.5, std. dev = 1.85). They were all university students recruited through social media. All of them had previous experience with the Multi-tap method (one year and more).

Fig. 1. Screen capture of the experimental software

Fig. 2. Keypad used in the experiment

The experiment was designed as a within-subject 5×5 factorial experiment, with the response time (RT) and keystroke loss ratio (KLR) as factors (independent variables). RT was a within-subject factor (0, 50, 100, 200, 500 ms), and KLR was a within-subject factor (0, 1, 2, 5, 10%). The dependent variables were entry rate, error rate, and subjective rating.

In order to measure the effect, we implemented the Multi-tap method and an interface for copying phrases (see Fig. 1). The software was capable of modeling the response time and keystroke loss ratio. The software was executed on a PC connected to a 40" TV screen with Full HD resolution located in a living room context simulated in a laboratory. Since using a real remote control could increase the response time or induce random keystroke losses, we used a modified numeric keypad, connected over USB (see Fig. 2) as a reliable and responsive input device.

Each experiment session was 45 minutes long. First, the participants were trained in writing for 10 minutes with dummy phrases. Then, they performed a standard text-copy task of English phrases from MacKenzie's phrase set for evaluating text entry methods [15]. All participants were Czech native speakers with good knowledge of English. Writing in foreign language has usually an impact on the type rate. According to Isokoski and Linden [16], the decrease is approximately 16%. This was, however, considered as a constant variable with insignificant influence on the study results. In order to compensate for a learning effect, the order of conditions for both independent variables, as well as the copied phrases, were counterbalanced. After entering a phrase, the participants were asked to rate subjectively the writing conditions. Approximately 50 phrases were copied during one session resulting in two phrases per participant and condition on average. The participants were instructed to type as fast and as accurate as possible. Error corrections were allowed by deleting the last typed character. They were also told to rest at their discretion between phrases (this time was not counted within their performance), but to proceed expeditiously through a phrase once the first character was entered.

The dependent variables were entry rate (words per minute; WPM), error rate (minimum string distance error rate; MSD error rate), and subjective rating. Both, WPM and MSD error rate, are standard measures described in detail in work by Wobbrock [17]. The subjective rating of comfort was measured using a five-point Likert item ("*The writing was comfortable.*"; 1 – Strongly Agree, 2 – Agree, 3 – Neutral, 4 – Disagree, 5 – Strongly Disagree) directly after entering each phrase.

4   Results

The data collected from 15 participants contained 541 copied phrases, approximately 22 phrases for each condition on average. In the following data analysis, the independent variables response time and keystroke loss ratio are abbreviated as RT and KLR. In the rest of the paper, data considered as significant are at p <0.01 level.

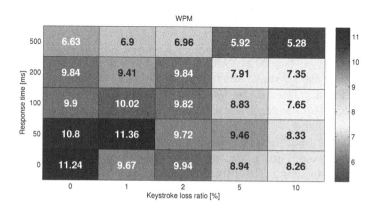

Fig. 3. Mean WPM entry rate for all conditions

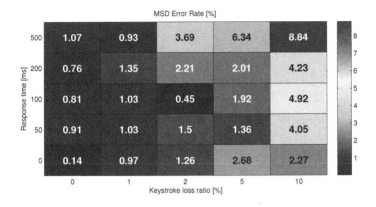

Fig. 4. Mean MSD error rate for all conditions

Entry Rate. The mean WPM entry rate for each condition is depicted in Figure 3. A two-factor ANOVA showed no significant interaction between RT and KLR (F(16,515) = 1.0494, p >0.05). However, a significant main effect was found for RT factor (F(4,532) = 56.7, p <0.001) and KLR (F(4,532) = 26.9, p <0.001). Subsequent post-hoc pairwise comparisons using Tukey's Honestly Significant Differences (HSD) test revealed that 5% and 10% of KLR as well as the 500 ms RT cause a significant decrease of the entry rate. Thus, we may conclude that the efficiency of the Multi-tap method is not significantly affected if the response time is 200 ms or lower, or if no more then 2% of keystroke loses are encountered.

Error Rate. The mean minimum string distance (MSD) error rate for each condition is shown in Figure 4. A two-factor ANOVA showed no significant interaction between RT and KLR (F(16,515) = 1.5434, p >0.05). However, a significant main effect for RT (F(4,532) = 8.45, p<0.001) and KLR (F(4,532) = 18.12, p<0.001) was found for the MSD error rate. The post-hoc pairwise comparisons using Tukey's HSD revealed that 5% and 10% KLR as well as 500 ms RT have a significant effect on the increased MSD error rate. As the MSD error rate express the accuracy after text entry, we may say that such high values of keystroke loss ratio and response time hinder participants from correcting errors.

Subjective Rating. Figure 5 shows a summary of the subjective responses of participants regarding perceived comfort. The median values are shown from one five-point Likert item. Even though Likert data are ordinal, parametric statistic tests (e.g., ANOVA) can be utilized as discussed, for example, in research by Norman [18]. A two-factor ANOVA showed a significant interaction between the RT and the KLR (F(16,516) = 2.113, p <0.01). Thus, we could not split both factors as in the case of the entry and error rates. We analyzed the effect of the KLR separately for each RT level and vice versa using ANOVA and Tukey's HSD for post-hoc comparisons. The effect of KLR for each RT level is shown in Table 1(a). KLR at 5% and 10% were rated as significantly less comfortable in the most cases. KLR at 2% was rated significantly less comfortable

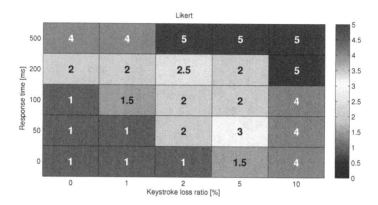

Fig. 5. Median values of subjective responses for all conditions

than 0% at RT of 200 ms. Similarly, the effect of RT was analyzed separately for each KLR level as shown in Table 1(b). The RT of 500 ms was rated significantly less comfortable then other levels of the KLR, while the RT of 200 ms was found significantly less comfortable only on two levels of the KLR factor. Thus, we may conclude that the response time of 500 ms and keystroke loss ratio at 5% and 10% were not comfortable enough for the participants, while response time of 200 ms and keystroke loss ratio at 2% were comfortable to a certain extent. Participants did not perceive any significant difference between lower values of both factors and zero response time and keystroke loss ratio in terms of comfort.

5 Discussion

An optimal remote control should not introduce any response time or keystroke losses. The reality, however, can be sometimes quite far. The aim of the experiment was to find limits for these two parameters in the context of text entry. The results show that the entry rate and error rate are significantly affected by the response time of 500 ms and longer, and by the keystroke loss ratio of 5%. Similarly, participants detected a significant decrease in subjective comfort on response time of 500 ms and keystroke loss ratio of 5%. Less significant decrease in comfort was found for response time of 200 ms and keystroke loss ratio of 2%.

Table 1. Significantly different pairs in subjective rating split by (b) response time (RT) and (a) keystroke loss ratio (KLR). The significantly different pairs are denoted by less than sign.

(a) Significant differences of KLR on all RT levels

RT [ms]	KLR [%]
0	$\{0, 1, 2\} < 10;$
50	$\{0, 1, 2\} < 10; \{0,1\} < 5$
100	$\{0, 1, 2\} < 10;$
200	$\{0, 1, 2, 5\} < 10; 0 < \{2, 5\}$
500	$1 < \{5, 10\}$

(b) Significant differences of RT on all KLR levels

KLR [%]	RT [ms]
0	$\{0, 50, 100, 200\} < 500$
1	$\{0, 50, 100, 200\} < 500$
2	$\{0, 50, 100, 200\} < 500; 0 < 200$
5	$\{0, 50, 100, 200\} < 500$
10	$\{0, 50, 100\} < 500; \{0, 50\} < 200$

Table 2. Levels of response time and keystroke loss ratio examined in the experiment classified into three categories according to their acceptability.

	"acceptable"	"noticeable"	"unacceptable"
response time	0, 50, 100 ms	200 ms	500 ms
keystroke loss ratio	0%, 1%	2%	5%, 10%

Based on these results, we classified the levels of both parameters into three categories (see Table 2): "acceptable", "noticeable", and "unacceptable". The "acceptable" category contains levels of both parameters which affect neither objective nor subjective performance significantly. "Noticeable" are levels that do not affect the objective performance significantly, but cause a significant decrease in subjective performance. "Unacceptable" are levels which affect both objective and subjective performance significantly.

6   Conclusion

In this paper, we analyzed the effect of the response time and the keystroke loss ratio on text entry on Interactive Digital Television. We conducted a controlled experiment with 15 participants, who typed on a model of a remote control using the Multi-tap method. We found that the subjective rating implied stronger performance limits than the objective rating. The limit values were determined to 200 ms response time and 2% keystroke loss ratio when only objective rating was taken into account, and 100 ms response time and 1% keystroke loss ratio when both objective and subjective ratings were taken into account.

The future work will focus on investigating other parameters that influence text input, such as visual, tactile, and audio feedback of keystrokes as well as the method itself. The results of this work will be taken in consideration in our further efforts to develop text input techniques for people with disabilities. Another interesting question is the influence of such parameters on other text entry methods on IDTV systems such as virtual QWERTY, dictionary-based T9 method, or predictive text input.

Acknowledgements. This research has been supported by projects TextAble (LH12070; Ministry of Education, Youth and Sports of the Czech Rep.) and Automatically Generated UIs in Nomadic Applications (SGS10/290/OHK3/3T/13; FIS 10-802900).

References

1. Pavlovych, A., Stuerzlinger, W.: Less-Tap: A fast and easy-to-learn text input technique for phones. In: Graphics Interface, pp. 97–104 (2003)
2. MacKenzie, I.S.: KSPC (Keystrokes per Character) as a Characteristic of Text Entry Techniques. In: Proceedings of the 4th International Symposium on Mobile Human-Computer Interaction, Mobile HCI 2002, pp. 195–210. Springer, London (2002)

3. MacKenzie, I.S., Kober, H., Smith, D., Jones, T., Skepner, E.: LetterWise: prefix-based disambiguation for mobile text input. In: Proceedings of the 14th Annual ACM Symposium on User Interface Software and Technology, UIST 2001, pp. 111–120. ACM, New York (2001)
4. Grover, D.L., King, M.T., Kushler, C.A.: Reduced keyboard disambiguating computer (1998)
5. Gutowitz, H.: Barriers to adoption of dictionary-based text-entry methods: a field study. In: Proc. 2003 EACL Workshop on Language Modeling for Text Entry Methods, TextEntry 2003, pp. 33–41. ACL, Stroudsburg (2003)
6. Ingmarsson, M., Dinka, D., Zhai, S.: TNT: a numeric keypad based text input method. In: Proceedings of the SIGCHI Conference on Human Factors in Computing Systems, CHI 2004, pp. 639–646. ACM, New York (2004)
7. Kandogan, E., Zhai, S.: Two-key input per character text entry apparatus and method (2004)
8. Sporka, A.J., Polacek, O., Slavik, P.: Comparison of two text entry methods on interactive tv. In: Proceedings of the 10th European Conference on Interactive TV and Video, EuroiTV 2012, pp. 49–52. ACM, New York (2012)
9. Aoki, R., Maeda, A., Watanabe, T., Kobayashi, M., Abe, M.: Twist tap: text entry for TV remotes using easy-to-learn wrist motion and key operation. IEEE Transactions on Consumer Electronics 56, 161–168 (2010)
10. Vega-Oliveros, D.A., de Carvalho Pedrosa, D., da Graça Campos Pimentel, M., de Mattos Fortes, R.P.: An approach based on multiple text input modes for interactive digital TV applications. In: Proceedings of the 28th ACM International Conference on Design of Communication, SIGDOC 2010, pp. 191–198. ACM, New York (2010)
11. Iatrino, A., Modeo, S.: Text editing in digital terrestrial television: A comparison of three interfaces. In: Proceedings of EuroITV 2006 (2006)
12. Geleijnse, G., Aliakseyeu, D., Sarroukh, E.: Comparing text entry methods for interactive television applications. In: Proc. 7th European Conf. on European Interactive TV, EuroITV 2009, pp. 145–148. ACM, New York (2009)
13. Miller, R.B.: Response time in man-computer conversational transactions. In: Proceedings of the Fall Joint Computer Conference, Part I, AFIPS 1968 (Fall, Part I), December 9-11, pp. 267–277. ACM, New York (1968)
14. Dabrowski, J.R., Munson, E.V.: Is 100 Milliseconds Too Fast? In: CHI 2001 Extended Abstracts on Human Factors in Computing Systems, CHI EA 2001, pp. 317–318. ACM, New York (2001)
15. MacKenzie, I.S., Soukoreff, R.W.: Phrase sets for evaluating text entry techniques. In: CHI 2003 Extended Abstracts on Human Factors in Computing Systems, CHI EA 2003, pp. 754–755. ACM, New York (2003)
16. Isokoski, P., Linden, T.: Effect of foreign language on text transcription performance: Finns writing english. In: Proceedings of the Third Nordic Conference on Human-Computer Interaction, NordiCHI 2004, pp. 109–112. ACM, New York (2004)
17. Wobbrock, J.: Measures of text entry performance. In: MacKenzie, I.S., Tanaka-Ishii, K. (eds.) Text Entry Systems: Mobility, Accessibility, Universality, pp. 47–74. Morgan Kaufmann (2007)
18. Norman, G.: Likert scales, levels of measurement and the "laws" of statistics. Advances in Health Sciences Education 15, 625–632 (2010)

English Nominal Compound Detection
with Wikipedia-Based Methods[*]

István Nagy T.[1] and Veronika Vincze[2]

[1] University of Szeged, Department of Informatics,
6720 Szeged, Árpád tér 2., Hungary
[2] MTA-SZTE Research Group on Artificial Intelligence,
6720 Szeged, Tisza Lajos krt. 103., Hungary
{nistvan,vinczev}@inf.u-szeged.hu

Abstract. Nominal compounds (NCs) are lexical units that consist of two or more elements that exist on their own, function as a noun and have a special added meaning. Here, we present the results of our experiments on how the growth of Wikipedia added to the performance of our dictionary labeling methods to detecting NCs. We also investigated how the size of an automatically generated silver standard corpus can affect the performance of our machine learning-based method. The results we obtained demonstrate that the bigger the dataset, the better the performance will be.

Keywords: Wikipedia, multiword expressions, nominal compounds, MWE detection, silver standard corpus.

1 Introduction

In natural language processing, multiword expressions (MWEs) have been receiving special interest. Nominal compounds (NCs) form a subtype of multiword expressions: they form one unit the parts of which are meaningful units on their own, the unit functions as a noun and it usually has some extra meaning component compared with the meanings of the original parts [1]. The semantic relation between the parts of the nominal compound may vary: it may express a "made of" relation (*apple juice*), a "location" relation (*neck pain*) or a "made for" relation (*hand cream*) just to name a few. Thus, nominal compounds encode some important meaning components that can be fruitfully applied by e.g. information extraction systems. However, such applications like these require that nominal compounds should be previously known to the system.

Nominal compounds occur frequently in everyday English (in the Wiki50 corpus [2], 67.3% of the sentences on average contain a nominal compound). Furthermore, they are productive: new nominal compounds are entering the language all the time, hence they cannot be exhaustively listed and appropriate methods should be implemented for their identification.

It is also important to emphasize that a nominal compound candidate does not always function as a nominal compound. Take, for instance, *tall boy*: when it refers to a can of

[*] This work was supported in part by the European Union and the European Social Fund through the project FuturICT.hu (grant no.: TÁMOP-4.2.2.C-11/1/KONV-2012-0013).

I. Habernal and V. Matousek (Eds.): TSD 2013, LNAI 8082, pp. 225–232, 2013.

beer, it is an MWE, but when it refers to a young male of somewhat unusual height, it is simply a productive combination of an adjective and a noun and does not constitute an MWE. Thus, nominal compounds should be identified in context, i.e. in running texts, and we will follow this approach in our investigations.

2   Related Work

The identification of MWEs or more specifically nominal compounds has been received considerable attention. Bonin et al. [3] use contrastive filtering in extracting multiword terminology (mostly nominal compounds) from scientific, Wikipedia and legal texts: term candidates are ranked according to their belonging to the general language or the sub-language of the domain. Caseli et al. [4] use alignment-based techniques to extract multiword expressions from parallel corpora in the pediatrics domain. Nagy T. et al. [5] describe a rule-based method, which heavily relies on morphological information to identify nominal compounds in Wikipedia texts. The machine learning-based tool mwetoolkit is designed to extract MWEs from texts, which is illustrated by extracting English nominal compounds from the Genia and Europarl corpora and from general texts [6,7]. In this paper, we present our experiments to automatically detect English nominal compounds in running texts with Wikipedia-based machine learning methods similar to [8] and investigate how the extension of Wikipedia contributes to the process.

3   Experiments

For the evaluation of our models, we made use of two corpora. First, we used Wiki50 [2], in which several types of multiword expressions (including nominal compounds) and Named Entities were marked. This corpus consists of 50 Wikipedia pages, and contains 2929 occurrences of nominal compounds. We also investigated approaches on the 1000-sentence dataset from the British National Corpus that contains 485 two-part nominal compounds [9]. The dataset includes texts from various domains such as literary work, essays, newspaper articles etc. Statistical data on the corpora can be seen in Table 1.

Table 1. Corpora used for evaluation with the number of tokens of the nominal compounds, based on their length

Corpus	Sentences	Tokens	Nominal Compounds	2	3	4≤
Wiki50	4,350	114,570	2929	2442	386	101
BNC dataset	1,000	21,631	485	436	40	9

3.1   Wikipedia-Based Method for Detecting Nominal Compounds

To identify nominal compounds we used a Wikipedia-based approach similar to Vincze et al. [2]. They collected lowercase n-grams from English Wikipedia links, and automatically filtered the non-English terms, Named Entities and non-nominal compounds.

They combined three methods in the following way: a candidate was marked as a nominal compound if it occurred in the list of n-grams. The second method involved the merging of two possible nominal compounds; namely if A B and B C both occurred in the list, A B C was also accepted as a nominal compound. Third, a nominal compound candidate was marked if it occurred in the list and its Part of Speech (POS)-tag sequence matched one of the previously defined patterns (e.g. `adjective + noun`). POS-tags were determined by the Stanford POS Tagger [10]. Finally, they combined these three methods, and this combined approach proved to be the most successful. This is why we applied this method later on.

Vincze et al. [2] investigated the performance of their Wikipedia-based method only on an actual Wikipedia state. However, we thought it interesting to examine this approach from the beginning of Wikipedia and to investigate how the size of Wikipedia influences the results. Hence, we collected the above mentioned nominal compound list from the actual Wikipedia state of the beginning of each year. The English Wikipedia was launched in 2001, so the first list was collected from the state of 1 January 2002.

3.2 Machine Learning Approaches

In order to automatically identify nominal compounds, we also applied a machine learning-based method [2]. The tool uses the MALLET implementations [11] of the Conditional Random Fields (CRF) classifier [12]. Identifying multiword Named Entities and nominal compounds can be carried out in a similar way as both nominal compounds and multiword Named Entities consist of more than one words. They form one semantic unit and thus, they should be treated as one unit in NLP systems [8]. Therefore the feature set employed was developed on the basis of a general Named Entity feature set, which includes the following categories: **orthographical features:** capitalization, word length, bit information about the word form (contains a digit or not, has an uppercase character inside the word, etc.), character level bi/trigrams, suffixes; **dictionaries** of first names, company types, denominators of locations; **frequency information:** frequency of the token, the ratio of the token's capitalized and lowercase occurrences, the ratio of capitalized and sentence beginning frequencies of the token, which was derived from the Gigaword dataset; **shallow linguistic information:** part of speech; **contextual information:** sentence position, trigger words (the most frequent and unambiguous tokens in a window around the word) from the training database and the word between quotes.

This basic feature set was extended with features adapted to nominal compounds. The **dictionaries** were extended with different nominal compound lists. We collected a nominal compound list from the state of Wikipedia on 1 January 2013 and sorted it according to frequency of occurrence. The components with different frequencies were included in different dictionaries. In addition, the training and test sets of Task 9 of the SemEval 2010 [13] were used as dictionaries. The shallow linguistic features were extended with the **POS-rules**, so if the POS-tag sequence in the text matched one pattern typical of nominal compounds (e.g. `noun − plural noun`), the sequence tags were marked as *true*, otherwise *false*. Furthermore, the **other entities** were also specified in the sentence, like Named Entities (NEs) or Light Verb Constructions (LVCs), which were also used as features. To identify Named Entities, the Stanford Named Entity Recognition (NER) tool was applied [14] and we detected Light Verb Constructions similar to the method described in [5].

We trained the first-order linear chain CRF classifier with the above mentioned feature set and evaluated it on the two corpora in a 10-fold cross-validation setting at the sentence level. We trained the CRF models with the default settings in Mallet for 200 iterations or until convergence was reached. We applied the above mentioned dictionary-based method to automatically generate a silver standard corpus. In this case, the training set consisted of randomly selected Wikipedia pages, which do not contain lists, tables or other structured texts. These documents were not manually annotated, so the dictionary-based nominal compound labeling was treated as the silver standard. The resulting dataset is much bigger than the available manually annotated corpora, but the annotation is less reliable. In this case, we would like to exploit the fact of the big training data with less accurate annotation.

The CRF model was trained on the silver standard dataset with the above presented feature set. We investigated how the size of the automatically labeled training set influenced the performance of CRF. First, we analyzed the results when the training set only consisted of 10 Wikipedia pages. After, we gradually increased the automatically labeled training set with randomly selected Wikipedia pages.

As we used randomly selected Wikipedia pages to train our CRF model, we investigated how the random selection affected the results. We automatically generated ten different training sets. One set consisted of ten thousand randomly selected Wikipedia pages, where dictionary-based labeling was used as the silver standard and a CRF model was trained with the above described feature set.

We also compared the results achieved by the supervised leave-one-document-out model, the model trained on the automatically generated dataset, and the dictionary-based method on the Wiki50 corpus.

4   Results

Table 2 shows the results obtained by the dictionary-based approach, the number of Wikipedia pages and the size of the collected lists, depending on the years and the actual state of Wikipedia.

After the first year, the English Wikipedia only consisted of 13,200 pages, and we were able to extract 5,892 potential nominal compounds from the links and the dictionary-based method, which yielded an F-score of 9.52 on the Wiki50 corpus. At the beginning of 2013, the English Wikipedia consisted of 9,914,544 pages, the potential NC list contains 687,574 elements and the approach achieved an F-score of 56.59. As Table 2 shows, with the expansion of Wikipedia, the method managed to produce better results, but the rate of improvement is negligible after 2007. Moreover, in 2013 the dictionary-based method yielded an F-score that was 0.15 lower than that in 2012.

We also investigated how the training set size affected the results of the model trained on the automatically generated dataset. As Figure 1 shows, with an increased training set the machine learning approach could achieve better results, but the improvement was smaller. The method produced an F-score of 46.69 when the training set just consisted

Table 2. The results of applying the Wikipedia-based dictionary labeling method, depending on the expansion of Wikipedia in terms of recall, precision, and F-score. **WikiPages:** the number of Wikipedia pages. **NC list:** the size of the lists collected from the Wikipedia links.

Year	WikiPages	NC list	Recall	Precision	F-score	Diff.
2002	13,200	5,892	5.12	68.42	9.52	-
2003	124,229	25,431	16.22	59.05	25.45	+15.93
2004	271,160	58,696	24.99	71.69	37.06	+11.61
2005	752,239	120,028	33.81	69.57	45.50	+8.44
2006	1,611,876	211,802	40.11	66.20	49.96	+4.46
2007	2,988,703	322,918	44.42	64.15	52.49	+2.53
2008	4,432,034	405,635	46.91	63.35	53.90	+1.41
2009	5,281,708	459,544	48.51	62.82	54.74	+0.84
2010	6,009,776	511,303	49.33	62.45	55.12	+0.38
2011	7,167,621	567,288	50.69	62.66	56.04	+0.92
2012	9,007,810	640,879	53.36	60.58	56.74	+0.7
2013	9,914,544	687,574	53.67	59.84	56.59	-0.15

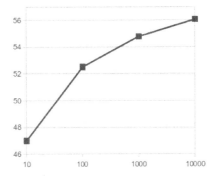

Fig. 1. Results of the machine learning approach depending on the automatically generated training set size (the number of Wikipedia pages)

of 10 Wikipedia pages and an F-score of 56.06 when it was constructed from 10,000 Wikipedia pages.

Table 3 lists ten different CRF model results, trained on ten different automatically generated datasets. The average F-score of ten runs was 55.99 and the standard deviation was 0.3237. Table 4 gives the results of the different approaches for the Wiki50 and BNC datasets.

To perform an error analysis, we examined the length of nominal compounds in the corpora. As Table 1 shows, the Wiki50 corpus contains 83.37% (2442 occurrences) two-part, 13.17% (386 occurrences) three-part nominal compounds and only 3.46% (101 occurrences) are longer than three tokens. As for the BNC dataset, there are 436 (89.89%) two-part nominal compounds, and only 8.25% (40 occurrences) are three-part, while 1.86% (9 occurrences) contain more than three tokens. Table 4 shows that all the methods got their best results on the two-part nominal compounds. Longer nominal compounds yielded worse results in the case of each method and corpus.

Table 3. Machine learning results obtained on different automatically generated training sets in terms of recall, precision, and F-score in Wiki50

	Recall	Precision	F-score
1	57.02	55.21	56.1
2	56.74	55.38	56.05
3	57.26	55.73	56.48
4	56.64	55.02	55.82
5	57.46	55.25	56.33
6	56.88	55.61	56.24
7	56.98	55.03	55.99
8	56.2	54.94	55.56
9	57.08	53.73	55.36
10	56.85	55.04	55.93
avg.:	56.91	55.1	55.99

Table 4. Results of different methods for nominal compounds in terms of recall, precision, and F-score in Wiki50 corpus and BNC dataset. **LOO:** evaluated in the leave-one-document-out scheme. **WikiTrain:** CRF model trained on the automatically generated dataset. **Dict:** Wikipedia-based dictionary labeling.

	LOO Wiki50	WikiTrain Wiki50	Dict. Wiki50	Wikitrain BNC	Dict. BNC
2	69.12/79.62/74.00	64.86/60.14/62.41	61.14/64.66/62.85	40.60/45.04/42.70	33.49/45.06/38.42
3	52.33/62.93/57.14	29.02/47.86/36.13	30.05/49.79/37.48	20.00/22.86/21.33	17.50/17.95/17.72
4≤	24.73/45.10/31.94	8.60/40.00/14.16	6.45/75.00/11.88	0.00/0.00/0.00	0.00/0.00/0.00
All	64.39/72.40/68.16	56.57/55.57/56.06	53.67/59.84/56.59	38.02/41.53/39.70	31.40/40.75/35.47

Table 4 also reveals that on the Wiki50 corpus the CRF model evaluated with the leave-one-document-out scheme yielded the best results with an F-score of 68.16. The CRF model trained on the automatically generated dataset and the Wikipedia-based dictionary labeling method achieved the same F-score on the Wiki50 corpus with different recall and precision scores. The machine learning-based method yielded a higher recall with a lower precision. Moreover, this approach yielded an F-score that was 4.23 higher on the BNC dataset than the dictionary labeling method.

5 Discussion

Due to the dynamic expansion of Wikipedia, the dictionary-based method was able to extract bigger potential nominal compound lists from Wikipedia links and achieved better recall scores for each year. At the same time, while the automatically extracted list was noisy, the precision score continuously decreased over the years, but the F-score value increased up to 2013. Then in 2013 the rise in recall was less than the decrease in precision, hence the F-score value was lower for 2013 than that for 2012. As Table 2 shows after 2009 the F-score improvement was less than 1. However, we found that the dynamic expansion of Wikipedia had a positive effect on the recall score, so in order to improve the precision score, we should define stronger rules.

Next, we evaluated the machine learning-based model with the leave-one-document-out scheme on the Wiki50 corpus. This approach achieved the highest F-score value since we used a supervised model here. As we applied a silver standard corpus, we had a less accurate but much bigger training dataset where the automatic (therefore noisy) labeling was used as a silver standard. This method had a detrimental effect on the precision scores for the CRF model, but recall scores improved because the model had access to more labeled nominal compounds. We examined how the training set size influenced the performance of this machine learning-based approach and we found that the size of training data had a large impact on the performance of the method when we exploited the automatically generated training data. We also wanted to see how the random selection of Wikipedia pages affected the performance. The method proved to be sufficiently robust as the standard variation of F-score values was 0.3237.

We also examined the nature of English nominal compounds, and we found that the majority of nominal compounds are two-part and the investigated approaches performed well on the two-part compounds as opposed to longer compounds, which is probably due to the fact that automatically labeled examples contained fewer instances of longer compounds.

On the BNC dataset the machine learning method proved to be more effective than the dictionary-based method. Due to the BNC paper [9] they annotated sequences of two nouns. However, we found 40 three-part, and 9 longer nominal compounds too in the data. On the other hand, some of the errors are related to annotation errors, for instance, marking nominal compounds that contain a proper noun, e.g. *Belfast primary school headmaster*, as simple nominal compounds instead of proper nouns (as they should be according to the guidelines). These differences can be responsible for the weaker performance of our methods on the BNC dataset.

6 Conclusions

Here, we examined dictionary and machine learning-based methods for identifying nominal compounds in two corpora. These approaches made intensive use of Wikipedia data. The dictionary-based approach applied a list automatically collected from Wikipedia. We examined the results of this method that depended on the expansion of Wikipedia over the years. We found that the growth of Wikipedia improved the performance, especially the recall score, but the rate of improvement is decreased over time. We also looked at the effectiveness of the machine learning-based method when it was trained on an automatically generated silver standard corpus and we demonstrated that this approach can also provide acceptable results. In the future, we would like to improve the precision of automatic labeling as this will have a positive effect on the performance of both the machine learning approach and the dictionary labeling method.

References

1. Sag, I.A., Baldwin, T., Bond, F., Copestake, A., Flickinger, D.: Multiword Expressions: A Pain in the Neck for NLP. In: Gelbukh, A. (ed.) CICLing 2002. LNCS, vol. 2276, pp. 1–15. Springer, Heidelberg (2002)
2. Vincze, V., Nagy T., I., Berend, G.: Multiword expressions and named entities in the Wiki50 corpus. In: Proceedings of RANLP 2011, Hissar, Bulgaria (2011)

3. Bonin, F., Dell'Orletta, F., Venturi, G., Montemagni, S.: Contrastive filtering of domain-specific multi-word terms from different types of corpora. In: Proceedings of the 2010 Workshop on Multiword Expressions: from Theory to Applications, pp. 77–80. Coling 2010 Organizing Committee, Beijing (2010)

4. de Medeiros Caseli, H., Villavicencio, A., Machado, A., Finatto, M.J.: Statistically-driven alignment-based multiword expression identification for technical domains. In: Proceedings of the Workshop on Multiword Expressions: Identification, Interpretation, Disambiguation and Applications, pp. 1–8. ACL, Singapore (2009)

5. Nagy T., I., Vincze, V., Berend, G.: Domain-dependent identification of multiword expressions. In: Proceedings of the International Conference on Recent Advances in Natural Language Processing 2011, pp. 622–627. RANLP 2011 Organising Committee, Hissar (2011)

6. Ramisch, C., Villavicencio, A., Boitet, C.: mwetoolkit: a framework for multiword expression identification. In: Proceedings of LREC 2010. ELRA, Valletta (2010)

7. Ramisch, C., Villavicencio, A., Boitet, C.: Web-based and combined language models: A case study on noun compound identification. In: Coling 2010: Posters, Beijing, China, pp. 1041–1049 (2010)

8. Nagy T., I., Berend, G., Vincze, V.: Noun compound and named entity recognition and their usability in keyphrase extraction. In: Proceedings of the International Conference on Recent Advances in Natural Language Processing 2011, pp. 162–169. RANLP 2011 Organising Committee, Hissar (2011)

9. Nicholson, J., Baldwin, T.: Interpreting Compound Nominalisations. In: LREC 2008 Workshop: Towards a Shared Task for Multiword Expressions (MWE 2008), Marrakech, Morocco, pp. 43–45 (2008)

10. Toutanova, K., Manning, C.D.: Enriching the knowledge sources used in a maximum entropy part-of-speech tagger. In: Proceedings of EMNLP 2000, pp. 63–70. ACL, Stroudsburg (2000)

11. McCallum, A.K.: Mallet: A machine learning for language toolkit (2002),
    http://mallet.cs.umass.edu

12. Lafferty, J.D., McCallum, A., Pereira, F.C.N.: Conditional random fields: Probabilistic models for segmenting and labeling sequence data. In: Proceedings of the Eighteenth International Conference on Machine Learning, ICML 2001, pp. 282–289. Morgan Kaufmann Publishers Inc., San Francisco (2001)

13. Erk, K., Strapparava, C. (eds.): Proceedings of the 5th International Workshop on Semantic Evaluation. ACL, Uppsala (2010)

14. Finkel, J.R., Grenager, T., Manning, C.: Incorporating non-local information into information extraction systems by gibbs sampling. In: Proceedings of the 43rd Annual Meeting on Association for Computational Linguistics, ACL 2005, pp. 363–370. Association for Computational Linguistics, Stroudsburg (2005)

Evaluating Voice Quality and Speech Synthesis Using Crowdsourcing

Jeanne Parson[1], Daniela Braga[1], Michael Tjalve[1,2], and Jieun Oh[3]

[1] Microsoft, USA
[2] University of Washington, Seattle, USA
{jeannepa,dbraga,mitjalve}@microsoft.com
[3] CCRMA, Stanford University, Stanford, USA
jieun5@ccrma.stanford.edu

Abstract. One of the key aspects of creating high quality synthetic speech is the validation process. Establishing validation processes that are reliable and scalable is challenging. Today, the maturity of the crowdsourcing infrastructure along with better techniques for validating the data gathered through crowdsourcing have made it possible to perform reliable speech synthesis validation at a larger scale. In this paper, we present a study of voice quality evaluation using the crowdsourcing platform. We investigate voice gender preference across eight locales for three typical TTS scenarios. We also examine to which degree speaker adaptation can carry over certain voice qualities, such as mood, of the target speaker to the adapted TTS. Based on an existing full TTS font, adaptation is carried out on a smaller amount of speech data from a target speaker. Finally, we show how crowdsourcing contributes to objective assessment when dealing with voice preference in voice talent selection.

Keywords: voice quality evaluation, speech synthesis, Text-to-Speech (TTS), crowdsourcing (CS), voice preference, gender preference.

1 Introduction

Online crowdsourcing (CS) marketplaces provide an environment for fast turn-around and cost-effective distributed outsourcing, at a statistically meaningful scale, leveraging human intelligence, judgment, and intuition. Such services are typically used by businesses to clean data, categorize items, moderate content and improve relevancy in search engines. However, over the past several years, the Speech Science community has also adopted them as a novel platform for conducting research that offers a more scalable means of: a) measuring speech intelligibility [1] and naturalness [2], b) of collecting data [3] and c) of processing that same data, either through annotation for natural language tasks [4] or audio transcriptions for automatic speech recognition [5]. For a complete review on the use of CS for speech-related tasks and anticipated challenges for the future of CS for speech processing, see [6-7]. In all the described experiments, we used Microsofts Universal Human Relevance System (UHRS) as the crowdsourcing platform. UHRS is a marketplace that connects a large worker pool with human intelligence tasks. Tasks can be distributed to workers within a specific country. The UHRS workers are

I. Habernal and V. Matousek (Eds.): TSD 2013, LNAI 8082, pp. 233–240, 2013.
© Springer-Verlag Berlin Heidelberg 2013

provided by several vendors across world-wide markets, providing many thousands of unique workers. Through these studies, we demonstrate the potential of CS to obtain subjective evaluation of real and synthesized speech. This paper details how CS can be used to evaluate human voice quality and synthesized speech in the context of text-to-speech (TTS), using subjective ratings from listeners and users. We survey voice gender preference (Section 2), examine the effects of voice adaptation technology on the perception of synthesized speech (Section 3), and evaluate voice preference to select the "best" voice from recordings of several voice talents (Section 4).

2   Surveying Voice Gender Preference

We had two goals with this experiment. One was to understand end users' voice gender preference when using Text-to-Speech (TTS) systems on mobile phones across three different scenarios: 1) instructions and confirmations, 2) read-out of text (SMS and/or tweets), and 3) driving directions, and covering eight locales. The second goal was to ascertain if we could extract reliable data for speech validation using CS. We are aware that there is considerable discussion around the effect of gender vis-a-vis users' preferences in all types of voice response systems [8], however it was not a goal in this experiment to investigate beyond the scope of our scenario.

We first framed the experience with a statement intended to ensure, as much as possible, that the crowd judges understood the meaning of TTS and its general usage on a mobile phone: *"You probably know that many smartphones have the ability to talk back to you. For example, if you ask to "Call Anna", the phone might talk back to you and say "Call Anna, at home or on the mobile phone?" Or the phone might have GPS and could read driving directions to you. For this survey, we will use "TTS" to refer to the phone's ability to talk back to you."*

We then presented the judge with a short series of questions: *1) Do you use a mobile phone that has TTS? 1.a. If "yes", do you use the TTS feature?; 1.b. If your previous answer is "sometimes", when do you use the TTS feature?; 2) If the TTS is confirming actions for you, do you prefer a male or female voice?; 3) If the TTS is reading a text message or tweet to you, do you prefer a male or female voice? 4) If the TTS is reading driving directions, do you prefer a male or female voice?*

We also asked the judges' to briefly describe their impression of TTS voices, and we asked the judges' gender. The text-input responses for question 1b and the follow-up question asking for a brief description of judges' impression of TTS served as a spam filter, and also provided opportunities to validate if judges' who did not have a mobile phone with TTS understood what TTS is. As with all CS, we also knew that it was not possible to control the balance between male and female respondents and that results would need to be normalized for gender.

The total crowd size for 6 of the 8 locales was large enough to provide solid data. For en-CA and it-IT, with less than 29 judges, the text comments validated that the responses from these locales were still highly valuable (the comments from Italian judges were especially robust and insightful). As expected with CS, the gender distribution of

Table 1. Demographics distribution per locale

Locale	Total Crowd	Female	Male
en-US	100	69	31
en-CA	23	14	9
en-GB	54	28	26
de-DE	38	21	17
fr-FR	30	12	18
es-ES	31	6	25
it-IT	18	5	13
es-MX	29	9	20

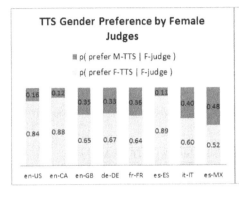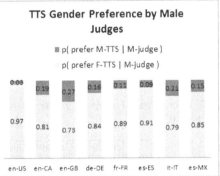

Fig. 1. Gender preference by female (left) and male (right) judges. Light gray denotes preferring female TTS voice, and dark gray denotes preferring male TTS voice.

the judges was uneven (Table 1). When answers to questions 2-4 were aggregated, we observed that both female and male judges in all 8 locales tended to prefer a female voice to a male voice, although the extent to which female voice was preferred was stronger by male judges than by female judges, for all locales except en-CA (Fig. 1).

Judges' responses to questions 2 and 3 varied by less than 2 except for de-DE and es-MX, which varied by 3 and 4 respectively. Because the responses for these two scenarios (confirming actions and reading a text message) were so aligned, we combined the results for both questions. For all locales, there was a strong preference for a female TTS voice in both of these scenarios (Table 2). For driving directions, there was more variance in the gender preference gap for some locales (Table 4), and especially for judges from Mexico. Since Mexico also had the largest gap for questions 2 and 3, it is worth looking at Mexico separately (Table 3).

For the Mexican market, there were only 9 female judges. Only 3 of the female judges preferred a female TTS voice for the driving directions scenario, but 7 preferred female for confirming actions, and 4 preferred female for reading text. These numbers reflect the overall pattern, regardless of the gender of the judge, indicating no preference based on judge's gender.

Table 2. Gender preference for confirming actions and reading text messages

Locale	Total Crowd	% prefer Female
en-US	100	89.0%
en-CA	23	91.1%
en-GB	54	70.3%
de-DE	38	73.7%
fr-FR	30	80.0%
es-ES	31	96.8%
it-IT	18	77.8%
es-MX	29	82.3%

Table 3. Mexico gender preference by scenario

(TTS gender preferred)	Confirm Actions	Read Text	Driving Directions
Female	26	22	17
Male	3	7	12

Table 4. Preference for female gender for confirming actions/reading text [Q2-3] vs. driving directions [Q4]

	en-US	en-CA	en-GB	de-DE	fr-FR	es-ES	it-IT	es-MX
Q2-3	89.0%	91.1%	70.3%	73.7%	80.0%	96.8%	77.8%	82.3%
Q4	85.0%	73.9%	68.5%	78.9%	76.7%	80.6%	72.2%	58.6%

3   Perception of Speaker Mood in Adapted TTS Fonts

3.1   Overview and Methodology

Voice Adaptation is a technique in text-to-speech (TTS) that generates a new voice (target voice) based on the training of a source voice [9]. The adaptation technology takes an existing TTS font and "adapts" it to the voice of a new target speaker based on a smaller quantity of speech data of the new speaker. Thus, we designed a study to better understand the extent to which voice qualities of the human target speaker impact that of the adapted font; our underlying motivation was to estimate the potential of improving certain problematic voice qualities such as friendliness or mood of existing TTS fonts through adaptation. We employed two types of listeners: non-expert crowd workers through UHRS, and language experts who are validated native speakers in the language. This study investigated the impact of adaptation on the perceived voice qualities of a newly generated TTS font in the es-ES locale. Four voices (2 human and their respective synthetic voices) were used: *Helena-human* was used to generate *Helena-TTS*, and *Laura-human* was the target speaker for generating the adapted font, *Laura-TTS*.

We conducted analysis in terms of the following four **voice qualities**: Listener perception on speaker's naturalness of prosody, Listener perception of speakers mood, voice color (timbre) preference by listener, and General preference by listener.

We generated 12 unique tasks using different scripts. In choosing our scripts, we ensured that the script was available in both of the human voice recordings, and that the semantic content of the script was neutral. We deployed the tasks using 1) crowd-sourced non-expert workers and 2) recruited language experts (LEs). For CS, we created two versions of each of the 12 unique HITs (tasks), reversing the order of speech stimuli. We allowed for up to 12 worker responses per HIT-order for a total of 288 HITs available in the marketplace. Similarly, for deployment to LEs, we created two versions of 12 unique surveys (tasks), reversing the order of speech stimuli. Twenty medium-skilled LEs each completed the 12 surveys, resulting in 10 LE responses per order-pattern. The LE surveys were implemented as a web application which matched the format of the CS surveys.

We designed a task consisting of three sets. In Set 1, listeners compared between the two human recordings, *Helena-human* and *Laura-human*. In Set 2, listeners compared between two synthesized speech, *Helena-TTS* and *Laura-TTS*. Set 3, in which listeners compared between *Helena-human* and its derived font *Helena-TTS*, was designed as a catch trial to check on the quality of workers responses, with the assumption that a non-spam response would show a clear preference for the human recording over the synthesized font. For each set, two speech stimuli under comparison were presented, followed by 11 questions (translated into Spanish) addressing the perception of our four voice qualities: prosody, mood, timbre, and general preference.

We hypothesized that adaptation would result in a font (*Laura-TTS*) whose voice qualities match that of the target speaker (*Laura-human*) rather than the parent speaker (*Helena-human*). That is, for a given voice quality, we hypothesized that (1) if a listener prefers *Laura-human* over *Helena-human*, then s/he would prefer *Laura-TTS* over *Helena-TTS*; and inversely, (2) if a listener prefers *Helena-human* over *Laura-human*, then s/he would prefer *Helena-TTS* over *Laura-TTS*.

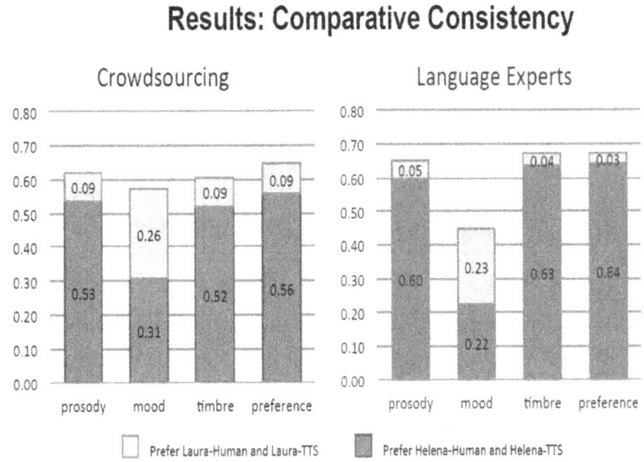

Results: Comparative Consistency

Fig. 2. Results for crowd-sourced workers and language experts

Fig. 2 summarizes the percentage of responses that follow our hypothesis, categorized by four voice qualities. The upper bar, in light gray, represents responses preferring *Laura-human* and *Laura-TTS*; the lower bar, in dark gray, represents responses preferring *Helena-human* and *Helena-TTS*. The sum of two bars represents the percentage of responses that met our hypothesis, which was shown to be greater than 50 percent for all of the voice qualities, with the exception of "mood" ratings by LEs. More interestingly, speaker's mood was shown to be a voice quality most positively impacted by adaptation in our experimental context; judges perceived *Laura-human* and *Laura-TTS* to be in a better "mood", as shown by the height of the light-gray bar for "mood". Though the trained LEs responses are slightly more pronounced, the graphed results clearly show that results from the crowd are consistent with LE results. This validates that crowd-sourcing can be a reliable resource for perception of voice qualities.

4   Evaluating Voice Preference

4.1   From Top 10 to Top 3

The goal of this study was to contribute to a decision to narrow a pool of 8-10 voice talent candidates (voices) down to 2 or 3 finalists. A listening survey was deployed to 100 judges via a customizable tool for audio listening and judgment collection from the crowd. The survey had two main sections: 1) the judge listened to sets of 3 voices and chose a favorite, and 2) the judge chose his/her favorite and least favorite voice overall, and provided comments as to why s/he made that choice. To ensure no bias based on listening order in the first section, comparison sets were pre-mapped on a matrix such that each voice was: 1) heard three times by each judge 2) compared in different orders (e.g., first, middle and last) and 3) compared an equal number of times against each other voice. For all listening samples, we used a recording of the human voice speaking the same sentence. It was not expected that the results of the first section would exactly match the results of the second section because a judge may have had to choose between 3 voices in a set, none of which were their favorite. Judge's comments offered insight into their favorite and least favorite voice.

Table 5. Results of the crowds

Voice	1	2	3	4	5	6	7	8	9	10
Total votes	57	51	109	78	74	114	79	84	93	111
Favorite	6	6	11	9	11	15	3	4	9	11
Least Favorite	13	32	3	6	9	5	5	3	7	5

In total votes from the first section, Voices 3, 6 and 10 were the most preferred, with a clean margin between the 3rd choice and 4th choice (Voice 9). When voting for a single favorite (second section of the survey), users gave Voice 6 the most votes, however Voices 3, 5 and 10 tied for 2nd place. In terms of the least favorite, Voice 2, there was consistency between the least favorite score (e.g. 32) and the lowest score in total votes (e.g. 51). In the favorite category, Voice 2 did receive more votes (e.g. 7) than Voices 7 and 8. However, a score of less than 7 in the Favorite category marks the mid-to-low range. This survey was used as one of three streams of input for selecting 3 top finalists for recording a new TTS font. It was coupled effectively with objective analysis from TTS developers who ranked each voice using algorithmic measurements in a separate study, and expert opinion from audio designers experienced in TTS voice production.

4.2   Voice Talent Selection - From Top 3 to the Best

This study had two types of speech assets available for review: 1) utterances from a 1500 sentence recorded corpus of each candidate, and 2) a prototype unit selection based TTS font (font). A pair of surveys was created: Survey #1 compared the human voices, while Survey #2 compared samples generated from the fonts. The surveys were identical with the exception of whether the source was human or TTS. Each CS survey contained two listening sets, and each set consisted of one sentence with matched content spoken by each (human) voice or font (presented in varying order). Judges were asked to choose which voice they preferred for each set, choose an overall favorite and least favorite, and also to give reasons for their preference. 100 judges were polled for each survey. A few surveys came back with incomplete results. For spam detection, we checked two things: a) we reviewed all comments and b) we made sure no judges chose the same voice as both favorite and least favorite. Although unusual, we felt confident that there were no spam responses (judges' comments support this conclusion). A total of 95/100 and 93/100 complete responses were collected for each survey. We examined responses for sets 1 and 2 in relation to the choice for favorite overall and found strong correlation: the mean score for Set 1 + Set 2 is ≤ 5 of the overall favorite score. We found this correlation to be consistent enough that we used only the scores for favorite and least favorite for conclusions.

Table 6. Set preferences compared to overall favorite for both voice and font

Voice/TTS	Voice 1	Voice 2	Voice 3	TTS 1	TTS 2	TTS 3
Set 1	43%	46%	11%	22%	43%	35%
Set 2	42%	45%	13%	28%	58%	14%
Overall Fav.	46%	47%	7%	26%	49%	25%

Table 7. Crowd results: favorite and least favorite

	Voice 1	Voice 2	Voice 3
Favorite / human	46%	47%	7%
Favorite / TTS	26%	50%	24%
Least favorite / human	30%	28%	42%
Least favorite / TTS	33%	17%	50%

The first trend to note is that of least favorite: there is a clear least favorite in this study - Voice 3. Next, looking across the scores for favorite and least favorite TTS, it is evident that the font from Voice 1 was not as well-liked as the human Voice 1. In terms of picking a "best" voice, when both human and TTS versions of the voice are considered, the crowd chose Voice 2. Like the previous study, this survey was used as just one of three inputs for selecting a new TTS voice talent. Its input was combined with analysis from TTS developers and the insights of audio designers to make the final decision.

5 Conclusions

In this paper, we have presented results from three studies on evaluating human voice talents and synthesized speech. The focus of each study was quite different, but all three are unified by the use of CS. The first study shows the power of CS to query not one but many countries about voice gender preference. The second one shows how CS helped to uncover one voice quality ("mood") which is more strongly carried over by TTS adaptation technology. The third study demonstrates 2 methodologies to effectively query general voice preference, depending on the size of the sample set. Considering the variety of focus across the 3 studies and the meaningful and relevant data uncovered by each, it is a reasonable conclusion that CS holds a wealth of potential feedback for developers of TTS voices and other applications of voice output.

Executing and reporting on a variety of CS experiment types helps us understand the strengths and weaknesses of CS apropos to research on human and synthesized speech. In this way, we can build reliable and repeatable experiment templates for use in crowdsourcing and tap the power of the crowd to hone and improve voice talent selection, steer gender or other country-dependent application decisions, and identify which areas of TTS technology innovation have the greatest impact with end-users.

References

1. Wolters, M., Isaac, K., Renalds, S.: Evaluating Speech Synthesis intelligibility using Amazon's Mechanical Turk. In: Proc. 7th Speech Synthesis Workshop, SSW7 (2010)
2. King, S., Karaiskos, V.: The Blizzard Challenge 2012. In: Proc. Blizzard Challenge Workshop 2012, Portland, OR, USA (2012)
3. Lane, I., Waibel, A., Eck, M., Rottman, K.: Tools for Collecting Speech Corpora via Mechanical-Turk. In: Proc. of Creating Speech and Language Data with Amazon's Mechanical Turk, pp. 185–187 (2010)
4. Snow, R., O'Connor, B., Jurafsky, D., Ng, A.Y.: Cheap and fast—but is it good?: evaluating non-expert annotations for natural language tasks. In: Proc. of the Conference on Empirical Methods in Natural Language Processing, pp. 254–263. Association for Computational Linguistics (2008)
5. Marge, M., Banerjee, S., Rudnicky, A.: Using the Amazon Mechanical Turk for transcription of spoken language. In: Proc. IEEE-ICASSP (2010)
6. Parent, G., Eskenazi, M.: Speaking to the Crowd: looking at past achievements in using crowdsourcing for speech and predicting future challenges. In: Proc. of INTERSPEECH 2011, pp. 3037–3040 (2011)
7. Cooke, M., Barker, J., Lecumberri, M.: Crowdsourcing in Speech Perception. In: Eskenazi, M., Levow, G., Meng, H., Parent, G., Suendermann, D. (eds.) Crowdsourcing for Speech Processing: Applications to Data Collection, Transcription and Assessment, pp. 137–172. Wiley, West Sussex (2013)
8. Lee, J., Nass, C., Brave, S.: Can computer-generated speech have gender?: an experimental test of gender stereotype. In: Proc. CHI EA 2000, CHI 2000 Extended Abstracts on Human Factors in Computing Systems, pp. 289–290. ACM, New York (2000)
9. Masuko, T., Tokuda, K., Kobayashi, T., Imai, S.: Voice characteristics conversion for HMM-based speech synthesis system. In: Proc. of ICASSP, pp. 1611–1614 (1997)

Experiment with Evaluation of Quality of the Synthetic Speech by the GMM Classifier[*]

Jiří Přibil[1,2], Anna Přibilová[3], and Jindřich Matoušek[1]

[1] University of West Bohemia, Faculty of Applied Sciences, Dept. of Cybernetics,
Univerzitní 8, 306 14 Plzeň, Czech Republic
jmatouse@kky.zcu.cz
[2] SAS, Institute of Measurement Science, Dúbravská cesta 9, SK-841 04 Bratislava, Slovakia
Jiri.Pribil@savba.sk
[3] Slovak University of Technology, Faculty of Electrical Engineering & Information
Technology, Institute of Electronics and Photonics, Ilkovičova 3, SK-812 19 Bratislava, Slovakia
Anna.Pribilova@stuba.sk

Abstract. This paper describes our experiment with using the Gaussian mixture models (GMM) for evaluation of the speech quality produced by different methods of speech synthesis and parameterization. In addition, the paper analyzes and compares influence of different types of features and different number of mixtures used for GMM evaluation. Finally, the GMM evaluation scores are compared with the results obtained by the conventional listening tests based on the mean opinion score (MOS) evaluations. Results of evaluations obtained by these two ways are in correspondence.

Keywords: GMM classifier, spectral and prosodic features of speech, synthetic speech evaluation.

1 Introduction

At present, the requirements on the quality of produced synthetic speech are rapidly increasing because the proper, quick, and easy understanding is a basic condition for effectivity and suitable strategy of dialogue management in the voice communication systems with the human-machine interface. For that reason the evaluation of the synthetic speech quality –first of all, intelligibility and naturalness – must often be performed. Several subjective and objective methods are used to verify the quality of produced synthetic speech. As regards subjective approaches [1], [2] the listening tests are usually used for giving the feedback information about user's opinion. On the other hand, especially in the case of intelligibility, the objective method based on automatic speech recognition (ASR) system can be used, where the final result represents the recognition score [3]. These recognition systems are often based on neural networks, hidden Markov models [4], or Gaussian mixture models (GMM) [5] in the case of speech emotion classification. The main advantage of this statistical evaluation method is that it works automatically without human interaction and the obtained results can be numerically judged.

---

[*] The work has been supported by the Technology Agency of the Czech Republic, project No. TA01030476, the Grant Agency of the Slovak Academy of Sciences (VEGA 2/0090/11), and the Ministry of Education of the Slovak Republic (VEGA 1/0987/12).

I. Habernal and V. Matousek (Eds.): TSD 2013, LNAI 8082, pp. 241–248, 2013.

This paper describes our experiment with using the GMM for evaluation of the speech quality produced by different methods of speech synthesis and parametric description that are often used for the speech production by the text-to-speech (TTS) systems. The GMM were trained with a corpus consisting of the original male and female Czech speech. The original sentences were resynthesized with different setting of segmentation method, type of speech modelling, as well as with different number of parameters of the vocal tract model. For the GMM evaluation, the basic and complementary spectral properties [6] including the supra-segmental parameters [7], were determined from the synthesized sentences and used in the input feature vector. The paper next analyzes and compares influence of different types of spectral features and supra-segmental parameters used for GMM evaluation as well as the influence of the number of used GMM mixture components.

Motivation of our work was to find an alternative approach to the standard listening tests, especially when audible differences were too small or hardly recognizable by listeners, in problems with their collective realization, etc.

2   Subject and Method

The Gaussian mixture models can be defined as a linear combination of multiple Gaussian probability distribution functions of the input data vector. For GMM creation, it is necessary to determine the covariance matrix, the vector of mean values, and the weighting parameters from the input training data. Using the expectation-maximization (EM) iteration algorithm, the maximum likelihood function of GMM is found [8]. For control of the EM algorithm, the N_{gmix} represents the number of used mixtures in each of the GMM models. In standard use of the GMM classifier, the resulting score is given by the maximum overall probability for the given class using the $score(T, i)$ representing the probability value of the GMM classifier for the models trained for current i-th class in the evaluation process, and an input vector T of the features obtained from the tested sentence [5]. For our purpose, only one model is created and trained in dependence on the speaker voice (male/female) with the help of the input feature vectors from the original sentences. In the classification phase, we obtain the scores using the input feature vectors from the tested sentences synthesized by various methods. These scores are sorted by the absolute size and quantized to N levels corresponding to N output classes. It means that the obtained highest score represents the synthesized sentences having the speech features that are most similar to those obtained from the original sentences used for GMM model training; the minimum score corresponds to the tested sentence with the greatest differences in comparison to the originals. To obtain correspondence (comparable values) with the mean opinion score (MOS) evaluation method where the perceived quality is scaled from "5" representing the best quality to "1" corresponding to poor quality; finally we use five output classes: in the score discriminator block (see Fig. 1) the highest obtained score is assigned to the value 5, the lowest score to the value 1. To obtain speaker independent GMM classification, the data k-fold cross-validation method [8] can be applied during the training and the

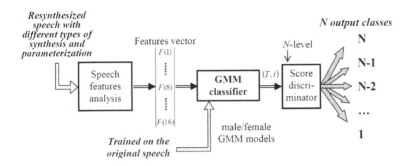

Fig. 1. Block diagram of the GMM classifier used for speech quality evaluation

testing processes. In our case this approach is not currently applied, so the classification works speaker-dependently. In the first stage of development – the GMM speech quality classifier has only one-level structure. This simple architecture expects that the gender of the voice (male/female) was correctly recognized in the previous process – manually, by listening tests, etc. Subsequently, the quality of the speech is identified for each of two gender classes.

2.1 Spectral Features and Prosodic Properties Determination

Spectral analysis of the speech signal is performed in the following way: from the input samples, after segmentation and weighting by a Hamming window, the absolute values of the fast Fourier transform, the power spectrum, and the smoothed spectral envelope (using the mean Welch's periodogram) are calculated for further use in the feature determination process. The spread of the signal around the higher harmonics has consequences on spectral tilt measurements. On the other hand, the harmonics-to-noise ratio (HNR) provides an indication of the overall periodicity of the speech signal. Specifically, it quantifies the ratio between the periodic and aperiodic components in the signal. The spectral centroid (SC) is a centre of gravity of the power spectrum and represents an average frequency weighted by the values of the normalized energy of each frequency component in the spectrum. The spectral flatness measure (SFM) can be used to determine the degree of periodicity in the signal. This spectral feature can be calculated as a ratio of the geometric and the arithmetic mean values of the power spectrum. The spectral entropy (SE) is a measure of spectral distribution. It quantifies a degree of randomness of spectral probability density represented by normalized frequency components of the spectrum.

Microintonation component of speech melody given by F0 contour can be supposed to be a random, band-pass signal described by its spectrum and statistical parameters. We use differential microintonation signal $F0_{DIFF}$ obtained by subtraction of mean F0 values and linear trends. The jitter related to microvariations of a pitch curve is computed as a relative number of zero crossings (ZCR) of a derivative pitch curve normalized by utterance duration. For calculation of the absolute jitter values, the average absolute difference between consecutive pitch periods measured in samples was used. In the case of the shimmer measure determination, a period-to-period variability of amplitudes of the speech signal was used.

3   Material, Experiments, and Results

In our experiment we use the speech material which consists of sentences with duration from 1.5 to 10.5 seconds, resampled at 16 kHz, representing speech in a neutral (declarative) state, and yes/no questions uttered by one male and one female speaker. The original sentences were analyzed and subsequently resynthesized using five different types of speech modelling (cepstral [9], harmonic [10]), segmentation, and with different number of applied coefficients N_{coeff} – see detailed specifications in Table 1. These synthesis methods are currently used in the Czech TTS system based on the HMM approach for synthetic speech personification or expressive speech production [11]. The whole test speech corpus includes $180 + 180$ sentences consisting of the originals and five types of resynthesis. For the listening test evaluation only $80 + 80$ sentences (male + female) were selected from the main test corpus. The processed speech material originates from speakers with different mean F0 values so different parameter settings for analysis – frame (window) length W_L and window overlapping W_O – were applied. The F0 values (pitch contours) were determined by autocorrelation analysis method with experimentally chosen pitch ranges as follows: $55 - 250$ Hz for male voices, and $105 - 350$ Hz for female ones.

Table 1. Detailed specification of used speech analysis and synthesis method

Type	Model	Specification	N_{coeff}
1 (h48)	harmonic	Spectral envelope smoothed by B-Splines	48[*)]
2 (i80)	cepstral	Impulse response of the real cepstrum	80[*)]
3 (k64)	cepstral	Minimum phase of real cepstrum, mixed excitation	64[*)]
4 (s64)	cepstral	Real cepstrum, excitation by Hilbert impulse	64[**)]
5 (o50)	cepstral	Optimized structure of cascade approximation filter	50[*)]

[*)] W_L=12/10 ms, W_O=W_L/2 for male/female voices. [**)] W_L=24/20 ms, W_O=W_L/4.

Two basic types of experiments were performed for comparison in this paper:

1. Objective – automatic evaluation of the synthetic speech quality using the statistical method based on the GMM approach,
2. Subjective – manually performed listening test using the MOS evaluation method.

In GMM evaluation the analysis and comparison was aimed at investigation of:

- influence of the number of used mixtures on GMM evaluation (in the range of $N_{gmix} \in < 1 - 3 >$),
- influence of the used type of the feature set on GMM evaluation score (sets P1-3),
- influence of the feature order in the input data vector on GMM score; the set P1 was used with the reversed order of features giving thus the set called P4.

The feature set of 16 values as the input data vector for GMM training and classification containing the features determined from the spectral envelopes (skewness, kurtosis, spread, and tilt), the complementary spectral parameters (HNR, SC, SFM, SE), and supra-segmental parameters (F0, jitter, and shimmer) was used, as shown Table 2. In the case of the spectral features, the basic statistical parameters-mean values and standard deviations (std)-were used as the representative values in the feature vectors for GMM evaluation. For implementation of the supra-segmental parameters of speech, the statistical types – median values, range of values, std, and/or relative maximum and minimum we used in the feature vectors. A simple diagonal covariance matrix of the GMM was applied in this first evaluation experiment. The basic functions from the Ian T. Nabney "Netlab" pattern analysis toolbox [12] were used for creation of the GMM models, data training, and classification.

Table 2. Detailed specification of used speech analysis and synthesis method

No	Feature name			Statistical value		
	P1	P2	P3	P1	P2	P3
1	HNR	Spec. envelope	F0	Mean	Skewness	Median
2	HNR	Spec. envelope	F0	Std	Kurtosis	Std
3	Spec. tilt	Spec. centroid	$F0_{DIFF}$	Min	Mean	Median
4	SC	Spec. spread	$F0_{DIFF}$	Mean	Std	Std
5	SC	Spec. tilt	$F0_{DIFF}$	Std	Min	Rel. max
6	SFM	SFM	$F0_{DIFF}$	Mean	Mean	Rel. min
7	SFM	F0	$F0_{ZCR}$	Std	Std	Median
8	SE	F0	$F0_{ZCR}$	Mean	Rel. max	Std
9	SE	$F0_{DIFF}$	$F0_{ZCR}$	Std	Median	Rel. max
10	Signal ZCR	$F0_{DIFF}$	$F0_{ZCR}$	Median	Std	Rel. min
11	Signal ZCR	$F0_{ZCR}$	Jitter	Std	Median	Median
12	$F0_{DIFF}$	$F0_{ZCR}$	Jitter	Rel. max	Rel. max	Std
13	Jitter	Jitter	Jitter	Median	Median	Rel. max
14	Shimmer	Jitter	Shimmer	Max	Rel. max	Median
15	Shimmer	Shimmer	Shimmer	Median	Median	Std
16	Shimmer	Shimmer	Shimmer	Rel. max	Rel. max	Rel. max

Subjective evaluation was realized by the conventional listening test called "Speech quality evaluation – male/female voice" located on the web page http://www/lef.um.savba.sk/scripts/itstpos12.dll. This listening test program in the form of MS ISAPI DLL script including the testing speech runs on the server PC and communicates with the user via the HTTP protocol by WEB pages with frames in the HTML language. The currently used type of the listening test is based on the MOS evaluation for naturalness and intelligibility of the synthetic speech example. By reason of differentiation for creation and training of the GMM models, the evaluation must be performed separately for male and female voices. The complete test consists of 10 evaluation sets using sentences selected randomly from the speech corpus. In addition, the listening test program generates also the test protocol with time marks, so we can determine the duration of the performed test.

3.1   Obtained Results

Obtained results of performed GMM evaluation experiment are presented in graphical form for visual comparison by the boxplot of basic statistical parameters (see Fig. 2), or by the bar graphs (see Figures 3 and 4) separately in dependence on the speaker's voice. Twenty two listeners (5 women and 17 men) took part in our listening test experiment: 20 listening tests of male voice and 20 tests of female voice were executed, 40 tests in total. Evaluation of the listening test results was realized in dependence on listener's sex for all synthesis method categories, see Fig. 5. The final comparison of evaluation results based on GMM approach and MOS listening tests separately for male/female voice are shown in Table 3.

Table 3. Final comparison of evaluation results based on GMM approach and MOS listening test separately for the male/female voice and the applied type of the synthesis method

Evaluation method	Male voice					Female voice				
	h48	i80	k64	s64	o50	h48	i80	k64	s64	o50
GMM-full corpus[*]	2.45	4.75	3.05	2.95	1.50	2.75	4.05	3.75	3.20	1.25
GMM-limit. corpus[*]	2.06	4.87	3.01	3.13	1.95	2.86	4.06	3.32	3.12	1.44
MOS listening test	2.75	3.65	2.70	3.01	2.34	2.27	3.55	3.06	2.31	2.19

[*] $N_{gmix}=1$, feature set P1

4   Discussion and Conclusion

As follows from our first comparison experiment, increasing the number of the used mixtures brings not always positive effect to the GMM evaluation – it holds especially in the case of female speech, where differentiation of the discriminated GMM score is worse for two or three mixtures than for only one mixture – see Fig. 3. Therefore, in further analysis we use setting $N_{gmix} = 1$; but in this case there are actually compared averages of features in the input vector from natural and synthesized speech using the Mahalanobis distance measure. Application of the proper type of the input features for GMM evaluation is very important – as demonstrated by the results of our second experiment (see Fig. 4): the best results are produced by the feature set P1 consisting of a mix of spectral and prosodic features, the worst results correspond to the set P3 when only supra-segmental features were used. These results consider the fact, that the resynthesized sentences have the similar distribution of F0 values as the original ones. The differences can be caused only by the used type of pitch-period detection and the segmentation method (processing) for signal analysis as it is documented by Table 1.

On the other hand, the order of the parameters in the input feature vector has minimal influence on the GMM score – values for the sets P1 and P4 are practically the same. Comparison of the GMM scores obtained with the help of the full speech corpus and using only the sentences applied in the listening test shows that they are similar. The summary results of the MOS test are also in correspondence – compare values in Table 3 and Fig. 5. Due to disproportion in the groups of listener's gender (the female group is smaller than the male one) the score differentiation in the case of the female listeners is not as objective as in the male listeners.

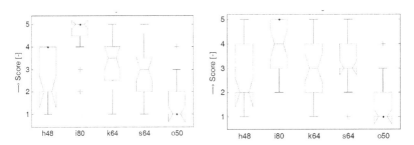

Fig. 2. The boxplot of the basic statistical parameters of the discriminated score values: for male (left) and female (right) voices; $N_{gmix} = 1$, feature set P1

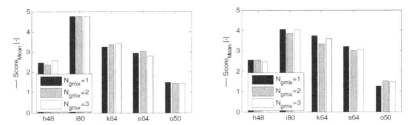

Fig. 3. Influence of the number of used mixtures on the discriminated GMM score: mean values for male (left) and female (right) voices; feature set P1

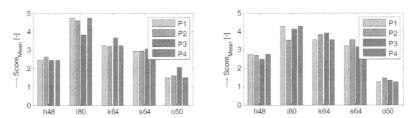

Fig. 4. Influence of the type of the feature vector on the discriminated GMM score: mean values for male (left) and female (right) voices; $N_{gmix} = 1$

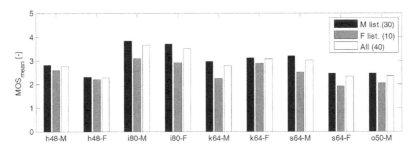

Fig. 5. Summary results of the listening test MOS evaluation in dependence on listener's categories joined for male and female voices

The observed quality of the synthetic speech with the female voice was lower than that with the male voice, so the score values were also lower and worse differentiated. At the same time, recognizable differences were observed in the resynthesis quality of the declarative sentences and yes/no questions, which reached the lowest MOS score. From the listener's feedback information follows that the listening test constructed by this way is very difficult: several listeners executed only one type of the test; some of the addressed persons tried the test but they could not recognize differences so the test was not finished. Therefore, this is the right case when the GMM based evaluation can be applied more effectively and it can bring more accurate results than the conventional listening tests.

In near future, we plan testing of the GMM evaluation using the larger speech databases as well as the ones spoken in other languages, and also make a comparison with other widely applied quality assessment schemes used for telephony speech signals and speech codecs evaluation. Increase of the GMM score can be expected if the full covariance matrix is used, so both approaches will be compared in future. We would also like to try to use the GMM classification for quality evaluation of the voice conversion (male/female/child) or the expressive style transformation of the synthesized speech.

References

1. Audibert, N., Vincent, D., Aubergé, V., Rosec, O.: Evaluation of Expresive Speech Resynthesis. In: Proceedings of LREC 2006 Workshop on Emotional Corpora, Gènes, pp. 37–40 (2006)
2. Iriondo, I., Planet, S., Socoró, J.C., Martínez, E., Alías, F., Monzo, C.: Automatic Refinement of an Expressive Speech Corpus Assembling Subjective Perception and Automatic Classification. Speech Communication 51, 744–758 (2009)
3. Takano, Y., Kondo, K.: Estimation of Speech Intelligibility Using Speech Recognition Systems. IEICE Transactions on Information and Systems E93D(12), 3368–3376 (2010)
4. Vích, R., Nouza, J., Vondra, M.: Automatic Speech Recognition Used for Intelligibility Assessment of Text-to-Speech Systems. In: Esposito, A., Bourbakis, N.G., Avouris, N., Hatzilygeroudis, I. (eds.) HH and HM Interaction. LNCS (LNAI), vol. 5042, pp. 136–148. Springer, Heidelberg (2008)
5. Yun, S., Yoo, C.D.: Loss-Scaled Large-Margin Gaussian Mixture Models for Speech Emotion Classification. IEEE Transactions on Audio, Speech, and Language Processing 20(2), 585–598 (2012)
6. Hosseinzadeh, D., Krishnan, S.: On the Use of Complementary Spectral Features for Speaker Recognition. EURASIP Journal on Advances in Signal Processing 2008, Article ID 258184, 10 pages (2008)
7. Lu, Y., Cooke, M.: The Contribution of Changes in F0 and Spectral Tilt to Increased Intelligibility of Speech Produced in Noise. Speech Communication 51(12), 1253–1262 (2009)
8. Reynolds, D.A., Rose, R.C.: Robust Text-Independent Speaker Identification Using Gaussian Mixture Speaker Models. IEEE Transactions on Speech and Audio Processing 3, 72–83 (1995)
9. Vích, R.: Cepstral Speech Model, Padé Approximation, Excitation, and Gain Matching in Cepstral Speech Synthesis. In: Proceedings of the 15th Biennial EURASIP Conference Biosignal 2000, Brno, Czech Republic, pp. 77–82 (2000)
10. Madlová, A.: Autoregressive and Cepstral Parametrization in Harmonic Speech Modelling. Journal of Electrical Engineering 53, 46–49 (2002)
11. Grůber, M., Hanzlíček, Z.: Czech Expressive Speech Synthesis in Limited Domain Comparison of Unit Selection and HMM-Based Approaches. In: Sojka, P., Horák, A., Kopeček, I., Pala, K. (eds.) TSD 2012. LNCS, vol. 7499, pp. 656–664. Springer, Heidelberg (2012)
12. Bishop, C.M., Nabney, I.T.: NETLAB Online Reference Documentation (accessed February 16, 2012), http://www.fizyka.umk.pl/netlab/

Experiments on Reducing Footprint
of Unit Selection TTS System*

Zdeněk Hanzlíček, Jindřich Matoušek, and Daniel Tihelka

Department of Cybernetics, Faculty of Applied Sciences, University of West Bohemia,
Univerzitní 8, 306 14 Plzeň, Czech Republic
{zhanzlic,jmatouse,dtihelka}@kky.zcu.cz
http://www.kky.zcu.cz/en

Abstract. The quality of speech produced by modern TTS systems utilizing the unit selection approach is very high. However, the system demands are enormous. The storage requirements are directly proportional to the size of speech unit inventory from which the units are selected during the synthesis process. This paper presents the analysis and reduction experiments performed on two large speech corpora employed by a unit selection TTS system for the Czech language. A procedure for exclusion of utterances from the default speech corpus based on statistics of the usage of particular speech units was proposed. The exclusion of whole utterances from the corpus was preferred over the exclusion of individual speech units in order to preserve the fundamental feature of the unit selection method – selection of possibly longest sequences of speech units. Experiments were performed for several reduction levels. Resulting synthetic speech was evaluated by a proposed statistics based on the concatenation points density. Moreover, the speech quality was evaluated in listening tests. All reduced versions of TTS system were evaluated as similar or slightly worse than the baseline system.

Keywords: speech synthesis, TTS, unit selection, reducing footprint.

1 Introduction

The current trend in speech synthesis is using large carefully prepared speech corpora containing a lot of instances of each speech unit[1]. Speech is created by the selection of an optimal sequence of speech units and their concatenation.

The quality of generated speech is very high [1]. However, the system demands (memory and computational requirements) are enormous. The size of required memory (so-called footprint) is usually hundreds of megabytes. It prevents to use this technology on less powerful or low-resource devices like pocket computers, cellphones etc.,

---

* This work was supported by the Ministry of Industry and Trade of the Czech Republic, project No. MPO FR-TI1/518. The access to computing and storage facilities owned by parties and projects contributing to the National Grid Infrastructure MetaCentrum, provided under the program *"Projects of Large Infrastructure for Research, Development, and Innovations"* (LM2010005) is highly appreciated.

[1] The objective of this work is the unit selection method. Some assertions may not be fully valid for other approaches to speech synthesis, e.g. HMM-based synthesis.

I. Habernal and V. Matousek (Eds.): TSD 2013, LNAI 8082, pp. 249–256, 2013.

or even on server-like systems where more voices are to be produced. This work is focused mainly on the reduction of memory requirements (footprint). However, both storage and computational demands are somehow directly proportional to the size of speech unit inventory from which the units are selected during the synthesis process. Thus, reducing the number of speech units in the inventory decreases actually the computational requirements, too.

This paper presents the analysis and reduction experiments performed on two large speech corpora (with male and female voice) employed by a unit selection TTS system ARTIC [4] for the Czech language. A special procedure for exclusion of utterances from the speech corpus is proposed. This procedure is based on statistics of the utilization of speech units from particular utterances contained in the default speech corpus. The exclusion of whole utterances from the corpus was preferred over the exclusion of individual speech units. The main motivation is to preserve the fundamental feature of the unit selection method – selection of possibly longest sequences of speech units.

2   Experiments

There are several methods dealing with the reduction of the footprint of a TTS system described in the literature. Some of them employs various speech coding techniques [2,3]. Those approaches preserve all the instances of speech units in the inventory.

In our experiments, besides the ADPCM speech coding technique, the system footprint is reduced by excluding some speech unit instances. Specifically, instead of particular isolated speech units as proposed e.g. in [5,6], the whole utterances were excluded. The main motivation was to preserve the fundamental feature of the unit selection framework – to select as longest sequence of contiguous speech units as possible.

2.1   Baseline TTS System

In our experiments TTS system ARTIC [4] together with two speech corpora (male and female voice – referred to as AJ and KI, respectively) were employed. Each corpus contains about 15 hours of speech. The utterances were carefully selected in order to be both phonetically and prosodically balanced [7]. The composition of both corpora is briefly described in Table 1. The overall footprint of the original system is about 620 MB for speaker AJ and 640 MB for speaker KI.

Table 1. Composition of original corpora – types of utterances

type of utterancs	speaker AJ	speaker KI
declarative sentences	10,035	9,936
yes/no questions	1,002	996
wh-questions	700	700
application-based sentences	531	531

2.2   The Reduction Process

The fundamental idea of the proposed reduction process is quite simple: The size of the resulting acoustic unit inventory is directly proportional to the size of the utilized speech corpus, i.e. the fewer utterances are present in the training corpus, the smaller is the final unit inventory.

For the best result, a new corpus should be designed with regards to the required size of the final unit inventory. However this approach is absolutely impractical because the process of obtaining a new speech corpus is very lengthy and expensive. A more viable way is recording of one sufficiently large corpus, which can be additionally reduced to the required size. The proposed algorithm works in three main steps:

Step 1: The Exclusion of Special Utterances. In the first step, all application-based utterances (useful for applications in special domains such as railway stations, call centers etc.) and wh-questions (non-polar questions) were excluded. Those utterances are supposed to be superfluous in a general-purpose TTS system. Moreover, the wh-questions are prosodically similar to declarative sentences (in the Czech language) and no additional specific speech units are needed for their synthesis.

Step 2: The Exclusion of Utterances with Suspicious Segmentation. Speech units with badly segmented boundaries can significantly degrade the quality of synthetic speech. Moreover, the bad segmentation usually afflicts several neighboring units. For the detection of potentialy badly segmented utterances that are suitable candidates for exclusion from the corpus, speech units matching the following criteria (based on our previous experiments) were sought

- phones with a suspiciously long duration (more than 400 ms)
- phones with a suspiciously short duration (less than 12 ms)
- phones with a low segmentation score [8] (less than -120 log-probability)

Utterances with at least one speech unit with suspiciously segmented boundaries were excluded. Because corpra employed in our experiments were already verified and corrected within former experiments and applications, this step had only a trivial effect on sizes of resulting speech inventories. Detailed information is presented in Table 2.

Table 2. Number of removed utterances with suspicious segmentation

speaker	AJ		KI	
type of utterances	declar. sent.	yes/no quest.	declar. sent.	yes/no quest.
initial amount	10,035	1,001	9,936	996
removed	20	3	121	6
final amount	10,015	998	9,815	990

Step 3: The Exclusion of Rarely Employed Utterances. In this step, the usage of units from the remaining utterances during the synthesis was analyzed. Statistics were obtained by processing of a huge set of sentences (approximately 524,000); hereinafter this set will be referred to as development data. The synthesis of those sentences was simulated and a record of how many times each utterance from the source speech corpus was employed during the synthesis was stored. The most rarely used utterances were excluded from the speech corpus. The numbers of excluded and kept utterances were determined to reach the desired size of the resulting speech inventory.

This reduction step was performed independently for declarative sentences and yes/no questions. In the case of reducing all the sentences together, a significant part of yes/no questions would be removed from the corpus. As a result, some specific speech units would be probably missing and the TTS system would be unable to synthesize that kind of question correctly. The reason is a rare occurrence of questions in the development data set. It contains mostly the declarative sentences, which is a common composition of a standard written text. Moreover, questions are naturally significantly shorter than declarative sentences. And because the utterance length is directly proportional to the probability of selecting a unit from it, yes/no questions have worse utilization statistics than declarative sentences. Thus an independent exclusion is convenient.

Detailed information is presented[2] in Table 3. The size of the question part of the footprint was ad hoc determined as 30, 20 and 10 MB for particular reduction levels; it correspond to 100%, 2/3 and 1/3 of the original size of the question part.

Table 3. Number of utterances left in the reduced speech corpra

speaker	AJ		KI	
type of utterances	declar. sent.	yes/no quest.	declar. sent.	yes/no quest.
O	10,015	998	9,815	990
R – 400 MB	5,104	998	4,891	990
R – 200 MB	2,116	598	2,047	568
R – 100 MB	907	253	878	251

2.3 Objective Evaluation

To quantify the extent of the corpus reduction so called *coverage rate* was introduced. It expresses the utilization rate of speech units from the new unit inventory within the original (full) unit inventory, i.e. how often the speech units remaining in the reduced inventory were used by the default system. This quantity can be also directly used to control the corpus reduction process. However, the size of the resulting unit inventory seems to be more practical. Results for particular reduction levels are presented in Table 4.

As an inevitable consequence of corpus reduction the number of concatenations of originally non-neighbouring speech units increases. Those concatenation points will be referred to as cross-points. At the cross-points the transition between units may not

---

[2] Hereinafter, the original a reduced systems are briefly referred to as O and R, respectively.

Table 4. Coverage rate [%] for particular levels of reduction

speaker	AJ		KI	
type of utterances	declar. sent.	yes/no quest.	declar. sent.	yes/no quest.
O	100.0	100.0	100.0	100.0
R – 400 MB	74.0	100.0	75.2	100.0
R – 200 MB	42.5	92.0	43.9	87.8
R – 100 MB	21.8	54.2	22.4	43.5

be fluent and some audible skips or other artifacts can appear. The more cross-points are present in the synthetic utterance the higher is the probability that some artifacts cause the quality degradation. Thus, number of cross-points could be use as a objective criterion for evaluation of the synthetic speech quality. To allow for the length of the utterance, we introduce the *concatenation density* $\mathcal{CD}$ defined as

$$\mathcal{CD} = \frac{\text{number of crosspoints}}{\text{number of all concatenation points}} \tag{1}$$

Statistical evaluation of original and reduced TTS systems (obtained by synthesis of all development utterances) is presented in Table 5 and Figure 1.

Table 5. Concatenation density for particular TTS systems

TTS system	speaker AJ	speaker KI
O	0.260 ± 0.070	0.271 ± 0.069
R – 400 MB	0.278 ± 0.071	0.286 ± 0.070
R – 200 MB	0.311 ± 0.075	0.316 ± 0.074
R – 100 MB	0.352 ± 0.079	0.356 ± 0.078

2.4 Listening Tests

The quality of synthetic speech produced by the original and reduced TTS systems was also evaluated by means of listening tests. The test contained 60 pairs of utterances for each speaker. Pairs were composed of one sentence synthesized by the original TTS system and one synthesized by a reduced version of the system. To keep the extent of listening tests reasonably large, particular versions of reduced system were not directly compared against each other. The length of utterances was limited to 5–8 words.

For each speaker and reduction level, the test contained a set of 20 queries. To cover various prosodic forms, all sets contained 10 declarative sentences, 5 yes/no questions and 5 complex sentences. The criterion for selection of sentences for the listening test was the possibly biggest difference between utterances synthesized by compared TTS systems. To quantify the difference between utterances U_1 and U_2 independently on their length, the normalized measure $\mathcal{D}$ was defined

$$\mathcal{D}(U_1, U_2) = \frac{\text{number of different units}}{\text{number of all units}} \tag{2}$$

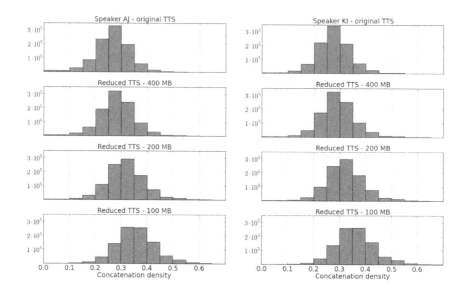

Fig. 1. Histograms of concatenation density for particular TTS systems

By the analysis of aforementioned large set of synthesized sentences, pairs with the highest value of normalized difference were selected. Statistics for particular levels of reduction are presented in Table 6 and Figure 2.

Table 6. Relative difference between synthesized utterances (average value ± standard deviation)

compared systems	speaker AJ	speaker KI
O vs R (100 MB)	0.833 ± 0.168	0.817 ± 0.175
O vs R (200 MB)	0.685 ± 0.224	0.654 ± 0.227
O vs R (400 MB)	0.410 ± 0.266	0.397 ± 0.249

13 listeners took part in the test; the following normalized scale was employed[3]:

-1.0 ... utterance generated by the original TTS system is significantly better
-0.5 ... utterance generated by the original TTS system is slightly better
0.0 ... both utterances are of a similar quality
$+0.5$ ... utterance generated by the original TTS system is slightly worse
$+1.0$ ... utterance generated by the original TTS system is significantly worse

Then, the average normalized score ranged from -1 to $+1$ is calculated; the limit values correspond to cases when utterances produced by the default TTS-system were always evaluated as significantly better (-1) or always significantly worse $(+1)$. The results are presented in Table 7 and Figure 3.

---

[3] Naturally, the random order of utterances in particular pairs is also taken into account.

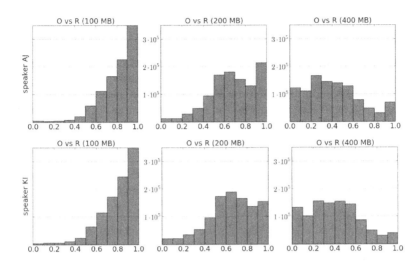

Fig. 2. Histograms of relative difference between synthesized utterances

Table 7. Results of preference listening tests (average normalized score ± standard deviation)

compared systems	speaker AJ	speaker KI
O vs R (100 MB)	-0.227 ± 0.522	-0.158 ± 0.433
O vs R (200 MB)	-0.225 ± 0.441	-0.075 ± 0.493
O vs R (400 MB)	-0.222 ± 0.478	-0.106 ± 0.444

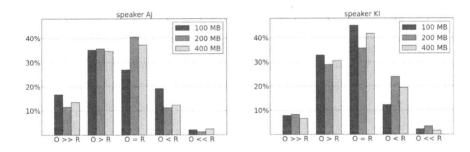

Fig. 3. Results of listening tests

3   Conclusion

This paper presented several experiments on the reduction of the footprint of the Czech unit-selection TTS system ARTIC. The reduction was performed by excluding specially selected utterances from the source speech corpus. Excluded utterances were selected according to their usage during synthesis of a huge amount of collected sentences. From the initial size 620–640 MB the footprint was reduced to 400, 200 and 100 MB.

The quality of synthetic speech generated by particular versions of reduced system was evaluated by using concatenation density (relative number of potentially influent concatenation points in synthesized utterances). As expected, the concatenation density increased with the number of excluded utterances. However the rise of the density was quite low, e.g. for the reduction to 100 MB, the average value raised from 0.260 to 0.352 for the male speaker and from 0.271 to 0.356 for the female speaker.

The quality of synthetic speech was also evaluated by means of comparison listening tests. All versions of the reduced system were assessed (on average) as similar or slightly worse than the baseline system. Thus, all levels of footprint reduction can be applied when the system/device resources are restricted.

References

1. Dutoit, T.: Corpus-based Speech Synthesis. In: Benesty, H., Sondhi, M., Huang, Y. (eds.) Springer Handbook of Speech Processing, pp. 437–455. Springer, Dordrecht (2008)
2. Chazan, D., Hoory, R., Kons, Z., Sagi, A., Shechtman, S., Sorin, A.: Small Footprint Concatenative Text-to-Speech Synthesis System using Complex Spectral Envelope Modeling. In: Proc. of Interspeech 2005, Lisbon, Portugal, pp. 2569–2572 (2005)
3. Strecha, G., Eichner, M., Hoffmann, R.: Line Cepstral Quefrencies and Their Use for Acoustic Inventory Coding. In: Proc. of Interspeech 2007, Antwerp, Belgium, pp. 2873–2876 (2007)
4. Matoušek, J., Tihelka, D., Romportl, J.: Current State of Czech Text-to-Speech System ARTIC. In: Sojka, P., Kopeček, I., Pala, K. (eds.) TSD 2006. LNCS (LNAI), vol. 4188, pp. 439–446. Springer, Heidelberg (2006)
5. Kominek, J., Black, A.W.: Impact of durational outlier removal from unit selection catalogs. In: Proc. of the 5th ISCA Speech Synthesis Workshop, Pittsburgh, USA, pp. 155–160 (2004)
6. Tihelka, D.: Corpus-based Approach to Unit Selection Speech Unit Inventory Reduction in ARTIC TTS. In: Proc. of 17th Czech-German Workshop on Speech Processing, pp. 160–167. Institute of Photonics and Electronics AS CR, Prague (2007)
7. Matoušek, J., Tihelka, D., Romportl, J.: Building of a Speech Corpus Optimised for Unit Selection TTS Synthesis. In: Proc. of LREC 2008, Marrakech, Morocco (2008)
8. Young, S.: The HTK Book (for HTK version 3.4). Cambridge University, UK (2009)

Expressive Speech Synthesis for Urgent Warning Messages Generation in Romani and Slovak

Milan Rusko, Marián Trnka, Sakhia Darjaa, and Marian Ritomský

Slovak Academy of Sciences, Institute of Informatics, Bratislava, Slovakia
{milan.rusko,trnka,utrrsach,marian.ritomsky}@savba.sk

Abstract. Warnings generated by a specially designed speech synthesizer can be used to inform, warn, instruct and navigate people in dangerous and critical situations. The paper presents the design of the speech synthesizer capable of generating warning messages with different urgency levels in Slovak and also in Romani - the under-resourced and digitally endangered language of the Slovak Roma. An original three-step method is proposed for creating expressive speech databases. Expressive synthesizers trained on these databases and capable of generating Romani and Slovak synthetic warning speech and messages in three levels of urgency are presented.

Keywords: Expressive speech synthesis, speech synthesis in Romani, speech synthesis in Slovak, crisis management, urgent warning messages generation.

1 Introduction

Previous attempts at building synthesizers in Slovak and the language of Slovak Roma ("Serviko Romaňi čhib") [1] were limited to emotionally neutral speech and used diphone concatenative and unit-selection speech synthesis. Present work is aimed at generating expressive synthetic speech - warning messages with different degrees of urgency.

Almost all up-to-date speech synthesis methods apply large speech databases to acquire the data necessary for their development, training, and testing. Specialized expressive speech databases have to be built for every particular language. A flexible type of speech synthesizer is needed to study different aspects of the synthesis process with the ability to change the speech characteristics independently. The statistical parametric synthesis was therefore chosen for this work.

The European Structural Funds project "Research and Development of New Information Technologies for Prediction and Solution of Critical Situations of Inhabitants" CRISIS (ITMS 26240220060) is aimed at predicting and solving critical situations when the inhabitants are endangered, and the environment has to be protected.

One of the activities of the project is aimed at the design of extremely expressive speech synthesis for urgent warning messages generation in two languages - Slovak [2] and Romani. Roma people represent the second biggest minority group living in the Slovak Republic estimated to include as many as 380 000 people [3].

The synthesizer that is being developed will be applicable to generate warning system messages for the Romani speaking people in case of fire, flood, state security threats, or other crisis situations.

I. Habernal and V. Matousek (Eds.): TSD 2013, LNAI 8082, pp. 257–264, 2013.

2   Text Resources

For the Romani language, the quantity and quality of texts available is insufficient. Moreover, many of the texts that have been published are written in a local dialect of their author and not in the standardized form of the language.

To have at least some amount of texts for initial efforts, we used the archive of the Slovak Romani newspaper "Romano nevo ĺil" from the years 2003 to 2010, unpublished Romani fairy tales by Vladimír Zeman and several tenths of pages of texts that we were provided by PaeDr. Stanislav Cina (SC) who is our Romani language expert is also experienced in voice recordings. We obtained only about 600 kB of texts in total from these sources, and this forms our corpus of Romani texts. These were analyzed and a basic set of 1574 phonetically rich sentences was selected for the recording of the emotionally neutral part of the Romani speech database. We refer to these texts as "neutral" although they were not checked for the expressivity of their semantic content. In fact, some of the sentences can be considered emotionally slightly loaded based on their semantic or pragmatic meaning, but the database as a whole can be considered as emotionally neutral.

The texts chosen for the neutral part of the Slovak SC database represent a subset of the collection of phonetically rich sentences used in our earlier works [1]. The method of preparation of the set of the Slovak warning sentences was described in [2]. These sentences were translated to Romani by our expert, who is a bilingual Romani-Slovak speaker of Romani nationality. The set contains 97 prompts, which is about 160 sentences in total, as some of the messages contain more than one sentence.

3   Speech Database

The speech synthesis database consists of several partial speech databases recorded by the same voice (SC).

- Emotionally neutral speech database in Romani (*Neutral_SC-ro*)
- Emotionally neutral speech database in Slovak (*Neutral_SC-sk*)
- Expressive speech database in Romani (*Crisis_SC-ro*)
- Expressive speech database in Slovak (*Crisis_SC-sk*)

The speech databases were recorded in the acoustically treated studio. Rode K2 microphone was used for recording due to its wide dynamic range.

3.1   Emotionally Neutral Speech Databases in Romani and Slovak

Emotionally neutral speech database Neutral_SC-ro consists of 1574 phonetically rich sentences read by one speaker. The sentences were taken from newspapers and fairy tales. Emotionally neutral speech database Neutral_SC-sk consists of 800 phonetically rich sentences read by one speaker. The sentences were taken from newspapers and fiction. The speaker read the sentences from a list and he was instructed to read it in his most natural way and not to "act" it as an actor. The style of this recording is similar to reading news.

3.2 The Three Steps Method of Recording the Expressive Speech Databases

One of the biggest problems with recording an acted expressive speech database is that the actor is often unable to keep the level of portrayed emotion consistent for a longer time interval. We have therefore proposed a new method of elicitation. The speaker does not try to maintain the same level of expressivity during an extended interval of recording, but rather, he varies emotions on one dimension producing triples of identical utterances and trying to keep equidistant steps between three levels of expressive load.

The speaker is instructed to utter the prompted message once in a neutral way, which is natural and comfortable for him, and to ignore the warning semantic content of the text. We assume that the collection of the recordings of this level (the first level of expressivity) reflects the neutral state of the speaker at that particular recording session. The same message is then uttered in the second level of expressivity with higher imperativeness, functioning as a serious command or directive. The speaker should imagine that he has to make a strong statement, an announcement of a very serious fact to big audience. Then the same message is recorded with extremely high urgency, pragmatically functioning as a warning that can save lives. The speaker should imagine that people around are directly endangered and there is little time to warn them (the third level of expressivity).

After recording the message in the third level, the speaker relaxes for a short moment and then he starts with a new message and utters it at the first, second and third level of expressivity respectively.

We assume this approach produces three emotionally consistent sets of utterances per language with three degrees of arousal. Furthermore, we believe that an amateur in voice recording is not able to distinguish and display more then three levels of expressivity consistently in one session. As the highly urgent warning speech uses high vocal effort, the untrained voice becomes tired soon and the recording session should not last longer than 40 to 60 minutes. The number of sentences recorded in one session should therefore be about 450 (150 per level). The speech material in the Crisis speech database consists of warning messages with lengths ranging from one word to four sentences.

4 Acoustic Properties of the Databases

We investigated several acoustic characteristics of the databases recorded in the way described above. Many investigators observed that vowel durations (including diphthongs and syllabic consonants) increase, while consonant durations become shorter in loud speech [4]. Figures 1 to 4 show the average segmental lengths in the Crisis_SC expressive databases. We observe that our data confirm this trend. As there are only short vowels in the orthographic form of Romani and the language does have diphthongs, we present the mean of segmental durations only for a, e, i, o, u vowels (Figure 1). For Slovak we present the whole range of vowels, diphthongs and syllabic consonants that follow the abovementioned trend (Figure 2). Figures 3 and 4 present the segmental lengths of consonants that are getting shorter with increasing vocal effort in Romani and Slovak respectively. The vowels and diphthongs are lengthened to about 105% at the second level and 110% at the third level. The aspirated consonants Ch, Kh, Ph, Th, that are typical for Romani are (similarly to other consonants) shortened with increasing expressive level. The second investigated feature was the speech rate (in syllables per second).

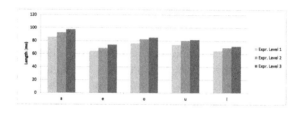

Fig. 1. Lengths of Romani vowels with increasing voice effort, Crisis_SC-ro

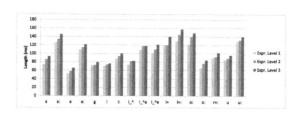

Fig. 2. Lengths of Slovak syllabic nuclei (vowels, diphthongs and syllabic consonants) in increasing voice effort, Crisis_SC-sk

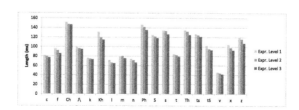

Fig. 3. Lengths of Romani consonants with increasing voice effort, Crisis_SC-ro

Fig. 4. Lengths of Slovak consonants with increasing voice effort, Crisis_SC-sk

Table 1. Average speech rate of the recorded databases in syllables per second

	AVERAGE SPEECHRATE (SYLL/SEC)			
	Neutral	Expr. Level 1	Expr. Level 2	Expr. Level 3
Romani	5.16	5.27	5.12	5.09
Slovak	5.56	5.37	5.14	5.01

The results in Table 1 show, that the rate increases with the increasing expressive load. Moreover it can be observed, that the Romani Neutral database was read slower than the corresponding Slovak Neutral database. This is probably due to the fact that reading aloud in Romani is a very infrequent activity. Furthermore, the subject is a skilled radio broadcaster in Slovak and he was probably tired - and subconsciously wanted to finish the recording of this last database as soon as possible.

The fundamental frequency (F0) was measured on the voiced parts of the database, and its mean and range was calculated. To represent these characteristics we used a statistical model employing normal distribution in the form:

$$f(x) = k \frac{1}{\sigma\sqrt{2\pi}} e^{-\frac{(x-\mu)^2}{2\sigma^2}} \tag{1}$$

where μ is the mean of the distribution, and σ is its standard deviation and k is a constant related to amount of samples [5]. For the normal distribution two standard deviations

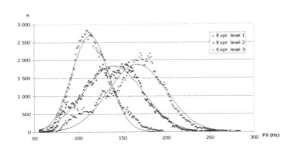

Fig. 5. The distribution of the measured F0 in Crisis_SC-ro

from the mean (in both sides) account for about 95.449% of the values, therefore we used the interval $\mu - 2\sigma \leq x \leq \mu + 2\sigma$ to define the F0 range in the database. Pearson product-moment correlation coefficient r [6] is used to express how well the measured data fit to the normal distribution. Figure 5 shows the measured F0 data and predicted normal distributions for Romani expressive databases. The results for all databases are presented in Table 2. The data indicate that the speaker was able to achieve bigger F0 range between levels Slovak than in Romani. Listening to the recordings showed, however, that the three levels of urgency are well distinguishable in both languages.

Table 2. F0 data for Romani and Slovak databases (speaker SC)

F0 (Hz)	Neutral-ro	Crisis-ro level 1	Crisis-ro level 2	Crisis-ro level 3	Neutral-sk	Crisis-sk level 1	Crisis-sk level 2	Crisis-sk level 3
μ	118	114	140	167	130	122	159	207
σ	25	21	34	33	22	18	29	24
$\mu - 2\sigma$	69	73	73	101	87	86	102	158
$\mu + 2\sigma$	167	156	208	234	174	158	217	256
range	98	83	135	132	86	71	115	97
r	0.992	0.990	0.986	0.968	0.992	0.993	0.987	0.983

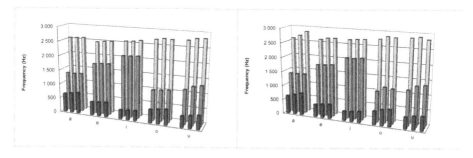

Fig. 6. The first three formants of a,e,i,o,u vocals in three levels of urgency in Crisis_SC-ro (left) and Crisis_SC-sk (right)

The changes in the positions of the first three formants of a,e,i,o,u vocals were measured too on the expressive databases and the results can be seen in Figure 6. The positive correlation of the formant positions with the level of urgency is observable - nevertheless we would need much more data to interpret the results objectively.

5 The Expressive Speech Synthesizers in Romani and Slovak

The HTS system [7] was used for experiments in speech synthesis. The neutral HMM-TTS voice was trained from the Neutral_SC database in the corresponding language. This voice was then adapted to the three levels of expressivity using Crisis_SC databases and Constrained Structural Maximum A-Posteriori Linear Regression (CSMAPLR) technique [8].

According to our informal listening tests the synthesized speech keeps the voicequality, rhythm, pitch features, and the resulting expressive load from the source recordings very well. The rising urgency is reliably distinguishable across the three versions (level 1, 2, and 3) in both languages.

Figures 7 and 8 show the spectrograms and oscillograms of the expressive utterance Imminent danger of gas leak! in Romani - "Sthoďa pes plinoskero čalaviben!" and in Slovak - "Hrozí únik plynu!" - synthesized at three levels of urgency.

It can be seen, that the range in amplitude is bigger in the Slovak version, than in Romani. If even bigger urgency is needed, some of the post-processing methods can

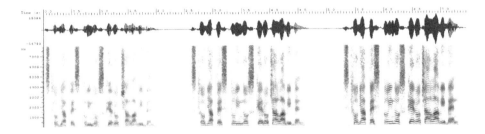

Fig. 7. The spectrogram and oscillogram of the Romani utterance "Sthoďa pes plinoskero čalaviben!" (Imminent danger of gas leak!) synthesized at three levels of urgency

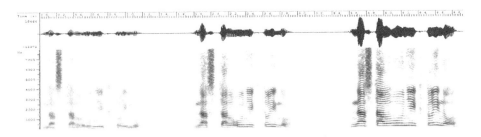

Fig. 8. The spectrogram and oscillogram of the Slovak utterance "Hrozí únik plynu!" (Imminent danger of gas leak!) synthesized at three levels of urgency

be used to increase the volume of the signal, for instance using the compression of dynamics. The method of building expressive voices by adapting the neutral voice using a small expressive database has proven its potential. The amount of about 150 expressive sentences seems to be sufficient to achieve a reliable effect of expressive load on the set of the urgent messages in the new voice.

6   Conclusion

We introduced an approach to eliciting speech resources that can be used for the development of bilingual expressive speech synthesis in Slovak and in highly underresourced language of the Slovak Roma minority - the Serviko Romani language. The produced speech synthesizer is intended for use in the public safety domain.

The expressive speech databases consisting of the annotated recordings of 90 prompted short warning messages - 160 sentences per level of urgency and per language - were designed.

An original three-step method of recording expressive speech database was proposed and successfully employed. The sentences were uttered by one male bilingual speaker in three levels of urgency. The first one represents neutral speech and serves mainly as a base comparison and reference level to the higher two levels. The second level represents assertive warnings or commands, and in the third level the speaker utters the messages in extremely intense and urgent way as if human lives were directly endangered and he should try to save them.

Larger neutral speech databases were recorded by the same speaker in both languages to create higher quality neutral voices, that can be later adapted to warning voices.

Preliminary experiments with HMM speech synthesis were performed. Their results suggest that the proposed method of expressive speech database development is suitable for gathering a good quality expressive and hyper-expressive speech database for the design of speech synthesizers for emergency situations.

The samples of the synthesized speech can be downloaded from [9].

Acknowledgements. This publication is the result of the project implementation: CRISIS, ITMS 26240220060 supported by the Research & Development Operational Programme funded by the ERDF.

References

1. Rusko, M., Darjaa, S., Trnka, M., Zeman, V., Glovňa, J.: Making Speech Technologies Available in (Serviko) Romani Language. In: Sojka, P., Horák, A., Kopeček, I., Pala, K. (eds.) TSD 2008. LNCS (LNAI), vol. 5246, pp. 501–508. Springer, Heidelberg (2008)
2. Rusko, M., Darjaa, S., Trnka, M., Cerňak, M.: Expressive speech synthesis database for emergent messages and warnings generation in critical situations. In: Language Resources for Public Security Workshop (LRPS 2012) at LREC 2012 Proc., Istambul, pp. 50–53 (2012)
3. Hübschmannová, M., et al.: Rules of Romani Orthography, State Paedagogical Institute, Bratislava (2006) (in Slovak)
4. Geumann, A.: Segmental durations in loud speech. In: Proceedings of the ITRW on Temporal Integration in Perception of Speech, Aix-en-Provence, France (2002)
5. Hazewinkel, M.: Normal distribution. Encyclopedia of Mathematics. Springer (2001) ISBN 978-1-55608-010-4
6. Rodgers, J.L., Nicewander, W.A.: Thirteen ways to look at the correlation coefficient. The American Statistician 42(1), 59–66 (1988)
7. Zen, H., Nose, T., Yamagishi, J., Sako, S., Masuko, T., Black, A.W., Tokuda, K.: The HMM-based speech synthesis system version 2.0. In: Proc. of ISCA SSW6, Bonn, Germany (2007)
8. Nakano, Y., Makoto, T., Yamagishi, J., Kobayashi, T.: Constrained Structural Maximum A Posteriori Linear Regression for Average-Voice-Based Speech Synthesis. In: Proc. of ICSLP 2006 (2006)
9. http://speech.savba.sk/ROMANI

Extracting Relations between Arabic Named Entities

Abdullah Alotayq

Imam Muhammad ibn Saud Islamic University, Riyadh, Saudi Arabia
asalotayq@imamu.edu.sa

Abstract. In this paper, a machine learning based Relation Extraction experiments for data from Arabic language are presented. There were 6 Types of relations involved in the experiments and 18 Sub-Types. By this work, a baseline for statistical Relation Extraction for Arabic is being created, with .85 acc., testing on 10-folds for the ACE 2005 Multilingual Training Data V6.0. Several factors contributed to the enhancements of accuracy for the testing data, especially the morphological and POS information using MaxEnt classifiers.

Keywords: Arabic, Classification, Relation, Named Entity, Information Extraction, Natural Language.

1 Introduction

Relation Extraction (RE) is a sub-field of Information Extraction (IE). IE is the field that aims to automatically extract structured information from unstructured and/or semi structured machine-redable documents. The Named Entity Recognition (NER) part of it greatly benefits from the approaches in IE, data mining and NLP in general. In our case, we are extracting information about relations that exist between Named Entities. This is essentially a next step to the previous NER task. Pattern Recognition Techniques are considered the backbone for IE and so for RE. Much of the success in this task relies on the success of capturing all representative patterns.

2 Relations

A relation is an aspect or quality that connects two or more things or parts as being, belonging, working together, or as being of the same kind. In our case, the relation is what connects two things together (Things with independent, separate, or self-contained existence, aka Named Entities). So, in formal, we can define Relation Extraction as the process of recognizing the type of relation that connects two or more Named Entities. Examples of such entities include: *names of persons, organizations, locations, expressions of times, quantities, monetary values, percentages,* etc.

In fact, relations are too many to be counted, but relations involved in the experiment we have were: **Physical, Part-Whole, Personal-Social, ORG-Affiliation, Agent-Artifact, GEN-Affiliation**. In addition, there were many Sub-Types such as: Geographical for **Part-whole**, Business or Family for **Personal-Social**. Some brief explanation for each of the six Types follows:

I. Habernal and V. Matousek (Eds.): TSD 2013, LNAI 8082, pp. 265–271, 2013.

1. **Physical:** Such as Located in, or Near a physical entity.
2. **Part-Whole:** Can be either Geographical, Subsidiary, or Artifact link between a part and a whole entities.
3. **Personal-Social:** Describes the relationship between people.
4. **ORG-Affiliation:** Describes Employment, Ownership, Membership etc.
5. **Agent-Artifact:** Such as User, Owner, or Manufacturer (of an Artifact).
6. **GEN-Affiliation:** links such as: Citizen of, Resident of, Religion, Origin etc.

3 Data

Automatic Content Extraction (ACE) is a program created by NIST in 1999 [1] to develop some advanced information extraction techniques. ACE-creators' challenge was to detect Entities, Relations, and Events. Their research was committed to collecting and annotating data appropriate for each of these tasks. However, their work was also to create tools to evaluate and support research on these three tasks.

Data was taken from 3 different sources: Newswire, Broadcast news, and Web blogs. The first two ones consisted of 80% of the data with 40% for each, while the last one made up the rest (20%). Dual first pass (Complete annotation) was picked among other types of annotations available. Types of annotations available from ACE were: 1P (first pass), DUAL (dual first pass), ADJ (discrepancy resolution/adjudication), and NORM (TIMEX2 normalization) [1]. Annotation provided the following information: *Modality, Tense, Relation-Extent, Syntactic class, Time stamping, Types,* and *Sub-Types.* Table 1 shows the number of instances available for each Type.

Table 1. Number of Relation mentions for each Type

TYPE	Number of Instances
PHYS	494
PRT-WLE	981
PER-SOC	679
ORG-AFF	1567
AGT-ART	378
GEN-AFF	1424
Total	**5523**

Annotation provided essential information such as *Type, Sub-Type, Tense,* and *Modality* for every Relation. *Lexical-Condition* was provided for Relation-Mentions, and *Role* for Relation-Mention Argument. *Tense* was divided amongst four categories: *Future, Past, Present,* and *Unspecified.* Moreover, *Present* was the dominant in terms of number of relations. *Modality* was identified as either *Asserted* or *Other* with clear skew towards *Asserted,* knowing that *Other* occurred in only 30 Relation-Mentions.

Lexical-Condition refers to one out of many syntactic classes, in which these classes were found frequent in Relation-Mention extents. *Lexical-Condition* comprises of 8 classes including: *Coordination, Formulaic, Participial, Possessive, PreMod, Preposition, Verbal,* and *Other*. *PreMod* annotated 3760 of the entire data while *Coordination* occurred only twice. *Role, Argument-1* and *Argument-2* were annotated and summed up to 8928 each, given that a relation can be of multiple *Types*. *Role* values were mostly about Time (*Before, Ending, Holds, Starting, Within*) with *Within* being the most occurring (111 times).

4   Related Work

Much research has been done in the Relation Extraction domain. Works in this domain can be classified into 3 categories (Feature-based, Kernel-based, and Logic-based) according to the approach used.

4.1   Feature-Based Approaches

Knowledge/Rule-based is considered the first sub-approach for this category. It follows the construction of transducers which were constructed based on extracted patterns (Libraries). The pattern library is what determines the capturing of relations expressed in the mined text. One work by Abdelmajid Ben Hamadou et. al. [2] implemented a transducer-based system that transforms patterns into transducers. They used a journalistic corpus and some data from Wikipedia as their training data (from which they extracted the patterns). They used a system called NooJ which provided many dictionaries (used for processing). For example, they had used dictionaries for: *Titles, First Names, Last Names, Geographical Names, Adjectives, Demnonym Adjectives* etc. In their experimentation they scored 63% Precision and 78% Recall for Sports Venues data.

Another approach under this category is the *un-supervised* approach which requires large amount of un-labeled data. This is considered a fully automated approach amongst all other approaches that deal with the relation extraction problem and many other NLP tasks too. Un-supervised approach requires high frequent entity pairs, which is clearly not the case in most/all of relations expressed in running texts. A few works has been done in the un-supervised approach and they mostly focus on clustering and computing similarities between features that mark relations. As an example, Min Zhang et. al. [3] worked on computing parse tree similarity. And their approach follows three steps:

1. Computing the similarity between two parse trees using some similarity function.
2. Clustering using a hierarchal clustering algorithm.
3. Labeling each cluster, and pruning bad clusters.

They reported 90% Precision and 84% Recall on a .29 threshold. They set up this threshold to eliminate un-reliable clusters. This elimination is done through pruning clusters whose NE pair number is below a pre-defined threshold. Another work here was done by Hasegawa et al. [4]. They proposed a method that deals with analyzing contexts. The method assumes that relations of the same type share similar contexts. So, computing similarity between contexts and creating clusters was their method. There were also some methods that use massive clustering (ensemble, or hybrid). Those approaches use output from linguistic analysis and semantic typing information taken from a knowledge base.

The problem in this approach was recovered by the *semi-supervised* approach which uses a set of seeds. Seeds are either target-relation instances or sample of linguistic patterns. Seeds are particularly used to acquire more elements, till it reaches a stage where all the target-relations are discovered. This process is commonly referred to as Bootstrapping. One important work was presented by Zhu Zhang [5]. This work performed both *semi-supervised* and *supervised* experiments on ACE corpus for English (which is considered be to compared to this work). He used a Multi-stage Bootstrapping and SVM classifiers to preform his experiments. He achieved 78.5% Precision and 69.7% Recall for the best performing type he had (ROLE).

Chris Welty et. al. [6] used large amount of instance-level background knowledge in their work. A corpus and a knowledge base were used as a pipeline. However, They used English Gigaword corpus, Wikipedia, news-resources, IMDB, DBPedia, and Freebase (KB) as pipeline. Their method had three steps:

1. Gathering patterns from a knowledge base.
2. Associating patterns with a set of relations.
3. Recognizing patterns and extracting relations from text.

This approach seems a little different from most of the work in relation extraction since it focuses on association techniques. Another work here is the one by Jinxiu et. al. [7]. They used Label-Propagation instead of Bootstrapping to resolve the relation detection problem but most of the semi-supervised works are still Boostrapping-based.

The last approach here is the *supervised*. This approach assumes the presence of fully annotated data. Works in this approach can be also divided into two categories: Feature-based and Kernel-based. In the Feature-based, Zhu Zhang [5] reported 85.7% Precision and 82.4% Recall (for the same relation Type mentioned earlier).

4.2 Kernel-Based Approaches

Kernel-based Approaches try to solve the problem of data representation by implicitly calculating feature vector dot-products in very high dimensionality spaces without having to make each vector explicit. This is justified by the fact that much of data in natural language processing cannot be explicitly represented. To define what kernel-function means, we can say that it's a kernel function over an object space X. This is a symmetric, positive semi-definite binary function that assigns a similarity score between two instances of X. And necessarily, every dot-product between two vectors is a kernel-function.

4.3 Logic-Based Approach

Some recent Logic-based approach was done by Horvath et. al. [8]. It considers the dependency trees of sentences as relational-structures, and Examples of the target-relation as ground atoms of a target-predicate. They represent each Example by a definite first-order Horn-clause, and picked Plotkin's-LGG operator as their choice. A divide-and-conquer algorithm for listing a certain set of LGGs was used. However, it uses the LGGs to generate binary features. The hypothesis is computed by applying SVM to the feature vectors resulted. However, they claim that their approach is comparable to state-of-art approaches.

4.4   Relation Extraction for Arabic

To the best of my knowledge, the only available system and work thats been done so far with regard to Arabic relation extraction is the one by Abdelmajd Ben Hamadou, Odil piton, and Hela Fehri [2]. This work uses the platform NooJ as a resource to providing dictionaries for patterns (uses other tools too, such as NER, Stemmer and Tokenizer). It's important to note that this work is limited to PER-ORG relation. Patterns in this work were extracted from a corpus, and then transformed into Transducers (implemented in NooJ format). The NER and RE systems used were basically a collection of dictionaries; *Title, First-name,* and *Last-Name* for PER recognition. *Geographical-Names, Type-Institution, Adjective* Dictionaries for the ORG recognition.

5   Experiments

In order to have stable numbers, a 10-folds setting was used for every experiment performed. Experimentation started with the minimum information available about the instance. So, experimentation started with the lexical items and some info provided by the annotation (+TENSE, +MODALITY). That did not show great results with only 30% testing accuracy. A next step was to add more info regarding the instance.

Stanford Tagger [9] was used to tag the data and provide POS information for each word in the instance. This step improved the accuracy by 8%. Another Stanford tool was used to provide syntactic info about data. Stanford parser [9] was used to perform parsing and increased accuracy by 3%. One more syntactic feature that was incorporated is the Lexical-Condition for Relation mention's syntactic class and showed 4% increase.

Syntactic features summed up all to 7% increase. But, experimentation continued and the number of Types involved was reduced to 4, and later to 2. However, it was important to try some extra method to improve the classification. The author attempted to do morphological analysis using Darwish's Morphological Analyzer [10]. The Analyzer converts Arabic characters into Latin characters with some special treatment for vowels. As a next step, the analyzer performs one of two tasks: Stemming or Rooting. Clearly, Rooting seemed to be contributing a lot to the classification results. It in fact, increased the accuracy by 18% for the selected two relation types.

The Best performing Types at this stage were GEN-AFF and ORG-AFF. This great performance for these two types is largely linked to the large number of labeled data for these two types. Training data for these two Relations were higher than 1,400 instance for each. By adding the features mentioned earlier and reducing the number of labels, a testing accuracy of **85%** was reached for the selected two Relations. However, adding

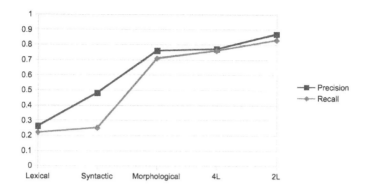

Fig. 1. Precision and Recall behavior

Table 2. Classification Results

Num. of Labels	Category	Sub-cat.	Acc.	P	R	F
6	Lexical	+TENSE +MODALITY	0.30	0.28	0.18	0.22
		+POS	0.38	0.26	0.22	0.24
	+Syntactic	+Chunks +LEXICAL-CONDITION	0.45	0.48	0.25	0.33
4	+Morphological	+Roots +Stems	0.71	0.67	0.71	0.69
			0.78	0.77	0.76	0.77
2		-Stems	**0.85**	0.87	0.83	0.85
		-Roots +Stems	0.84	0.82	0.86	0.84

2 more Relations (PER-SOC, PHYS) gave us 78%. This went down to 71% by adding remaining 2 Types (Total: 6). For the record, using the Sub-Type information did not seem helpful both as labels or just features.

It was clear that Morphological information was the most helpful part in dealing with Arabic data due to the rich Morphology Arabic expresses. However, unlike other languages, Arabic did not show that much improvement by doing the Syntactic chunking. Classification results are showed in Table 2. and Figure 1 shows the improvement resulted from label-reduction, and features added in terms of Precision and Recall.

6 Conclusion

It can be concluded that the most beneficial features tested were the Morphological, POS and the Syntactic, for relations expressed in Arabic. However, increasing the number of annotated instances for each type in the training data seemed crucial to training the classifier properly and get reasonable classification results.

As noticed earlier, most of this presented work is *supervised* and reducing/eliminating annotation cost means going towards more raw data. Therefore, a next step to this study would be doing both *semi* and *un-supervised* classification to this data, and this is what being considered for future work.

References

1. Walker, C., et al.: ACE 2005 Multilingual Training Corpus Linguistic Data Consortium, Philadelphia (2006)
2. Hamadou, A.B., Piton, O., Fehri, H.: Multilingual Extraction of functional relations between Arabic Named Entities using NooJ platform. hal-00547940 - version 1 (2010)
3. Zhang, M., Su, J., Wang, D., Zhou, G., Tan, C.L.: Discovering relations between named entities from a large raw corpus using tree similarity-based clustering. In: Dale, R., Wong, K.-F., Su, J., Kwong, O.Y. (eds.) IJCNLP 2005. LNCS (LNAI), vol. 3651, pp. 378–389. Springer, Heidelberg (2005)
4. Hasegawa, T., Sekine, S., Grishman, R.: Discovering relations among named entities from large corpora. In: Proceedings of the 42nd Annual Meeting on Association for Computational Linguistics (2004)
5. Zhang, Z.: Weekly-supervised relation classification for Information Extraction. In: Proceedings of ACM 13th Conference on Information and Knowledge Management (CIKM 2004), Washington D.C., USA, November 8-13 (2004)
6. Welty, C., Fan, J., Gondek, D., Schlaikje, A.: Large scale relation detection. In: Proceedings of the NAACL HLT 2010, First International Workshop on Formalisms and Methodology for Learning by Reading. ACL (2010)
7. Jinxiu, C.: Automatic relation extraction among named entities from text contents. PhD thesis, University of Singapore (2007)
8. Horváth, T., Paass, G., Reichartz, F., Wrobel, S.: A Logic-Based Approach to Relation Extraction from Texts. In: De Raedt, L. (ed.) ILP 2009. LNCS, vol. 5989, pp. 34–48. Springer, Heidelberg (2010)
9. Green, S., Manning, C.: Better Arabic Parsing: Baselines, Evaluations, and Analysis. In: COLING (2010)
10. Darwish, K.: Building a shallow Arabic morphological analyser in one day. In: ACL 2002 Workshop on Computational Approaches to Semitic Languages, Philadelpia, PA. ACL (2003)

Foot Detection in Czech
Using Pitch Information and HMM

Jan Bartošek and Václav Hanžl

Department of Circuit Theory, FEE CTU in Prague,
Technická 2, 166 27 Praha 6 - Dejvice, Czech Republic
{bartoj11,hanzl}@fel.cvut.cz
http://obvody.fel.cvut.cz

Abstract. In the presented work we are dealing with modelling and detection of lexical stress-group (foot) for Czech language. Detection of foot as one type of supra-segmental (prosody) information nearly corresponds to detection of word boundaries. Every native speaker is able to distinguish the feet in continuous speech, but on the other hand there are still no obvious connections between the sound qualities (pitch, intensity, syllable length) and foot prominence realization in Czech. In the experiment we tried to train the Hidden Markov Models (HMM) for Czech feet representation using only pitch information in the syllable nuclei. The most of Czech SPEECON database was used as an experiment source database. A necessary part of the presented system is a tool that transforms given Czech text into the foot units according to the known linguistic rules.

Keywords: prosody, stressed-group detection, foot, pitch, clitics absorption, ASR, HMM.

1   Introduction

Prosodic information is still not sufficiently used in nowadays automatic speech recognition (ASR) systems. Although this kind of system can generally benefit from the use of prosody, it is commonly lost during the parametrization process. Besides modality detection of the sentence, the lost prosodic information should be also able to give a cue about borders of stress-group units (feet). This could in the end help the ASR to decide in special cases where syllable chain of the utterance is ambiguous without foot placement decision (typical Czech examples are "proti vnějším" vs. "protivnějším" or "světlo v ní mají" vs. "světlo vnímají").

There have been several attempts to find stressed syllables in the utterance. Probably one of the first was study [1] which was dealing with relevant factors as prominence indicators. For fixed stressed language there is study [2] developing word boundary detection system for Hungarian with trained HMM on F0 and energy as prosodic features obtained in regular time interval. In Hungarian, all acoustic qualities realizing lexical stress correspond and presented success rate was about 77% for word boundary detection task. Study [3] engaged in detection of emphasized words for Czech, but in that case it was sentential stress detection rather than lexical stress we are dealing with. They claim that stressed syllables in Czech are generally characterized mostly by intensity

I. Habernal and V. Matousek (Eds.): TSD 2013, LNAI 8082, pp. 272–279, 2013.

increase, plus there is some increase in duration and minor increase in pitch. Authors then detect emphasized word in the utterance by simply summing the normalized contours of all of these qualities and the syllable with highest peak is then considered as the beginning of the emphasized word in the utterance. Their system achieves overall score of 91% in the task of emphasized word detection out of 180 sentences recorded specially for this experiment. The most significant feature alone was found to be relative word prolongation (86% score). In [4] there was introduced a special system of stressed/unstressed vowels for Dutch acoustic modelling in their ASR, but the final decrease of WER was not observed. Work [5] followed up the stress detection using the spectral slope as the feature. They recorded three-syllable pseudo words with different option of stress placement. By computing the energy ratios of bands 350-1100Hz and 2300-5500Hz they observed that all stressed vowels show less spectral scope value, they are more spectral-balanced than the unstressed. Unfortunately, their approach is dependent on a knowledge of particular vowels.

According to known research [6,7], the Czech prosodic system for lexical stress realization is very unique and mentioned approaches [2] are not directly applicable for Czech. We do not know about any research that explores detection success rate of Czech lexical stress (or foot units) on larger data corpus based on various acoustic qualities. In contrast to all mentioned works for Czech, we operate on huge set of utterances from Czech SPEECON database used commonly for training of ASR systems for Czech. In this initial experiment we focus on the possibility of stress-group (foot) unit detection based purely on melody (pitch) information. We also focus on development of an automatic syntactic-level ASR module with ability of utterance division into stressed-groups. Our Czech feet modelling approach is based on collaboration with the ASR that provides specific information (especially syllable nuclei centers time-stamps obtained by force-alignment process) in combination with utilizing the raw acoustic data for the feature extraction. In this paper we tried to find how pitch information as the only feature used can model feet in Czech.

The article is organized as follows: Section 2 brings some theory about Czech feet. Used data set is described in Section 3, used features and their normalizations in Section 4. An overview of the whole training phase of the system for Czech feet detection is presented in Section 5. Experimental setup is closely presented in Section 6 followed by achieved results (Section 7), which are discussed together with possible future work.

2   Facts about Czech Language and Feet Realizations

A foot (stressed group) is a unit of speech binding one prominent syllable and certain count of non-prominent syllables. Czech is a fixed-stress language with stressed first syllable. From the speech rhythm point of view Czech is "syllable-timed" language (syllable is considered as basic perceived time resolution step of the speech), which means there is usually no speeding up or slowing down of the speech according to the length of the foot. In [6] it is claimed that foot is really perceived as a basic unit of the utterance. Prominence in Czech is probably achieved by mixture of sound qualities - melody contour, speech dynamic and also durational features of syllables (especially their vowels), but there still has not been discovered definitive description of how these sound quality features correspond together when creating the prominence (in contrast with [2] where speech dynamic and also pitch have similar contours with peaks corresponding to prominent syllables). Description of each type of prominence realization follows:

a) When prominence is realized by pitch change, it can be done by both rising or lowering the pitch in Czech. There are known typical musical intervals for both cases that represent valid pitch prominence (measured on synthetic isolated word database): up - by one semitone, down - by four semitones (musical third). When there is longer chain of syllables in the foot, the pitch changes are even less (level of musical quartertones). This fact involves the need of a precious pitch detection algorithm (PDA) when studying foot intonation contours. Pitch prominence is probably used most often as word-level prominence indicator, but greater pitch changes than denoted are probably perceived as key sentence melody events. This involves the problem - sentence modality is distinguished by the melody contour and this occurs also across the foot prominence realized also by the pitch. Resulting pitch contour of the sentence is thus a combination of foot pitch prominences and sentence melody and when examining the pitch contour, we are dealing with both sentence and word-level segmentation. One of the latest studies [7] about sound qualities of the feet prominence and the role of the pitch in determining foot boundaries for Czech confirms one of the former theories. The strongest tendency found is that the pitch contour with clear F0 drop in the middle does not represent an acceptable form of a foot.

b) In the past it was thought that the prominence in Czech is done only by the dynamic. This has been proven to be true for isolated words synthetic material, but according to the later studies aimed at real complex utterances it very often occurs, that the syllable with prominence is less dynamic than the rest of the foot. That is why the dynamic is apparently not the driving feature of the prominence.

c) Length of syllables: In Czech it is not allowed to evidently prolong or shorten the syllable duration, because of its functional meaning. Nevertheless, very slight changes in the duration might also play a role in determining the prominence.

In continuous speech Czech multi-syllable words very often keep their independence and they create own feet. On the other hand single syllabic words very often lose their autonomy and create one foot together with the other neighbouring word. These single syllabic words are called clitics. There exist grammatical rules [8] how particular word categories behave - if they join their predecessor (then they are marked as enclitics) or successor (proclitics). On the basis of these rules was created lexical module capable of text-to-foot conversion [9].

3   Used Data Set

Firstly, we considered to use the Czech Audiobooks as the corpus for the whole system but we decided not to do this. The main reason was the professionalism of the speakers which lead into excessive intonation over the whole data set. This fact can be beneficial for sentence modality detection, but for foot detection this can be inconvenient. Rather we decided to use part of the standard Czech SPEECON speech database with recordings of 550 adult non-professional speakers (16kHz, 16-bit, mono). We used cleaned subset of sentences and further filtered diftongs because of their ambiguous syllabification in Czech. We finally obtained final subset of 10,022 sentences which were then processed to obtain their foot division and features. Sentences were then divided into training and testing subset (ratio 9/1).

4   Used Features and Their Normalizations

In this experiment we limited on using this set of features computed at each syllable nucleus: pitch, Δpitch and $\Delta\Delta$pitch. The features are not extracted equidistantly on the time axis, but are driven by the occurrence of the syllable nucleus. To obtain pitch in musical units we firstly converted fundamental frequencies from Hz into semitones scale related to the mean pitch frequency for the utterance measured across voiced regions only (Eq.1). We further refer to this type of normalization as "norm0".

$$SemitoneDifference = 12 \log_2(\frac{F0}{F0_{mean}}) \tag{1}$$

Second type of normalization ("norm1") is based on the knowledge of utterance division into the feet and relates computed pitch to the mean pitch of given foot. Third type of normalization ("norm2") is based on fitting the "norm0" values with the 2nd order polynomial function and computing the difference of each pitch point to the corresponding function value. Unfortunately, the very last foot (as being often very decreasing in pitch) tends to make the curve more convex for the other parts of the utterance. This is why we introduced "norm3" where we do the same process as for "norm2" but with removed last foot of the utterance from all data structures.

5   System Overview

The process of training (and also testing) data preparation can be seen in the Fig.1. Czech SPEECON sentence subset is the only information source for further processing and feature extraction. Firstly, fundamental frequency (F0) of all audio files are obtained using Praat [10] autocorrelation function with lowered voice/unvoiced threshold (set to value 0.2). This threshold is quite low for ordinary use, but is more convenient for our case as we know exactly where the syllable nuclei should be located (and thus we allow some VUV errors outside our regions of interest). Praat F0 output is converted into the form with 1ms timestep for convenient further processing.

Another step is a retrieval of the .NUC files with timestamps of syllable nuclei. Using HTK we created and trained standard context dependent triphone acoustic models (three states per model) on the same dataset that were further used for force-alignment of all the utterances. This allowed us to obtain syllable nuclei time-stamps as the aligned times of triphone middle states. Stressed and non-stressed vowels were not anyhow distinguished in the phoneme alphabet used for force-alignment.

Having both .F0 and .NUC files we can create feature vector (.TRAIN) files for model training. A production of these feature files is not limited to just picking the

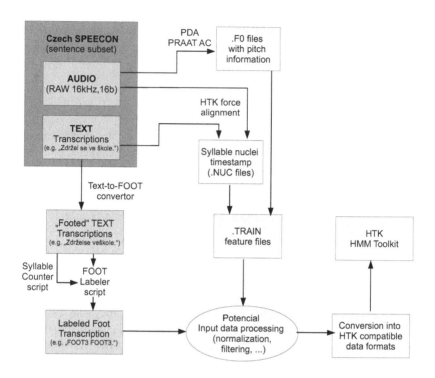

Fig. 1. Scheme of the training data preparation process

corresponding pitch in the time of nucleus center occurrence. It does also contain a logic for repairing the estimates of these centers by seeking for the contiguous regions of pitch around nucleus center (influencing the range for delta and double delta features computation) and also logic for approximation of missing pitch (frame is denoted as unvoiced by PDA). Moreover, various types of pitch normalization can occur during and after creation of .TRAIN files.

On the lexical side of the chain utterance text transcriptions enter the Text-to-Foot (TTF) module [9]. Basically, it follows the strong rules for Czech clitics absorption [8]. The transcription is converted by the module into the output text containing stressed groups division (an example of Czech sentence on the input and its foot-oriented conversion can be seen in the Fig.1). It has been proven [9], that used TTF module generates (after transformation of the whole dataset) very similar distribution of feet with given lengths as it is referenced by respected literature [6] for Czech speech. Obtained stress-grouped transcription enters the FOOT labeller which in connection with the syllable counter module produce final FOOT-labeled utterance transcription. One of possible outputs can be e.g, "FOOT2 FOOT3." (one 2-syllable foot followed by one 3-syllable foot). The syllable counter can manage special diftongs as well as syllable-creating consonants 'l' and 'r'.

6   Experiment Setup

To model our problem we utilized Hidden Markov Model (HMM) approach, because our task is similar to standard speech recognition tasks, where HMM framework is widely adopted. All the experiments were performed using HTK Speech Recognition Toolkit [11]. We trained HMMs for various feet lengths (1-8 syllables) on the training subset. Our models of feet are in comparison with standard speech triphone models without state self-loops and backward state transitions. Thus, we do not allow the model to stay in any state and state-flow of our system is strictly forward with coming input features. We illustrate HMM model for two-syllable stressed-group (label FOOT2) in the Fig.2(a). Each emitting HMM state was modelled using one mean and variance (no mixtures). Each utterance is generally modelled by the grammar depicted in the Fig.2(b). The a priori probability was not modified for any of the models and thus is equal for all of them.

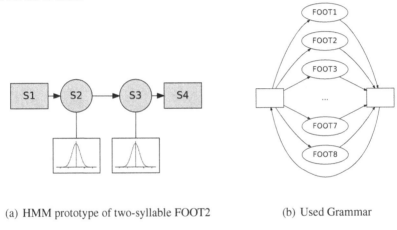

(a) HMM prototype of two-syllable FOOT2   (b) Used Grammar

Fig. 2. Illustration of used HMM modelling

We trained and tested the models using various forms of pitch data, but main experiments were related to four suggested types of normalization. As for testing with "norm1" data, we realized that in real system these data will not be available, because the division of the utterance into feet will be unknown. This is why we decided to make experiments with feet models trained on "norm1" data, but tested with different normalization types available in real situation. We also prepared version of experiment with filtered final feet of utterances. We expected this could improve the accuracy of the system, because we would remove the feet most influenced by sentence intonation. Besides, in another version we filtered out all the sentences that contained comma indirect speech, because their feet should be mostly affected by complex sentence intonation.

7   Results, Discussion and Future Work

We performed various versions of experiment, but only those most valuable are quoted in Tab.1. We actually found that neither filtering of the last foot nor filtering the complex-sentences out of the dataset did not improve the accuracy. By using "norm1" feature

Table 1. Results from HResults for foot detection using pitch information

Training data	Testing data	Corr	Acc	D	S	I
norm0	norm0	34.2	18.2	2326	2718	1223
norm1	norm1	53.4	46.1	1739	1755	561
norm2	norm2	30.4	18.7	2659	2563	876
norm3	norm3	32.1	20.4	2240	2154	759
norm1	norm3	38.7	32.1	1460	3145	493

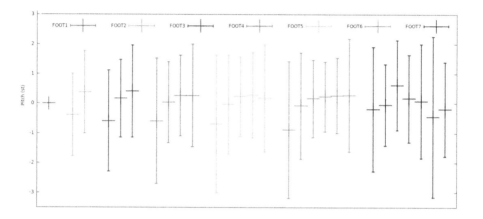

Fig. 3. Pitch means and STD for "norm1" trained HMM models of Czech feet

files, we were able to obtain accuracy up to 46%, which denotes rightness of this type of normalization for our task. In the Fig.3 we can see these "norm1" trained models FOOT1-FOOT7 with plotted pitch values of emitting states. Models for FOOT2-FOOT6 very well satisfy the theoretical condition for foot existence declared in [7] – that intonational contour with pitch drop in the middle does not create acceptable form of foot. In more real scenario using norm3 features as testing data while still keeping HMM models trained on norm1 features, we were able to recognize feet in utterances with 32% accuracy.

Although reached results are not much impressive, there is still a lot of possible future work and improvements pending:

1. Collaboration with Czech language specialists (phonetics), which can lead into careful verification of used features and corresponding labels and also improvements of the used text-to-foot module
2. Choice of features - except the pitch there are another features that are obtainable and worth to try (intensity, durations of syllables or vowels and spectral slope [5]).

8   Conclusions

We have created the framework for the experiments dealing with the automatic Czech feet detection based on the subset of the Czech SPEECON database. On the lexical level we utilize necessary module that converts given sentence into feet units. In this experiment, stressed-groups HMM models were trained using only pitch information, but with various types of normalizations. Trained models are in strong accordance with results of other phonetic studies examining Czech feet pitch behaviour [7]. They statistically confirm the fact that most of lexical stress is realized by pitch drop on the first syllable of the foot. Unfortunately, on our testing set we were able to achieve only 32% foot recognition accuracy. To achieve better scores, it seems that additional features extracted from the acoustic signal would be needed.

Acknowledgments. This work has been supported by the GA of the Czech Technical University in Prague, grant No. SGS12/143/OHK3/2T/13.

References

1. Fry, D.B.: Experiments in the perception of stress. Language and Speech 1, 126–152 (1958)
2. Vicsi, K., Szaszák, G.: Using prosody to improve automatic speech recognition. Speech Commun. 52, 413–426 (2010)
3. Kroul, M.: Automatic detection of emphasized words for performance enhancement of a czech asr system. In: Proceedings of 13th International Conference Speech and Computer (Specom 2009), Petersburg, Russia, pp. 470–473 (2009)
4. Kuijk, D.V., van den Heuvel, H., Boves, L.: Using lexical stress in continuous speech recognition for dutch. In: Proc. ICSLP 1996, pp. 1736–1739 (1996)
5. Volín, J., Zimmermann, J.: Spectral slope parameters and detection of word stress. In: Proceedings of Technical Computing Prague (Humusoft), Praha, pp. 125–129 (2011)
6. Palková, Z.: Fonetika a fonologie češtiny (Phonetics and phonology of Czech). Karolinum, Praha (1994)
7. Palková, Z., Volín, J.: The role of f0 contours in determining foot boundaries in czech. In: Proceedings of the 15th ICPhS, Barcelona, vol. 2, pp. 1783–1786 (2003)
8. Hauser, P.: Základy skladby češtiny. Masarykova Univerzita, Brno (2003)
9. Bartošek, J.: Czech text-to-foot converter. In: POSTER 2013 (CD-ROM) (2013)
10. Boersma, P.: Praat, a system for doing phonetics by computer. Glot International 5, 341–345 (2001)
11. Young, S.: The htk hidden markov model toolkit: Design and philosophy. Entropic Cambridge Research Laboratory, Ltd. 2, 2–44 (1994)

Improved Hungarian Morphological Disambiguation with Tagger Combination

György Orosz[1,2], László János Laki[1,2], Attila Novák[1,2], Borbála Siklósi[1], and Nóra Wenszky[2]

[1] Pázmány Péter Catholic University, Faculty of Information Technology
50/a Práter street, Budapest, Hungary
{oroszgy,laki.laszlo,novak.attila,siklosi.borbala}@itk.ppke.hu
http://itk.ppke.hu
[2] MTA-PPKE Natural Language Processing Group
50/a Práter street, Budapest, Hungary
wenszkynora@gmail.com
http://nlpg.itk.ppke.hu

Abstract. In case of morphologically rich languages, full morphological disambiguation is a fundamental task that is more difficult than just providing PoS tags. In our paper, we overview Hungarian morphological disambiguation tools, and evaluate some common tagger combination techniques in order to improve annotation accuracy. Following an error analysis of the existing tools, we introduce a method that independently selects the proper tag and lemma and harmonizes them achieving a 28.90% error rate reduction compared to PurePos, a state-of-the-art Hungarian morphological annotation tool.

Keywords: morphological disambiguation, machine learning, PoS tagging, machine translation.

1 Introduction

Part-of-speech tagging is one of the basic and most studied tasks of computational linguistics. There are several freely available tools and algorithms that work with high precision. However, assigning PoS tags is only a subtask of morphological disambiguation. It is also crucial to identify the lemma, which is not a trivial task for languages having a rich morphology like Turkish or Hungarian. Nevertheless, most of the currently available tools only deal with disambiguating morphosyntactic labels; there are only few that do the whole job. Robust and accurate operation of these tools is important, since they are usually parts of larger linguistic processing chains. Thus errors propagating from this level affect the performance of systems performing more complex language processing tasks.

In our paper, we survey taggers that perform full morphological disambiguation for Hungarian, investigating and comparing their common errors. Lessons learned from the error analysis help us to combine them successfully to gain better performance.

I. Habernal and V. Matousek (Eds.): TSD 2013, LNAI 8082, pp. 280–287, 2013.

2   Background

First we give a brief overview of full morphological annotation tools for Hungarian. After comparing them, we overview commonly used tagger combination techniques. The experiments described in this paper were performed on the Hungarian Szeged Corpus [3] with PoS annotation automatically converted to morphosyntactic tags used by the Hungarian HuMor morphological analyzer [10,9]. 10% of the corpus was separated for testing and another 10% is used for development and tuning purposes. Each set contains about 7,100 sentences, while the rest, about 57,000 sentences, were used for training the systems.

2.1   Morphological Annotation Tools

PurePos [8] is an open source hybrid system for full morphological disambiguation. It is based on hidden Markov models, but it can use an integrated morphological analyzer (MA) module as well to tag unseen words and to assign lemmas. The tool uses well-known trigram tagging algorithms, but what distinguishes it from its predecessors is the complete integration of a morphological analyzer, which results in a further boost in its PoS tagging accuracy and also makes high precision lemmatization possible.

HuLaPos [7] is a purely statistical annotation tool based on an SMT[1] decoder. An advantage of applying this methodology to PoS tagging is that it can consider the context in both directions. Moreover, HuLaPos uses a higher order language model than PurePos. On the other hand, HuLaPos has an inferior performance on unseen words, although it utilizes a simple smoothing algorithm that enables it to handle such words to some extent.

Magyarlanc [13], another commonly used tool, is a full processing chain, consisting of a sentence splitter, tokenizer, part-of-speech tagger, lemmatizer, and its latest version even contains a dependency parser. It also contains a built-in morphological analyzer based on morphdb.hu [11]. As a tagger, it is reported to attain 96.33% precision on a random 4:1 split of the Szeged Corpus.

2.2   Tagger Combination Schemes

The design process of a combined system of classification or annotation tools involves several steps. First, it needs to be examined whether the errors of each system to be combined are different enough for the aggregate system to be likely to outperform the best individual system significantly. Then an appropriate combining algorithm must be found.

A basic combining scheme, which is often used as a baseline, is majority voting. Other, more advanced, combining schemes involve training a top-level classifier for the task of generating the output of the combined system based on outputs of the individual embedded systems. This class of combination schemes is commonly referred to as stacking learners. The top-level classifier may use various features of both the input and

---

[1] statistical machine translation.

the outputs of the bottom-level classifiers when making its decision. The set of features used may have a significant impact on the performance of the combined system.

Finally, decisions to be made by the top-level classifier can be of at least two sorts: it can either always select the output of one of the bottom-level systems, or it can generate an output of its own that may differ from the output of each individual embedded system. When applying the former solution, the errors of the embedded systems determine a theoretical upper limit on the accuracy of the combined system (it can never generate the expected output whenever neither of the embedded classifiers generate it), thus the latter solution seems more beneficial in theory. However, complexity of the annotation task to be performed and the available training data may have an influence on which of these options is feasible and how they perform in practice. If the cardinality of the output annotation and of the features involved in training the classifier is high, there may be either data sparseness or performance problems with the combining classifier.

One of the first attempts of combining English PoS taggers was done by Brill and Wu [2]. They propose a memory-based learning system for tagger combination that employs contextual and lexical clues. In their experiments, the solution where the top-level learner always selects the output of one of the embedded taggers outperformed the more general scheme that allowed the output differ from either of the proposed tags.

A comprehensive study by van Halteren et al. [6] presents detailed overview of previous combination attempts using mainly machine learning techniques. Several combination methods are compared and evaluated systematically in the paper. The authors show that cross-validation can be used to train the top-level classifier for an optimal utilization of the training corpus. They found a scheme perform best in their experiments that they characterize as generalized voting, although it is a scheme that can output annotation that may differ from the output of either of the embedded taggers and thus can also be interpreted as a stacking method. However, the cardinality of the tag set and the dimensionality of the feature space was modest compared to that in our case.

A system of different architecture is presented in e.g. Hajič et al. [4]: in contrast to the parallel and hierarchical architecture of the systems above, it employs a serial combination of annotators starting with a rule-based morphological analyzer, followed by constraint-based filters feeding a statistical tagger at the end of the chain.

3 Error Analysis

As we mentioned above, it is useful to start the design process with an error analysis of the systems to be combined in order to see whether a system combination is likely to improve performance. We present tagging accuracy values of PurePos (PP) and Hu-LaPos (HLP) measured on the development set in Table 1.[2] Unfortunately, magyarlanc is not directly comparable with the others above, since its built-in annotation scheme is not compatible with the HuMor scheme used by the two other tools.

It may not be evident from these values why and how combining these tools can boost performance, but deeper investigation on common errors suggests that chances for success are good.

---

[2] All other measurements in Sections 3 and 4 were also made on the development set.

Table 1. Base system accuracies

	Tagging	Lemmatization	Full disambig.
PurePos	98.57%	99.58%	98.43%
HuLaPos	97.61%	98.11%	97.03%

Table 2. Comparison of PurePos and HuLaPos

	Tagging	Lemmatization	Full disambig.
$OER(PP, HL)$	22.41%	11.66%	21.16%
$OER(HLP, PP)$	53.58%	80.21%	58.24%
Agreement rate	97.60%	98.02%	96.92%
They are right when they agree	99.30%	99.85%	99.29%
One is right when they disagree	97.53%	98.89%	97.14%
Oracle	99.26%	99.83%	99.22%

We use the metric[3] $OER(A, B) = (\#errors\ of\ A\ only/\#all\ errors)$ that measures the percentage of the cases where tagger A is wrong but B is correct in proportion of all errors that were made by either A or B. We do not use the complementarity formula proposed by Brill et al.[2], because that gives hard-to-interpret unlimited negative values in cases where there is a significant overlap between the errors made by the two taggers. Although HuLaPos makes more errors than PurePos, own error rates (Table 2) indicate that error distribution is fairly balanced between the two tools for tagging and full disambiguation. In addition, we calculated the agreement rate of the tools and the relative percentage of times they agree on the right morphological annotation. Table 2 also shows that one of them assigns the right annotation most of the time they disagree. Assuming a hypothetical oracle that can always select the better annotation output, the performance of the better tagger can be increased by more than 0.6% corresponding to 72.73% relative error rate reduction on the development set. These results encourage us to combine the two tools.

4 Annotation Tool Combinations

It was shown previously [2,6] that stacking of classifiers can improve PoS tagging accuracy. For an optimal utilization of the training material, we applied training with cross-validation in our experiments, as it was suggested by e.g. [6]. The training set was split into 5 equal-sized parts and level-0 taggers (PurePos and HuLaPos) were trained 5 times using 4/5 of the corpus, and the rest was annotated by both taggers in each round. The union of these automatically annotated parts of the training corpus was used to train the top-level (or level-1) metalearner. Thus the full training data was available for level-1 training, yet separating the two phases of the training process. In addition, this workflow made the full training material available also for the level-0 learners.

---

[3] OER=own error rate.

As the use of a "relatively global, smooth" level-1 learner is suggested in [12], we investigated the naïve Bayes (NB) classifier and instance-based (IB) learners[4] [1], which, in addition to be simple, were shown to perform well in sequence classifier combination tasks. Given the high agreement rate of the level-0 taggers on correct events, we decided to apply all metalearners only in cases of disagreement. We used the instance-based combinators as follows. After extracting features for a word on the annotation of which the tools disagree, the classifier finds the most similar previously seen case(s) based on Manhattan distance, and it selects the output of the annotation tool that most often generated the correct output in the most similar case(s). We decided to use the tagger-picking approach, since Hungarian has a tag set with a cardinality of over a thousand and an almost unlimited vocabulary, which suggests that the tag-picking approach would not be feasible.[5]

Brill et al. [2] proposed a simple but powerful feature set (FS1 in Table 3) that consists of the word to be tagged, its immediate neighbours and all their suggested tags. Since many others (e.g. [6]) applied these features successfully, we intended to extend them systematically in order to better fit languages with a very productive morphology. Since wordforms in Hungarian are composed of a lemma and numerous affixes, we incorporated longer suffix properties in the feature sets to handle data sparseness issues. To manage the free word order nature of the language, wider context features are also examined. Several experiments were run[6] in order to investigate whether using word shape (FS2,FS4), word suffix (FS3,FS4,FS6) or wider contextual features (FS5) can improve the performance of tagger selection (see Table 3).

Table 3. Feature sets used in the experiments

Feature set	Base FS	Additional features
FS1	Brill-Wu	—
FS2	FS1	whether the word contains a dot or hyphen
FS3	FS1	use at most 5-character suffixes instead of the word form
FS4	FS2, FS3	—
FS5	FS1	guessed tags for the second word both to the right and left
FS6	FS4	use at most 10-character suffixes instead of the word form

As expected, the naïve Bayes classifier (NB) performed significantly worse than the instance-based learners (IB) even when using seemingly independent features. Moreover, lemmatizer combination turned out to be an almost insoluble task for it, as error rate reduction data in Table 4 show. It is also interesting that accommodating word shape features (FS2) always improved the tagging accuracy, while the increased contexts (FS5) were generally useless. The results show that using longer suffix features is beneficial in cases where assigning a lemma is part of the task. However, for combining part-of-speech taggers only, omitting the word form and using at most five-character-long suffix features gives the best result.

---

[4] The C4.5 decision tree algorithm was also tested, but it was unable to handle the large amount of feature data involved in our experiments.

[5] We plan to verify this assumption in the future. Nevertheless, all experiments described in this paper used the tagger-picking model.

[6] The WEKA [5] machine learning software was used in the experiments.

Table 4. Error rate reduction using metalearners with different feature sets sets

Task:	Tagging		Lemmatization		Full annotation	
Feature set	NB	IB	NB	IB	NB	IB
FS1	19.03%	24.65%	-6.21%	22.24%	5.06%	22.89%
FS2	18.91%	24.82%	-0.80%	23.85%	4.95%	23.16%
FS3	21.04%	27.60%	0.80%	26.65%	18.42%	25.31%
FS4	20.92%	_27.90%_	4.01%	26.65%	18.96%	25.20%
FS5	16.37%	17.55%	-19.24%	16.03%	-0.70%	18.47%
FS6	19.27%	27.30%	-17.03%	_26.85%_	16.16%	_25.79%_

We describe below how these combinations were applied to improve automatic Hungarian morphological annotation quality, reporting on relative error rate reduction compared to that of PurePos on the development set.

4.1 Full Disambiguator Combination

The most straightforward combination method is to treat annotations as atomic and simply use the full output of the embedded tool selected by the metalearner. Results in Table 4 show that the best combination for this approach is the instance-based learner with FS6. Comparison of relative error rate reduction achieved on the development set using this model and alternative models described below is shown in Table 5.

Table 5. Relative error rate reduction on the development set

System	Tagging	Lemmatization	Full disamb.
Disamb. combination	23.05%	18.64%	25.79%
Tagger combination	27.90%	6.81%	25.26%
Multiple metalearners	29.85%	30.06%	32.42%

4.2 Combining Taggers

Another plausible scheme is to combine the morphosyntactic tagger subsystems of the annotation tools. However, in this case, one has to deal with lemmatization as well. The PurePos lemmatizer performs better than the one built into HuLaPos, so we chose to use that to generate the lemma corresponding to the selected PoS tag. The highest accuracy values for tag selection were achieved by extending the baseline feature set with word shape and at most five-character-long suffix features (FS4).

This algorithm allowed us to achieve higher error rate reduction (see Table 5) for the morphosyntactic tags, however, the gain in lemma accuracy is much lower, thus the overall accuracy improvement is inferior to that achieved using the previous method.

4.3 Multiple Metalearners

It is possible to benefit from the strengths of both types of combination using two level-1 learners: one is trained to choose the better lemmatizer and the other to select the optimal tagger for the given case. The combination scheme with best accuracy for lemmatization was the IB classifier with FS6, and that for tagging was the same algorithm with FS4. This configuration may yield incompatible tag-lemma pairs[7]. We used the HuMor analyzer to find and fix these cases: if the tag is found to be correct, the lemma is provided by the analyzer and vice versa. With this enhancement, we achieved higher relative error rate reduction and better overall accuracy than with any of the previous two methods.

5   Evaluation

We present the performance of the best combinations on the unseen test set in Table 6.

Table 6. Relative error rate reduction on the test set

System	Tagging	Lemmatization	Full disamb.
Oracle	48.60%	59.42%	51.53%
Disamb. combination	23.23%	23.55%	26.86%
Tagger combination	22.76%	13.77%	23.81%
Multiple metalearners	25.07%	29.89%	28.90%

In accordance with our results on the development set, the multiple combination method achieved the best performance with over 98.90% full disambiguation accuracy; that is a 28.90% overall relative error rate reduction. Taking a closer look at the output, we found that the best compound annotation system can partly deal with error types that are typical for PurePos. This combination scheme attained 56.08% of the possible improvement that could be achieved by a perfect oracle.

6   Conclusion

In this paper, we presented a combination of two automatic morphological annotation tools for Hungarian reducing the error rate of the better system by 28.90%. The tools combined, one based on an SMT decoder and the other an HMM-based tool, were found to complement each other well. The combination is based on a machine learning algorithm, but we use a technique that allows the whole training data to be utilized for training all level-1 and level-0 models at the same time. The described combination scheme benefits from utilization of a morphological analyzer during the whole process. The combined system outperforms the known best system for Hungarian, thus it can be used in cases where very high disambiguation accuracy is crucial.

Acknowledgement. This work was partially supported by TAMOP – 4.2.1.B – 11/2/ KMR-2011-0002 and TAMOP – 4.2.2/B – 10/1–2010–0014.

---

[7] A lemma and a tag for a word is incompatible if the MA can analyze the word, but no analysis contains both the lemma and the morphosyntactic label.

References

1. Aha, D.W., Kibler, D., Albert, M.K.: Instance-based learning algorithms. Machine Learning 6(1), 37–66 (1991)
2. Brill, E., Wu, J.: Classifier combination for improved lexical disambiguation. In: Proceedings of the 36th Annual Meeting of the Association for Computational Linguistics and 17th International Conference on Computational Linguistics, vol. 1, pp. 191–195. Association for Computational Linguistics, Stroudsburg (1998)
3. Csendes, D., Csirik, J., Gyimóthy, T.: The Szeged Corpus: A POS tagged and syntactically annotated Hungarian natural language corpus. In: Proceedings of the 5th International Workshop on Linguistically Interpreted Corpora LINC 2004 at The 20th International Conference on Computational Linguistics COLING 2004, pp. 19–23 (2004)
4. Hajič, J., Krbec, P., Květoň, Oliva, K., Petkevič, V.: Serial combination of rules and statistics: A case study in Czech tagging. In: Proceedings of the 39th Annual Meeting on Association for Computational Linguistics, pp. 268–275. Association for Computational Linguistics (2001)
5. Hall, M., Frank, E., Holmes, G., Pfahringer, B., Reutemann, P., Witten, I.H.: The WEKA data mining software. ACM SIGKDD Explorations Newsletter 11(1), 10 (2009)
6. Van Halteren, H., Zavrel, J., Daelemans, W.: Improving Accuracy in Word Class Tagging through the Combination of Machine Learning Systems. Computational Linguistics 27(2), 199–229 (2001)
7. Laki, L.: Investigating the Possibilities of Using SMT for Text Annotation. In: Simões, A., Queirós, R., da Cruz, D. (eds.) 1st Symposium on Languages, Applications and Technologies. OpenAccess Series in Informatics (OASIcs), vol. 21, pp. 267–283. Schloss Dagstuhl–Leibniz-Zentrum fuer Informatik, Dagstuhl (2012)
8. Orosz, G., Novák, A.: PurePos – an open source morphological disambiguator. In: Sharp, B., Zock, M. (eds.) Proceedings of the 9th International Workshop on Natural Language Processing and Cognitive Science, Wroclaw, pp. 53–63 (2012)
9. Prószéky, G.: Industrial applications of unification morphology. Association for Computational Linguistics, Morristown (1994)
10. Prószéky, G., Novák, A.: Computational Morphologies for Small Uralic Languages. In: Inquiries into Words, Constraints and Contexts, Stanford, California, pp. 150–157 (2005)
11. Trón, V., Halácsy, P., Rebrus, P., Rung, A., Vajda, P., Simon, E.: Morphdb.hu: Hungarian lexical database and morphological grammar. In: Proceedings of the Fifth Conference on International Language Resources and Evaluation, Genoa, pp. 1670–1673 (2006)
12. Witten, I.H., Frank, E., Hall, M.A.: Data Mining: Practical Machine Learning Tools and Techniques, 3rd edn. (2011)
13. Zsibrita, J., Vincze, V., Farkas, R.: magyarlanc 2.0: szintaktikai elemzés és felgyorsított szófaji egyértelműsítés. In: IX. Magyar Számítógépes Nyelvészeti Konferencia, pp. 238–374. Szegedi Tudományegyetem, Szeged (2013)

Improving Dependency Parsing
by Filtering Linguistic Noise

Tomáš Jelínek

Charles University in Prague, Faculty of Arts
Institute of Theoretical and Computational Linguistics

Abstract. In this paper, we describe a way to improve stochastic dependency parsing by simplifying both the training data and new text to be parsed. Many parsing errors are due to limited size of the training data, where most of the words of a given language occur seldom or not at all, thus the parser cannot learn their syntactic properties. By defining narrow classes of words with identical syntactic properties and replacing members of these classes by one representative, we facilitate language modeling done by the parser and improve its accuracy. In our experiment, a 17.8% decrease in forms variability in the training data of the Czech dependency treebank PDT led to a 8.1% relative error reduction.

Keywords: syntax, Czech, dependency parsing, text pre-processing.

1 Introduction

In recent years, much effort has been devoted to improving stochastic dependency parsing, which is used in many fields of NLP, but the error rate remains high. It could be reduced by further development of parsing algorithms and more sophisticated parser settings, but alternative ways to improve parsing are sought as well: grammar-based pre- and post-processing, hybrid parsing methods etc. In this paper, we describe a procedure that allowed us to significantly improve the parsing accuracy of Czech. It is language-specific, but the main ideas can be easily applied to other languages, tree-banks and parsers.

Our analysis of the results of stochastic parsing of Czech showed that parsers make most mistakes in less frequent structures and with less frequent words which they didn't encounter with sufficient frequency in the training data. From the syntactic point of view, however, the choice of specific word forms is often irrelevant, as this choice may be influenced by preferences of the speaker or extralinguistic reality. Insofar as this variability does not change the syntactic interpretation of a structure, we label it as linguistic "noise": for the parser, it makes the understanding of language structures more difficult. This "noise" can be filtered out: forms or lemmas with identical syntactic properties can be replaced by one representative; thereby language modeling done by the parser is simplified and the parser's efficacy (speed and accuracy) increases.

The two most thoroughly tested stochastic dependency parsers for Czech with best results are MaltParser [6] and MSTParser [5]. We experiment with both of these parsers in this paper. However, we do not describe further attempts to improve their settings. We achieve a significant improvement of their results by modifying their input data instead.

I. Habernal and V. Matousek (Eds.): TSD 2013, LNAI 8082, pp. 288–294, 2013.

The paper is organized as follows: Section 1 presents the treebank and previous work in Czech parsing. Section 2 explains why it is necessary to reduce the variability of forms and lemmas in the data and how it is achieved. In Section 3, we propose a more sophisticated feature setting for the parsers. Section 4 shows the results of the experiment. A brief conclusion is presented in Section 5.

2 Previous Work on Dependency Parsing of Czech

2.1 Prague Dependency Treebank

In this experiment, we use data from the Prague Dependency Treebank (PDT, see [3]). PDT includes approximately 1.5 million words (88,000 sentences) with a surface syntactic dependency annotation, divided into training data (80%), and two sets of test data (d-test and e-test, 10% each).

Every word or punctuation in the sentence is represented as a node in a dependency tree, each node is assigned a syntactic function, as shown in Fig. 1, representing the sentence *Nemáme jediný výbor z Bretona.* 'We don't have a single anthology of Breton.' The text is lemmatized and morphologically annotated with a positional morphological tagset (15 positions, see [2]).

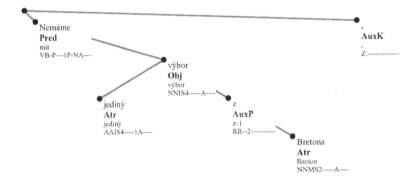

Fig. 1. Example of a dependency tree in PDT

2.2 Automatic Parsing of PDT

Based on the PDT data, many experiments of automatic parsing were undertaken. So far, the best published results were achieved with a combination of several parsers [4].[1]

[1] http://ufal.mff.cuni.cz/czech-parsing/ records the results of experiments. We take into account only the results of tests carried out on the full training data and tested on the d-test and/or e-test of PDT. We provide two measures of accuracy: unlabeled attachment score (*UAS*, percentage of words with correct head) and labeled attachment score (*LAS*, correct head and syntactic label).

2.3 MaltParser

The best result for the MaltParser has been achieved by Daniel Zeman (unpublished; 86.1% UAS/ 79.8% LAS) with the *stacklazy* algorithm and LibSVM learner. The parser uses all the non-zero values of the positional morphological tag as single features, as shown in Tab. 1. For example, the morphological tag VB-P---1P-NA--- of the first word in the sentence is divided into POS and 6 features: SP, Nu, Per, Ten, Neg and Vo. Other positions in the tag, with the value "-", are omitted. In this example, the values of the features mean: POS=V (part of speech: verb), SP=B (subtype of POS: verbal present form), Nu=P (number: plural), Per=1 (person: first), Ten=P (tense: present), Neg=N (negation: negative), Vo=A (voice: active).

Table 1. CONLL format used for MaltParser (simplified)

ID	FORM	LEMMA	POS	FEATS	HEAD	DEPR
1	Nemáme	mít	V	SP=B|Nu=P|Per=1|Ten=P|Neg=N|Vo=A	0	Pred
2	jediný	jediný	A	SP=A|Gen=I|Nu=S|Cas=4|Gr=1|Neg=A	3	Atr
3	výbor	výbor	N	SP=N|Gen=I|Nu=S|Cas=4|Neg=A	1	Obj
4	z	z	R	SP=R|Cas=2	3	AuxP
5	Bretona	Breton	N	SP=N|Gen=M|Nu=S|Cas=2|Neg=A	4	Atr
6	.	.	Z	SP=:	0	AuxK

2.4 MSTParser

The best result for the MSTParser was reached by Miroslav Spousta (unpublished; 85.9%/ 78.8%). The parser uses only reduced morphological tags (POS + case or POS + SubPOS) as they were proposed in [1]. Wider tags (including, e.g., gender and number) paradoxically decrease the parser's accuracy. Tab. 2 presents the format used by the parser:

Table 2. MCD format used for MSTParser

Nemáme	jediný	výbor	z	Bretona	.
VB	A4	N4	R2	N2	Z:
Pred	Atr	Obj	AuxP	Atr	AuxK
0	3	1	3	4	0

Both parsers were trained on data annotated morphologically by automatic tools[2] instead of the gold data (manually annotated), so that the parsers are better able to cope with errors in morphological annotation in unknown data and achieve a slightly better accuracy.

---

[2] Lemmatization and morphological annotation performed by the Featurama tagger, http://sourceforge.net/projects/featurama/, with an accuracy of 95.65% on the d-test, 95.39% on the e-test and 98.39% on the training data (which the tagger itself used for its training).

3   Filtering Linguistic "Noise"

3.1   Linguistic "Noise"

Both MSTParser and MaltParser use word forms, in addition to automatically assigned morphological tags (MaltParser also uses lemmas). When creating models, however, they can rely only on the limited training data in which most lemmas and forms appear only once or several times[3], and most words from the Czech vocabulary do not appear at all. Based on these rare occurrences, the parser cannot generate any reliable prediction of their behavior in syntactic structures.

In a similar way, formal variations of more frequent words (such as prepositions) render the task of language modeling more difficult, too. Several prepositions in Czech have a vocalized variant, used for easier pronunciation before a word starting with a similar consonant (*přijel z Německa/ přijel ze Slovenska* 'he came from Germany / Slovakia'). It is a phonetic, syntactically insignificant phenomenon, but it fragments the co-occurrences of forms (or forms and tags) and thus reduces the efficiency of the parser.

From the syntactic point of view, a large part of language variability is irrelevant: cardinal numerals *pět* 'five', *šest* 'six' ... *devět* 'nine' have exactly the same syntactic properties. Stylistic variants *mohu* / *můžu* 'I can' behave in the same way in syntactic structures etc. We call this variability linguistic "noise": it does not provide any useful information to the parser, it just renders the understanding of the structures more difficult.

The solution is to automatically filter out what is irrelevant for parsing: different forms and lemmas with the same syntactic properties can be replaced by a single proxy form or lemma: for example, ordinal numerals *five* to *nine* will be replaced by one, arbitrarily chosen numeral (*five*), stylistic variants will be replaced by the basic form and so on.

3.2   Processing the Data

Filtering of linguistic "noise" is performed automatically by pre-processing the data. Before any training or parsing is done, the original text (morphologically analyzed, and, in case of training data, with the "gold", i.e. manual syntactic analysis) is processed sentence by sentence. Whenever a form or lemma of the appropriate type is found, it is replaced by a proxy form or lemma, while the original text is stored in a separate backup file. The parser is then trained on the simplified data, and can be used to parse new data processed in the same way. After the parsing is complete, the original forms and lemmas are retrieved from the backup file.

So far, members of some 80 word classes (lemmas or forms) are sought and replaced, and more can be added. The selection of such word classes suitable for simplification is performed manually, based on linguistic intuition and frequency statistics from a large Czech textual corpus, which is not related to the treebank.

We use, for example, a list of about 300 female first names (*Jana* 'Jane', *Marie* 'Mary', *Eva* 'Eve' ...), with identical syntactic properties. For any such proper name in

---

[3] 53.3% word forms and 43.4% lemmas occur exactly once in the training data.

the text, its lemma is replaced by one arbitrary lemma (*Alena*) and its form is replaced by the corresponding form. In a similar way we treat masculine first names, masculine and feminine surnames, etc. Thus the sentence *Leona Machálková přišla s Vendulou Svobodovou* 'Leona Machálková came with Vendula Svobodová' is simplified as *Alena Nováková přišla s Alenou Novákovou*. Understanding of the text is disabled, but the syntactic structure has not changed, and the parser deals with structures populated by fewer lemmas and forms.

Linguistic data needed for the filtering of noise (lists of words with identical properties) come from three sources: a linguistic database in a project of a rule-based automatic morphological disambiguation [7], a valency dictionary PDT-Vallex [8] and data directly extracted from a Czech textual corpus SYN (1.3 billion words, http://www.korpus.cz). No data were obtained directly from the training or test data of PDT. The use of a large corpus and external resources ensure that new text (forms and lemmas not occurring in the training data) are treated in the same way as the training data.

3.3   Some Examples of Simplification and Statistics of Variability Reduction

The following paragraphs will present a few more examples of the filtering of linguistic "noise", i.e. the replacement of equivalent forms by a single representative, using examples of adjectives, numerals and verbs.

Several groups of adjectives are simplified: deverbal adjectives, possessive adjectives, some composed adjectives etc. Deverbal adjectives are divided into subclasses corresponding to their types (active, passive; reflexive...) and their potential valency frame. For example *obávající se* 'fearing', *štítící se* 'loathing', etc.: reflexive adjectives with a genitive valency, will be replaced by one lemma.

Among numerals (both numerals written as words and numbers written with digits), a significant reduction in variability is possible: the specific numerical value plays only a minor role and the variability is very high. There are 260 different lemmas of numerals written with words and 3,304 lemmas using digits (*1984, – 4.5*) in PDT; 1,300 word lemmas and 380,000 different numbers in the whole SYN corpus. We sort the numerals by their type and value and replace their lemmas and forms accordingly, the result is a simplification down to about twenty lemmas and corresponding forms.

For the simplification of verbal lemmas or forms, we proceed carefully, considering the various obligatory and optional valency frames of Czech verbs (important for correct syntactic analysis). We only replace less frequent transitive verbs, which have no other valency frame besides an object in accusative in PDT-Vallex, nor do they co-occur frequently with nouns in other cases or with prepositions in the SYN corpus.

On the whole, the procedure reduces the variability of forms and lemmas in the training data by nearly 20%, as shown in Tab. 3:

Table 3. Numbers of lemmas and forms in the training data (original/simplified)

train PDT	original data	simplified data	number decrease
forms	125369	103016	17.8%
lemmas	54610	44126	19.2%

4 "Syntactic" Features for Parsers

Besides the filtering of linguistic noise in the data, we modify the way in which the parsers use the information from morphological tags, the "features". The morphological tagging of PDT is not well adapted for stochastic parsing, mainly because the tags contain too much information unrelated to syntax. Several POS classes are meticulously divided into many subtypes, but this classification does not reflect the syntactic properties of the words. There is, for example, a subtype of "negative" pronouns, comprising both syntactic nouns as *nic* 'nothing', *nikdo* 'nobody', and syntactic adjectives *žádný* 'none', *ničí* 'no-one's'. If, therefore, we supply the parser with features taken from unmodified morphological tags (as in the "original" setting of MaltParser), the parser has to create its own system for syntactic interpretation of these tags, which makes errors more probable.

Even worse, if the parser is given only reduced morphological tags (POS+case or POS+SubPOS, as in MSTParser), it cannot base its decisions on the tag only, it must combine it with the word form or risk misinterpreting the structure: for example, a reduced tag *P4* (pronoun in accusative) describes both syntactic nouns as *ji* 'her' or *sebe* 'herself' and syntactic adjectives, such as *můj* 'my' or *nějaký* 'some'.

We solve this problem by introducing a "syntactic" variable. It differentiates, for example, between safe syntactic nouns as *já* 'I', *kdo* 'who', safe syntactic adjectives as *můj* 'my', *čí* 'whose' and words, which can be both syntactic nouns and adjectives as *všichni* 'all', *který* 'which'. For the MSTParser, we expand the reduced morphological tags by this variable. For the MaltParser, we introduce this variable as a separate feature; we also modify the morphological features it uses (we remove unnecessary features and adjust others).

5 Results of the Experiment

Thanks to the filtering of linguistic "noise" and more sophisticated morphosyntactic features, both parsers achieved improvements in UAS and LAS accuracy. Tab. 4 shows the contribution of both parts of the procedure compared to the original settings described in Section 2. For both MaltParser and MSTParser, four experiments of training and testing were run: original settings; original features with "noise" filtering; better morphosyntactic features with the original text and the complete procedure (noise filtering, better features). The results were obtained with the d-test of PDT.[4]

For MaltParser, a higher error rate reduction (a relative error reduction of 5.6% UAS / 8.1% LAS) was achieved compared to the MSTParser (2.4% / 1.8%), while both parsers achieved results exceeding the state-of-the-art accuracy in Czech parsing. The procedure increases also the speed of training and parsing for both parsers (by 10% to 25%). The tool (a Perl program) for Czech parsing is available on the website: www.utkl.ff.cuni.cz/simplify_data_for_parsers.

---

[4] Results achieved on the e-test are similar, but slightly worse.

Table 4. UAS/LAS accuracy of the MaltParser and the MSTParser on PDT d-test

MaltParser	original settings	noise filtering
original settings	86.11 / 79.80	86.41 / 80.55
morphosynt. features	86.37 / 80.46	86.85 / 81.31

MSTParser	original settings	noise filtering
original settings	85.93 / 78.80	86.17 / 79.10
morphosynt. features	86.04 / 78.95	86.26 / 79.18

6 Conclusion

In this paper, we have shown that reducing the variability of forms and lemmas in the training data and in newly processed text can lead to a noticeable improvement in the accuracy of the parser. The filtering of linguistic "noise", i.e. simplification of forms and lemmas, is performed very carefully: only well-defined, narrow word classes are simplified. Further research is needed to determine whether a more radical simplification would lead to better results.

Acknowledgments. This research was supported by grant no. 13-27184S of the Grant Agency of the Czech Republic.

References

1. Collins, M., Hajič, J., Ramshaw, L., Tillmann, C.: A statistical parser for Czech. In: Proceedings of the 37th Annual Meeting of the ACL, College Park, MD, USA, pp. 505–512 (1999)
2. Hajič, J.: Disambiguation of Rich Inflection, Computational Morphology of Czech. Karolinum, Charles Univeristy Press, Prague, Czech Republic (2004)
3. Hajič, J.: Complex Corpus Annotation: The Prague Dependency Treebank. In: Šimková, M. (ed.) Insight into the Slovak and Czech Corpus Linguistics, pp. 54–73. Veda, Bratislava (2006)
4. Holan, T., Žabokrtský, Z.: Combining Czech Dependency Parsers. In: Sojka, P., Kopeček, I., Pala, K. (eds.) TSD 2006. LNCS (LNAI), vol. 4188, pp. 95–102. Springer, Heidelberg (2006)
5. McDonald, R., Pereira, F., Ribarov, K., Hajic, J.: Non-projective Dependency Parsing using Spanning Tree Algorithms. In: Proceedings of HLT/EMNLP, pp. 523–530. ACL, Vancouver (2005)
6. Nivre, J., Hall, J., Nilsson, J.: MaltParser: A Data-Driven Parser-Generator for Dependency Parsing. In: Proceedings of the Fifth International Conference on Language Resources and Evaluation (LREC 2006), Genoa, Italy, pp. 2216–2219 (2006)
7. Petkevič, V.: Reliable Morphological Disambiguation of Czech: Rule-Based Approach is Necessary. In: Šimková, M. (ed.) Insight into the Slovak and Czech Corpus Linguistics, pp. 26–44. Veda, Bratislava (2006)
8. Urešová, Z.: Building the PDT-VALLEX valency lexicon. In: On-line Proceedings of the Fifth Corpus Linguistics Conference. University of Liverpool, UK (2009)

Improving Speech Recognition by Detecting Foreign Inclusions and Generating Pronunciations

Jan Lehečka and Jan Švec

Department of Cybernetics, Faculty of Applied Sciences, University of West Bohemia,
Univerzitní 8, 306 14 Plzeň, Czech Republic
{jlehecka,honzas}@kky.zcu.cz
www.kky.zcu.cz

Abstract. The aim of this paper is to improve speech recognition by enriching language models with automatically detected foreign inclusions from a training text. The enriching is restricted only to foreign, proper-noun inclusions which are typically a dominant part of miss-recognized words. In our suggested approach, character-based n-gram language models are used for detection of foreign, single-word inclusions and for a language identification, and finite state transducers are used to generate foreign pronunciations. Results of this paper show that by enriching language model with English proper nouns found in Czech training text, the recognition of a speech containing English inclusions can be improved by 9.4% relative reduction of WER.

Keywords: Language Identification, G2P, ASR.

1 Introduction

Automatic Language Identification is the process of detecting and determining in which language or languages a given piece of text is written [1]. The past decades have seen the rapid development of automatic language identification in many succesful aplications. Various approaches have been developed and used in text-to-speech (TTS) systems and automatic-speech-recognition (ASR) systems.

There is a large volume of published studies describing automatic language identification on text basis, e.g. [2], [3]. Succesful aproach using phoneme recognition followed by a language model was described in [4], [5], [6] and many others.

Also some approaches using sub-word units for short-text language identification have been introduced in [7], [8], but only little attention has been paid to automatic language identification on single-word basis. Typically in a speech and written text, people are using inclusions from other languages (e.g. proper nouns) embedded in sentences from their mother tongue, so around the foreign-language entity, there is no context which can be used for better language indentification. That is why the text-based language identification can not be used in order to detect foreign inclusions, but a language of each isolated single word has to be identified separately.

The correct synthesis and recognition of foreign inclusions is an increasingly important area in TTS and ASR systems. There have been several attempts to solve this

I. Habernal and V. Matousek (Eds.): TSD 2013, LNAI 8082, pp. 295–302, 2013.

problem. In [7], an approach of analyzing and ranking n-grams of characters in different languages to detect foreign entities has been described. In [9], a combination of short-text language identification and n-gram-based grapheme-to-phoneme (G2P) conversion has been used to solve mixed linguality in ASR systems. However, no attempt to detect foreign, single-word inclusions and generate the pronunciation by a modern G2P convertor (i.e. a weighted finite state transducer) has been published so far.

In this paper, a statistical approach to detect foreign, single-word language inclusions in a written text, and a WFST-based approach to generate pronunciations of foreign words is presented. The improvement of these approaches is demonstrated on an ASR experiment.

2   Single-Word Language Identification

Many of the language identification methods are based on n-gram language models. These language models can be trained on word-units or sub-words units such as syllables, phonemes or characters. The word-level language models are unusable in our task because the foreign inclusions are typically out-of-vocabulary (OOV) words and we classify isolated words in the ASR pronunciation lexicon. Therefore the word-level classifier cannot use the word context to classify the word into target classes.

In our task we use character-level language modeling which has a number of advantages:

- a probability $P(C|L)$ of an out-of-vocabulary word can be well-estimated (see below),
- language-specific sequences of characters can be observed in the training data and used for better language identification,
- higher order of language models can be used because of a very small vocabulary (size of the vocabulary is equal to the number of different characters in the training texts which is typically less then 50),
- the training text for each language model can be easily obtained and used in training robust language-specific n-gram model.

We can formalize the single-word language identification task. We assume that we have a set of n target languages $\mathcal{L} = \{L_1, L_2, \ldots, L_n\}$. In the training phase we use the training data consisting of N pairs $T = (w_k, l_k)_{k=1}^N$, where w_k is a training word which has assigned a target language $l_k \in L$. The word w_k consists of m_k characters $c^i : w_k = (c_k^1, c_k^2, \ldots, c_k^{m_k})$. Using the training data we are able to construct language-specific training data T_L containing only character sequences from the given language L:

$$T_L = \{(c_k^1, c_k^2, \ldots, c_k^{m_k}) : (w_k, l_k) \in T, l_k = L\} \tag{1}$$

The language-specific training data for language L are used to train character-level n-gram language model $P(C|L)$. It models the probability of observing character sequence C (i.e. word decomposed into characters) given the language L. We used the SRI Language Modelling toolkit (SRILM) [10] to estimate the back-off n-gram language model $P(C|L)$ for each language in $\mathcal{L}$.

Then the novel word $\hat{w} = (\hat{c}^1, \hat{c}^2, \ldots, \hat{c}^{\hat{m}})$ can be classified using the maximum a posteriori (MAP) estimate:

$$L^*_{\text{MAP}} = \arg \max_L P(C|L)P(L) \tag{2}$$

where L^* is the predicted language of word $\hat{w}$. In our experiments the prior distribution $P(L)$ over the set $\mathcal{L}$ was unknown because we used the language-specific training data T_L sampled from different distributions [11]. Therefore we used the maximum likelihood estimate (MLE):

$$L^*_{\text{MLE}} = \arg \max_L P(C|L) \tag{3}$$

Difficulties arise, however, when an attempt is made to identify language of a word which belongs to more then one language. This phenomenon is typical for very short words (e.g. the word *die* belongs to both English and German) and for adopted words (e.g. the word *minus* adopted from Latin has different pronunciation in English and in German). This is a serious problem but not even a human is able to decide about the language and pronunciation of such a word if he sees it isolated without any context.

3   Grapheme to Phoneme Conversion

Let's assume that the word $\hat{w}$ is classified as a belonging to a language L^*. To use this word in an ASR lexicon it is necessary to convert it into a sequence (or sequences) of phoneme pronunciations. In our task we assume that the pronunciation of the main language is defined, i.e. we have a main, probably large pronunciation lexicon over the phoneme alphabet $\mathcal{P}$. The goal is to perform phonetic transcription of the novel word $\hat{w}$ from language L^* using the same phoneme alphabet $\mathcal{P}$.

We will assume that the pronunciation lexicon over $\mathcal{P}$ for language L^* is available and it does not necessarily include the word $\hat{w}$. To be able to transcribe out-of-vocabulary words in language L^* we use statistical grapheme-to-phoneme mapping trained on the pronunciation lexicon for language L^*.

The goal of grapheme-to-phoneme (G2P) mapping is to predict the pronunciation of a novel word given only its orthography. We used a modified G2P method which employs joint sequence modeling using an n-gram model and multiple-to-multiple sequence alignment [12]. We used the implementation of this algorithm called Phonetisaurus. The training algorithm consists of three steps:

1. The pronunciation lexicon of L^* is used to generate an *alignment model*. The alignment model uses several EM-training iterations to find the proper aligned pairs {graphemes:phonemes}. The algorithm allows only m-to-1 and 1-to-m mapping to achieve better balance between the pair length and data sparsity problem which appears in the training of n-gram model in subsequent steps.
2. The *n-gram model* is trained to predict the joint probability of some alignment sequence. The alignment sequence consists of pairs {graphemes:phonemes} generated in the first step.

3. The trained n-gram model is converted to a weighted finite state transducer M (WFST) with grapheme input and phoneme output labels.

The phonetic transcription of the novel word is generated by an algorithm which can be described using standard WFST operations. The input finite state acceptor (FSA) representing the all possible grapheme alignments is created. For example the word *right* can be converted to following sequences of graphemes: *r-i-gh-t, r-i-g-h-t, r-i-g-ht, r-ig-h-t, r-ig-ht,* etc. Then, the set of possible grapheme subsequences is generated from the alignment model. The grapheme sequences are encoded in the input FSA W and the G2P decoding process can be defined as:

$$H = \text{shortest\_path} \left[\det \left[\Pi_2 \left[W \circ M \right] \right] \right] \tag{4}$$

where $\text{shortest\_path}[\cdot]$ denotes the application of n-shortest paths algorithm, $\det[\cdot]$ denotes the determinization, $\Pi_2[\cdot]$ projection of WFST to output labels (phoneme alphabet in this case) and $\circ$ operator represents transducer composition.

The main idea of using G2P for improving speech recognition is as follows. After collecting a text for training speech-recognition language model, all novel words are found and for each one of them the automatic language identification is used to detect if the word belongs to some other language. If it does, the word is sent as an input sequence of graphemes to the WFST trained from phonetic dictionary of the same language. An output sequence of corresponding phonemes is then returned from the WFST and used in the speech recognizer as a pronunciation of the novel word.

4   Language Identification Experiments

This paper is focused on speech recognition of Czech, therefore, Czech is considered as a base language and words from all other languages are considered as foreign inclusions. Of course, it is impossible to take all foreign languages into acount, so three main languages with the highest amount of adopted words into Czech were empiricaly chosen, namely Latin, English and German. For each language (including Czech), a character-based back-off language model $P(C|L)$ was trained on word lexicons created from a few books written in the corresponding language L. We used the books freely available from the Project Gutenberg [11] as a source of training words. Prior to language model training, all words were converted to lower case because of the reduction of a vocabulary size and better probability estimation of language-specific sequences.

To prove that this is a valid method for single-word language identification, the following experiment had been prepared. For each chosen language, a character-based language model was trained as described in section 2. Then, from each language, a short text (about 300 words per language) not appearing in the training data was chosen for the test data. Then the single-word language identification was performed on all words appearing in these texts. Once language of all single words in the test data had been identified, the predicted language and the original language (i.e. the language of a text, from which the word came to the test data) were compared and correct-language-identification accuracy was computed. The experiment had been repeated 9x with different order of n-gram models to see which order was suitable for this task. Results obtained from the experiment are shown in Figure 1.

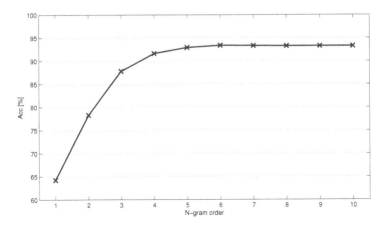

Fig. 1. Accuracy of single-word language identification with different orders of character-based n-gram language models

As can be seen, if only unigram language models were used, i.e. only pure statistics about the frequency of single characters in each language were computed from the training data, language of more then 64% words would be identified correctly. Best accuracy, however, was achieved by using 6-grams and further increasing of the n-gram order did not show any improvement. When 6-grams were used, the language of Czech words was correctly identified with accuracy 96.8%, English with 88%, German with 93.5% and the accuracy of correct language identification of Latin words was 95.1%, which all together made 93.38% accuracy. Based on this experiment, 6-gram character-based back-off language models were used as a language identifier in following sections.

5   Speech Recognition Experiment

To evaluate the impact of single-word language identification and G2P conversion on speech recognition, a text with some foreign language inclusions had been chosen. Then the text was read by four speakers. Once the text had been dictated, the speech was recognized using general language model to get the baseline accuracy. Then the speech recognition was repeated but this time with the lexicon and language model enriched by automatically identified foreign words with automatically generated pronunciations.

As a source of texts for the language modeling, large web-mined corpus [13], which had been primarily focused on automatic topic identification and data filtering [14], was used. The text database from which the general language model had been trained contained texts from various sources, e.g. automatically downloaded articles from various Czech web servers, manual transcripts from TV broadcasts, movie subtitles, etc. All texts had been automatically cleaned for purpose of language modelling but, of course, there was a lot of errors in these texts. To avoid words with errors to be recognized in the speech, a vocabulary of the most frequent Czech words which had been manually checked was created. The size of this vocabulary, which is still continuously supplemented with novel words, was about 3 million words. In most of our ASR applications, language models are limited by this checked vocabulary, so most of foreign inclusions are OOV words although they can appear in the training text.

However, novel foreign inclusions are usually not appearing even in the training data, so it can not be added into the lexicon in order to recognize OOVs correctly. In the text corpora [13], which was used to build language models, this issue is solved by storing texts organized by the time of its release. When some older text is to be recognized, only texts released before this time are used to build language models, so "cheating" by looking into the future is avoided. By respecting this time-release restriction, most of the foreign inclusions appearing in the speech can be usually found in the training data, especially in recent texts.

To train the general language model, roughly 850 million words (more then 5 GB of text) were used and limited by the vocabulary with checked words. All unigrams, all bigrams and only those trigrams that appeared in the text more than once were involved in the language model. Since foreign inclusions were expected to be a proper nouns, some restrictions for novel foreign words were defined to lower the amount of words to be added into the lexicon. Foreign novel words were added to the lexicon only if the frequency of the word in the text was at least 2 and if the word started with a capital letter.

To simplify the experiment, inclusions only from one language, namely from English, were investigated. Since only English inclusions were to be detected, only Czech-or-English language identification was used and only English G2P had been trained from the CMU dictionary [15]. An approximate mapping from the English phonetic alphabet used by CMU to the Czech phonetic alphabet used in the speech-recognition lexicon had to be defined in order to add new pronunciations. Because foreign languages typically use different phones which have no alternatives in Czech and vice versa, in Czech are phones with no alternatives in foreign languages, each English phoneme was mapped to the most similar phoneme in our Czech phonetic alphabet. Using this approximate mapping, the pronunciation of how would Czech speaker pronounce the foreign word by using only Czech phones was obtained, which is exactly how foreign inclusions are usually pronounced in the Czech speech.

The large-vocabulary continuous speech recognition (LVCSR) with the acoustic model which had been originally created for purposes of subtitling live TV programs [16] was used. Typically, the task of capturing live TV programs suffers a lot from foreign language inclusions not appearing in the pronunciation lexicon, e.g. foreign names and geographic locations from actual cases, which are analyzed in TV broadcasts.

As for the speech to be recognized, a news article about hurricane Sandy from the date 2.11.2012 was dictated by four non-professional speakers. This article was chosen because it contained many typical English inclusions, e.g. unknown places where the hurricane had stroke, unknown names of local mayors and residents, etc. In this article, there were 774 words and 23 of them were OOV words (English inclusions).

The vocabulary size of the general language model built only from texts released before the date 2.11.2012, in which only checked Czech words were involved, was 1.08 million words and the size of the vocabulary enriched by English words was 1.27 million, so less then 200k unknown words had been identified as an English inclusion in the training text.

Table 1. Speech recognition accuracy with two different language models: the first one is the general Czech LM (baseline), the second LM is the baseline enriched with automatically detected English words with automatically generated pronunciation

	baseline	baseline+eng	improvement	p-value
speaker 1	85.22	87.00	+1.78	$2.14 \cdot 10^{-2}$
speaker 2	87.64	87.77	+0.13	$4.30 \cdot 10^{-1}$
speaker 3	87.01	88.79	+1.78	$5.92 \cdot 10^{-3}$
speaker 4	86.11	87.52	+1.41	$7.86 \cdot 10^{-3}$
total	86.50	87.77	+1.27	$2.65 \cdot 10^{-4}$

6 Conclusion

Results of the experiment described in section 5 are tabulated in Tab. 1. Since each tested speaker had different voice, speaker-independent acoustic model was used. Nevertheless, the improvement is obvious: the speech recognition accuracy went up in average by 1.27% which corresponds to 9.4% relative reduction of WER. A statistical test confirmed that obtained results are statistically significant with p-value $2.65 \cdot 10^{-4}$.

The perplexity of the general language model with the reference text, i.e. the text which had been read to be recognized, was 593.9 and there were 23 OOVs. After enriching language model with English words, the perplexity was reduced to 566.3 which corresponds to 4.64% relative reduction of perplexity and number of OOVs was reduced to 6 (73.9% relative reduction of OOVs).

Results of this paper state that the general language model of one language can be enriched with automatically detected English proper nouns in order to improve the recognition of a speech containing English inclusions. In the experiment, it was shown that by using single-word language identification to find English inclusions in a text and by using WFST G2P to generate English pronunciation, WER of the speech recognition can be significantly decreased.

Acknowledgments. This research was supported by the Technology Agency of the Czech Republic, project No. TE01020197 and by the European Regional Development Fund (ERDF), project "New Technologies for Information Society" (NTIS), European Centre of Excellence, ED1.1.00/02.0090, and by the grant of the University of West Bohemia, project No. SGS-2013-032.

References

1. Yang, X., Liang, W.: An N-Gram-and-Wikipedia joint approach to Natural Language Identification. In: 2010 4th International Universal Communication Symposium (IUCS), pp. 332–339 (2010)
2. Martins, B., Silva, M.J.: Language identification in web pages. In: Proceedings of the 2005 ACM Symposium on Applied Computing, SAC 2005, pp. 764–768. ACM, New York (2005)

3. Zissman, M.A., Singer, E.: Automatic language identification of telephone speech messages using phoneme recognition and n-gram modeling. In: 1994 IEEE International Conference on Acoustics, Speech, and Signal Processing, ICASSP 1994, vol. 1, pp. I/305–I/308 (1994)
4. Zissman, M.A.: Language identification using phoneme recognition and phonotactic language modeling. In: 1995 International Conference on Acoustics, Speech, and Signal Processing, ICASSP 1995, vol. 5, pp. 3503–3506 (1995)
5. Yan, E.Y., Barnard: An approach to automatic language identification based on language-dependent phone recognition. In: 1995 International Conference on Acoustics, Speech, and Signal Processing, ICASSP 1995, vol. 5, pp. 3511–3514 (1995)
6. Matejka, P., Schwarz, P., Cernocký, J., Chytil, P.: Phonotactic language identification using high quality phoneme recognition. In: INTERSPEECH, pp. 2237–2240 (2005)
7. Ahmed, B., Cha, S.H., Tappert, C.: Detection of Foreign Entities in Native Text Using N-gram Based Cumulative Frequency Addition. In: Proceedings of CSIS Research Day. Pace University, New York (2005)
8. Hakkinen, J., Tian, J.: N-gram and decision tree based language identification for written words. In: IEEE Workshop on Automatic Speech Recognition and Understanding, ASRU 2001, pp. 335–338 (2001)
9. Maison, B., Chen, S., Cohen, P.S.: Pronunciation modeling for names of foreign origin. In: 2003 IEEE Workshop on Automatic Speech Recognition and Understanding, ASRU 2003, pp. 429–434 (2003)
10. Stolcke, A.: Srilm-an extensible language modeling toolkit. In: Proceedings International Conference on Spoken Language Processing, pp. 257–286 (November 2002)
11. Project Gutenberg, http://www.gutenberg.org
12. Novak, J., Dixon, P.: Improving WFST-based G2P conversion with alignment constraints and RNNLM N-best rescoring. In: Proceedings of International Conference on Spoken Language Processing Interspeech 2012 (2012)
13. Švec, J., Hoidekr, J., Soutner, D., Vavruška, J.: Web text data mining for building large scale language modelling corpus. In: Habernal, I., Matoušek, V. (eds.) TSD 2011. LNCS, vol. 6836, pp. 356–363. Springer, Heidelberg (2011)
14. Skorkovská, L., Ircing, P., Pražák, A., Lehečka, J.: Automatic Topic Identification for Large Scale Language Modeling Data Filtering. In: Habernal, I., Matoušek, V. (eds.) TSD 2011. LNCS, vol. 6836, pp. 64–71. Springer, Heidelberg (2011)
15. The CMU Pronouncing Dictionary, http://www.speech.cs.cmu.edu/cgi-bin/cmudict
16. Pražák, A., Loose, Z., Trmal, J., Psutka, J.V., Psutka, J.: Captioning of live TV programs through speech recognition and re-speaking. In: Sojka, P., Horák, A., Kopeček, I., Pala, K. (eds.) TSD 2012. LNCS, vol. 7499, pp. 513–519. Springer, Heidelberg (2012)

Intensifying Verb Prefix Patterns
in Czech and Russian*

Jaroslava Hlaváčová and Anna Nedoluzhko

Charles University in Prague
Institute of Formal and Applied Linguistics
Faculty of Mathematics and Physics
{hlava,nedoluzko}@ufal.mff.cuni.cz

Abstract. In Czech and in Russian there is a set of prefixes changing the meaning of imperfective verbs always in the same manner. The change often (in Czech always) demands adding a reflexive morpheme. This feature can be used for automatic recognition of words, without the need to store them in morphological dictionaries.

1   Introduction

An automatic text processing usually starts with a morphological analysis trying to recognize all the words and to assign morphological features to them. In practice, morphological dictionaries are used for that purpose. However, regardless of dictionary size, we meet words which are not included in the dictionary. There are several kinds of such words. So called guessers are used for their recognition. In the present paper, we will examine a special group of verbal prefixes that change imperfective verbs always in the same manner. This feature allows an easy and reliable recognition and subsequent analysis of such verbs. We will present our results on two representatives of Slavic languages — Czech and Russian.

2   Verb Intensification in Czech and Russian

The majority of Czech and Russian imperfective verbs have an interesting ability to be associated with one of several prefixes that change the meaning of the verb in a way that might be called intensification. With one exception on the Russian side, those prefixal verbs must be modified with a reflexive morpheme *se/si* in Czech and *-ся/-сь* in Russian. The prefixes are: *roz-, po-, za-, na-, vy-* and *u-* for Czech and *раз-(рас-)*, *по-, за-, на-, вы-* and *у-* for Russian, the set could be extended with the prefix *из-* for Russian. This prefix has also a Czech equivalent, namely *z-*, but not in the context we are going to present now.

The order we introduced the prefixes is not accidental. With a certain tolerance, both sets can be viewed as sequences ordered according to an intensity of the process

---

* This paper is a result of the project supported by the grant number P406/2010/0875 of the Grant Agency of the Czech republic (GAČR).

I. Habernal and V. Matousek (Eds.): TSD 2013, LNAI 8082, pp. 303–310, 2013.

described by the verb attached. This is the reason why we call this type of prefixation a **"verb intensification"**.

The meaning of the prefixes could be derived from short stories in both languages. These stories are not authentic, they were compiled by the authors. The English translations are not very good because it is really hard to translate the prefixal verbs of such kind into English properly. In majority of cases, the pure verb would be a better translation, but we wanted to stress that the meaning of the sentence lies primarily in the prefixal verb.

The Czech story with the intensification of the verb *lyžovat* (*to ski*):

Czech *V zimě jsem jela do Itálie **lyžovat**.*
English *(In winter, I went to **ski** to Italy.)*
Czech *Marie **si** chtěla se mnou **zalyžovat**.*
English *(Marie wanted to **enjoy skiing** with me.)*
Czech ***Rozlyžovaly** jsme **se** poměrně rychle.*
English *(We **got into skiing** fairly quickly.)*
Czech *První den jsme **si** jen tak zvolna **polyžovaly**.*
English *(The first day we have just **skied a little**.)*
Czech *Další dny jsme **se** už **nalyžovaly** víc.*
English *(Other days we've **skied more**.)*
Czech *Poslední den jsem už měla pocit, že jsme **se vylyžovaly** hodně.*
English *(The last day I have felt that we have been **skiing lot**.)*
Czech *Už jsme musely odpočívat, abychom **se neulyžovaly**.*
English *(We had to rest no to be completely **exhausted by skiing**.)*

The Russian story with the intensification of the verb *плавать* (*to swim*):

Russian *Этим летом я ездил в Италию на месяц **плавать**.*
English *(In summer, I went to Italy to **swim** for a month.)*
Russian *Первая неделя — это только **расплаваться, поплавать** себе в своё удовольствие.*
English *(The first week is good just to **get into swimming, to enjoy the swimming**.*
Russian *Однажды я так **заплавался**, что даже пообедать забыл.*
English *(Once I **forgot myself in swimming** so much, that I missed my dinner.)*
Russian *К концу третьей недели я наконец **наплавался**, а к середине четвёртой уже **уплавался** до изнеможения.*
English *(By the end of the third week I've **swimmed to my heart's content** and by the middle of the forths week I was **completely tired of swimming**.)*
Russian *В конце месяца у меня было такое ощущение, что я уже весь **изплавался**, просто **выплавался** напрочь и море просто видеть не могу.*
English *(By the end of the month, I felt **totally exhausted by swimming, swimming has completely drained my spirit** and the sea already ticked me off.)*

The figures 1 and 2 illustrate the distribution of meanings of the prefixal verbs with intensification meanings in both languages.

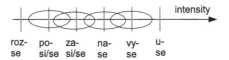

Fig. 1. Czech intensifying prefixes

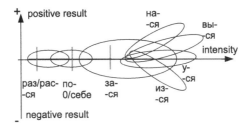

Fig. 2. Russian intensifying prefixes

In the Czech case, there are two strict points at both ends of the scale. They belong to the prefixes *roz-* and *u-*. The former one represents a starting stage of a process, while the latter one expresses its final stage. Other prefixes cannot be placed on the axis so unambiguously, which is indicated by the overlapping elipses.

In Russian, the situation is similar but not the same. The first three prefixes correspond with their counterparts in Czech, though the prefix *за-* stretches beyond common limits. Also the prefix *раз-(рас-)* is not placed so strictly as in Czech, it is more vague. The last four prefixes (*вы-, на-, у-, из-*) not only overlap in their intensities, but have another dimension of meaning. Although their intensity is roughly the same, their semantic and pragmatic properties differ. This fact is schematically described by the perpendicular axis expressing a scale of positivness / negativness of the process. However, the meanings of individual prefixes are also vague and overlapping. The picture shows the meanigs of prefixes only schematically. There is a more detailed description of the mentioned prefixes and their connection with reflexive morphemes in [1].

The following section presents a complete overview of the prefixes, their meanings and connection with reflexive morphemes. Every pair of prefixes (for Czech and for Russian) is represented by a word-forming pattern. The letter X stands for an imperfective verb. Adding the given prefix and reflexive morpheme, we derive the appropriate intensified verb. We always explain the meaning and, if needed, also a difference between the two languages. The meaning explanation is usually quite vague. A better understanding is usually visible from examples that accompany every prefix in both languages. The Czech examples come from the corpus SYN [2], the Russian ones were found in the Russian National Corpus [3] and on the internet.

2.1   roz-X se, раз-Х-ся

Meaning: to start X, in Russian also possibly to amplify the intensity.

Examples:

*Z jara se člověk nejdříve musí **rozlétat**.*
*(In spring, one must first **get used to flying**.)*
*Просто **расслушаться** надо, тогда почувствуешь.*
*(You should just **begin to listen** carefully and then you will feel it.)*

2.2 po-X se/si, по-X 0/себе

Meaning: to X during a certain time period
  Czech + usually rather calmly, enjoy X-ing
  Russian + usually relatively short time
Note: The difference *0/себе* is rather stylistic — the reflexive morpheme is not obligatory, it has an informal tincture.
Examples:

*Pak vyslechl Aliciny kritické poznámky, pár minut **si popřemítal**, načež prohlásil, že všecko souhlasí.*
  *(Then, he listen to Alice's critical remarks, he **was reflecting** several minutes and then declared that all agree.)*
*Анатоль Максимович укрепился на своихъ ногахъ, еще немного **пообду-мывалъ** ситуацию, а затемъ, очень внезапно, началъ проявлять повышенную активность.*
  *(Anatol Maximovič fixed himself on his legs, **spent some time on considering** the situation and then suddenly began to behave very actively.)*

2.3 za-X se/si, за-X-ся

Meaning: to X during a longer period
  Czech + enjoy X-ing
  Russian + longer than usually, or even too long
Examples:

*Loupil tady i loupežník Klempera a **zaloupežil si** tu i slavný lupič Babinský.*
  *(The bandit Klempera robbed here and also the famous robber Babinsky **liked to rob** here.)*
*Избавляемся от вещичекъ, которые давно **зависелись** в нашемъ шкафчике!*
  *(We are getting rid of old stuff, that **is (lit. is hanging) too long** in our cupboards!)*

2.4 na-X se, на-X-ся

Meaning: to X a lot
  Russian + completely, often with a slightly positive result
Examples:

*Přestože profesor napsal tolik dopisů ... a **napředstavoval se** v desítkách kanceláří, jeho úsilí se bohužel nesetkalo s úspěchem.*
  *(Although the professor wrote so many letters ... and **many times presented himself** in dozens of offices, his efforts unfortunately have not been successful.)*

*Она руку привязала, чтобы не позвонить Новикову и не спросить, каки-
ми же это чудесными именами он Евгению Гордееву **напредставлялся**.*

*(She has tied up her hand, in order not to call Novikov and to ask him, which won-
derful names he **used (lit. was using, i.e. many times) to introduce himself** to Eugene
Gordeev.)*

2.5   vy-X se, вы-X-ся

Meaning: very much X
  Czech + with more or less positive result
  Russian + until exhaustion
Examples:
  *Francis Kennedy se tam uchýlil, aby **se vytruchlil** ze smrti své ženy.*
  *(Francis Kennedy took refuge there to **have enough time to grieve** over the death of
his wife.)*
  *Я много **выстрадался** с тех пор, как расстался с вами в Петербурге.*
  *(I've **suffered really much** from the time, when we parted in St.Petersburg.)*

2.6   u-X se, у-X-ся

Meaning: until exhaustion
Examples:
  *Zemřel na těžkou bronchitidu. Prostě **se ukašlal!***
  *(He died of severe bronchitis. He simply **died of coughing**.)*
  *Всем смешно, а я уже весь **укашлялся и усопливился**...*
  *(Everybody finds it funny, but I'm **completely exhausted by coughing and having
the sniffles**.)*

2.7   —, из- X-ся (Only Russian)

Meaning: great intensity, with more or less negative result
Example:
  *За 40 с лишним лет рисования он не **изрисовался**.*
  *(For more than 40 years, he **hasn't weared himself out with painting**.)*

3   Intensified Verbs in the Context of a Common Vocabulary

The previous overview and all the examples show that the situation in Russian and
Czech is very similar. The languages differ in productivity, but mostly in shades of
meanings of prefixes investigated. That will be the subject of further, more linguistically
oriented research.
  The described formation of new meanings is quite rare and is used only in cases
where an author wants to stress a specific intensity of the basic word. They are created
only occasionally, like the majority of examples presented so far. However, some words

that were formed according to the method described above, are a normal part of the language. Some of them have exactly the same meaning that was described in the previous section, for instance the word *rozesmát se* (in Czech) and *рассмеяться* (in Russian) with the meaning *to start laughing*.

Others have a different meaning. The verb compound *usmát se* is an example. Its common meaning is described in [4] as "to express joy with a smile", as for instance in *Hezky se na mě usmál. (He smiled at me nicely.)*. In our intensified usage, it does not mean a smile, but a laugh, namely its very high, almost exhausting intensity. If we remove the prefix -*u*, we get the verb *smát se (to laugh)*. The following example is not authentic, as it is very difficult to look for a context where the word has different (moreover, very rare) meaning.

Czech: *Celý večer jsme se něčemu smáli, až jsme se skoro usmáli.*

English: *(We have been laughing all the evening, until we almost died from laugh.)*

4   Use in Automatic Language Processing

4.1   Morphological Dictionaries and Guessers

At the beginning of any automatic language processing task, one should detect the words in the text. For this purpose, the morphological analysis is used, which assigns each word its morphological characteristics. These are e.g. lemma, part of speech and according to the part of speech some other morphological values, such as grammatical gender, number, case, time and so on. These values are usually expressed by means of morphological tags. The lemma for verbs in Czech and Russian (but also in many other flective languages) is infinitive.

Today, many languages have a morphological dictionary that should contain all commonly used words. These dictionaries are used for word detection and analysis. Words that are not included in the dictionaries must be detected and tagged according to some heuristics. These heuristic tools are called guessers. Commonly, a guesser may use any possible rule observed in different texts of the language in question. The observation may be either "manual" (rule-based guessers) or automatic (statistical guessers). Our finding about verb intensification may contribute to better word recognition and analysis. The observation described in the previous sections is considered to be rather a rule-based approach. Due to the high regularity of intensified verbs, it is neither necessary nor possible, to include all of them into a dictionary. In most cases, frequently used intensified verbs are already present in the dictionary (see, for example, the verb *rozesmát se (begin to laugh)* described in the previous section); the rest can be detected very reliably.

4.2   Homonymy of Basic and Intensified Verbs

In cases of homonymy between basic and intensified verbs (see e.g. the example above with the Czech verb *usmát se (smile)* or *(to be completely exhausted by laughing)*), the most correct approach should be to analyze both meanings. However, due to the low frequency of such cases, we decided not to do so.

Looking at the examples, we have found some factors which can be useful for recognition of different meanings of homonymous verbs. First of all, in the sentences with intensified verbs, verb modifications are generally missing. In the example above, the Czech verb *usmát se* in the sense of smiling is commonly used with the object (*usmát se na něco/na někoho* (*smile at somebody/something*)). If used as an intensified verb, the object is scarcely expressed. This is true for the most verbs which have complements in their basic meaning (either objects or adverbials).

Another sign that could help recognizing intensified verbs in the sentence, is use of certain intensifiers. These intensifiers for Czech are, for example, *do sytosti, dosyta* (*to one's heart's content*), *úplně* (*completely*), *k smrti* (*to death*), *do (úplného) vyčerpání* (*to (complete) exhaustion*) and so on, for Russian there are *совсем* (*entirely, totally*), *досыта* (*to one's heart's content*), *до смерти* (*to death*) and so on.

4.3   Lemma of Intensified Verbs

At first glance, the lemmatization doesn't seem to be problematic. By adding a prefix, a new word is created. It seems to be reasonable, to assign this word a new infinitive with the prefix, in the same way as it is done for all prefixal verbs. In linguistics, this kind of derivation is considered to be a case of word formation. However, due to the high regularity of this phenomenon, the intensified verbs may be also observed as a result of a morphological derivation. All prefixes in this case are simply modifying the value of a basic verb, as shown above. For this reason, we consider it to be more natural, to lemmatize these verbs without the prefix. Thus, for example, the Czech verb *rozsténat se* (*begin to moan*) will have the lemma *sténat* (*moan*), not *rozsténat se* (*begin to moan*). The same is true for Russian.

In addition, it would significantly facilitate the distinction between homonymous verbs, examples of which we mentioned in the previous sections. For the homonymy of the Czech verb *usmát se* (*smile* and *to be completely exhausted by laughing*), two lemmas will be assigned: *smát se* for an intensified meaning and *usmát se* for *smile*.

Such lemmatization is also easier to use in further processes of automatic text processing. For example, in machine translation, it is very unlikely that an intensified verb occurs in a translation dictionary, because, as we pointed out above, these words are mostly casual. According to prefixes, the rules for translation would be created. For example, the Czech prefix *roz-* could be translated to English as the construction *start / begin to* + lemma. If we apply this rule to the verb *rozsténat se* (*begin to moan*), the dictionary will translate its lemma as *sténat* (*moan*) and the correct translation into English will be *start / begin to moan*:
Czech: *James Stidham, který to vše sledoval z nosítek,* se tiše **rozsténal** *hrůzou.*
Russian: (*James Stidham, who watched all this from a stretcher,* **began to moan** *quietly with horror.*)
Similar rules can be applied for all other prefixes which are analyzed here.

4.4   Some Remarks to the Morphological Analysis of Intensified Verbs

Although the evidence concerning the verb intensification presented here is being now just under investigation, a prefix guesser is already being used in the Czech morphology, namely in the Morfo project [5]. However, the lemmatization in the project has not been implemented as described in the previous sections yet. Verbs, as well as all other words, are lemmatized together with their prefixes.

The recognition procedure of intensified verbs would be useful to apply on all verbs with the presented prefixes, so that in cases of homonymy it could be detected which kind of verb is used and thus which lemma (with or without prefix) should be assigned to it. However, as intensified verbs are not frequent in texts, it seems to be sufficient (at least for the time being), to analyze only the verbs which are not recognized by a morphological analyzer using a morphological dictionary. Due to low frequency of intensified verbs in texts, the risk of an error is relatively small.

The low frequency of the cases in question does not allow us to provide any statistics of how the success rate of morphological analysis improves; the results would be negligible. However, the more precise description of word-formative and morphological characteristics will always contribute to a more precise language processing.

5   Conclusion

In this paper, we presented productive word-formative patterns consisting of a prefix (*roz-, po-, za-, na-, vy-, u-*, in Russian also *iz-*) and the reflexive morpheme. Using the examples from Czech and Russian, we have showed that being applied to almost any imperfective verb, these patterns always change its meaning in the similar way. For the time being, this regularity should be regarded as just an observation. To effectively use this property in NLP, it is necessary to analyse the formal grammar and semantic characteristics of these structures in more detail.

In further research, we also plan to look at other (mainly Slavic) languages, where we also assume the existence of intensified verbs.

References

1. Evgen'eva, A.P.: Malyj akademičeskij slovar'. Institut lingvističeskich issledovanij, Moscow (1999); Slovar' russkogo jazyka v 4-ch tomach
2. Czech national corpus (2012), http://www.korpus.cz
3. Russian national corpus (2011), http://www.ruscorpora.ru
4. Havránek: Slovník spisovného jazyka českého. Praha, Academia (1989)
5. Hlaváčová, J., Kolovratník, D.: Morfologie češtiny znovu a lépe. In: Vojtáš, P. (ed.) Informačné Technológie Aplikácie a Teória.Zborník Príspevkov, ITAT 2008, Seňa, Slovakia, Univerzita Pavla Jozefa Šafárika, PONT s.r.o., pp. 43–47 (2008)

K-Component Adaptive Recurrent Neural Network Language Models

Yangyang Shi[1], Martha Larson[1], Pascal Wiggers[2], and Catholijn M. Jonker[1]

[1] Intelligent System Department, Delft University of Technology
Mekelweg 4, 2628CD, Netherlands
{yangyang.shi,m.larson,c.m.jonker}@tudelft.nl
[2] CREATE-IT Applied Research, Amsterdam University of Applied Sciences (HvA)
p.wiggers@hva.nl

Abstract. Conventional n-gram language models for automatic speech recognition are insufficient in capturing long-distance dependencies and brittle with respect to changes in the input domain. We propose a k-component recurrent neural network language model (KARNNLM) that addresses these limitations by exploiting the long-distance modeling ability of recurrent neural networks and by making use of k different sub-models trained on different contextual domains. Our approach uses Latent Dirichlet Allocation to automatically discover k subsets of the training data, that are used to train k component models. Our experiments first use a Dutch-language corpus to confirm the ability of KARNNLM to automatically choose the appropriate component. Then, we use a standard benchmark set (Wall Street Journal) to perform N-best list rescoring experiments. Results show that KARNNLM improves performance over the RNNLM baseline; the best performance is achieved when KARNNLM is combined with the general model using a novel iterative alternating N-best rescoring strategy.

Keywords: Recurrent Neural Networks, Latent Dirichlet Allocation, N-best rescoring.

1 Introduction

The language model plays a crucial role in automatic speech recognition. It is responsible for constraining the sequence of recognized words to sequences occurring in natural language. Conventional n-gram language models calculate the probability of each word based on a history of the preceding $n - 1$ words. The length of the history is limited by number of times individual $n - 1$ word sequences appear in the training data; as the history grows in length, the n-gram language model faces a quickly increasing data sparseness challenge. In addition to their inadequacy in modeling long-distance dependencies, n-gram models are known for cross-domain brittleness. Within a single domain this shortcoming manifest itself as lack of robustness to variations in the input speech [1].

Recent studies have shown that a recurrent neural network language model (RNNLM) can outperform n-grams [2]. One claim about the performance of the RNNLM is that it projects the high dimensional vocabulary into a low dimensional continuous space.

I. Habernal and V. Matousek (Eds.): TSD 2013, LNAI 8082, pp. 311–318, 2013.

In this way, an RNNLM helps to relieve data sparseness issues. Simultaneously, the recurrent procedure in the RNNLM equips the language models with the memory to address the long-distance dependency insufficiency.

The k-component adaptive recurrent neural network language model (KARNNLM) proposed in this paper adopts the framework of RNNLMs with their capability to model long-distance word dependencies, and extends them to tackle brittleness to variation. Rather than using one single model as in conventional RNNLMs, the KARNNLM contains k component models, each trained on a specific domain that has been discovered in the training data. KARNNLM first uses Latent Dirichlet Location to automatically construct k contextual domains in the training data and divide the data into partitions corresponding to the k-components. Domain models are trained on the individual partitions and combined with a general model trained on the entire corpus. The resulting model is used to rescore N-best lists, the standard procedure for applying the RNNLMs in speech recognition. We propose three methods for combining domain models: KARNNLM, that makes a hard decision for the correct k-component model at the word level, MIX, that creates a linear combination of the sentence-level probabilities of all k-component models and an alternative rescoring strategy.

Taking the fact that word usage pattern varies among contextual domains, a novel approach for exploiting context domains within the RNNLM framework is proposed in this paper. We demonstrate that the proposed method can improve performance over both conventional n-gram language models and generic RNNLM models.

The rest of the paper is organized as follows. Section 2 discusses the related work. In Section 3, we introduce our KARNNLM approach, including the construction of the contextual domains and the k-component adaptation strategy. Section 4 presents experimental results. The final section gives the conclusion and outlook.

2   Related Work

Our work is related to previous work that has been carried out in three topic areas: recurrent neural networks language modeling, topics based language modeling and two-pass rescoring, each of which is covered in this section in turn.

The recurrent neural network language model RNNLM, originally proposed by [2,3], incorporates the time dimension by expanding the input layer, which represents the current input word, with the previous hidden layer. Theoretically, recurrent neural networks can store relevant information from previous time steps for an arbitrarily long period of time, making it possible to learn long-term dependencies. In [4,5], RNNLMs are used to model the long-term context information by directly treating the contextual information as input to the networks. In this paper, we use k-component strategy together with RNNLMs to take advantage of context domains in language modeling.

The potential of integrating context information into conventional n-gram language models has been long well-established by the speech recognition community [6,7,8,9]. Our work extends previous work on adapting language models to topical context, by looking not only at topic, but rather at general context domains. Such domains are more subtle since they may involve factors such as style, with less marked lexical distribution patterns.

Conventionally, language models that aim to capture context are not applied during the first decoding pass, but rather are used for rescoring [9]. In this paper, we apply standard rescoring that optimizes WER [10]. We use a iterative alternating N-best rescoring approach, motivated by a desire to avoid local optima during the rescoring process.

In our experiments, we test two patterns that we use for combining general models and context models. The first is sentence level mixtures, which is promising since it has previously demonstrated its usefulness within conventional n-gram language modeling framework [6,7]. The second is word-conditioned combinations, which can be seen as related to the approach of [11], that proposes a model in which each word is treated as a topic mixture model for predicting further word occurrences.

3 K-Component Adaptation Recurrent Neural Networks Language Models

3.1   Recurrent Neural Network Language Models

The recurrent neural network adopted in our work originated with [2]. It has three layers: an input layer x, a hidden layer h and an output layer y. It is characterized by the loop between the input layer and hidden layer. At each time t, the input vector $x(t)$ is constituted by the current word vector $w(t)$ as well as a copy $h(t-1)$ from the previous hidden neurons. The sigmoid function and softmax function are used as the activation functions in the hidden layer and output layer, respectively. The weight matrix is estimated by backpropagation-through-time (BPTT)[3].

3.2 K-Component Model and Domain Adaptation

In order to create the k-component domains, we cluster the training data on the sentence level using Latent Dirichlet Allocation (LDA) [12]. We chose LDA since it is a state-of-the-art method for clustering text data; other clustering algorithms can be expected to work as well. The LDA is a probabilistic generative model that represents each sentence as a combination of latent topics. Each sentence is assigned to a cluster based on its largest latent component.

We realize language model adaptation, by interpolating the general RNNLM models with specific contextual component domain RNNLM models:

$$p(w_t|h_t) = \mu_g p_g(w_t|h_t) + \mu_k p_k(w_t|h_t), \tag{1}$$

where p_g is the general model trained on the complete training set and p_k the k-component model. The interpolation weights μ_g and μ_k are tuned using the development data.

Language Model Adaptation. During rescoring, no information concerning the identification of the 'correct' component model is available. We propose two approaches for selection of the appropriate component model, thereby adapting the overall language model. In the first model (KARNNLM), we assign a sentence probability by applying a hard maximum likelihood decision at the word level.

$$p_{krnnlm}(s) = \prod_i \max_k p(w_i|h_i, M_k), \tag{2}$$

where s is a sentence under evaluation. $p(w_i|h_i, M_k)$ represents the conditional probability of the current word w_i given the component language model M_k and history h_i. In the second model (MIX), we do not use a hard decision, but rather create a sentence-level linear combination of the sentence of all k-component models to assign the sentence probability, expressed by:

$$p_{mix}(s) = \sum_k \lambda_k p(s|M_k) = \sum_k \lambda_k \prod_i p(w_i|h_i, M_k), \tag{3}$$

where λ_k is the interpolation weight of component model k. Notice that within each sentence, the w_i is dependent on the same model M_k. The interpolation weight is determined by the N-best list held out data.

3.3 Iterative Alternating N-Best Rescoring

We experiment with a further combination of the general model and models specific to the k-component using an iterative alternating N-best rescoring approach. In the N-best rescoring paradigm [10], N-best hypotheses are generated from one model, which are rescored by other models. In general, the combination weight is learned from the held out data, which can correspond to a local optimum. Here, we propose a simple iterative rescoring strategy to extend the standard strategy. It works as follows:

For two different language models m_1 and m_2, N-best hypotheses N and ratio $\alpha \in (0, 1)$.

1. The language model m_1 is used to rescore the N-best hypotheses N, of which a α portion of hypotheses are selected. $N := \alpha * N$.
2. The hypotheses selected in previous step are rescored by m_2. The N-best list is further reduced to a new list $N := \alpha * N$.
3. If $N == 1$, stop. Otherwise repeat from step 1.

This strategy approaches the optimization problem by exploiting a filter method. In each iteration, the size of N-best list is first reduced by the general RNNLM and then by the KARNNLM or the MIX. The result of the refined N-best list is used into next iteration until the best hypothesis is obtained. This method can contribute to preventing the different combinations of scores from falling into a local optimum.

4 Experimental Evaluation

We conduct two types of comparisons in our experiments. The first compares the general RNNLM with the KARNNLM specific to a particular contextual domain under the oracle situation in which the specific domain is known. The second compares the RNNLM, the KARNNLM and the mixture RNNLM in a speech recognition experiment involving N-best list rescoring.

Table 1. Word prediction accuracy (WPA) results by context domain (CGN data), comparing conventional (RNNLM) trained on the whole data set with domain-specific recurrent neural networks language models (context domain known). Substantial improvement in word prediction accuracy is indicated in bold. '-' indicates the component is too small, random selection did not select data from it.

comp	socio-situational settings	words	RNNLM	KRNNLM
a	Spontaneous conversations ('face-to-face')	2,626,172	24.0	24.2
b	Interviews with teachers of Dutch	565,433	20.2	19.4
c&d	Spontaneous telephone dialogues	2,062,004	25.5	25.4
e	Simulated business negotiations	136,461	24.5	25.0
f	Interviews/ discussions/debates	790,269	18.6	18.3
g	(political) Discussions/debates/ meetings	360,328	**15.9**	**16.5**
h	Lessons recorded in the classroom	405,409	20.9	20.5
i	Live (eg sports) commentaries (broadcast)	208,399	**16.5**	**19.9**
j	Newsreports/reportages (broadcast)	186,072	17.3	16.6
k	News (broadcast)	368,153	**14.5**	**20.0**
l	Commentaries/columns/reviews (broadcast)	145,553	15.2	13.7
m	Ceremonious speeches/sermons	18,075	-	-
n	Lectures/seminars	140,901	14.8	13.0
o	Reading speech	903,043	**14.2**	**15.6**
overall			20.6	21.3

4.1 Data

Spoken Dutch Corpus. This corpus, referred to in Dutch as *Corpus Gesproken Nederlands* and abbreviated CGN, [13,14] is an 8 million word corpus of contemporary Dutch as it is spoken in Flanders and the Netherlands. It consists of 14 components, each associated with a type of socio-situational setting, as shown in Table 1. The socio-situational setting is related to speech style, which is in turn related to the situation in which the speech is produced. Settings range from informal spontaneous conversation to formal read speech. Of the CGN data, 80% is randomly selected for training, 10% for development testing and 10% for evaluation. We selected a vocabulary with 45K words by choosing word types that occur more than once in the training data. In the test data, words which are not in the vocabulary are replaced by an out-of-vocabulary token (OOV rate is 3.8%).

Wall Street Journal. This corpus, abbreviated here WSJ is drawn from the DARPA WSJ '92 and WSJ'93 data sets. We chose to use the same issue of the data set, and the same N-best lists, as used by [2,15]. The training corpus contains 37M words of running text from the NYT section of English Gigaword. The held-out 230K words is used for testing. A part of the N-best list rescoring data is used as development data for tuning the weight for interpolation, language model score, acoustic model score and word insertion penalty. The rest of them are used for evaluation.

4.2   Word Prediction Accuracy Results

Word prediction has many applications in natural language processing, such as augmentative and alternative communication, spelling correction, word and sentence auto completion, etc. Typically word prediction provides one word or a list of words which fit the context best. It actually provides a measurement of the performance of language models [16].

Table 1 provides the comparison, in terms of WPA, between the general RNNLM and the specific component RNNLM over each component. The hidden layer has a size of 300 neurons, the class size is 100 and we train using 5 iterations of backpropagation through time (BPTT), with a block size of 10. We used the same parameter settings for the general and the component models [3].

The results in Table 1 indicate that if the context domain is known, in general, the domain-specific model outperforms the general model. Especially, in the component "News (broadcast)", the domain-specific model improves the WPA of the RNNLM by almost 38%. However, the component RNNLM is a balance of speciality and reliability. For some components which have much similar characteristics with other components, the specialized training can degrade the performance. For example, the component "Commentaries/columns/reviews (broadcast)", the component model gets almost 10% relative WPA reduction. From [17], we can find that this component also got the lowest classification accuracy. In other words, this component is not strictly discriminative with other components.

4.3   Word Error Rate Results

In WPA results, using a hand-labeled Dutch-language corpus, we confirm the ability of KARNNLM to automatically choose the appropriate component. However, in practice, the component information is not available before hand. The experiment on WSJ is intends to deal with such situation.

Table 2 shows the comparison of RNNLM, KARNNLM and MIX in N-best rescoring experiments performed on the the WSJ data set in terms of word error rate 'WER' under the setting –hidden layer size of 100 neurons, class size of 100, 5 iterations of BPTT, BPTT size of 10. The MIX represents mixture k-component RNNLM. T stands for latent topics. The final column is the WER for iterative alternating N-best rescoring, where a pass is carried out with RNNLM before KARNNLM or MIX are applied.

The best performance is achieved by KARNNLM with 10 latent topics. Beyond that, more topics lead to reduced performance. At the same time, we notice that the performance of KARNNLM faithfully tracks the number of topics indicating that determining the optimal number of topics is critical for applying the approach. The sentence-level mixture MIX is not as effective as KARNNLM, and the RNNLM framework appears to do well exploiting the hard decision between components imposed by KARNNLM. The final column shows that both the KARNNLM and MIX achieve additional improvement with the iterative alternating N-best rescoring strategy. The best KARNNLM reduces the WER of RNNLM by absolute 0.70%.

Table 2. The WER results of Kneser-Ney 5-gram, conventional RNNLM, KARNNLM and k component mixture RNNLM

model	WER(%) rescore	WER(%) iterative
KN5	17.30	-
RNNLM	16.83	-
KRNNLM+5T	16.60	16.25
KRNNLM+10T	16.34	16.13
KRNNLM+15T	17.13	16.57
KRNNLM+20T	17.15	16.56
MIX+5T	16.59	16.31
MIX+10T	16.55	16.21
MIX+15T	16.94	16.53
MIX+20T	17.09	16.52

5 Conclusion

We proposed a k-component recurrent neural network language model for speech recognition N-best list rescoring. Our approach addresses the shortcomings of conventional language models with its ability to capture long-distance dependencies and robustness to variations in domain. Each k component language model is a combination of a general RNNLM with a dedicated RNNLM trained on its associated contextual domain. The experiment on the CGN data set demonstrated the ability of k-component models to outperform general RNNLM under the oracle situation in terms of WPA. In N-best list rescoring of the WSJ data set, the KARNNLM with 10 latent topics reduced WER by 0.49% absolute. In order to reduce the risk of overfitting, we used a novel iterative alternating N-best rescoring strategy, which resulted in an absolute WER reduction of 0.70% over the RNNLM.

By demonstrating the potential of component language models in the recurrent neural network language modeling framework, we have set the stage for future work. In [2], it is observed that the performance improves when RNNLM with different architectures are combined. Our results suggest that diverse component models can be selected or combined to strengthen RNNLM. Future work will involve understanding how they can be further improved by combining different architectures. Further, although KARNNLM and RNNLM demonstrate large improvements over n-gram language models, they remain computationally expensive. An immediate next step will focus on methods for reducing computational cost.

Acknowledgement. Thank you to Tomas Mikolov for making the RNNLM Toolkit publicly available and for helpful discussion.

References

1. Rosenfeld, R.: Two decades of statistical language modeling: where do we go from here? Proceedings of the IEEE 88, 1270–1278 (2000)

2. Mikolov, T., Karafiát, M., Burget, L., Cernocký, J., Khudanpur, S.: Recurrent neural network based language model. In: INTERSPEECH, pp. 1045–1048 (2010)

3. Mikolov, T., Kombrink, S., Burget, L., Cernocky, J., Khudanpur, S.: Extensions of recurrent neural network language model. In: 2011 IEEE International Conference on Acoustics, Speech and Signal Processing (ICASSP), pp. 5528–5531 (2011)

4. Mikolov, T., Zweig, G.: Context dependent recurrent neural network language model. In: SLT, pp. 234–239 (2012)

5. Shi, Y., Wiggers, P., Jonker, C.M.: Towards recurrent neural networks language models with linguistic and contextual features. In: 13th Annual Conference of the International Speech Communication Association (2012)

6. Iyer, R., Ostendorf, M., Rohlicek, J.R.: Language modeling with sentence-level mixtures. In: Proceedings of the Workshop on Human Language Technology, pp. 82–87. Association for Computational Linguistics, Morristown (1994)

7. Iyer, R., Ostendorf, M.: Modeling long distance dependence in language: Topic mixtures vs. dynamic cache models. In: Proc. ICSLP 1996, Philadelphia, PA, vol. 1, pp. 236–239 (1996)

8. Kneser, R., Peters, J.: Semantic clustering for adaptive language modeling. In: 1997 IEEE International Conference on Acoustics, Speech, and Signal Processing, ICASSP 1997, vol. 2, pp. 779–782 (1997)

9. Clarkson, P., Robinson, A.: Language model adaptation using mixtures and an exponentially decaying cache. In: 1997 IEEE International Conference on Acoustics, Speech, and Signal Processing, ICASSP 1997, vol. 2, pp. 799–802 (1997)

10. Ostendorf, M., Kannan, A., Austin, S., Kimball, O., Schwartz, R., Rohlicek, J.R.: Integration of diverse recognition methodologies through reevaluation of n-best sentence hypotheses. In: Proceedings of the Workshop on Speech and Natural Language, HLT 1991, pp. 83–87. Association for Computational Linguistics, Stroudsburg (1991)

11. Chin, H.S., Chen, B.: Word topical mixture models for dynamic language model adaptation. In: IEEE International Conference on Acoustics, Speech and Signal Processing, ICASSP 2007, vol. 4, pp. IV-169 –IV-172 (2007)

12. Blei, D.M., Ng, A.Y., Jordan, M.I.: Latent dirichlet allocation. J. Mach. Learn. Res. 3, 993–1022 (2003)

13. Hoekstra, H., Moortgat, M., Schuurman, I., van der Wouden, T.: Syntactic annotation for the Spoken Dutch Corpus Project (CGN). In: Computational Linguistics in the Netherlands 2000, pp. 73–87 (2001)

14. Nelleke, O.N., Wim, G.W., Van Eynde, F., Boves, L., Martens, J.P., Moortgat, M., Baayen, H.: Experiences from the Spoken Dutch Corpus project. In: Proceedings of the Third International Conference on Language Resources and Evaluation, LREC 2002, pp. 340–347 (2002)

15. Wang, W., Harper, M.P.: The SuperARV language model: Investigating the effectiveness of tightly integrating multiple knowledge sources. In: Proceedings of the ACL 2002 Conference on Empirical Methods in Natural Language Processing, EMNLP 2002, vol. 2, pp. 238–247 (2002)

16. den Bosch, V.: Scalable classification-based word prediction and confusible correction. Traitement Automatique des Langues 46, 39–63 (2006)

17. Shi, Y., Wiggers, P., Jonker, C.M.: Socio-situational setting classification based on language use. In: IEEE Workshop on Automatic Speech Recognition and Understanding, pp. 455–460 (2011)

LIUM ASR System for ETAPE French Evaluation Campaign: Experiments on System Combination Using Open-Source Recognizers

Fethi Bougares*, Paul Deléglise, Yannick Estève, and Mickael Rouvier

LUNAM University, LIUM laboratory, Le Mans France
http://www-lium.univ-lemans.fr
{firstname.lastname}@lium.univ-lemans.fr

Abstract. In this paper, we report the LIUM participation in the ETAPE [1] (Évaluations en Traitement Automatique de la Parole) evaluation campaign, on the rich transcription task for French track. After describing the ETAPE goals and guidelines, we present our ASR system, which ranked first in the ETAPE evaluation campaign. Two ASR systems were used for our participation in ETAPE 2011. In addition to the LIUM ASR system based on CMU Sphinx project, we utilized an additional open-source ASR system based on the RASR toolkit. We evaluate, in this paper, the gain obtained with various acoustics modeling and adaptation techniques for each of the two systems, as well as with various system combination techniques. The combination of two different ASR systems allows a significant WER reduction, from 23.6% for the best single ASR system to **22.6%** for the combination.

Keywords: automatic speech recognition, cross-system combination, evaluation campaign.

1 Introduction

Interest in Automatic Speech Recognition (ASR) has grown over the last years, and various ASR toolkits are developed and made freely available on the Internet. These publicly available toolkits, generally published under an open-source license, facilitate the setting up of new transcription systems. The LIUM laboratory was involved in all French speech transcription evaluation campaigns organized by the AFCP since 2003. These campaigns present opportunities to enhance and measure the improvements obtained by the continuous development of the LIUM ASR system [2]. During the ETAPE evaluation campaign, we introduced a new open-source ASR system used jointly with our system. Both systems were first used individually and afterwards combined with different techniques. The present article focuses on the performance gains obtained by each technique used in each ASR system individually. Therefore we analyze the improvements with different combination techniques.

---

* This research was supported by the ANR (Agence Nationale de la Recherche) under contracts number ANR-09-BLAN-0161 (ASH) and ANR-08-CORD-026 (PORT-MEDIA).

I. Habernal and V. Matousek (Eds.): TSD 2013, LNAI 8082, pp. 319–326, 2013.

In the remainder of this paper we describe the LIUM participation in the ETAPE evaluation campaign. Section 2 depicts the ETAPE campaign. The third and fourth sections present the used multi-pass ASR systems and highlight the performance gain for each system, depending on decoding pass and adaptation techniques. Before concluding, section 5 presents the combination techniques that we tested, and the gain achieved with each one.

2   ETAPE Evaluation Campaign

The ETAPE evaluation campaign was organized within the framework of a project funded by the French National Research Agency (ANR). The project brings together national experts in the organization of such campaigns under the scientific leadership of the French-speaking Speech Communication Association (AFCP). The AFCP has organize several evaluation campaigns since 2003, in order to evaluate automatic speech transcription systems for the French language. The ETAPE 2011 campaign followed the ESTER 1 and 2 [3] evaluation campaigns organized respectively in 2003 and 2009. Unlike the ESTER campaigns, the ETAPE 2011 campaign focused on TV material with various levels of spontaneous speech and multiple overlapping speech.

3   ASR Systems

In addition to the LIUM ASR system, we built a new ASR system based on the open-source RASR toolkit. In this section we describe the development of both systems.

3.1   Speaker Diarization

Speaker diarization is carried out using the LIUM open-source speaker diarization toolkit. This toolkit includes hierarchical agglomerative clustering methods using well-known measures such as BIC (*Bayesian Information Criterion*) and CLR (*Cross Likelihood Ratio*). In order to remove music and jingle regions, a segmentation into speech / non-speech is obtained using a Viterbi decoding with 8 one-state HMMs. Detection of gender and bandwidth is done using a GMM (with 128 diagonal components) for each of the 4 combinations of gender and bandwidth [4].

3.2   LIUM ASR System

The LIUM ASR system is an expansion of the best open-source ASR system participating in the ESTER 2 evaluation campaign [2].

Feature Extraction. The transcription decoding process is based on multi-pass decoding using two types of acoustic features. The first set is composed of 39-dimensional PLP features (13 PLP with energy, delta, and double-delta). These features are computed for each show, corresponding to broadband and narrowband analysis. The second type of features if probabilistic features produced by a Multi Layer Perceptron (MLP), trained using the ICSI QuickNet libraries [5]. The input speech representation of our MLP is a concatenation of nine frames of 39 MFCC coefficients (twelve MFCC features, energy, Δ and $\Delta\Delta$). The topology of the MLP is as follows: the first hidden layer is composed of 4000 neurons; the second one, used as the decoding output, of 40 neurons; and the third one, used for training, of 102 neurons (34 phonemes included 1 phone for all fillers, 3 states per phoneme). The MLP features were decorrelated by a PCA transformation. The second feature vector has 79 parameters resulting from the concatenation of the PLP and MLP (39 PLP + 40 MLP).

Acoustic Models. Acoustic models are trained using a set of data from distinct sources. The training corpus is composed of 511 hours of wide-band and 60 hours of narrow-band training data (167.5h from ESTER 1 and 2 campaign, 60h from EPAC, 17.5h from ETAPE, 227h of internal Broadcast news and 99h of imperfect transcript of Podcast). Using these audio data, two sets of gender- and bandwidth-dependent acoustic models are trained. All models are adapted using MAP adaptation of means, covariances and weights. First set of models are used for first pass decoding and composed of 2500 tied states, each state being modeled by a mixture of 22 diagonal Gaussians. Others models are composed of 10000 tied states with a mixture of 48 Gaussians trained in a MPE framework applied over the SAT-CMLLR models and used after the first decoding pass.

Vocabulary. To build the vocabulary, we generate a unigram model as a linear interpolation of unigram models trained on the various training data sources detailed in next section. The linear interpolation was optimized on the ETAPE dev set so as to minimize the perplexity. Then, we extract the 159k most probable words from this language model. Phonetic transcriptions for the vocabulary are taken from the BDLEX database if found there, or generated by the rule-based, grapheme-to-phoneme tool LIA_PHON [6] otherwise.

Language Models. Different textual data from multiple sources are used to train trigram and quadrigram language models. The training corpus contain 1626 M words and composed of the audio corpus transcript, the GoogleNews corpus, french gigaWord and additional data collected from the Web. To estimate and interpolate these models, the SRILM [7] toolkit is employed using the modified Kneser-Ney discounting method without cut-off.

Decoding Process. The decoding strategy is close to that used in the LIUM'08 system with an additional decoding pass. The involved passes are as follows:

1. The first pass uses gender- and bandwidth-depended acoustic models and a 3-gram language model.
2. The word-graphs output of the first pass are used to compute a CMLLR transformation for each speaker. This second pass is performed using SAT and Minimum Phone Error (MPE) acoustic models with CMLLR transformations. Segments boundaries are also adjusted according to the one-best output of the previous pass in order to avoid cuts within words.

3. In the third pass, the word-graphs of the previous pass are used to drive a graph-decoding with full 3-phone context with a better acoustic precision, particularly in inter-word areas. This pass generates new word-graphs.
4. The linguistic scores of the third pass word-graphs are updated using a 4-gram language model.
5. The output of the previous pass are used to recompute the linguistic scores with a 5-gram language model.
6. The last pass generates a confusion network from the word-graphs and applies the consensus method to extract the final one-best hypothesis.

3.3   RASR ASR System

RASR is the RWTH Aachen University open-source speech recognition toolkit [8]. The toolkit includes state of the art speech recognition technology for acoustic model training and decoding. It includes speaker adaptation techniques, speaker adaptive training, unsupervised training and post-processing tools. Here is a description of the RASR system used during ETAPE campaign. This system uses the same language model and vocabulary as the ones used for the LIUM ASR system and described in sections 3.2 and 3.2.

Feature Extraction. The acoustic waveform is parameterized as 15 Mel-frequency cepstral coefficients (MFCCs) with energy. Afterwards, consecutive feature vectors in a sliding window of size 9 are concatenated. 135-dimentional feature vectors are obtained, which are then projected down to 45 components using an LDA transformation.

Acoustics Models. For the RASR system, acoustic modeling for 35 phonemes and 5 kinds of fillers is based on across-word triphone states represented by left-to-right, three-state Hidden Markov Models (HMMs). Training data are the same as used for the LIUM ASR system and described in section 3.2. The number of triphone states is 15000, generalized via decision tree clustering based on linguistic question. The emission probabilities of the remaining states are modeled by Gaussian mixtures with a globally shared covariance matrix and 2,8 millions densities. Contrary to the LIUM ASR system, the RASR acoustic model is gender and bandwidth independent.

Decoding Process. The RASR system involves four decoding passes. The first two passes are as follows:

1. The first pass is a decoding using a 3-gram language model.
2. The best hypotheses generated by pass 1 are used to compute a CMLLR and MLLR transformation for each speaker.

The two final passes are identical to passes 4 (recomputing linguistic scores with 4-gram language model) and 6 (consensus decoding) of the LIUM ASR system.

4   ASR System Initial Performance

4.1   Experimental Data

Systems are tuned using a development set of 8.5 hours made available to participants for benchmarking purposes. In this paper, we report the results obtained on the official test data, which contain 9.5 hours from various sources and of various sizes (BFM TV, LCP, TV8 Mont Blanc and France Inter).

4.2   System Performance

The ETAPE test set was first transcribed individually by each system. The performance of each ASR system are detailed in this section.

LIUM ASR System. Table 1 reports the WER of the LIUM ASR system after each decoding pass using the MLP feature vectors.

Table 1. LIUM ASR WER for each pass on ETAPE test set

Pass	WER
Pass 1	33.98 %
SAT-CMLLR-MPE decoding	29.36 %
Full 3-phone decoding	25.55 %
4-G LM rescoring	24.01 %
5-G LM rescoring	23.85 %
CN-decoding	23.60 %

RASR ASR System. The ETAPE test set was also transcribed using the RASR ASR system. Results are presented in table 2

Table 2. RASR performance on ETAPE test set

Pass	WER
Pass 1	33.44 %
RASR-CMLLR-MLLR	31.52 %
4-G LM rescoring (RASR-4G)	31.31 %
CN-decoding	29.45 %

5   ASR System Combination

Given multiple ASR systems, the aim of system combination is to exploit complementary information encoded by each system in order to minimize the final WER. This section describes various combination techniques that we explored in this evaluation campaign.

5.1   Cross-System Combination

In multi-pass speech recognition systems, it is common practice to apply an unsupervised acoustic model adaptation in-between passes. Cross-system adaptation consists in adapting a system $S2$ based on the output of a different system $S1$ [9,10]. This schema can be used in order to get higher gains out of adaptation compared to adapting in-between passes. Cross-system adaptation is applied into the RASR system to perform CMLLR and MLLR adaptation using the output of the fourth pass of the LIUM system which yields 24.01% of WER. In addition, the acoustic scores of the output lattices of this cross-system adaptation are updated using the LIUM ASR acoustic models.

Table 3. Cross-system combination on ETAPE test set

System	WER
LIUM Pass 4	24.01 %
RASR-CMLLR-MLLR	28.88 %
Acoustic re-scoring	28.81 %
4-G LM rescoring (RASR-4G)	27.90 %
CN-decoding	27.49 %

Table 3 shows the improvements obtained with cross-system combination. Cross-system adaptation allows to decrease the WER of RASR system by almost 2 points.

5.2   Lattice Combination

Instead of combining the one best hypotheses, the lattice combination method exploits the benefit of multiple hypothesis output by performing the combination using the lattices generated from the various systems. This combination method achieves consistent improvement in recognition accuracy compared one-best combination techniques [11]. Posterior scores are first calculated for each lattice involved in the combination. Afterwards, lattices from different systems are equally weighted and merged as follows: all the lattice starting nodes are linked to a new starting node S and all the ending nodes are linked to a new ending node E. The resulting lattice is finally used as input to final pass of the LIUM ASR system described previously.

In order to perform lattice combination, various lattices are extracted by varying the feature set and language model used for graph rescoring (passes 4 and 5 of the LIUM ASR system). Table 4 presents the combined graphs, their final WER (after CN-decoding) and the achieved improvements after lattice combination. The MPE-CSLM and MLP-CSLM lattices are respectively those obtained using the first feature set (PLP) and the second feature set (MLP) described in section 3.2, with 5-gram continuous space language model [12] for graph rescoring in pass 5 of the LIUM system. The MPE-4G are lattices obtained before the 5-gram rescoring in the LIUM system. Lattice combination permits to improve the overall WER by 0.66 point. Its provides an additional output to be integrated into the ROVER schema described in the next section.

Table 4. WER obtained after lattice combination technique

System	WER
(1) MPE-CSLM	24.43 %
(2) MLP-CSLM	24.02 %
(3) MLP-5G	23.60 %
(4) MPE-4G	24.05 %
(5) RASR-4G	27.49 %
Lattice combination-1-2-3-4-5	22,94 %

5.3 ROVER combination

The LIUM primary submission for the ETAPE evaluation campaign is a ROVER [13] combination of the various outputs presented in table 5.

Table 5. LIUM primary submission on the ETAPE campaign

System	WER
(a) Lattice combination-1-2-3-4-5	22,94 %
(b) MLP-5G	23.60 %
(c) MLP-4G	23.75 %
(d) MLP-4G-segbase	23.98 %
(e) MPE-CSLM	24.43 %
(f) RASR-4G	27.49 %
ROVER-a-b-c-d-e-f	**22.26 %**

The best result obtained during ETAPE evaluation campaign is 22.26%, which is ROVER over the six best outputs presented on table 5 using two open-source ASR systems. The MLP-4G-segbase output is obtained without the segment boundary modification and the 5-G rescoring in the LIUM system.

6   Conclusions

The work presented in this paper gives an experimental feedback about the performance of two open-source ASR systems within the well-defined framework of an evaluation campaign. The LIUM primary submission was the best ASR system participating in the ETAPE evaluation campaign in 2012. Experiments were made with a couple of open-source ASR systems. These toolkits, combined following various schemes, were used in a well-defined context with unprepared speech, and they allowed to reduce significantly the final WER (1.34 point WER reduction compared to the LIUM ASR system without additional training data).

References

1. Gravier, G., Adda, G., Paulsson, N., Carré, M., Giraudel, A., Galibert, O.: The ETAPE corpus for the evaluation of speech-based TV content processing in the French language. In: International Conference on Language Resources, Evaluation and Corpora, LREC (May 2012)
2. Deléglise, P., Estève, Y., Meignier, S., Merlin, T.: Improvements to the LIUM French ASR system based on CMU Sphinx: what helps to significantly reduce the word error rate? In: Interspeech (2009)
3. Galliano, S., Geoffrois, E., Mostefa, D., Choukri, K., Bonastre, J.F., Gravier, G.: The ES-TER phase II evaluation campaign for the rich transcription of French broadcast news. In: Eurospeech, European Conference on Speech Communication and Technology, Lisbon, Portugal (September 2005)
4. Meignier, S., Merlin, T.: LIUM SpkDiarization: an open source toolkit for diarization. In: ASRU (2010)
5. Zhu, Q., Chen, B., Morgan, N., Stolcke, A.: On using MLP features in LVCSR. In: Proc. ICSLP, Jeju, Korea, pp. 921–924 (2004)
6. Béchet, F.: LIA_PHON: un système complet de phonétisation de textes. Traitement Automatique des Langues – TAL 42, 47–67 (2001)
7. Stolcke, A.: SRILM – An extensible language modeling toolkit (2002)
8. Rybach, D., Gollan, C., Heigold, G., Hoffmeister, B., Lööf, J., Schlüter, R., Hermann, N.: The RWTH Aachen university open source speech recognition system. In: Interspeech, pp. 2111–2114 (2009)
9. Stüker, S., Fügen, C., Burger, S., Wölfel, M.: Cross-system adaptation and combination for continuous speech recognition: The influence of phoneme set and acoustic front-end?(2006)
10. Giuliani, D., Brugnara, F.: Experiments on cross-system acoustic model adaptation. In: ASRU, pp. 117–122 (2007)
11. Li, X., Singh, R., Stern, R.M.: Lattice combination for improved speech recogniton. In: Interspeech (2002)
12. Schwenk, H.: CSLM - A modular Open-Source Continuous Space Language Modeling Toolkit. In: Interspeech (2013)
13. Fiscus, J.G.: A post-processing system to yield reduced word error rates: Recognizer Output Voting Error Reduction (ROVER). In: IEEE Workshop on Automatic Speech Recognition and Understanding, pp. 347–352 (1997)

Lexical Stress-Based Morphological Decomposition and Its Application for Ukrainian Speech Recognition

Mykola Sazhok[1,2] and Valentyna Robeiko[2]

[1] Hlushkov Institute of Cybernetics, Kyiv, Ukraine
mykola@cybermova.com
[2] International Research/Training Center for Information Technology and Systems, Kyiv, Ukraine
{sazhok,valia.robeiko}@gmail.com

Abstract. This paper presents an approach to word morphological decomposition based on lexical stress modeling. Word segmentation quality is estimated by a hidden variable that assigns the lexical stress. The formulated segmentation criterion is based on a training set of words with manually pointed stresses and a large text corpus. The described search algorithm finds one or more segmentations with the best likelihood. Given arguments confirm the necessity to distinguish stressed and unstressed vowels in the phoneme alphabet for Ukrainian speech recognition systems. The developed tool allows to assign primary lexical stress in unknown words. Experimental research is described and results are discussed.

Keywords: lexical stress, morphological decomposition, speech recogntition, Ukrainian.

1 Introduction

The phenomenon of lexical stress plays significant role for many languages. Prosodic features like duration, pitch, and loudness are used to describe phonetic distinctions for stressed word segments. So any text-to-speech system must implement lexical stress prediction. Letter-to-sound rules practically always work for vowels under lexical stress even for highly spontaneous pronunciation manner. And this property might be useful for spontaneous speech recognition tasks.

In [1] authors assume that morphological decomposition is required for lexical stress prediction particularly for cases where a local context is insufficient. Moreover, presenting words as a sequence of reasonable segments or morphemes is a key to model the word formation and to strive beyond vocabulary limitations particularly in speech understanding systems. Known methods of morphological decomposition relies solely on orthography [2], [3]. In our research, lexical stress prediction and morphological decomposition are considered as a result of the same process through which phonetic, syntactic and semantic hidden features can be discovered from word spelling.

In the next section we consider motives, prerequisites and possible applications, in Section 3 we describe segmentation procedure formalization and implementation, Section 4 is devoted to data description, in Section 5 we report and discuss experimental results followed by Conclusion.

I. Habernal and V. Matousek (Eds.): TSD 2013, LNAI 8082, pp. 327–334, 2013.

2   Motivation

In Ukrainian, stress position is irregular and it can be changed even within forms of the same word. Fortunately, the available electronic lexicography system contains more than 1.8 million words with manually assigned lexical stress covering all valid word forms [4]. The web-based basic text corpus contains 275 million unchecked word samples, which makes a vocabulary for about two million words. Relatively to the lexicography system, half of which is detected in the basic corpus, OOV words make 2.5%. At least 200 thousand more words allow for OOV reduction to less than 0.5%. So stress prediction is a way to assign lexical stress for the large amount of both new and known words.

The reason to introduce stressed vowels to the text-to-speech system is obvious due to necessity to meet a human perception of duration, pitch, and loudness. In speech recognition, feature extraction models are mostly invariant to prosodic features. However, we believe that introduction of both stressed and unstressed vowels to the phoneme alphabet, at least for Ukrainian, is essential due to phonetic, lexical, and acoustical facts. Stressed vowels normally act as distinctive phonemes changing word grammatical function and meaning that we observe for more than 5% of words in the basic text corpus. In English language, stress shift, normally, causes changes in phonetic content of the pronunciation (compare: récord and recórd). Therefore, such argumentation might be rather inapplicable for a specific language.

Grapheme-to-phoneme conversion methods like [5] could be directly used also for modeling stress, however they have no provisions to account for the structural properties of stress. Here, rather than modifying an existing algorithm we prefer to construct a model concentrated on stress properties and then convert the stressed text to phoneme sequences by means that allow for counting specific pronunciation properties provided by the method that needs about 30 find-replace-and-step rules for Ukrainian [6].

2.1   Application to Speech Recognition

To explore the acoustical side of lexical stress we estimated phoneme model parameters considering stressed and unstressed vowels as different phonemes and inspected dissimilarities particularly by means of the HMM visualization tool [7]. Following Fig. 1 we can see the difference between models for unstressed and stressed phonemes of "a" and "ï" trained on 40 hour multi-speaker speech corpus for Ukrainian [8].

The presented central state contains 32 GMMs estimated in MFCC feature space accomplished with energy coefficient and mean subtraction that makes total 13 coefficients. The dotted line corresponds to a zero value. Visually, a stressed model looks like a subset for most coefficients. Overlaps rather than inclusions with respective coefficients in the stressed model are proper in cases like the 5th coefficient for "a" and the first coefficient for "ï". More HMMs are available from the tool's web-page.

Analyzing transition matrices, we can see that diagonal values corresponded to emitting states are 1.52 times greater for stressed models that confirms the essential difference in duration.

Introducing both stressed and unstressed vowels is relatively small overhead for Ukrainian language since only 6 vowel phonemes are distinguished: a, e, y, i, o, and u.

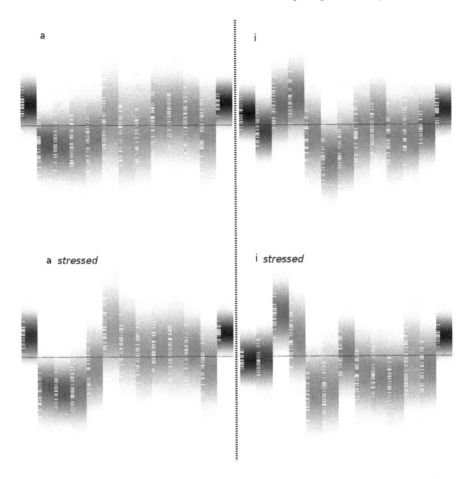

Fig. 1. HMM visualization for Ukrainian unstressed and stressed monophones a and i

However, such extension is more expensive for phoneme alphabets containing significantly more vowels (due to diphthongs, nasalization, etc.).

Perhaps, the most convincing argument for stressed vowel models became the analysis of Ukrainian speech recognition preliminary experiments. A multi-speaker and a single speaker > 40 hour training sets were used to learn models for both versions of the phoneme alphabet (49 and 55 monophones). For word recognition we applied a bi- and a trigram language models and a free order grammar was used for syllable and phoneme recognition. To make the results comparable we ignored the stress information from recognized word and phoneme sequences. In all cases we observed a $12 - 23\%$ improvement relatively to the word/phoneme error rate. Therefore, our concern about possible confusion of stressed and unstressed phonemes was rather exaggerated.

The benefit of morphological decomposition is a possibility to present the entire lexicon with a limited set of morpheme level segments irrespective to their correspondence to morphemes in a sense of classical linguistics.

3   Methodology

3.1   Lexical Stress-Based Word Segmentation Model

A training vocabulary W contains words with assigned attributes like a lexical stress. Each word w from the vocabulary W can be decomposed into a sequence of symbols $q^{(w)} = (q_1, q_2, \ldots, q_{K_w})$ taken from an alphabet of letters or phonemes Q.

We consider subsequences of $q^{(w)}$ as segments of a certain segmentation $s^{(w)}$ among all valid segmentations $S^{(w)}$ for the word w. The i-th segment of segmentation $s^{(w)}$

$$s_i^{(w)} = \left(s_{i1}, s_{i2}, \ldots, s_{iL_i^{(w)}} \right) \tag{1}$$

together with other segments of $s^{(w)}$ cover the entire $q^{(w)}$ without overlaps that is for any $w \in W$

$$\sum_i L_i^{(w)} = K_w, \ 1 \leqslant L_i^{(w)} \leqslant \min \{ L_{\max}, K_w \}, \tag{2}$$

$$I(s_{11}^{(w)}) = 1 \ \text{and} \ I(s_{i1}^{(w)}) = I\left(s_{(i-1)L_{i-1}^{(w)}}^{(w)} \right) + 1, \ i > 1, \tag{3}$$

where $I(\cdot)$ returns a segment item index within $q^{(w)}$. The constrain for the largest segment length, $L_{\max}$, determines a model order. More segmentation constrains might be introduced, e.g. restriction on running primarily stressed syllables.

Uniting all segments of valid segmentations for all words belonging to the vocabulary W we form a training set of segments

$$S = \bigcup_{w \in W, s^{(w)}, i} s_i^{(w)} \tag{4}$$

and consider each segment in this set, s_i, with no relation to words.

A stress level, $\theta_k^{(w)} = \{0, 1, 2\}$, assigned to each symbol forms a corresponding attribute sequence $\theta^{(w)} = (\theta_1, \theta_1, \ldots, \theta_k, \ldots, \theta_{K_w})$. We assume the stress level other than zero can be assigned to symbols that introduce a syllable, at least potentially. Normally, these are vowels that may be accomplished with specific consonants like "r" in Slovenian [3]. For other symbols the stress level is not applicable and is always equal to zero. Stress level values may be limited only to 0 and 1 that means only a primary lexical stress is considered. On the contrary, we may introduce more values corresponding to symbol attributes that might be hidden in spelling like reduction, lengthening and modifiers, as well as symbol attribute combinations. Thus, we generally refer to $\theta^{(w)}$ as an attribute sequence for corresponding symbols in w.

Obviously, the returned index in (3) is the same within $\theta^{(w)}$, which subsequences are assigned to $s_i^{(w)}$. Attribute sequences assigned to each segment of $s^{(w)}$ segmentation, in turn, form a set $\Theta^{(w)}$.

We can estimate a probability of the attribute sequence θ given a segment s_i observable in the training set:

$$P(\theta|s_i) \approx \frac{c(s_i, \theta)}{c(s_i)} \tag{5}$$

where $c(s_i, \theta)$ is a count of segments s_i with stress assignment defined by a stress indication vector θ and $c(s_i)$ is a number of s_i occurrence. All counts are taken from the text corpus for words included in the stress vocabulary. For the segments with low occurrence a smoothing technique should be applied.

Finally, we search through all valid segmentations $s^{(w)}$ and attribute sequences θ that satisfy the expression:

$$\left(\hat{s}^{(w)}, \hat{\Theta}\right) = \underset{s^{(w)}, \Theta^{(w)}}{\arg\max} \prod_{i,\theta} P\left(\theta | s_i^{(w)}\right) \tag{6}$$

For known words, θ is determined for each segment $s_i^{(w)}$ uniquely, otherwise, all valid attribute sequences are being searched.

Thus, to carry through morphological decomposition we introduced a segmentation model based on features, hidden in word spelling, like lexical stress. However, not all obtained segments are valid morphemes due to potentially more strict morpheme constrains like presence of at least one vowel. Therefore, unifying such a segment with adjacent one is a way to compose a valid morpheme.

3.2 Segmentation Graph Analysis

We constructed a dynamic programming graph where finding the shortest trajectory is equivalent to the search (5). Each input symbol introduces a set of valid pairs (segment, attributes), which are nodes on the graph where the partial criterion is accumulated. Connections between nodes correspond to valid segmentations. Memorizing N prospective arrows entering to nodes we can extract N-best word segmentations.

In Fig. 2 an example of one-best stress prediction search (6) is shown for a proper name Obama missing from the basic Ukrainian vocabulary. The word is represented as concatenation of all valid character segments where the largest segment length is limited to five. Input items are down-cased letters accomplished with the word boundary symbol "—". Each input item introduces a set of valid segments with attributes. For compact presentation we show only the result of the attribute application. Thus, a notation "obAm" in column 5 means the segment (o, b, a, m) with the attribute vector (0, 0, 1, 0). Potentially optimal arcs are either shown or coded with the name of previous node. Indicated partial criteria are log probability based. The optimal trajectory, respective nodes and criteria are bold.

In the example we illustrate two running stressed syllables forbidding: in column 7 the segment "mA" follows the segment "a" rather than "obA". As far no constrains for the segment content are introduced, segments may contain a single consonant like "b" in column 3. This is the way to guarantee a successful search (6) for any word. The system may decide that such a segment and an adjacent segment may belong to the same morpheme depending on constrains given by the expert. In accordance to morphology knowledge, for Ukrainian language, only the first and the last morphemes may consist of consonants solely. To compose a formally valid morpheme we may append the segment "b" to the preceding segment preferring more frequent morphemes and coming to Ob-áma. We can see that a foreign word has been approximated with native morphemes. The model updated with rather correctly stressed new word samples

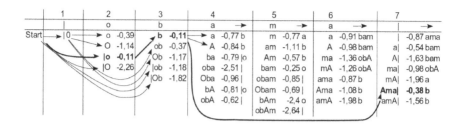

Fig. 2. Stress prediction for an out-of-vocabulary word "obama"

learns new morphemes, which make linguistically more justified decomposition of the considered word and its forms possible: Obám-a, Obám-y, etc.

3.3 Implementation

To implement the described word segmentation algorithm a set of basic three tools has been developed. Currently, these tools operate only with primary stress information.

The first tool, *putstress*, prepares the data necessary for estimation of probabilities eq:stressAttrBySegmentProb by input data and knowledge base, word frequency vocabulary and, optionally, corrected homograph frequency proportions. For each word the tool tries to retrieve records about stress position and stores found words accomplished with stress-mark and frequency. Found homographs are saved with frequencies updated in accordance to their corrected proportions. An expert may correct more proportions and run the tool again.

The second tool, *guessstress*, implements the search procedure (6) extracting $N-$best sequences of segments with corresponding attributes. A frequency vocabulary of words with assigned stresses is the source to estimate probabilities for hypothetical symbol subsequences.

Finally, the *prep_stressvcb* tool forms a stress vocabulary by the extracted segments. Several additional tools allow for extracting various information from input data, estimated models, and segmentations. All modules are written in Perl language.

4   Data

The stress dictionary is extracted from the electronic lexicography system subset containing 151 962 lemmas, including over 10 thousand names, that makes 1.90 million word forms [4]. Due to the shared spelling an actual word form the vocabulary consists of 1.83 million words that have different either spelling or primary lexical stress position.

The basic text corpus is derived from a hypertext data downloaded from several websites containing samples of news and publicity (60%), literature (8%), encyclopedic articles (24%), and legal and forensic domain (8%). To be noted that the data downloaded from news websites contains numerous user comments and reviews, which we consider as text samples of the spontaneous speech. A text filter, used for text corpus processing, provides conversion of numbers and symbolic characters to relevant letters, removing improper text segments and paragraph repetitions. Hereafter, we refer to the basic corpus as 275M corpus. In accordance to the corpus summary shown in Table 1, we observe 6.64 word forms per lemma in average, whereas this relation is twice greater, 12.3, within the dictionary [4]. Adding to the vocabulary 200 000 most frequent words we reduce OOV to less than 0.5%.

Table 1. Basic text corpus 275M summary

Words	Sentences	Vocabulary			OOV	Homographs
		All words	Known words	Known lemmas		
275 288 408	1 752 371	1 996 897	801 040	120 554	2.51%	16 729 476

Words that have two or more valid stress positions, referred as homographs, take 6% of the average text. However, homographs may occur with quite different frequency that may considerably affect occurrence for certain segments. Therefore, an expert may correct occurrence proportions in the homograph dictionary containing over 14 000 spellings.

5   Experiments

Known and OOV words were evaluated separately. The evaluation of known words was made in order to learn how a large part of the vocabulary can be coded without explicit lexical stress information. The largest model order, L_{max}, was set to five, and 4-best segmentations were analyzed to form a stress vocabulary. The expert has corrected occurrence proportions in the homograph dictionary for 500 most frequent spellings.

The system detected about one million pairs (segment, stress). Frequencies for different segment lengths are shown in Table 2.

Table 2. Detected segment counts

Segment length, L	1	2	3	4	5
Item count	46	1781	35280	233816	721575
Occurance (million)	2115.652	1848.766	1581.879	1314.993	1070.579

215 000 segments were used to predict a lexical stress for words in 275M text corpus. Stress position for less than 1% known words was detected incorrectly. Stress detection for 5 000 OOV words was incorrect in 21.1% words or 5.3% syllables. However, over 50% of incorrectly stressed words have strong foreign origins.

Perhaps, the most interesting is the case of stress moving with some morphological derivations. Checking derivations from photo/photography (fóto, fotóhraf, fotohráfija, fotohrafíchnyj, fotohrafuváty and their forms) we found that incorrect stress has been assigned only in one derivation (fotohráf).

6   Conclusion

The proposed morpheme level segmentation model allows for simultaneous extraction the features generally ignored in word spelling. However, these features are hypothetical and a word context must be used to choose either right stress position or letter modifiers or their absence that is typical for homographs. The segment level context introduction is a way to further model improvement.

The model refinement is possible in expectation-maximization manner, especially with a slight supervising for stress correction between iterations. Future research should also concern a model order, letter modifiers and phonemic input, as well as more languages should be considered.

References

1. Black, A., Lenzo, K., Pagel, V.: Issues in Building General Letter to Sound Rules. In: 3rd ESCA Workshop on Speech Synthesis, Jenolan Caves, Australia, pp. 77–80 (1998)
2. Creutz, M., Lagus, K.: Induction of a simple morphology for highly-inflecting languages. In: Proc. 7th Meeting of the ACL Special Interest Group in Computational Phonology (SIGPHON), Barcelona, pp. 43–51 (2004)
3. Sef, T., Skrjanc, M., Gams, M.: Automatic lexical stress assignment of unknown words for highly inflected Slovenian language. In: Sojka, P., Kopeček, I., Pala, K. (eds.) TSD 2002. LNCS (LNAI), vol. 2448, pp. 165–172. Springer, Heidelberg (2002)
4. http://lcorp.ulif.org.ua/dictua/
5. Bisani, M., Ney, H.: Joint-Sequence Models for Grapheme-to-Phoneme Conversion. Speech Communication 50(5), 434–451 (2008)
6. Robeiko, V., Sazhok, M.: Bidirectional Text-To-Pronunciation Conversion with Word Stress Prediction for Ukrainian. In: Proc. UkrObraz 2012, Kyiv, pp. 43–46 (2012)
7. http://www.cybermova.com/speech/visual-hmm.htm
8. Pylypenko, V., Robeiko, V., Sazhok, M., Vasylieva, N., Radoutsky, O.: Ukrainian Broadcast Speech Corpus Development. In: Proc. Specom 2011, Kazan, RF, pp. 244–247 (2011)

Motion Event in Croatian, English, German and Italian Concerning Path Prefixes and Prepositions

Nives Mikelić Preradović, Damir Boras, and Tomislava Lauc

Faculty of Humanities and Social Sciences, University of Zagreb
Ivana Lučića 3, 10000 Zagreb, Croatia
www.ffzg.unizg.hr
{nmikelic,dboras,tlauc}@ffzg.hr

Abstract. We present an approach to encode translocative meaning and explore not only the possible, but also the preferred patterns for expressing this phenomenon in four languages: Croatian, English, German and Italian. We manually developed verb subcategorization frames for translocative meaning of prefixes PRE- and PRO-, enlarged the Croatian part of the bilingual dictionaries with the appropriate subcategorization frames and created the frame transition rules for three language pairs. The analysis presented here should indicate the possibilities of using obtained results for the development of both monolingual and bilingual verb dictionaries to facilitate foreign language acquisition, but also to improve the native language proficiency.

Keywords: perlative meaning, translocative prefixes, verb subcategorization frame, Croatian language, English language, Italian language, German language.

1 Introduction

Linguistic treatment of the spatial domain is represented through diverse perspectives. In spite of diversity in which spatial features are coded in specific languages, the basic meaning of space can be divided into two types: firstly, it is a spatial construction denoting the location of the given object and secondly, it is a direction of an object in motion. Directional meaning can be divided into three meanings: (1) adlative that expresses direction "moving closer to"; (2) ablative that expresses direction "moving away from" and (3) perlative that expresses motion which is directionally unspecified and represents only the path of the moving object. If all basic types of spatial relationships are analyzed, we obtain locative (i.e. where), perlative (i.e. which way), adlative (i.e. where to), and ablative relationship (i.e. where from) [1]. Each of these relationships also has sub-meanings, such as intralocative (direction: into), extralocative (direction: out of) and translocative.

In this paper, we deal with translocative meaning that refers to route or path (i.e. trajectory in space tracing the points occupied by object that is moving from one location - source to another - destination). This is often referred to as "motion event", "translocation", "directed motion" or "translational motion", and the corresponding verbs are called "directed motion verbs", "change-of-location verbs", etc. [2]

I. Habernal and V. Matousek (Eds.): TSD 2013, LNAI 8082, pp. 335–342, 2013.
© Springer-Verlag Berlin Heidelberg 2013

Talmy defines six basic elements of motion events [3]: (1) presence or absence of translational motion (Motion); (2) moving entity (Figure); (3) object with respect to which Figure moves (Ground); (4) course followed by the Figure with respect to Ground (Path); (5) manner in which motion takes place (Manner) and (6) cause of its occurrence (Cause). Depending on how elements of "motion event" are mapped to different constituents in the clause, Talmy [3] formulates the basis for his motion-event typology in order to differentiate satellite-framed languages (S-languages where path is lexicalized as a "satellite" to the verb) and verb- framed languages (V-languages where path is lexicalized in the root of the motion verb) and also giving implications for second language acquisition.

In this paper we describe the translocative interlingual meaning, based on the analysis of 4 languages: Croatian, German, English and Italian, where only Italian belongs to V-languages, while the other three are S-languages that encode path as prefix added to the verb root (Croatian and German), or by combination of prefix and preposition (Croatian) or they exclusively employ prepositional phrases (English).

We aim to present translation patterns related to expressing PATH in four languages and illustrate them through verb subcategorization frames and rules for mapping the semantic component of motion event and path in these languages.

2   Motivation

Linguistic theory offers verb's lexical representation through its subcategorization frame specifying the number and types of complements corresponding to the participants in the event described by a verb. Verb's complements (arguments) can be identified as syntactic roles (e.g. Subject, Object) or semantic roles (e.g. Agent - entity initiating the action and Patient - entity suffering an action).

We agree with Stringer [4] that, in order to learn the second language (L2), it is not sufficient to learn the general rules for the expression of motion. L1 and L2 learners also need to acquire the argument structure associated with particular lexical items and be able to distinguish which elements denoting motion are mapped onto the satellites (i.e. prepositions, prefixes) or verb root and whether these elements occur in L2 constructions. Furthermore, in the evaluation of the language behavior of L2 learners, the potential influence of the predominant lexicalization pattern of the first language may be important for the investigation of the knowledge acquisition process. For example, the subcategorization frame of motion verb in Croatian language may contain PATH as a prepositional phrase (PP) argument (indirect object) or noun phrase (NP) argument (direct object). NP complement sets focus on achieving the goal with the object as center (*preploviti rijeku / to sail the river*), while PP emphasizes trajectory (*preploviti preko rijeke na suprotnu obalu / to sail across the river to the opposite bank*). The research [5] on 10 sentence pairs (the first one with NP and the second with PP complement) revealed that Croatian native speakers consider them semantically equal (out of 37 students, 52.6% preferred NP, while 47.4% preferred PP constructions).

Therefore, coding meanings of prefixes and subcategorization frames of motion verbs could be relevant for learners to facilitate understanding of path in L1 and L2. However, existing Croatian monolingual learner's dictionaries (MLD) are useful only for learners with high language competencies and not quite suitable for beginning learners of

Croatian as a foreign language. Our analysis should indicate the possibilities, as well as the importance of using obtained results for the development of both monolingual and bilingual verb dictionaries to facilitate L2 acquisition, but also to improve the native language proficiency.

3   Method

The importance of a prefix and its meaning(s) is reflected in the following: if a bilingual dictionary encodes **pre-** as a translocative prefix in the source language, then we can assume translocative meaning in the target language as well (realized as a prefix / prepositional phrase in S-language or as part of the root in V-language). Translocation refers to an object moving crosswise, through or above another (usually elongated) object, touching it or not.

As far as our approach is concerned, we systematically analyzed potential meaning for each member belonging to the set of 20 Croatian productive prefixes (do-, iz-, na-, nad-, o-/ob-, obez-, od-, po-, pod-, pre-, pred-, pri-, pro-, raz-, s-, su-, u-, uz-, za-). We manually developed verb subcategorization frames for each meaning of each prefix and discovered that the verbs sharing the same meaning of a prefix also have the same type of complement with the same or similar morphosyntactic form in their subcategorization frame.

Since our first goal was to encode and analyze the interlingual features of translocative meaning, we searched for verbs with prefix *pre- (over, across)* and *pro- (through)* in three bilingual dictionaries [6], [7], [8]. The reason for extracting prefixed verbs from dictionaries instead of corpus was twofold. Firstly, we wanted to obtain evidence for all meanings of two prefixes, even for those that are rarely present in the corpus. Secondly, we wanted to investigate the significance of prefixes, their meanings and subcategorization interlingual frames. Regarding the number of retrieved verbs: 153 verbs with prefix pre- and its translocative meaning were found in Croatia- English dictionary [6], 42 were found in Croatian-Italian dictionary [7], and 46 verbs in Croatian-German dictionary [8]. Also, 54 verbs with prefix pro- and its translocative meaning were found in [6], 35 were found in [7], while 42 were retrieved from [8].

In the next step, we performed the same search on PoS-tagged and lemmatised 1.2 billion word Croatian corpus hrWaC [9]. As a result, we found resultative meaning (completion of an action to the very end, to an effect or to the goal) as the most frequent meaning of the prefix pro- (verb *pronaći / to find* has 224567 tokens), while the translocative meaning comes in the 2nd place (verb *proći / to pass through* with 178535 tokens). Regarding the prefix pre-, its most frequent meaning in corpus is also resultative (verb *prenositi / to report* has 106026 tokens), while the translocative meaning comes in the 10th place (verb *prelaziti / to cross over* with 37219 tokens).

Finally, all verbs with prefix pre- and pro- extracted from bilingual dictionaries (124 in total) were given to ten annotators, who manually labeled each prefixed verb with one or more translocative meanings inherited from the prefix. They also examined the syntactic-semantic argument structure of each verb and assigned it to the predefined subcategorization frame(s) listed below. All complements (semantic roles) listed in each frame are obligatory and valid for both perfective and imperfective Croatian verbs.

Prefix pre- (over, across) frames:

AGT_1 **pre-V** PATH_preko+2 / On je **pretrčao preko** ceste / He ran **across** the street

AGT_1 **pre-V** PATH_4 / Dječak je **prešao cestu** / The boy crossed the street newline

AGT_1 **pre-V** PAT_4 PATH_preko+2 / Nosim je preko rijcke / I carry her **over** the river

Prefix pro- (through) frames:

AGT_1   **pro-V** PATH_kroz+4/7 / Kamion je **prošao kroz tunel/tunelom** / The truck passed **through** the tunnel

AGT_1   **pro-V** PATH_4 / Kamion je **prošao tunel** / The truck left the tunnel behind

AGT_1   **pro-V** PAT_4 PATH_kroz+4 / Kustos je **proveo** goste **galerijom** / Curator walked guests **through** the gallery

   Depending on the number of obligatory complements (2 or 3), apart from translocative meaning, these verbs also specify either MOTION (if they have 2 complements) or TRANSPORT (if they have 3 complements). Furthermore, verbs with 2 obligatory complements have different meanings, depending on the morphosyntactic structure of the 2nd complement. If second complement takes the form of accusative case (as in AGT_1 PATH_4), it denotes space as an object and prefixed verb becomes transitive despite the fact that verbs of movement do not possess transitivity as their distinctive quality. If 2nd complement takes the form of instrumental case or PP (as in AGT_1 PATH_kroz+4/7), it moves semantic focus from the object to the path in space. It is worth pointing out that verb *proći* / *to pass* (the most frequent verb in corpus) represents the only pro- prefixed verb that takes both the direct complement (with the *meaning to leave behind*) and indirect complement (denoting the path through the object). The corpus statistics for both NP and PP complements of pre- verbs showed that they are equally represented in web documents. In the next step, we enlarged the Croatian part of the bilingual dictionaries with the appropriate subcategorization frames and created the frame transition rules for three language pairs listed in Table1.

   Analyzing the frame transition rules we came to the following conclusion:

– Croatian prefixed verb and its PATH semantic role **(expressed with PP and additional instrumental case for pro-verbs)** is (regardless the aspect) translated with:

  • **English** non-prefixed verb and preposition *across* (for pre- verbs designating movement **from one side of an area or surface to the other side**)

    (1) Martin je **pre**-trčao preko ceste / Martin ran **across** the street

  • **English** non-prefixed verb and preposition *over* (for pre- verbs designating movement **above and across from one side of an area or surface to the other side**)

    (2) Ptice su **pre**-letjele preko planine / Birds flew **over** the mountain

  • **English** non-prefixed verb and preposition ***through*** (for pro- verbs)

    (3) Oni su pro-(i)šli kroz tunel / They passed **through** the tunnel

Table 1. Subcategorization Frame Transition Rules for Croatian-English, Croatian-German and Croatian-Italian[1]

Croatian frame	English frame	German frame	Italian frame
AGT_1 pre-V PATH_preko+2	AGT **V** PATHover / AGT **V** PATHacross	AGT_1 **V** PATH_über+4 /	AGT **V** PATH oltre
AGT_1 pre-V PATH_4	-	AGT_1 **über-V** PATH_4 / AGT_1 **durch-V** PATH_4	AGT **V** PATH
AGT_1 **pre-V** PAT_4 PATH_preko+2	AGT **V** PAT PATH_over / AGT **V** PAT PATH_across	AGT_1 PAT_4 **über-V** PATH_4 / AGT_1 PAT_4 **durch-V** PATH_4	AGT **V** PAT PATH_attraverso / AGT_1 **V** PAT PATH_oltre
AGT_1 pro-V PATH_kroz+4/7	AGT **V** PATH_through	AGT_1 **durch-V** PATH_4	-
AGT_1 pro-V PATH_4 (only for "proći")	AGT **V** PATH ("to leave behind" only)	AGT_1 **V** PATH_4 ("passieren" only)	AGT **V** PATH_attraverso
AGT_1 **pro-V** PAT_4 PATH_kroz+4/7	AGT **V** PAT PATH_through	AGT_1 **V** PAT4 PATH_durch	AGT **V** PAT PATH_attraverso

– Croatian prefixed verb and its PATH semantic role (**expressed with PP and additional instrumental case for pro-verbs**) is (regardless the aspect) translated with:

  • **German** non-prefixed verb and preposition **über** (for pre- verbs) or **durch** (for pro- verbs)

  (4) Martin je **pre**-skočio preko ograde / Martin sprang **über** den Zaun (Martin jumped over the fence)

  (5) Oni su **pro**-(i)šli kroz tunel / Sie gingen **durch** den Tunnel (They passed through the tunnel)

  • **Italian** non-prefixed verb and preposition **oltre** or **attraverso** (for pre- verbs) and **attraverso** (only for pro-verbs)

  (6) On je **pre**-letio preko Himalaja / Lui volò **oltre** l'Himalaya (He flew over the Himalaya Mountains)

  (7) On je **pre**-plivao preko rijeke / Lui ha nuotato **attraverso** il fiume (He swam across the river)

  (8) Dante je **pro**-šao kroz Pakao / Dante è passato **attraverso** l'Inferno (Dante passed through Hell)

– Croatian prefixed verb and its PATH semantic role with the **direct object in accusative case** is translated to:

  • **English** in the same way as the role expressed with the indirect object. Although in some cases (see sentences 9-10), English marks the distinction between these two types of objects, it is not a regular phenomenon:

---

[1] AGT(agent) = subject; PAT(patient) = direct object; PATH= route, way; pre-V= verb with prefix pre-; pro-V= verb with prefix pro-; 1= nominative case; 2= genitive case; 4= accusative case; 7= instrumental case; preko+2/over; kroz+4/through = prepositional phrase consisting of preposition and either genitive or accusative case.

(9) Martin je **pre**-plivao Atlantic / Martin swam **across** the Atlantic / **Martin swam the Atlantic** (note: *to swim the pool* is a colloquial expression)

(10) Martin je **pro**-šao tunel / Martin **passed the tunnel** (note: meaning in both languages is *to leave behind*)

- **German** with a **durch** prefixed verb (for pre- verbs expressing movement in a three dimensional space)

(11) Martina je **pre**-plivala La Manche / Martina **durch**schwamm den Ärmelkanal (Martina swam **across** La Manche)

- **German** with a **über** prefixed verb and the direct object in accusative (for pre-verbs)

(12) Martin je **pre**-skočio ogradu / Martin **über**sprang den Zaun (Martin jumped over the fence)

- **German** with a non-prefixed verb and the direct object in accusative (only for pro- verbs)

(13) Martin je **pro**-šao tunel / Martin passierte den Tunnel / Martin passed the tunnel (note: meaning in all 3 languages is *to leave behind*)

- **Italian** with *oltre-*, *attraverso-* or *tra-* prefixed verb and the direct object in accusative (which proves the hypothesis [10] that French, Italian and Spanish, although being V-languages, allow S-framed constructions in verbs with path preverbs)

(14) Mi smo **pro**-(i)šli kroz park / Noi abbiamo **attraversato** il parco (We crossed the park)

(15) Vijest je **pre**-šla more / La notizia ha **oltrepassato** l'oceano (News crossed the ocean)

(16) On je **pre**-letio ocean / Lui ha **trasvolato** l'oceano (He flew over the ocean)

Croatian prefixed verbs of TRANSPORT have 3 complements: AGENT, PATIENT and PATH and follow the same pattern in all 3 languages: PATH is always expressed as PP, but only Croatian has verb prefix that correlates with the head of PP.

(17) Martin je **pre**-nio dijete **preko** rijeke./ Martin hat das Kind **über** den Fluss getragen / Martin carried the child **over** the river

(18) On je **pro**-veo posjetitelje kroz izložbu / Er führte die Besucher **durch** die Ausstellung / He guided the visitors **through** the exhibition)

Regarding the Italian language, it usually expresses PATH within the root verb (which supports Talmy's [3] distinction between satellite and verb-framed languages).

(19) Dječak je **pre**-skočio zid / Il ragazzo ha **saltato** il recinto (The boy jumped over the wall)

(20) Lindberg je **pre**-letio Atlantski ocean / Lindberg *sorvolò l'Oceano* Atlantico (Lindberg flew **over** the Atlantic ocean)

However, Italian expresses the path of **TRANSPORT** verbs outside the root verb as a preposition, following the English and German pattern.

(21) On je **pre**-nio dječaka preko rijeke / Lui **portava** il bambino **attraverso** un fiume /He carried the child across a river (note: following these translation patterns L2 learner can easily distinguish between *prevesti-to carry over/take across* and *prevesti-translate*)

(22) Oni su **pre**-vezli eksploziv preko oceana/ Loro hanno **trasportato** l'esplosivo **oltre** l'occano (They transported the explosive over the occan)

4   Results

Analyzing these translation patterns, we discovered similarities between languages, as well as peculiarities. English, Italian and German *through* and *across* prepositions can specify movement on a path in a 3-dimensional space (*to swim through the river / durch den Fluss schwimmen/ nuotare attraverso il fiume; to swim across the sea / schwimm über den See*). While *across* emphasizes the straightness of movement, *through* emphasizes the path from one side of an enclosed space to the other side, or from end to end.

Besides, English distinguishes *across* movement (from one side of an area or surface to the other side in a rectilinear manner) and *over* movement (above and across from one end or side to the other). Depending on the features of the indirect object (its length/width or height/depth), English verbs take either *over* or *across* preposition. Therefore, the construction "to run across the street" is more accurate than "to run over the street". On the contrary, construction "to jump over the fence" is more accurate than "to jump across the fence". Furthermore, both constructions "to jump over the ditch" and "to jump across the ditch" are equally correct, because the first one meets the length/width criterion, while the second one meets the height/depth criterion. This difference is partially revealed in Italian language with prepositions **oltre** and **attraverso**, but it cannot be presented as a rule in a V-language. Neither *across* nor *over* are evidently conveyed by the Croatian prefix *pre-* and the preposition *preko* nor by the German prefix *über*.

We can conclude that the combination of a PRE- or PRO- - prefixed verb and prepositional phrase (or instrumental case and pro-prefixed verbs) represents PATH in Croatian. If a prefixed verb takes only the complement in accusative case instead, it does not emphasize trajectory, but rather achieving the goal with the object as center.

Our results show that the effect of prefixes on the argument structure of verbs in Croatian exhibits regularity that can be exploited to produce both a high-coverage subcategorization monolingual and bilingual lexicon for foreign language learning.

One can also notice the semantic harmony between prefixes, prepositions and motion verb complements. The meaning of the prefix is amplified by the meaning of the preposition. For instance, prefix **pre-** and preposition **preko** (over) share the meaning of PATH. Therefore, preposition **preko** (*over*) is always part of the verb obligatory complement (PATH_preko+2) that verbs with prefix pre- typically bind.

5   Conclusion

Combinations of prefixes PRE- and PRO- with the base motion verbs gave us insight into the close relationship between the syntax and the semantics of verbs sharing the same prefix, as well as idea of foreign language learning using additional set of rules that underlie a set of languages. In our experiment we introduced explicit syntactic and semantic features of each Croatian motion and transport verb with prefix pre- or pro- in the bilingual dictionary. We designed matching frames in both V- and S-languages that allow generalization over syntactic and semantic features of these verbs. As a result, each verb and its translation are described by semantic roles (Agent, Patient and Path) and selectional restrictions.

If grouped according to their prefix and syntactic complements, verbs create semantically and syntactically coherent class (such as MOTION or TRANSPORT) with members sharing a range of features, from the implementation of obligatory complements to morphologically related forms. As a result, it is possible to reduce the effort required to enlarge both the monolingual and bilingual dictionaries.

As our further step, we will give a set of Croatian prefixed verbs that were not found in the dictionaries to annotators to translate them and check if they might guess the meaning from the prefixes and the pattern around the verb.

Also, we plan to enlarge these dictionaries by extracting the most frequent adlative and ablative prefixed verbs (*with prefix do-/to; iz-/out*) from hrWaC corpus using predefined subcategorization patterns that rely on morphosyntactic cues. As a result, we expect to systematically encode SOURCE (expressed by ablative prefix) and GOAL (expressed by adlative prefix) roles that proved to have huge explanatory potential in the analysis of different linguistic phenomena. If Croatian Dependency Treebank becomes available for research purposes, we plan to use it for the extraction of specific instances of verb subcategorization frames directly from the Treebank instead of the tagged corpus. Finally, we will try to extract all other (non-motion) frequent prefixed verbs from hrWaC and implement predefined patterns for each of their meanings in order to fully discover comparable semantic features linked to productive verbal prefixes and prepositions in Croatian and other languages.

References

1. Sysak-Borońska, M.G.: Some remarks on the spatio-relative system in English and Polish. Poznan Studies in Contemporary Linguistics 3, 185–208 (1974)
2. Smit, V.: From Motion Events To(wards) a Semantics of Relocation. Paper delivered at New Directions in Cognitive Linguistics. First UK Cognitive Linguistics Conference, Brighton, UK, October 23-25 (2005)
3. Talmy, L.: Cognitive semantics, Cambridge (2000)
4. Stringer, D.: Paths in first language acquisition: motion through space in English, French and Japanese. PhD thesis, University of Durham (2005)
5. Brala-Vukanović, M., Rubinić, N.: Space prepositions and prefixes in Croatian language (Prostorni prijedlozi i prefiksi u hrvatskome jeziku). FLUMINENSIA 23(2), 21–37 (2011)
6. Filipović, R. (ed.): Englesko-hrvatski rječnik (English-Croatian Dictionary). Školska knjiga, Zagreb (1996)
7. Parcic, D.A.: Vocabolario croato-italiano (Rječnik hrvatsko-talijanski) compilato per cura di Carlo A. Parcic. 3. ed. coretta ed aumentata. Narodni list, Zara (1901)
8. Mažuranić, I., Užarević, J.: German-Croatian Dictionary (Deutsch-ilirisches Wörterbuch). Agram (1845)
9. Ljubešić, N., Erjavec, T.: hrWaC and slWac: Compiling Web Corpora for Croatian and Slovene. In: Habernal, I., Matoušek, V. (eds.) TSD 2011. LNCS (LNAI), vol. 6836, pp. 395–402. Springer, Heidelberg (2011)
10. Fortis, J.-M.: Space in language. Summer School on Linguistic Typology – Part III. Leipzig, Germany (2010)

Multi-label Document Classification in Czech

Michal Hrala[1] and Pavel Král[1,2]

[1] Dept. of Computer Science & Engineering
Faculty of Applied Sciences
University of West Bohemia
Plzeň, Czech Republic
[2] NTIS - New Technologies for the Information Society
Faculty of Applied Sciences
University of West Bohemia
Plzeň, Czech Republic
{hrala36,pkral}@kiv.zcu.cz

Abstract. This paper deals with multi-label automatic document classification in the context of a real application for the Czech news agency. The main goal of this work is to compare and evaluate three most promising multi-label document classification approaches on a Czech language. We show that the simple method based on a meta-classifier proposes by Zhu at al. outperforms significantly the other approaches. The classification error rate improvement is about 13%. The Czech document corpus is available for research purposes for free which is another contribution of this work.

Keywords: Czech, Czech News Agency, Maximal Entropy, Multi-label Document Classification, Naive Bayes, Maximal Entropy, Support Vector Machines.

1 Introduction

Automatic document classification becomes very important for information organization and storage because of the fast increasing amount of electronic text documents and the rapid growth of the World Wide Web. In this work, we focus on the *multi-label* document classification[1] in the context of the application for the Czech News Agency (CTK).[2] CTK produces daily about one thousand of text documents. These documents belong to different categories such as weather, politics, sport, etc. Nowadays, documents are manually annotated but this annotation is often not accurate enough. Moreover, the manual labeling represents a very time consuming and expensive task. Therefore, automatic document classification is very important.

In our previous work [1], we proposed a precise Czech document representation (lemmatization and POS tagging included) and evaluated five feature selection methods, namely document frequency, mutual information, information gain, Chi-square test and Gallavotti, Sebastiani & Simi metric on three classifiers (Naive Bayes (NB), Maximal

---

[1] One document is usually labeled with more than one label from a predefined set of labels.
[2] http://www.ctk.eu

I. Habernal and V. Matousek (Eds.): TSD 2013, LNAI 8082, pp. 343–351, 2013.

Entropy (ME) and Support Vector Machines (SVMs)) in order to build en efficient one class (sometimes also called single-label) Czech document classification[3] system.

The main goal of this work is to adapt our previously developed system to multi-label classification task. The main scientific contribution is to compare and evaluate three most promising multi-label document classification approaches on a Czech language in order to build an efficient Czech multi-label document classification system. Note that to the best of our knowledge, there is no comparative study on the multi-label document classification approaches evaluated on Czech documents. Another contribution of this work is the public availability of the Czech document corpus for the research purposes.

Section 2 presents a short review about the document classification approaches. Section 3 describes three document classification approaches that are compared. Section 4 deals with the realized experiments on the CTK corpus. In the last section, we discuss the research results and we propose some future research directions.

2   Related Work

The document classification task is basically treated as a supervised machine-learning problem, where the documents are projected into the so-called Vector Space Model (VSM), basically using the words as features. Various classification methods have been successfully applied [2,3], e.g. Bayesian classifiers, decision trees, k-Nearest Neighbour (kNN), rule learning algorithms, neural networks, fuzzy logic based algorithms, maximum entropy and support vector machines. However, the task suffers from the issue that the feature space in VSM is highly dimensional which negatively affects the performance of the classifiers.

To deal with this issue, techniques for feature selection or reduction have been proposed [4]. The successfully used classical feature selection approaches include document frequency, mutual information, information gain, Chi-square test or Gallavotti, Sebastiani & Simi metric [5,6].Furthermore, a better document representation may lead to decreasing the feature vector dimension, e.g. using lemmatization or stemming [7]. More recently, advanced techniques based on Principal Component Analysis (PCA) [8] incorporating semantic concepts [9] have been introduced.

Recently, multi-label document classification [10,11,12] becomes a popular research field, because it corresponds usually better to the needs of the real applications, than one class document classification. Unfortunately, it is much more complicated. *The choice of 1 class from the predefined set of N classes* becomes *the choice of M classes from N ones* (M value is unknown). Several approaches have been proposed as summarized for instance in a survey [13].

The most of the proposed methods (see above) deals with English and are usually evaluated on the Reuters,[4] TREC[5] or OHSUMED[6] data sets.

---

[3] One document is assigned exactly to one label from a predefined set of labels.

[4] http://www.daviddlewis.com/resources/testcollections/reuters21578

[5] http://trec.nist.gov/data.html

[6] http://davis.wpi.edu/xmdv/datasets/ohsumed.html

Only few work focuses on the document classification in other languages. Yaoyong et al. investigate in [14] learning algorithms for cross-language document classification and evaluate them on the Japanese-English NTCIR-3 patent retrieval test collection.[7] Olsson presents in [15] a Czech-English cross-language classification on the MALACH[8] data set. Wu et al. deals in [16] with a bilingual topic aspect classification of English and Chinese news articles from the Topic Detection and Tracking (TDT)[9] collection.

Unfortunately, to the best of our knowledge, there are no language-specific multi-label document classification method for documents written in Czech language. In such a case, the issues of large feature vectors become more significant due to the complexity of Czech language when compared to English.

3   Multi-label Document Classification

3.1   Preprocessing, Feature Selection and Classification

The same preprocessing as in our previous work [1] is used, i. e. a morphological analysis including *lemmatization* and *Part-Of-Speech (POS) tagging*. The lemmatization decreases the number of features by replacing a particular word form by its *lemma* (base form) without any negative impact to the classification accuracy.

The knowledge of the POS tags is used for the further feature vector reduction. We filter out the words that should not contribute to classification according to theirs POS tags. The words with the uniform distribution among all document classes are removed from the feature vector. After this filtration, only words with the POS tags noun, adjective or adverb remain in the feature vector.

As a feature selection, the mutual information method is used because it achieves the best results in our previous work.

Note, that the above described steps are very important, because irrelevant and redundant features can degrade the classification accuracy and the algorithm speed.

Three classifiers that are successfully used for document classification in the literature (see previous section) and in our previous work are used: Naive Bayes (NB), Maximal Entropy (ME) and Support Vector Machines (SVMs).

3.2   Multi-label Document Representation

The existing approaches can be divided into two groups: 1) problem transformation methods; and 2) algorithm adaptation methods. We focus here only on the first group. According to the authors of the survey [13], we have implemented two approaches that give the best classification scores. These approaches are described next. Then, we present a simple approach proposed by Zhu et al. in [17].

---

[7] http://research.nii.ac.jp/ntcir/permission/perm-en.html
[8] http://www.clsp.jhu.edu/research/malach/
[9] http://www.itl.nist.gov/iad/mig//tests/tdt/

Class and Complement. Let N be the number of the classes. The first approach uses N binary classifiers $C_{i=1}^{N} : x \to l, \neg l$, i. e. each binary classifier assigns the document x to the label l iff the label is included in the document, $\neg l$ otherwise.

The final classification result is given by:

$$C(x) = \cup_{i=1}^{N} : C_i(x) = l \tag{1}$$

The main drawback of this method is a very long training and classification time. This approach is hereafter called *Class & complement*.

Merged Categories. Let K be the number of the different sets of labels existing in the corpus. The second approach uses each different set of labels as a new single label:

$$L = \cup_{k=1}^{K} l_k \tag{2}$$

One class document classifier $C : x \to L$ is then used for the document classification. Authors of [13] state, that this approach brings the best classification results. The principal weakness of this method is the data sparsity, i.e. some new classes with few document occurrences are created. This approach is further called *Merged categories*.

Threshold Classification. In this approach, the corpus is transformed as follows: the document with K labels is considered as K one class documents for training. The same classifier C as in the one label document classification task is created. This classifier produces a sorted list of the N labels l_i according to their classification scores s_i.

The core of the method consists in building a meta-classifier C_M in order to separate K classes belonging to the document and the rest $\neg K$. In this work, we distinguish these two classes by a *threshold T*. The document x is associated with a label l_i iff:

$$s_i(x) > T \tag{3}$$

The resulting set of labels L is given by:

$$L = \cup l_i \leftrightarrow C_M : x \to l_i \tag{4}$$

The threshold value is determined experimentally on the development corpus.

Note, that this approach is very simple. Nevertheless, there are two main advantages of this method: 1) minimal adaptation of our previously developed system is necessary; 2) algorithm speed. This approach is hereafter called *Threshold classification*.

4   Experiments

4.1   Tools and Corpora

For lemmatization and POS tagging, we used the mate-tools.[10] The lemmatizer and POS tagger were trained on 5853 sentences (94.141 words) randomly taken from the PDT

---

[10] http://code.google.com/p/mate-tools/

2.0^{11} [18] corpus. The performance of the lemmatizer and POS tagger are evaluated on a different set of 5181 sentences (94.845 words) extracted from the same corpus. The accuracy of the lemmatizer is 81.09%, while the accuracy of our POS tagger is 99.99%. Our tag set contains 10 POS tags as shown in Table 1.

We used an adapted version of the MinorThird[12] tool for implementation of the document classification methods. This tool has been chosen mainly because the three evaluated classification algorithms were already implemented.

As mentioned previously, the results of this work will be used by the CTK. Therefore, for the following experiments we used the Czech text documents provided by the CTK. Table 1 shows the statistical information about the corpus. Figure 1 illustrates the distribution of the documents depending on the number of labels. This corpus is available only for research purposes for free at http://home.zcu.cz/~pkral/sw/ or upon request to the authors.

In all experiments, we used the five-folds cross validation procedure, where 20% of the corpus is reserved for the test. For evaluation of the classification accuracy, we used as frequently in some other studies a *Error Rate (ER)* metric. The resulting error rate has a confidence interval of $< 1\%$.

Table 1. Corpus statistical information

Unit name	Unit number	Unit name	Unit number
Document	11955	Numeral	216986
Category	60	Verb	366246
Word	2974040	Adverb	140726
Unique word	193399	Preposition	346690
Unique lemma	152462	Conjunction	144648
Noun	1243111	Particle	10983
Adjective	349932	Interjection	8
Pronoun	154232		

4.2 Class and Complement

The first section of the Table 2 shows the classification results of the *class & complement* approach. These results show clearly, that SVM and ME classifiers having comparable scores outperform significantly the NB.

Note that the ER metric is very strict, because the document is considered as classified incorrectly when only one label (from K) is not correct.

4.3 Merged Categories

As already stated, this approach suffers from the data sparsity problem. There are some classes with few document occurrences and a correct estimation of such models is very difficult. One solution is not to consider the classes with few occurrences and remove them from the classification.

---

[11] http://ufal.mff.cuni.cz/pdt2.0/
[12] http://sourceforge.net/apps/trac/minorthird

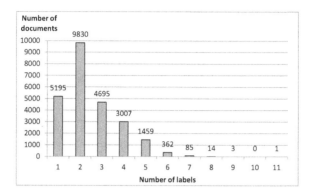

Fig. 1. Distribution of the documents depending on the number of labels

Table 2. Classification error rates [in %] of the all evaluated approaches using NB, SVM and ME classifiers

	Approach	Classifier		
		NB	SVM	ME
1	Class & complement	63.45	**36.62**	37.60
2	Merged categories	52.88	60.87	**32.26**
3	Threshold classification	38.40	25.67	**23.12**
4	Number of classes given	19.04	**9.44**	9.96

Figure 2 illustrates the classification error rates of the *merged categories* approach using NB, SVM and ME classifiers depending on the number of classes. This number is given by the value of the minimal number of documents per class. The figure demonstrates that the ME classifier brings better results than NB and SVM. The difference is most significant in the case when all classes are considered.

The second section of the Table 2 shows the error rates when all classes are used. The best error rate value is 32.26% which outperforms the best score of the previous experiment by 4% in the absolute value.

4.4 Threshold Classification

Figure 3 illustrates the classification error rates of the *threshold classification* approach using NB, SVM and ME classifiers when the different thresholds are used. The best error rates are reported in the third section of the Table 2. These results show that this approach outperforms significantly both previous methods. The best result is given by the ME classifier as in the previous approach.

4.5 Results Analysis

In this section, we would like to analyze the most accurate method from the two aspects:

1. Evaluation of the classification result where the correct number of the classes is given (see the fourth section of the Table 2). We can conclude that it is possible to

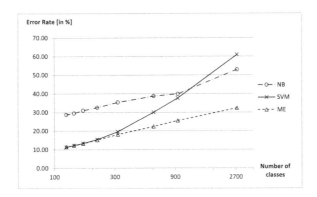

Fig. 2. Document classification error rates of the *merged categories* approach using NB, SVM and ME classifiers depending on the number of classes (x-axis in logarithmic scale)

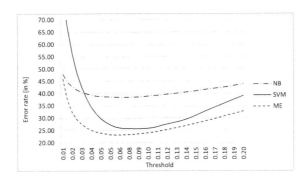

Fig. 3. Document classification error rates [in %] of the *threshold classification* approach

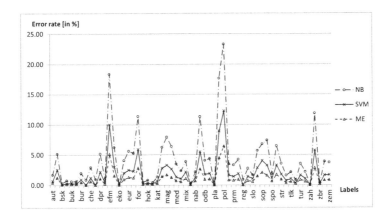

Fig. 4. Error rates of the separated classes (without any combination using a meta-classifier)

improve the classification score by 13% when the "ideal" meta-classifier is used. The ME and SVM give comparable results, while the error rate of the NB is significantly worse.

2. Evaluation of the single-label classification (see Figure 4), i. e. error rates of all classes separately without any combination. This analysis confirms that the single-label classification is much easier than the multi-label ones. Note, that the best global error rate is given by the SVM classifier and is 9.44%.

5   Conclusions and Future Work

In this work, we have implemented three promising multi-label document classification methods. Then, we have evaluated these methods on the Czech CTK corpus. We have shown that the simplest method based on the meta-classifier outperforms significantly both other approaches. The classification error rate improvement is about 13%.

The main perspective consists in proposing a more suitable document representation. For this task, we would like to study the impact of the syntactic structure of the sentence, semantic spaces, etc.

Acknowledgements. This work has been partly supported by the UWB grant SGS-2013-029 Advanced Computer and Information Sstems and by the European Regional Development Fund (ERDF), project "NTIS - New Technologies for Information Society", European Centre of Excellence, CZ.1.05/1.1.00/02.0090. We also would like to thank Czech New Agency (CTK) for support and for providing the data.

References

1. Hrala, M., Král, P.: Evaluation of the Document Classification Approaches. In: Burduk, R., Jackowski, K., Kurzynski, M., Wozniak, M., Zolnierek, A. (eds.) CORES 2013. AISC, vol. 226, pp. 875–885. Springer, Heidelberg (2013)
2. Bratko, A., Filipič, B.: Exploiting structural information for semi-structured document categorization. In: Information Processing and Management, pp. 679–694 (2004)
3. Della Pietra, S., Della Pietra, V., Lafferty, J.: Inducing features of random fields. IEEE Transactions on Pattern Analysis and Machine Intelligence 19, 380–393 (1997)
4. Forman, G.: An extensive empirical study of feature selection metrics for text classification. The Journal of Machine Learning Research 3, 1289–1305 (2003)
5. Yang, Y., Pedersen, J.O.: A comparative study on feature selection in text categorization. In: Proceedings of the Fourteenth International Conference on Machine Learning, ICML 1997, pp. 412–420. Morgan Kaufmann Publishers Inc., San Francisco (1997)
6. Galavotti, L., Sebastiani, F., Simi, M.: Experiments on the use of feature selection and negative evidence in automated text categorization. In: Borbinha, J.L., Baker, T. (eds.) ECDL 2000. LNCS, vol. 1923, pp. 59–68. Springer, Heidelberg (2000)
7. Lim, C.S., Lee, K.J., Kim, G.C.: Multiple sets of features for automatic genre classification of web documents. Information Processing and Management 41, 1263–1276 (2005)
8. Gomez, J.C., Moens, M.F.: Pca document reconstruction for email classification. Computer Statistics and Data Analysis 56, 741–751 (2012)

9. Yun, J., Jing, L., Yu, J., Huang, H.: A multi-layer text classification framework based on two-level representation model. Expert Systems with Applications 39, 2035–2046 (2012)
10. Novovičová, J., Malík, A., Pudil, P.: Feature selection using improved mutual information for text classification. In: Fred, A., Caelli, T.M., Duin, R.P.W., Campilho, A.C., de Ridder, D. (eds.) SSPR&SPR 2004. LNCS, vol. 3138, pp. 1010–1017. Springer, Heidelberg (2004)
11. Novovičová, J., Somol, P., Haindl, M., Pudil, P.: Conditional mutual information based feature selection for classification task. In: Rueda, L., Mery, D., Kittler, J. (eds.) CIARP 2007. LNCS, vol. 4756, pp. 417–426. Springer, Heidelberg (2007)
12. Forman, G., Guyon, I., Elisseeff, A.: An extensive empirical study of feature selection metrics for text classification. Journal of Machine Learning Research 3, 1289–1305 (2003)
13. Tsoumakas, G., Katakis, I.: Multi-label classification: An overview. International Journal of Data Warehousing and Mining (IJDWM) 3, 1–13 (2007)
14. Yaoyong, L., Shawe-Taylor, J.: Advanced learning algorithms for cross-language patent retrieval and classification. Information Processing & Management 43, 1183–1199 (2007)
15. Olsson, J.S.: Cross language text classification for malach (2004)
16. Wu, Y., Oard, D.W.: Bilingual topic aspect classification with a few training examples. In: Proceedings of the 31st Annual International ACM SIGIR Conference on Research and Development in Information Retrieval, pp. 203–210. ACM (2008)
17. Zhu, S., Ji, X., Xu, W., Gong, Y.: Multi-labelled classification using maximum entropy method. In: Proceedings of the 28th Annual International ACM SIGIR Conference on Research and Development in Information Retrieval, pp. 274–281. ACM (2005)
18. Hajič, J., Böhmová, A., Hajičová, E., Vidová-Hladká, B.: The Prague Dependency Treebank: A Three-Level Annotation Scenario. In: Abeillé, A. (ed.) Treebanks: Building and Using Parsed Corpora, pp. 103–127. Kluwer, Amsterdam (2000)

On Behaviour of PLDA Models in the Task of Speaker Recognition

Lukáš Machlica and Vlasta Radová

University of West Bohemia in Pilsen,
Faculty of Applied Sciences, Department of Cybernetics,
Univerzitní 22, 306 14 Pilsen
{machlica,radova}@kky.zcu.cz
http://www.kky.zcu.cz/en

Abstract. Nowadays, Factor analysis based techniques become part of state-of-the-art Speaker Recognition (SR) systems. These are the Joint Factor Analysis, its modified version called the concept of i-vectors, and the Probabilistic Linear Discriminant Analysis (PLDA). PLDA, as a generative statistical model, is usually used as the back end of a SR system, e.g. once i-vectors have been extracted, a PLDA model is used in the i-vector space to provide a verification score of two given i-vectors. In order to train the system huge amount of development data are utilized. In this paper the behaviour of the PLDA model is investigated. It is shown how does the amount of development data influence the system's performance. PLDA has several parameters to be tuned, i.e. dimensions of latent variables/subspaces, which represent the speaker and the channel variabilities. These will be examined too.

Keywords: PDLA, i-vectors, robustness, speaker recognition.

1   Introduction

The concept of i-vectors is closely related to the term SuperVectors (SVs). These are high dimensional mappings of a set of feature vectors extracted from one recording of a speech of one speaker. In order to construct SVs a Gaussian mixture model is trained in the feature space from a huge amount of data from a lot of development speakers recorded on several channels. Such a model is also called Universal Background Model (UBM). It is used to specify the acoustic conditions, in which the speaker recognition is carried out. Statistics related to a UBM and a recording are computed, and concatenated to a SV. Since such a SV is of significantly high dimension, often higher then the number of available recordings, techniques used to reduce its dimension are utilized. In this paper a system based on i-vectors and PLDA back end will be used [1]. I-vectors are low dimensional representations of supervectors. The main assumption is that these low dimensional representations are responsible for most of the variances in the high dimensional SV space. Since development corpora contain speakers recorded on several channels, such information is also utilized. However, rather than in the SV space it is handled in the i-vector space, because in [2] experiments were carried out proving that the channel space, obtained when decomposing SV space, does contain information on

I. Habernal and V. Matousek (Eds.): TSD 2013, LNAI 8082, pp. 352–359, 2013.

speaker's identity. Therefore, a PLDA model is estimated in the i-vector space, which does a decomposition of the i-vector space to a speaker and a channel subspace.

PLDA is a Factor Analysis (FA) based model, hence in the training process covariance matrices are decomposed in order to find the final subspaces [3]. More precisely, two covariance matrices are decomposed: between and within speaker covariances. Such a decomposition is very similar to an eigenvalue decomposition, where the estimated subspaces are those, in which the projected variance of training data is highest. In FA model, in addition noise corruption is treated so that subspaces with higher noise contamination are suppressed. Hence, having several acoustic/environment conditions present in the development set, some of them may be suppressed by higher variations in other conditions. If test and development conditions do not match the test environment, or if the test conditions are more diverse, this may lead to poor recognition rates.

Some investigations were already carried out in [4]. This paper aims to extend the results from [4] and take a closer look also on the dependence of such conclusions on the amount of development data or dimensions of subspaces generated by the PLDA model. The idea how to deal with diverse acoustic conditions is to treat available corpora individually, and compute the final verification score as a combination of results obtained from individual systems.

2 I-Vector Extraction

In order to specify an i-vector, concepts like Universal Background Model (UBM) and SuperVectors (SVs) have to be specified first. As already stated UBM is used to delimit the part of the feature space of interest. UBM is a Gaussian Mixture Model given by a set of parameters $\lambda_{\text{UBM}} = \{\omega_m, \boldsymbol{\mu}_m, \boldsymbol{C}_m\}_{m=1}^{M}$, where M is the number of Gaussians in the UBM, ω_m, $\boldsymbol{\mu}_m$, $\boldsymbol{C}_m$ are the weight, mean and covariance of the mth Gaussian, respectively. In order to construct a SV, statistics

$$f_{sm}^0 = \sum_t \gamma_m(\boldsymbol{x}_{st}), \quad \boldsymbol{f}_{sm}^1 = \sum_t \gamma_m(\boldsymbol{x}_{st})\boldsymbol{x}_{st} \tag{1}$$

for each Gaussian in the UBM and a set $\boldsymbol{X}_s = \{\boldsymbol{x}_{st}\}_{t=1}^{T_s}$ of feature vectors extracted from recording s are computed, where

$$\gamma_m(\boldsymbol{x}_t) = \frac{\omega_m \mathcal{N}(\boldsymbol{x}_t; \boldsymbol{\mu}_m, \boldsymbol{C}_m)}{\sum_{i=1}^{M} \omega_i \mathcal{N}(\boldsymbol{x}_t; \boldsymbol{\mu}_i, \boldsymbol{C}_i)} \tag{2}$$

is the probability that a feature vector $\boldsymbol{x}_{st} \in \mathbb{R}^D$ belongs to the mth Gaussian. Note that if $\boldsymbol{x}_t$ is not close to any of the Gaussians in the UBM ($\gamma_m(\boldsymbol{x}_t) \to 0$) it does not contribute to any of the statistics (1).

In order to construct a SV $\boldsymbol{\psi}_s$, statistics from individual Gaussians are concatenated:

$$\boldsymbol{\psi}_s = [(\boldsymbol{f}_{s1}^1/f_{s1}^0)^{\text{T}}, \ldots, (\boldsymbol{f}_{sM}^1/f_{sM}^0)^{\text{T}}]^{\text{T}}. \tag{3}$$

Note that $\hat{\boldsymbol{\mu}}_{sm} = \boldsymbol{f}_{sm}^1/f_{sm}^0$ is the maximum likelihood estimate of the mean $\boldsymbol{\mu}_m$ of mth Gaussian in the UBM given the dataset $\boldsymbol{X}_s$.

Now, to reduce the dimension of a SV $\psi_s \in \mathbb{R}^{DM}$ a Factor analysis based model is used. It has the form

$$\psi_s = \mu_\psi + T\nu_s + \epsilon, \tag{4}$$
$$\nu_s \sim \mathcal{N}(0, I), \quad \epsilon \sim \mathcal{N}(0, \Sigma),$$

where $\mu_\psi \in \mathbb{R}^{DM}$ is the mean of ψ_s (one can take a SV constructed from UBM means), $T \in \mathbb{R}^{DM \times K}$ is a low rank matrix of rank $\leq K$, $\nu_s \in \mathbb{R}^K$ is the i-vector and $\epsilon \in \mathbb{R}^{DM}$ is a residual noise term with Gaussian distribution and diagonal covariance matrix Σ.

In order to train T, a SV is extracted from each recording of each speaker contained in the development set. However, no relation is made between SVs of one speaker, each SV of a speaker is handled as if it would come from a different speaker. Therefore, the i-vector space contains both the speaker and the channel variability. That is why T is often denoted as the total variability space matrix. The estimation process of parameters of (4) is an extension to an ordinary FA training algorithm, where in addition dimensions related to individual Gaussians are weighted, for details see [5].

3 Probabilistic Linear Discriminant Analysis (PLDA)

PLDA [6] is a statistical generative model. It is based on Factor analysis and it is well suited as a back end for i-vector based systems. However, it is utilized as a back end also in speaker recognition systems based on Large Vocabulary Continues Speech Recognition (LVCSR) and SVs derived from adaptation matrices [7]. The PLDA model can be expressed as

$$\nu_{sh} = \mu_\nu + Fh_s + Gw_{sh} + \epsilon, \tag{5}$$
$$h_s, w_{sh} \sim \mathcal{N}(0, I),$$
$$\epsilon \sim \mathcal{N}(0, S),$$

where ν_{sh} is the i-vector to be decomposed, $h_s \in \mathbb{R}^{D_h}$, $w_{sh} \in \mathbb{R}^{D_w}$, and the index h relates to a session of speaker s. The similarity with (4) is obvious, substitute

$$\hat{T} = [F, G], \quad \hat{\nu}_{sh} = [h_s^T, w_{sh}^T]^T. \tag{6}$$

However, note that now an additional requirement is set on the form of $\hat{T}$. More precisely, it has to be composed of a subspace formed by columns of $F \in \mathbb{R}^{K \times D_h}$ responsible for speaker changes, and subspace generated by $G \in \mathbb{R}^{K \times D_w}$ responsible for channel changes. Thus, the decomposition of $\hat{\nu}_{sh}$ to speaker and channel factors is straightforward.

3.1 Verification

Once PLDA model parameters $\theta = \{\mu_\nu, F, G, S\}$ have been estimated, the model (5) can give predictions like e.g. $p(\nu|h_s, \theta)$ or $p(\nu_1, \nu_2|\theta)$. In the first case a speaker identity factor h_s and PLDA model parameters are given, and the probability that an unknown i-vector ν has the identity factor h_s is computed. In the latter case, we do not care about the decomposition of i-vectors ν_1, ν_2, we just want to know the probability that they were generated from the same source given a decomposition of the total variability space generated by matrices F and G. In speaker recognition usually the second probability is evaluated since only one recording of a target speaker is given, thus the estimate h_s becomes inaccurate. To be more specific a log-likelihood ratio

$$\text{LLR}(\nu_1, \nu_2) = \log \frac{p(\nu_1, \nu_2 | \theta)}{p(\nu_1 | \theta) p(\nu_2 | \theta)} \tag{7}$$

is evaluated, thus $p(\nu_1, \nu_2 | \theta)$ is normalized by the probability that ν_1 and ν_2 are independent.

4 Experiments

At first an i-vector system is built, which set-up will be described in Section 4.3. The focus is laid on the PLDA back end used in order to obtain the verification score. Following problems will be investigated:

- How does the change of dimensions of PLDA subspaces D_h and D_w change the recognition rates?
- How does the PLDA back end behave when amount of development data changes?
- Should several PLDA models be trained in order to make the system more flexible to variable test conditions?

The last item assumes that outputs of several subsystems will be fused at the end. For this purpose the FoCal toolkit [8] based on logistic linear regression will be used. Thus, the overall score will be given as linear combination of scores obtained from individual subsystems.

4.1 Used Corpora

Datasets from NIST SRE 2004, NIST SRE 2005, NIST SRE 2006, Switchboard 1 Release 2 (SW1), Switchboard 2 Phase 3 (SW2) and Switchboard Cellular Audio Part 1 and Part 2 (SWC) were utilized for development purposes, NIST SRE 2008 was used for calibration of fusion coefficients obtained by FoCal, and the results are reported on NIST SRE 2010. Only male telephone speech was used and only those development speakers who had more than 4 recorded sessions were assumed. Further, the development dataset was divided into 4 classes according to the corpora the data belong to[1]:

1. NIST040506 – containing 3787 recordings of 465 males of approximately 8 sessions for each male speaker,
2. SW1 – containing 2342 recordings of 211 males of approximately 11 sessions for each male speaker,

---

[1] see http://www.ldc.upenn.edu/Catalog/index.jsp

3. SW2 – containing 2183 recordings of 216 males of approximately 10 sessions for each male speaker,
4. SWC – containing 2707 recordings of 232 males of approximately 12 sessions for each male speaker,

Each of the recordings had approximately 5 minutes in duration including the silence.

All male telephone trials contained in "core-core trials" from NIST SRE 2010, yielding 74762 trials in total, were used to obtain the reported verification results. Results will be given in terms of Equal Error Rates (EERs) [%].

4.2   Feature Extraction

Hamming window of length 25 ms was shifted each 10 ms on a speech recording, 25 triangular filter banks were spread linearly across the frequency spectrum, 20 linear frequency cepstral coefficients were extracted and delta coefficients were added. Hence, the dimension of a feature vector was 40. Simple voice activity detection, based on detection of energies in filter banks located in the frequency domain, was carried out to discard non speech frames. At the end, feature vectors were normalized utilizing Feature warping with 3 sec. sliding window – distribution of feature vectors along each dimension was mapped to Gaussian.

4.3   System Set-Up

The number of Gaussians in the UBM was set to 1024, diagonal covariance matrices were used. UBM was trained on data from all the development corpora. Next, 7 Maximum Likelihood (ML) and 1 minimum divergence iterations were performed to train the total variability space matrix $T \in \mathbb{R}^{(1024*40) \times 800}$, thus the dimension of i-vectors was $K = 800$, T was trained on all development corpora too. And 20 ML iterations were done to estimate one PLDA model, used corpora will be specified in subsequent subsection.

In order to demonstrate the system performance in dependence on different dimensions of PLDA subspaces, the dimensions of speaker identity and channel subspace varied from $D_h = 100$, $D_w = 100$ to $D_h = 800$, $D_w = 800$ with step 100, respectively.

4.4   Results

At first, each dataset NIST040506, SW1, SW2, SWC was used alone to train a PLDA model. Results for each set are given in Table 1, these were computed for all specified values of D_h and D_w, but because of insufficient amount of space only minimum, maximum and median values of EER are given. Notice the difference between worst and best results obtained for NIST040506 and SW2, respectively. In Figure 1 and Figure 2 varying dimensions of PLDA subspaces are studied. In Figure 1 all development data are pooled and 1 PLDA model is trained, in the first plot results are given when all development data except SWC are used for the training, in the second one data from SWC were added. It is worth noting that even if the corpora SW2 is included in the training set, the EERs do not considerably decrease until SWC is added. This is the consequence of presence of NIST040506 (and partially of SW1) corpus in the training database, which does not suit the task very well regarding results given in Table 1. Variations in this dataset are more significant than those in SW2, and they become reflected in the estimated PLDA subspaces. The same may be stated when inspecting the left plot in Figure 1, where lower values of the identity space induce significant degradation in system's performance.

Table 1. EERs for speaker recognition systems with PLDA back ends trained only from a specific database NIST040506, SW1, SW2, SWC. EERs were computed for distinct dimensions of PLDA subspaces; minimal, maximal and median EERs are given.

	NIST040506	SW1	SW2	SWC
minEER[%]	10.21	8.84	7.05	7.58
maxEER[%]	12.21	10.11	8.63	8.95
medEER[%]	10.42	9.37	7.37	7.79

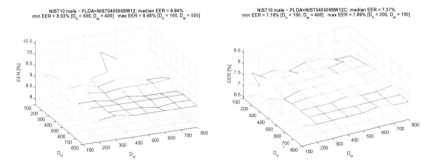

Fig. 1. EERs acquired for different dimensions of PLDA subspaces are shown. Available development data from corpora NIST040506, SW1, SW2, SWC were pooled and one PLDA model was trained. Left plot depicts the situation, where the corpus SWC was omitted from the training.

In Figure 2 a PLDA model is trained for each dataset NIST040506, SW1, SW2, SWC and results are fused at the end. Note that changing the dimensionality D_h of the speaker space does not change much the value of EER, it depends more on the value of D_w. The reason may be caused by the fact that a lower amount of data is used to train individual PLDA systems, and for higher values of D_w, more speaker variations are attributed to channel changes rather than to speaker changes.

Finally, in order to compare the pooling and fusing approaches one can compare Figure 1 and Figure 2. However, it is difficult to compare the differences for all values of D_h and D_w. Therefore Figure 3 was constructed. Here, EERs are taken from Figure 1 (pooled system) and Figure 2 (fused system), but instead of a grid plot, EERs obtained from the fused system are sorted (blue line) and for each pair (D_h, D_w) a respective value of EER from the pooled system is drawn (red line). Left plot depicts the situation, where the corpus SWC was omitted from the training. The vectors on the x-axis are couples $[D_h, D_w]^T$, but since the lines are sorted according to EER values of the fused

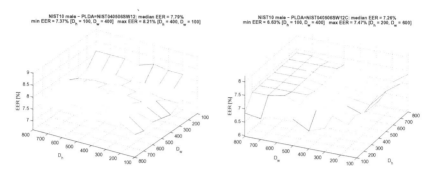

Fig. 2. EERs acquired for different dimensions of PLDA subspaces are shown. For each data class NIST040506, SW1, SW2, SWC one PLDA model was trained and verification results were fused at the end (fusion coefficients were trained on NIST08). Left plot depicts the situation, where the corpus SWC was omitted from the training.

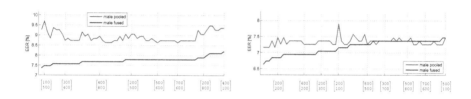

Fig. 3. Comparison of fusion and pooling of corpora in order to train a PLDA model

system, they are in any order. If the SWC is omitted from the training process the fused system does significantly outperform the pooled system for all the dimensions of subspaces. Loosely speaking, the fusion of individual systems assigns small weights to systems, which do not perform well (in this case PLDA system based on NIST040506). Thus, they do not spoil the performance. Once data from SWC were added (the plot on the right side of Figure 3), undesirable variations in NIST040506 are suppressed, and performances of the pooled and fused system become alike. However, for some values of D_h and D_w fused system performs still better and it is not outperformed in any of the cases. In practice the test conditions are often not clear, and one cannot count on a few of the development databases performing well on a development set. Therefore, it turns out to be more useful to train a PLDA model for each database (assuming sufficient amount of data in each of the databases to train a reliable PLDA model), and fuse the results of PLDA models at the end.

5   Conclusion

The behaviour of PLDA back end in the i-vector based speaker recognition system was studied. It was shown how the development data and meta parameters to be set influence the recognition rates. Conclusions of this paper may be beneficial mainly for systems dealing with diverse acoustic environment present in the operating conditions.

Acknowledgments. This research was supported by the Grant Agency of the Czech Republic, project No. GAČR P103/12/G084.

References

1. Matějka, P., Glembek, O., Castaldo, F., Alam, J., Plchot, O., Kenny, P., Burget, L., Černocký, J.: Full-covariance UBM and Heavy-tailed PLDA in I-Vector Speaker SVerification. In: ICASSP 2011, pp. 4828–4831 (2011)
2. Dehak, N., Kenny, P., Dehak, R., Dumouchel, P., Ouellet, P.: Front-End Factor Analysis For Speaker Verification. IEEE Transactions on Audio, Speech and Language Processing (2010)
3. Machlica, L., Zajíc, Z.: An Efficient Implementation of Probabilistic Linear Discriminant Analysis. In: ICASSP 2013 (2013)
4. Machlica, L., Zajíc, Z.: Analysis of the Influence of Speech Corpora in the PLDA Verification. In: Sojka, P., Horák, A., Kopeček, I., Pala, K. (eds.) TSD 2012. LNCS, vol. 7499, pp. 464–471. Springer, Heidelberg (2012)
5. Kenny, P., Ouellet, P., Dehak, N., Gupta, V., Dumouchel, P.: A Study of Interspeaker Variability in Speaker Verification. IEEE Transactions on Audio, Speech and Language Processing 16, 980–988 (2008)
6. Prince, S., Elder, J.: Probabilistic Linear Discriminant Analysis for Inferences About Identity. In: IEEE 11th International Conference on Computer Vision, pp. 1–8 (2007)
7. Scheffer, N., Lei, Y., Ferrer, L.: Factor Analysis Back Ends for MLLR Transforms in Speaker Recognition. In: Interspeech 2011, pp. 257–260 (2011)
8. Brümmer, N.: FoCal: Tools for Fusion and Calibration of Automatic Speaker Detection systems (2006), http://sites.google.com/site/nikobrummer/focal

On the Quantitative and Qualitative Speech Changes of the Czech Radio Broadcasts News within Years 1969–2005

Michaela Kuchařová[1], Svatava Škodová[2], Ladislav Šeps[1], Václav Lábus[2],
Jan Nouza[1], and Marek Boháč[1]

[1] Institute of Information Technology and Electronics, Technical University of Liberec,
Studentska 2, 461 17, Liberec, Czech Republic
{michaela.kucharova1,ladislav.seps,jan.nouza,marek.bohac}@tul.cz
[2] Department of the Czech Language and Literature, Technical University of Liberec,
Studentska 2, 461 17, Liberec, Czech Republic
{svatava.skodova,vaclav.labus}@tul.cz

Abstract. In this paper we introduce the quantitative and qualitative characteristics of the Czech Radio Broadcasts News during a period of significant political and social changes in the Czech Republic (1969 - 2005). The research is mainly focused on the quantitative features of speech that can be determined from the results of automatic speech recognition system. We describe the used archive transcription system and selected characteristics of the macro- and micro- structure of the Radio Broadcasts News; namely the changes in studio vs. out-of-studio speech ratio, distribution of speakers by male and female, moderators and guest-speakers, changes in the use of signature tunes (including jingles), approximate use of phrasal introductory and closing language specific for the time periods, speech speed changes, average silence length, coordinative vs. subordinate conjunctions ratio and the most frequent semantic words. The sample of data consists of 6,580 hours of news broadcasting and 48,721,952 lexical words.

Keywords: audio archive processing, spoken formal speech, radio broadcast news, non-speech events, automatic speech recognition.

1 Introduction

The language of news media, especially of radio news, offers a lot of potential for areas worth investigating [1]. Being tightly connected with the historic events significant during the period of their occurrence, the radio broadcasts news (BN) also represent linguistic and social norms changing over the time [2].

In the text, we suggest various possibilities for investigation based on the automatically transcribed speech data of the radio BN. All the data were automatically processed, recognized and made accessible as a part of a national research project "Disclosure of the Czech Radio archive for sophisticated search" supported by the Czech Ministry of Culture [3]. The project's ultimate goal is to automatically process historical and contemporary spoken documents in the archive (the oldest date back to the

I. Habernal and V. Matousek (Eds.): TSD 2013, LNAI 8082, pp. 360–368, 2013.
© Springer-Verlag Berlin Heidelberg 2013

1920s), to index the results and to make them publically accessible. Similar research project exists for English [4].

The archive contains several hundred thousands audio files, from which more than 75,000 have already been processed since 2011, when the project was launched.

The presented statistics, based on the automatic processing of the BN speech, are complemented by a partly manually gathered overview of some qualitative features.

2   Archive Transcription System

The system used for archive document transcription is the large-vocabulary continuous speech recognition (LVCSR) system, which has been developed and continually improved at the Technical University of Liberec since 2003. It has to meet specific needs of highly inflected languages which require very large lexicons.

The system processes audio documents converted into 16 kHz PCM format. The first step is parameterization into a stream of 39-dimensional MFCC (Mel-frequency Cepstral Coefficients) vectors computed every 10 ms. These are normalized by the CMS (Cepstral Mean Subtraction) technique within a 2-second long sliding window and forwarded to the decoder. The acoustic model uses context-dependent phoneme units (trigram). (For Czech language, these have been trained on 320 hours of speech.) The segmentation into speech and non-speech parts, as well as the detection of speaker turns is performed after the first pass through the decoder. We found this approach more robust because it can benefit from the already identified noise events and word boundaries. To make the first pass quick, usually a smaller subset of the lexicon (10 to 50 thousand words) is used. After that, the document is segmented into speech and non-speech parts, and the former ones are split into the parts belonging to different speakers. At this point, optional language identification (Czech or Slovak) is performed in case bilingual audio documents are expected. Next, each speech segment is passed to the decoder with appropriate lexicon and language model (LM). For Czech, the currently used lexicon contains 490,000 items (with 540,000 pronunciations). The LM is based on bigrams trained on a 11 GB corpus of multi-genre Czech texts. For the transcription of historical documents, the corpus has been complemented by texts extracted from scanned editions of *Rude Pravo*, the official newspapers of the Czechoslovak government and Czech communist party in the 1970 to 1989 period [5].

The output of the decoder provides substantial information about the document content. Besides the standard orthographic transcription, the system stores pronunciation of each word, start and end times (measured in milliseconds) as well as for each non-speech sound, like silence, large noise (in BN it is usually signature tunes), vocal noise, hesitation sounds, etc. Also each document segment has its own set of descriptors such as: a speech or non-speech label, a language label and a speaker label (a speaker name if known or at least his/her gender). Every item of this information is stored in a huge database and indexed for future use, either for searching or for statistical purposes.

In this contribution, we use the LVCSR output for most of the statistics. We are aware of the fact that every speech-to-text system can produce errors. The designers of the system put much effort to its development, performance tuning and proper evaluation. Large experiments showed that the typical values of the word-error-rate (WER) were about 11 % for contemporary BN and about 14 % for those from the 1969 to 1989 period. These values may seem large, but in our earlier work [6] we demonstrated that the majority of errors were either omissions of very short words (mainly 1-letter and 2-letter prepositions and conjunctions) or misrecognized word-endings (due to the rich morphology of Czech with many acoustically similar suffixes). Therefore, the impact of these errors on the statistics provided in this contribution can be considered to be insignificant.

3 Data

On the selected phenomena we would like to show some quantitative and qualitative changes in the macro- and micro- structure of BN.

The material we analyzed comes from the main news broadcasted every evening by the Czech Radio (or Czechoslovak Radio before 1993) from 1969 to 2005. Namely the survey deals with daily news summary called *Rozhlasove noviny* (1969 - 1993) and *Ozveny dne* (1993 - 2005), approximately 25-30 minutes long programs which sum up daily home and foreign affairs. These BN are not type-homogenous: although traditional studio read news prevails, besides that there is also out-of-studio speech, which contains spontaneous speech (reports and short interviews) and prepared speech (brief commentaries or recordings such as public speeches). It means that non-professional speakers occur in BN as well.

The amount of data used for the statistics in this contribution consists of a really large and representative sample of time-evolved speech (6,580 hours; 48,721,952 lexical words).

To illustrate the changes, we separated the period into regular 8 segments, each containing 5 years, except the last one containing the remaining two years. Our decision to divide the period into these segments is a compromise between the limits on the text range (too detailed segmentation would not allow us to present many BN features) and the possibility to show the continuity of changes (the 5 year segments are detailed enough to illustrate the process of changes during 42 years).

4 Macrostructure and Microstructure of the Radio News Broadcasts

The statistics illustrating the BN changes are divided into two main parts that we labeled as macro- and micro- structure. The macrostructure describes changes in the formal structure of BN; the microstructure concerns the issues connected with the language and speech production.

4.1 Macrostructure

The macrostructure includes the statistic comparison of studio vs. out-of-studio speech (Fig. 1); ratio of male vs. female speakers, including the difference between the moderators and guests (Fig. 2); changes in signature tunes (including jingles) (Fig. 3); and the summarization of typical beginnings and endings changing during the periods (Table 1).

Ratio of Studio and Out-of-studio Speech. Through the whole period there is a constant increase of out-of-studio speech and decrease of studio speech in BN (Fig. 1), where studio speech consists of the read speech, that represents the structures of written language, and out-of-studio speech, representing mostly spontaneous together with prepared speech. The prevailing proportion of the studio speech which was typical for the beginning of the period decreases by almost 30 % in the beginning of the nineties.

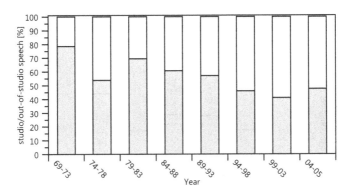

Fig. 1. Verbal distribution of BN: Studio (grey) vs. out-of-studio (white) speech ratio

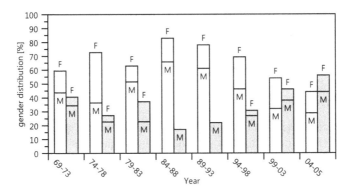

Fig. 2. Gender distribution of BN: Male (M) vs. female (F) speaker ratio and moderators (white column) vs. guest-speakers (grey column) ratio

Gender Distribution of BN. Fig. 2 illustrates ratio of male and female speakers in BN. White columns represent the moderators; except the period 1974 - 1978 the male moderators prevail. Grey columns represent guest-speakers; even up to our time there are very few female guest-speakers present in the BN.

Proportion of Signature Tunes. The graph in Fig. 3 shows signature tunes proportion. Until 1994, the signature tunes are nearly homogenous, being represented by an orchestra. From 1994 there is a rapid diversification and increase of signature tune types. We recognized 19 types; as far as they are not namely under the investigation, we do not list them. Fig. 3 shows the proportion of all types of signature tunes towards the length of a whole broadcasting.

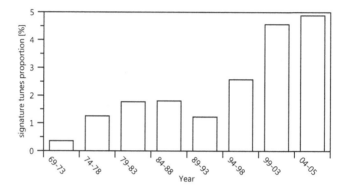

Fig. 3. Proportion of signature tunes towards the length of the whole broadcasting

Introductory and Closing Phrases. Table 1 summarizes regularities in the formal structure of BN concerning the introductory and closing phrasal language. Beginnings and endings as the only parts of BN represent a type of personal contact with the listeners; in comparison to the text of news, they have not only the informative function but also the phatic function. Their division into the periods in Table 1 follows the significant changes in the character of used introductory and closing phrases.

4.2   Microstructure

The microstructure includes the statistics of changing speech speed (Fig. 4); average silence length (Fig. 5); coordinative vs. subordinate conjunctions ratio (Fig. 6); and the occurrences of the most frequent semantic words over two compared periods (Table2).

Speech Speed. Fig. 4 shows the quantitative changes, constant increase in the amount of words pronounced per hour. The lowest amount of words per hour was 6,712 in the period 1969 - 1973 and the highest was 8,404 words in the last period 2004 - 2005. There is a growth of 1,692 words per hour through the periods observed; the constant growth pattern is evident after 1989.

Table 1. Characteristics of introductory and closing phrases

1969 - 1977
The period is marked by the strict use of phrases in the beginnings and ends of BN; the only variation is in the word order of the phrases. There is hardly any contact word signaling that the information is intended for the audience, we can find only the formal greeting phrase; there are no vocatives and personal pronouns signaling the orientation to the listener. The sentence perspective is formed from the point of view of the broadcasting itself, or from the perspective of a moderator.
1978 - 1988
Introductory and closing phrases through the period are identical to period 1969 - 1977, except the notions of names of moderators representing some personalization elements.
1989 - 1990
The shortest period is marked by structural chaos and complete omission of formal standardized phrases. All the elements of the introductions and conclusions are subjected to the personal point of view of a moderator. A typical feature of the period is an enumeration of many people cooperating on BN. The phrases oscillate between the formulations from a moderator or the listener perspective.
1991 - 1996
The penultimate period represents a return to the formulaic language of 1978 - 1988 as a compromise between the structural chaos of the preceeding period 1989 - 1990 and the non-personal language of 1969 - 1977. Compared to the period of 1969 - 1977 there is constant use of various means signaling listener-oriented BN.
1996 - 2005
The final period is marked by a return to neutral introductory phrases without contact means; this phatic function is represented only in closing phrases. It is standardised in its means in the use of vocatives and contact cases of personal pronouns.

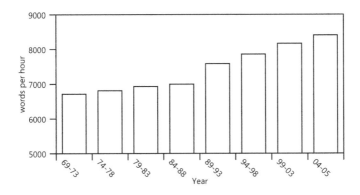

Fig. 4. Speech speed measured in words pronounced per hour

Silence Length. In general, verbal production in all media of audible type accelerates, together with the acceleration of word production per hour; it is possible to observe shortening of silence in speech. We define silence as a segment not containing either speech or noise events. Fig. 5 illustrates the average length of silence in BN.

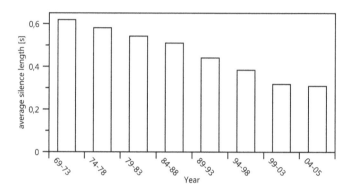

Fig. 5. Average silence length in BN

Coordination vs. Subordination Ratio. Fig. 6 shows the ratio changes of coordinative to subordinate conjunctions in BN. For the analysis, we have chosen the most frequent Czech one-word conjunctions according to the monograph [7]: 40 subordinate and 35 coordinative. The value 1 on the vertical axis means that number of coordinative conjunctions is equal to subordinate ones. Until 1989, the average ratio was approximately 2:1 (coordination: subordination), afterwards it rapidly decreased to 1.10:1. The average ratio during the whole period was 1.54:1.

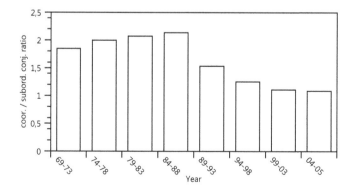

Fig. 6. Ratio of coordinative to subordinate conjunctions

The Most Frequent Words. As far as the lexical system of each language is very dynamic on its periphery, it reflexes changes of the lexical field depending on the political and social changes through the time flow. Over the whole period from 1969 to 2005, divided into the communist and post-communist phases, we have selected the most frequent 150 words. Table 2 shows, that the whole amount of semantic words among the first 100 was only 24 for the phase till 1989 and only 12 starting from this year. For comparison, we have added following 12 words according to the frequency rank. (It does not display the grammar words, deictic words and ordinary numbers.) The column Frequency rank shows their order among the most frequent words. Most of these words reflect the tight connection of BN with the political system.

Table 2. The most frequent semantic words over the period 1969 - 2005

1969 - 1989			1990 - 2005		
Word	**Translation**	**Freq. rank**	**Word**	**Translation**	**Freq. rank**
strany	*party*	14	dnes	*today*	22
dnes	*today*	18	České	*Czech*	58
výboru	*committee*	19	první	*first*	70
komunistické	*communistic*	22	řekl	*(he) said*	71
ústředního	*central*	27	jednání	*negotiation*	72
svazu	*federation*	30	republiky	*republic*	77
Československa	*Czechoslovak*	34	vlády (genitiv)	*government*	78
republiky	*republic*	35	prezident	*president*	79
další	*next*	42	ministr	*minister*	81
Sovětského	*Soviet*	46	rozhlasu	*radio*	82
socialistické	*socialistic*	49	strany	*party*	85
vlády	*government*	54	roku	*year*	98
práce	*work*	58	vláda (nominativ)	*government*	101
států	*states*	60	Český	*Czech*	103
zemí	*countries*	68	unie	*union*	111
jednání	*negotiation*	70	korun	*crowns*	112
první	*first*	73	stupňů	*degree*	114
národní	*national*	75	zpravodaj	*reporter*	116
Rady	*Council*	79	teploty	*temperature*	120
Praze	*Prague*	80	zahraničí	*abroad*	126
zasedání	*session*	83	lidí	*people*	127
předseda	*chairman*	94	myslím	*(I) think*	130
organizace	*organization*	95	let	*years*	132
Spojených	*United*	99	Evropské	*European*	133

5 Conclusions and Future Work

National archives of spoken documents represent a considerable corpus for linguistics research of spoken data. The processed part of the Czech Radio archive contains speech together with automatically recognized texts. That allows searching in spoken data and analyzing them. The value of the search is that one can find the information on the two linguistic levels: in the written form and the spoken form. This doubled information-storage is important especially for the comfortable retrieval of information and it extends diametrically the possibilities of work with the information. Finally, there is the benefit of search through the data of a long period of time.

Our investigation shows quantitative and qualitative speech changes of the Czech Radio BN. In the paper, we have chosen several features from the formal and linguistic sides of BN to show through the statistics methods the changes in the BN. We have pointed out (i) the increase of spontaneous and prepared speech compared to the read speech; (ii) surprising absence of female guest-speakers; (iii) increasing amount and variety of signature tunes; (iv) changing habits of introductory and closing phrasal language. The statistics covering the changes in the speech of BN itself comprise (i) the increase of speech speed, (ii) shortening silence, (iii) decrease of paratactic coordinative connectors, (iv) changes in occurrences of the most frequent semantic words.

This paper is meant as an introductory study that shows the large potential of the modern speech-to-text technology applied to the archive of historical spoken documents.

In this paper, we did not present a comprehensive topped research of spoken Czech in BN over nearly four decades but we maintained to introduce possible ways of investigation of the speech stored in the Czech Radio audio archive.

Acknowledgments. This work was supported by project no. DF11P01OVV013 provided by The Czech Ministry of culture in research program NAKI.

References

1. Sonkova, J.: Morfologie mluvene cestiny: Frekvencni analyza (Morphology of Czech: Frequency Analysis). NLN, Praha (2008)
2. Cmejrkova, S., Hoffmannova, J. (eds.): Mluvena cestina: hledani funkcniho rozpeti (Spoken Czech in Search of Functional Range). Academia, Praha (2011)
3. Nouza, J., et al.: Making Czech Historical Radio Archive Accessible and Searchable for Wide Public. Journal of Multimedia 7(2), 159–169 (2012)
4. Hansen, J.H.L., et al.: SpeechFind: Spoken document retrieval for a National Gallery of the Spoken Word. In: 6th Nordic Sig. Proc. Symposium, NORSIG 2004, pp. 1–4 (2004)
5. Chaloupka, J., Nouza, J., Červa, P., Málek, J.: Downdating lexicon and language model for automatic transcription of czech historical spoken documents. In: Habernal, I., Matousek, V. (eds.) TSD 2013. LNCS (LNAI), vol. 8082, pp. 201–208. Springer, Heidelberg (2013)
6. Boháč, M., Nouza, J., Blavka, K.: Investigation on Most Frequent Errors in Large-scale Speech Recognition Applications. In: Sojka, P., Horák, A., Kopeček, I., Pala, K. (eds.) TSD 2012. LNCS, vol. 7499, pp. 520–527. Springer, Heidelberg (2012)
7. Barton, T., Cvrcek, V., Cermak, F., Jelinek, T., Petkevic, V.: Statistiky cestiny (Statistics of Czech). NLN, Praha (2009)

On the Use of Phoneme Lattices
in Spoken Language Understanding

Jan Švec and Luboš Šmídl

[1] Department of Cybernetics
[2] NTIS - New Technologies for Information Society,
Faculty of Applied Sciences, University of West Bohemia, Czech Republic
{honzas,smidl}@kky.zcu.cz

Abstract. This paper presents a novel approach to spoken language understanding in dialogue systems. Unlike prevalent methods that use only the word lattices, the presented approach works with phoneme lattices generated by a phoneme recognizer. The hierarchical discriminative model for speech understanding was used together with modifications proposed in this paper. The method was experimentally evaluated using two semantic corpora and the results are presented.

Keywords: spoken language understanding, phoneme recognition.

1   Introduction

The automatic speech recognition (ASR) task is widely studied scientific problem. The commonly used speech recognizers employ probabilistic models based on the noisy channel approach. The probability of observing a word sequence W given some acoustic observation O is expressed as:

$$P(W|O) = \frac{P(O|W)P(W)}{P(O)} \propto P(O|W)P(W) \tag{1}$$

where the acoustic model $P(O|W)$ uses Hidden Markov Models and the language model $P(W)$ is approximated with an n-gram language model (LM). This scheme is widely used in current systems but it has a number of drawbacks. The following description of these drawbacks should serve as the motivation for the use of phoneme lattices in the SLU task:

– Training of the LM requires a huge amount of training data. The number of tokens required to train robust LM is quite large even for a simple bigram LM with the lexicon containing only severl thousand words. The collection of sufficient amount of data is expensive and time consuming task. It conflits with the requirements on the development phase of many spoken dialogue systems where rapid development and simplified maintenance of the system is preferred.

– The noisy channel model leads to a generative model, i.e. it models the joint probability $P(O, W)$. Therefore every input frame contained in the observation O has to be classified as some being part of some word from the recognition lexicon. Although the

I. Habernal and V. Matousek (Eds.): TSD 2013, LNAI 8082, pp. 369–377, 2013.

recognition lexicon can contain some non-speech events such as *hmm* or *laugh*, it is still necessary to model their prior probability in the language model $P(W)$.

– The background speech or noise can significantly degrade the recognition performance. Therefore many production-grade spoken dialogue systems use some type of voice activity detection or speech/non-speech classifier. These systems can filter out observations that do not fit to a statistic model of speech.

– The out-of-vocabulary (OOV) words cannot be recognized. In addition, the occurrence of OOV word frequently causes subsequent errors on surrounding words. The generic word-based generative model has no means of marking some subsequences of acoustic observation as unintelligible speech. This limitation cannot be overcome by the use of speech/non-speech classifier because the OOV words are surely speech but their lexical realisation is not contained in the recognition lexicon.

The mentioned problems have many solutions such as LM interpolation and adaptation, speech/non-speech classifiers or iterative adaptation of spoken dialogue systems but the requirement on the amount of annotated training data is reduced only slightly.

In this paper, we will focus on the spoken language understanding (SLU) task in the spoken dialogue systems. We consider the SLU module as a classifier which predicts the semantics of a given input utterance. The structure of semantics can vary from simple tags (i.e. call routing/classification task) to more complex set of attribute-value pairs or semantic trees. The fundamental feature of the SLU task is the fact that not all words of the utterance determine the semantics. The spontaneous speech usually contains a large number of filler words and words that are necessary to form grammatically correct sentence but does not influence the meaning of the sentence. Therefore the SLU task can be understood as an application of keyword spotting – the classifier searches for some patterns in the input and outputs the corresponding semantics.

2   SLU as Keyword Spotting Application

The idea of keyword spotting is contained in many SLU modules currently used. For example, the implementation of Hidden Vector State parser described in [1] uses vocabulary pruning to obtain robust lexical model. Then the set of words in the lexical model vocabulary can be viewed as a set of keywords. The Semantic Tuple Classifier [2] uses a set of discriminative classifiers to predict semantic tree corresponding the input utterance. The classifiers are based on Support Vector Machines (SVMs) and are trained from lexical features (counts of n-grams) derived from the input sentence. In this case the keywords are represented as support vectors of a classifier. Also the semantic interpretation approach described in [3] uses a filler model to filter out the words not belonging to a given named entity in speech lattice. The keywords consist of the input symbols of the word-to-concept transducer. Even in the case of full knowledge-based design of spoken language understanding (e.g. [4]), there is a need to model the filler words using dedicated context free grammar. Then the non-terminal symbols not contained in the filler grammar can be considered as keywords.

This interpretation of the SLU task as a keyword spotting application leads to the idea of using phoneme recognizer instead of word recognizer that is standardly employed in methods described in the previous section. The speech recognizer which uses phonemes as recognition units is widely used in tasks such as keyword spotting or spoken term detection. Since the phoneme inventory of most languages is usually very limited and fixed, the recognition lexicon of the phoneme recognizer is much smaller and closed. Consequently, the phoneme recognizer based on HMMs and n-gram LM can use longer n-gram history, because the data sparsity problem virtually disappers for phoneme sequences. The phoneme recognizer is also immune to OOV words because all words are modelled as phoneme sequence regardless of whether they belong to the vocabulary or not. And finally, the phoneme LM can be easily adapted to a new domain using unsupervised training. This feature is valuable in commercial applications since the customers often have a huge number of unannotated data available.

However, the statistical SLU module based on phoneme recognizer still needs the semantically annotated data. Therefore we present an approach where the SLU module is trained from unaligned data. This means that only the utterance recordings tagged with the reference semantics are needed, not the full lexical transcription nor lexical-semantic alignment. In addition, many use-cases of SLU in telecommunications can exploit the information assigned to the call in the call-center databases. This information can be used to predict the call routing or goal classification.

The remaining sections of this paper are organized as follows: Sec. 3 briefly describes the Hierarchical discriminative model which is used as an SLU module. Sec. 5 presents the experimental setup. This includes description of data and acoustic models used in experimental validation. Also the process of phoneme LM adaptation is described. The experimental results are shown in Sec. 6.

3 Hierarchical Discriminative Model

In this section we briefly describe the Hierarchical Discriminative Model (HDM) which was first introduced in [5]. The description of HDM uses the terminology of feed-forward neural networks – the *input layer* computes lexical features, the *hidden layer* transforms these features into a new feature space. The vectors in this feature space represent the presence or absence of a given semantic tuple (part of semantic tree). The *output layer* then predicts lexicalized probabilities. The probabilities are used to parameterize the generalized probabilistic context-free grammar (PCFG). The symbols used in this PCFG correspond with the set of domain-dependent *semantic concepts*. These concepts represent the atomic units of the meaning important in the given task – for example TIME, STATION, ACCEPT, CREATE etc. The structure of HDM is fully discriminative – it models directly the posterior probability distribution. In addition, it uses weighted finite state transducers (WFST) to represent the uncertainty of the ASR output. Since the training data consist of ASR lattices it allows to model ASR errors. In experiments presented in [5], the HDM outperforms both the generative (hidden vector state parser) and the discriminative (semantic tuple classifiers) models.

The input layer uses rational kernel theory to compute the values of a kernel function directly without the use of explicit feature vector extraction. The kernel computation is very fast, it is possible to compute the vector of 5,000 kernel function values in times of the order of tens of milliseconds. Kernel function values are used in the hidden layer classifiers. The hidden layer corresponds to a Semantic Tuple Classifiers (STC) model

[2]. It uses support vector machines (SVMs) with kernel values computed in the input layer. The predictions of these classifiers are not used directly, instead the distance to the decision boundary is used as an output of the hidden layer. The output layer employs a set of multi-class SVMs and uses the feature vector computed in the hidden layer to predict expansion probabilities of the PCFG. The rules of PCFG are inferred from training data. Therefore the HDM allows to predict only the expansions of non-terminals seen during the training phase. This does not limit the prediction power of the SLU – the HDM is still capable to predict the semantic tree not seen during the training phase. Since the input lattice is a generic WFST, the structure of the HDM classifier allows to simply change the lattice representation from word-level lattices to phoneme-level lattice.

4   Proposed HDM Modification

The HDM was designed to predict semantic trees from generic representation of the input utterance. This representation can include word-level ASR lattices but in this paper we will focus on the phoneme-level lattices and on the ability to decode the semantics from these sub-word units. The proposed modification of the HDM model consists of the modification of the *input layer* which computes the values of kernel functions used in the SVM classifiers in the hidden layer (originally the STC model).

We propose to change the kernel function to compute the dot-product of two vectors representing the expected counts of n-grams in the phoneme lattice. Since the unigram phonemes does not represent the sentence meaning very well, we need to raise the order of n-gram features to be able to decode the sentence meaning. In addition, with the higher order of n-gram features we also use the feature space normalization described below. In the following equations, we will denote the expected number of counts the substring x occur in the lattice A as $c(A, x)$. This number represents the "soft-counts" of x in A. If A is the deterministic string, then $c(A, x) \in \mathbb{N}$. In the case when A is a WFST which represents probability distribution over a set of all string accepted by A, the $c(A, x) \in \mathbb{R}$. The function $k_n(A, B)$ defined as:

$$k_n(A, B) = \sum_{|x|=n} c(A, x)c(B, x) \tag{2}$$

is a positive definite symmetric (PDS) kernel function [6]. Since the class of PDS kernel functions is closed under addition, the kernel function $K_{n,m}(A, B)$ defined over two WFSTs A and B:

$$K_{n,m}(A, B) = \sum_{i=n}^{m} k_i(A, B) \tag{3}$$

is also a PDS kernel function. In addition, the feature space normalization described in [7] can be used to obtain the kernel function $\bar{K}_{n,m}(A, B)$ defined as:

$$\bar{K}_{n,m}(A, B) = \frac{K_{n,m}(A, B)}{\sqrt{K_{n,m}(A, A)K_{n,m}(B, B)}} \tag{4}$$

Since $K_{n,m}$ is a rational kernel function [6], the computation of its values can be performed using WFST composition. The optimized algorithm based on factor transducer was presented in [8]. Using the definition of kernel functions $K_{n,m}$ and $\bar{K}_{n,m}$ we are able to modify the HDM to be able to parse utterances which are represented using phoneme lattices.

5 Experimental Setup

To evaluate the phoneme-level SLU performance, we used two semantically annotated corpora designed for spoken language understanding in a spontaneous dialog. The first one was the Human-Human Train-Timetable (HHTT) corpus [9] used in [10,1]. The corpus contains inquiries and answers about train connections. The second one was a Czech Intelligent Telephone Assistant (TIA), a corpus containing utterances about meeting planning, corporate resource sharing and conference call management. These corpora contain unaligned semantic trees together with word-level transcriptions. We have split the corpora into train, development (shortened to dev) and test data sets (72:8:20) at the dialog level, so that the speakers do not overlap. The corpora characteristics are summarized in Tab. 1 (a).

We used our in-house real-time decoder which was configured as a phoneme recognizer to process the input utterances. The recognizer was the same as in our previous work employing phoneme recognition [11]. The phoneme-level recognition accuracy can be improved using phoneme LM. This is caused by the fact that the HMM-based phoneme recognizer is a generative model which relies on the use of the LM. In our experiments we use 5-gram phoneme LMs. The 5-gram models are more robust than the models of lower orders and at the same time the decoder still allows to recognize the input utterances in real-time.

In the presented experiments, we trained a separate LM for each semantic corpus (HHTT and TIA). We use the following naming convention for LMs: phoneme LM trained from forced alignment (ph), phoneme LM trained from BH data (see bellow, ph-bh) and finally the ph-bh LM adapted to target domain (ph-ad). We also used a pseudo-phoneme lattices obtained by mapping the word lattices to phonemes using the pronunciation dictionary (ph-map). The BH data corpus contains a large Switchboard-like collection of spontaneous telephone dialogues collected using the toll free number. The corpus is described in more detail in [12]. The adaptation process is described in Sec. 5.1.

The phoneme-level recognition accuracies are summarized in Tab. 1 (b). The accuracy was evaluated using the word-level transcription of development and test sets. These transcriptions were first aligned with the audio signal to determine the correct pronunciation variant of words. Then the phoneme-level alignment was used as a reference. This leads to a bias in the phoneme-level accuracy because the pronunciation dictionary was generated using rule-based phonetic transcription. But whole groups of phonemes are often missing or changed in the spontaneous speech. Therefore the results of the ph LMs are biased because these models contain the statistics from in-domain words and alignment of in-domain words was used also as a reference.

Table 1. (a) Semantic corpora characteristics. (b) Recognition accuracies for different semantic corpora and different phoneme LM. The word-level LM is also presented.

<table>
<tr><td colspan="3" align="center">(a)</td></tr>
<tr><td></td><td>HHTT</td><td>TIA</td></tr>
<tr><td># train sentences</td><td>5240</td><td>4337</td></tr>
<tr><td># dev sentences</td><td>570</td><td>469</td></tr>
<tr><td># test sentences</td><td>1439</td><td>1073</td></tr>
<tr><td># different concepts</td><td>28</td><td>24</td></tr>
<tr><td># train concepts</td><td>8849</td><td>9027</td></tr>
<tr><td># dev concepts</td><td>989</td><td>1107</td></tr>
<tr><td># test concepts</td><td>2546</td><td>2305</td></tr>
</table>

(b)				
	HHTT $Acc[\%]$		TIA $Acc[\%]$	
LM	dev	test	dev	test
words	70.5	72.9	72.4	62.5
ph	74.6	76.1	79.1	70.1
ph-bh	65.4	67.5	59.4	51.8
ph-ad	71.8	73.5	70.7	62.8
ph-map	74.8	76.1	81.9	74.8

5.1 Adaptation of Phoneme Language Model

The 5-gram phoneme LMs used in this work offer a good recognition performance. On the other hand, the n-gram histories are already modelling the whole words. Therefore the 5-gram LMs are domain-dependent. To adapt the phoneme LM to a different domain, we use the fact that the set of all phonemes is finite and relatively small. We use an unsupervised approach – given a set of unannotated speech data from the target domain, we first recognize them with a generic phoneme LM (*ph-bh*) and then we use the 1-best recognition hypothesis to train a new adapted LM (*ph-ad*).

6   Results and Conclusion

To evaluate the SLU performance we used the *concept accuracy* measure defined as $cAcc = \frac{H-S-D-I}{N}$, where H is the number of correctly recognized concepts, N is the number of concepts in reference and S, D, I are the numbers of substitutions, deletions and insertions [10].

First of all, we analysed the influence of the kernel parameters n, m and kernel normalization on the semantic classifier performance. We used the HHTT training data and split them into train$_t$, train$_d$ and train$_e$ subsets. These subsets were used as train, development and test data in the HDM training process to evaluate the accuracy for different $K_{n,m}$ and $\bar{K}_{n,m}$ kernel functions. The concept accuracy of different configurations is depicted on Fig. 1. The first observation is that the kernel normalization significantly improves prediction accuracy, especially for the higher values of n and m. Based on the results of this experiment, we use the $\bar{K}_{1,5}$ kernel function defined by Eq. 4. In this case the kernel normalization improves the prediction accuracy from 64.7% to 71.5%. Another interesting observation is that a feature vector consisting of unigram phonemes ($n = m = 1$) still yields the accuracy above 50%.

The kernel function $\bar{K}_{1,5}$ was used in the input layer of an HDM and the model was trained on different data sets obtained by recognizing the utterances with different phoneme LMs. The concept accuracies for these experiments are show in Tab. 2. The results for word-level HDM are also included.

The *ph* LM outperforms the *ph-bh* and adapted *ph-ad* models. But this model is not very usable in the scenario where the word-level transcriptions are hard to obtain. The adapted model *ph-ad* is about 1% of accuracy worse than the *ph* model and the increase in accuracy in comparison with *ph-bh* model is 3-4%. This means that the phoneme LM adaptation process can be used to easily adapt high-order phoneme n-gram models to the target domain. The most interesting result is that the *ph-map* model outperforms the

Table 2. Concept accuracies for phoneme-level HDMs in comparison with word-level HDM

LM used	HHTT $cAcc[\%]$		TIA $cAcc[\%]$	
to generate ph. lattices	dev	test	dev	test
words	70.7	73.5	76.1	74.8
ph	67.0	70.9	73.8	70.9
ph-bh	61.5	66.5	67.5	65.7
ph-ad	68.8	69.8	70.7	69.6
ph-map	73.4	75.6	76.6	75.5

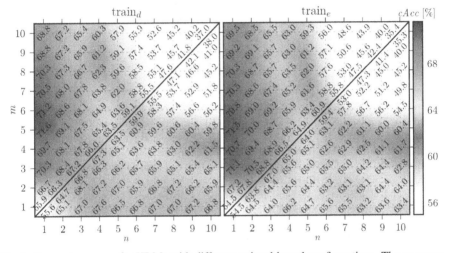

Fig. 1. Concept accuracy for HDMs with different rational kernel configurations. The accuracy was computed for different values of n and m with (above diagonal) and without (below diagonal) kernel normalization (Eq. 4).

word-level model although the pseudo-phoneme lattices was generated from the same word-level lattices as was used for the word-level HDM training. This result can be interpreted in two ways: (1) By having better phoneme recognizer, the phoneme-level HDM parsing performance can outperform the word-level one and (2) the mapping of words to phonemes and using them to train the phoneme-level HDM performs a kind of "linguistic" preprocessing. This preprocessing can be described as follows: if two different words are sharing some subsequence of phonemes (for example *arriving* and *arrival*) they are completely different at the word level but similar at the phoneme-level. And this phoneme-level similarity is effectively measured using the $\bar{K}_{1,5}$ kernel function used in this work.

In this paper, we presented a novel approach to spoken language understanding from phoneme lattices. The presented phoneme HDM cannot be directly used to obtain lexical realisation of semantic concepts. It only performs semantic tree assignment. To assign concrete values to semantic concepts some kind of spoken term detection methods must be used [13].

The results are nevertheless very interesting because the use of the adapted phoneme LM can support rapid development of spoken dialogue systems. The call routing and utterance classification tasks can also benefit from these results. The classification accuracy drop caused by using adapted phoneme LM (in comparison with the word-level HDM) is compensated by a faster deployment process – there is no need to collect and annotate data for word LM. And finally the use of pseudo-phoneme lattices obtained by mapping the word lattices to phonemes even outperform the word-level SLU model.

Acknowledgments. This research was supported by the Technology Agency of the Czech Republic, project No. TE01020197 and by the European Regional Development Fund (ERDF), project "New Technologies for Information Society" (NTIS), European Centre of Excellence, ED1.1.00/02.0090, and by the grant of the University of West Bohemia, project No. SGS-2013-032.

References

1. Švec, J., Jurčíček, F.: Extended Hidden Vector State Parser. In: Matoušek, V., Mautner, P. (eds.) TSD 2009. LNCS, vol. 5729, pp. 403–410. Springer, Heidelberg (2009)
2. Mairesse, F., Gašić, M., Jurčíček, F., Keizer, S., Thomson, B., Yu, K., Young, S.: Spoken language understanding from unaligned data using discriminative classification models. In: IEEE International Conference on Acoustics, Speech and Signal Processing, ICASSP 2009, Taipei, pp. 4749–4752. IEEE (2009)
3. Raymond, C., Béchet, F., De Mori, R., Damnati, G.: On the use of finite state transducers for semantic interpretation. Speech Communication 48(3-4), 288–304 (2006)
4. Valenta, T., Švec, J., Šmídl, L.: Spoken Dialogue System Design in 3 Weeks. In: Sojka, P., Horák, A., Kopeček, I., Pala, K. (eds.) TSD 2012. LNCS, vol. 7499, pp. 624–631. Springer, Heidelberg (2012)
5. Švec, J., Šmídl, L., Ircing, P.: Hierarchical Discriminative Model for Spoken Language Understanding. In: IEEE International Conference on Acoustics Speech and Signal Processing, pp. 8322–8326. IEEE, Vancouver (2013)
6. Cortes, C., Haffner, P.: Rational kernels: Theory and algorithms. The Journal of Machine Learning 5, 1035–1062 (2004)
7. Graf, A.B., Smola, A.J., Borer, S.: Classification in a normalized feature space using support vector machines. IEEE Transactions on Neural Networks 14(3), 597–605 (2003)
8. Švec, J., Ircing, P.: Efficient algorithm for rational kernel evaluation in large lattice sets. In: IEEE International Conference on Acoustics Speech and Signal Processing, pp. 3133–3137. IEEE, Vancouver (2013)
9. Jurčíček, F., Zahradil, J., Jelínek, L.: A human-human train timetable dialogue corpus. In: Proceedings of EUROSPEECH, Lisboa, pp. 1525–1528 (2005)
10. Jurčíček, F., Švec, J., Müller, L.: Extension of HVS semantic parser by allowing left-right branching. In: IEEE International Conference on Acoustics Speech and Signal Processing, vol. (1), pp. 4993–4996 (2008)

11. Psutka, J., Švec, J., Psutka, J.V., Vaněk, J., Pražák, A., Šmídl, L., Ircing, P.: System for fast lexical and phonetic spoken term detection in a Czech cultural heritage archive. EURASIP Journal on Audio, Speech, and Music Processing (1), 1–11 (2011)
12. Soutner, D., Loose, Z., Müller, L., Pražák, A.: Neural Network Language Model with Cache. In: Sojka, P., Horák, A., Kopeček, I., Pala, K. (eds.) TSD 2012. LNCS, vol. 7499, pp. 528–534. Springer, Heidelberg (2012)
13. Dogan, C., Saraclar, M.: Lattice Indexing for Spoken Term Detection. IEEE Transactions on Audio, Speech and Language Processing 19(8), 2338–2347 (2011)

Online Speaker Adaptation of an Acoustic Model Using Face Recognition

Pavel Campr[1], Aleš Pražák[2], Josef V. Psutka[2], and Josef Psutka[2]

[1] Center for Machine Perception, Department of Cybernetics, Faculty of Electrical Engineering,
Czech Technical University in Prague, Technická 2, 166 27 Prague 6, Czech Republic
camprpav@cmp.felk.cvut.cz
cmp.felk.cvut.cz
[2] Department of Cybernetics, Faculty of Applied Sciences,
University of West Bohemia in Pilsen, Univerzitní 8, 306 14 Pilsen, Czech Republic
{aprazak,psutka_j,psutka}@kky.zcu.cz
www.kky.zcu.cz

Abstract. We have proposed and evaluated a novel approach for online speaker adaptation of an acoustic model based on face recognition. Instead of traditionally used audio-based speaker identification we investigated the video modality for the task of speaker detection. A simulated on-line transcription created by a Large-Vocabulary Continuous Speech Recognition (LVCSR) system for online subtitling is evaluated utilizing speaker independent acoustic models, gender dependent models and models of particular speakers. In the experiment, the speaker dependent acoustic models were trained offline, and are switched online based on the decision of a face recognizer, which reduced Word Error Rate (WER) by 12% relatively compared to speaker independent baseline system.

Keywords: acoustic model, speaker adaptation, face recognition, multimodal processing, automatic speech recognition.

1 Introduction

Automatic speech recognition systems are used in many real world applications. An unpleasant problem is the frequent and sometimes very rapid change of speakers. This disallows to use an online speaker adaptation technique, which requires relatively long part of speech for adaptation. Special focus is given to real-time systems, such as automatic subtitling systems, where this problem is more visible. One solution is to enhance online speaker adaptation techniques, but here we study a different multimodal approach that uses video modality for rapid speaker change detection.

The proposed system uses a face recognizer to identify the speaker's identity and gender, in times lower than 100 ms. Based on the results, the pre-trained acoustic models in the LVCSR system can be switched, which leads to decrease in WER compared to the baseline system using only a speaker independent model. Advantages and disadvantages of such an approach, compared to traditional audio-only approach, are discussed. The experiment is carried out on TV broadcast of Czech parliamentary meetings.

The paper is organized as follows. Section 2 describes LVCSR system used in experiments. Section 3 describes face recognition system used for gender and identity

I. Habernal and V. Matousek (Eds.): TSD 2013, LNAI 8082, pp. 378–385, 2013.

estimation. Section 4 presents proposed system as a whole, with focus on an acoustic model selection from video stream. Section 5 presents the experiment and results. Final sections shortly discuss open problems for future work and conclude the paper.

2   Real-Time Automatic Speech Recognition System

One of the key applications for real-time LVCSR systems is the automatic online subtitling of live TV broadcasts. To increase the recognition accuracy, some speaker adaptation techniques with suitable fast online speaker change detection can be used [1]. Traditionally, the speaker change detections are based only on audio track analysis [2].

A common metric of the performance of a LVCSR system is Word Error Rate (WER), which is valuable for comparing different system and for evaluating improvements within one system. It was used in this paper for the evaluation.

In the following, the LVCSR system used for the evaluation of proposed system is described. The baseline LVCSR system uses speaker independent (SI) acoustic models. Gender dependent acoustic models are used if the gender of the speaker is known. Finally, speaker adaptation of an acoustic model is used to fine-tune the models for particular speakers. All models are trained offline before they are used in the online experiments. In a real-world application, the models could be trained and added to the bank of acoustic models in the course of time.

2.1   Language Model Details

The language model was trained on about 52M words of normalized Czech Parliament meeting transcriptions from different electoral periods. To allow captioning of an arbitrary (including future) electoral period, five classes for representative names in all grammatical cases were created and filled by current representatives. See [3] for details. To reduce out-of-vocabulary (OOV) word rate, additional texts (transcriptions of TV and radio broadcasting, newspaper articles etc.) were added. The final trigram language model with vocabulary size of 588923 words was trained using Kneser-Ney smoothing.

2.2   Acoustic Model Details

The acoustic model was trained on 585 hours of parliamentary speech recordings using automatic transcriptions. Since we trained the previous acoustic model on a much smaller amount of speech records (90 hours) using manual transcriptions, we tried to upgrade that model using parliamentary recordings collected during the real captioning [4]. These records were automatically recognized and reliable parts of automatic transcriptions were used for acoustic model training. Only words, which had confidence greater than 99% and their neighboring words had confidence greater than 99% too, were selected. Hence, so the real amount of training data was 440 hours. For details on the confidence measure see [5]. Such an enhanced acoustic model reduced WER by 20% relatively.

We used three-state HMMs and Gaussian mixture models with 44 of multivariate Gaussians for each state. The total number of 236940 Gaussians is used for the SI model. In addition, discriminative training techniques were used [6]. The analogue input speech signal is digitized at 44.1 kHz sampling rate and 16-bit resolution format. We use PLP feature extraction with 19 filters and 12 PLP cepstral coefficients, both delta and delta-delta sub-features were added. Feature vectors are computed at the rate of 100 frames per second (fps).

Gender dependent acoustic models were trained using an automatic speaker clustering method [5]. The initial split of the data was based on male/female markers obtained from the manual transcriptions utilizing previous acoustic model. The resulting acoustic models contained 29 Gaussians per state for men and 14 Gaussians for females. Discriminative training techniques were used for both acoustic models as well.

Acoustic model adaptation of specific speakers was carried out using unsupervised fMLLR adaptation [7]. Reliable parts of automatic transcriptions were chosen in the same way as for acoustic model training. Only one transformation matrix and shift vector was trained on available speech data (from 40 seconds to 150 minutes) for each speaker using gender dependent acoustic model.

3   Face Recognition

As shown in Figure 2, the goal of face recognition module is to detect and track faces in the image sequence, and to estimate their gender and identities, all in real time.

The task of real-time face detection and identification was widely studied and existing solutions are capable to solve this task with high accuracies [8] [9].

In this paper, we use a detector of facial landmarks based on Deformable Part Models, proposed by Uřičář [8]. In addition to the required position of the face in the image, this real-time face detector provides a set of facial landmarks like nose, mouth and canthi corners. Such landmark positions are used for normalized face image construction, which is used in the next recognition step.

Fig. 1. Example of a processed video frame. The circle denotes a face detection zone. The speaker's face is marked by the green rectangle, non-speaker face by the red rectangle.

The problem of gender and identity estimation are classification problems, for which we use multi-class Support Vector Machine (SVM) classifier, more precisely the implementation done by Franc [10]. The classifier uses image features, computed from normalized face images, based on Local Binary Patterns (LBP). For the gender classification task we used SVM learned on 5 million examples that was presented by Sonnenburg [9]. The results of classificators, which process single images, are accumulated for whole face tracks in the video. The decisions are accumulated and are provided as a global decision when the confidence is high enough.

Example depicted in Figure 1 presents results of the face detector, that is limited to process only highlighted circular area, where the speaker occurrence is expected. The results of gender and identity estimation are presented later in Experiment & Results section.

4   Proposed Multimodal LVCSR System

Figure 2 presents the schema of the proposed real-time automatic speech recognition system that uses face recognition to detect speaker changes. The main question is how to associate the face(s) coming from video modality with the speaker(s) coming from audio modality. This opened topic, sometimes referred as "audiovisual speaker diarization", is discussed for example in [11] or [12]. To our knowledge, no other work presented the use of speaker change detection from video modality for the use in real-time LVCSR system.

Our system is based on a speech recognizer as described in Section 2, on face recognizer as described in Section 3 and on acoustic model selector, that identifies the speaker based on the results from face recognizer, i.e. it answers the question "who is the speaker" based only on the video stream.

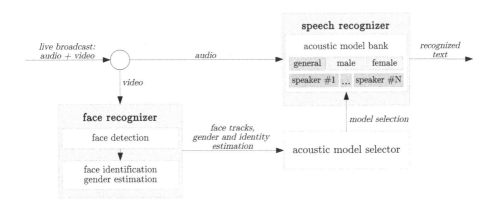

Fig. 2. Schema of an online automatic speech recognition system using multiple acoustic models, which are activated based on identity and gender estimations obtained from the face recognition module

4.1 Acoustic Model Selection

This module (see Figure 2) should identify the speaker, in real time, from the results of the face recognition module. Generally, the video stream can contain several or none faces, similarly to audio stream where the speakers can overlap or be quiet. Additionally, it is not expected that a speaker is always visible, and, vice versa, the face always speaks. A discussion about all the cases and possible solutions are presented in [11], but only for offline processing.

Here, for the real time processing, we impose some rules for the input broadcast that facilitate the speaker identification from the video. The rules are based on the type of input data, in our case for the parliamentary broadcasts. Sample image is shown in Figure 1, where the speaker is visible in the middle of the frame. Sometimes, the camera is switched and the frame contains a view of the whole room. Such knowledge about the broadcast allows to build a simple and real-time mapping between the faces and the speaker:

1. ignore small faces (we ignore faces smaller than 75 pixels)
2. the face whose position is closest to the center of the screen is denoted as current speaker
3. if the identity of the speaker's face is recognized, the acoustic model of this particular speaker is activated; otherwise go to step 4
4. if the identity is not recognized but the gender is, the gender dependent acoustic model is activated; otherwise go to step 5
5. if we have no knowledge about the speaker's face at all, the general acoustic model is activated

Such rule-based system is not general, but is sufficient for the first experiment and evaluation in this area, and it can be enhanced and generalized later, taking inspiration from offline variant of this task [11].

Additional decision must be done when all the faces are lost. We examined two strategies. The first is to immediately activate the SI model. The second is to keep the current acoustic model until the next face is found. In our experiment, the second strategy performed slightly better and is presented in the following section. This can be caused by occasional camera switches, that are present in the broadcast and show some graphics or the whole parliamentary room even if the speaker is still speaking.

5 Experiments and Results

Several experiments based on gender or speaker change detection were performed on one test record (200 minutes, 22286 words, 35 speakers, 105 speaker changes, 0.2% OOV words, perplexity of 315) simulating on-line transcriptions. The camera captured the speaker in 70.6% of the time, in the rest of the time the camera captured the whole parliamentary room or some graphics.

The face recognition module was able to identify the gender immediately in one frame, the median number of frames required to identity the face was 3, corresponding to 75 ms (with fps 25). The module identified 74 faces fully and in 34 cases only the gender was identified. The fails of identity recognition were mostly caused by the side orientation of the face or overlapping of the face with the microphone. The face classifier was trained on a closed set of speakers that were present in the test record.

The evaluation of the entire system follows. Firstly, only gender dependent acoustic models were switched during recognition based on speaker gender changes that were manually annotated. Relative WER reduction of 11% was achieved over baseline system with the SI acoustic model. Furthermore, by switching off-line prepared speaker adapted acoustic models, we reduced WER by 15% relatively over the baseline system (see Table 1). This represents a maximal improvement achievable by the proposed approach.

Next, the same experiments were performed based on gender and speaker changes detected automatically as described above. The WER reduction was slightly worse than in the case of manual annotation. To summarize, the baseline WER, our system WER and maximal achievable WER are 9.68%, 8.96% and 8.62%. for gender recognition only. For identity recognition, the results are 9.68%, 8.62% and 8.22%, with 11% relative WER reduction (which was 15% for the manual annotation).

In Table 1, the gender dependent results and results for two individual speakers are presented as well. The speaker "Schwarzenberg" is known for his speech disorder that leads to the worst WER of all speakers in the test record. The speaker "Němcová" acts as a chair of the meeting, so she often speaks during applauses or other types of speech noise, overlaps with other speakers, or is not visible in the camera when she speaks, so that an acoustic model of another speaker is used.

To conclude, the results show that decisions of a face recognizer can be used very efficiently for speaker change detection, which leads to improvement of the LVCSR system. Such approach has some advantages and disadvantages compared to audio-based detection, which is widely used nowadays. The advantage is the accurate and fast response of the face recognizer (usually less than 100 ms after the face is found). The disadvantage is the need to identify the speaker from the video, which is only possible when the speaker's face is visible.

Table 1. Results of LVCSR system: Word Error Rates (WER). Columns denoted as "gender" represent results for the case where only information about speaker's gender was used for acoustic model adaptation, columns "identity" relate to the case where both gender and identity were used. At the bottom, results for two worst performing speakers are presented: speaker "Schwarzenberg" leads to the worst absolute WER, speaker "Němcová" leads to increased WER with enabled adaptation.

| | | *without* | *with adaptation* | | | |
| | | *adaptation,* | *manual annotation* | | *face recognition* | |
speaker	*words*	*baseline*	*gender*	*identity*	*gender*	*identity*
all	22286	9.68%	8.62%	8.22%	8.96%	8.62%
men ♂	19635	9.91%	9.00%	8.51%	9.07%	8.71%
women ♀	2651	7.95%	5.82%	6.03%	7.93%	7.81%
Schwarzenberg	212	42.16%	35.20%	32.10%	35.20%	32.88%
Němcová	865	6.64%	5.31%	7.11%	9.34%	9.34%

6   Future Work

Many enhancements can be applied to the whole proposed system or to particular modules. In the speech recognition module, the speaker dependent acoustic models can be trained online as proposed in [1]. A speaker verification could be employed to verify the decisions of the face recognizer. In the face recognition module, the face models could be improved during the time from the new data. Lip activity detector or correlation between lip movements and the audio signal could be employed for a more robust identification of the speaker among found faces.

The speaker identification could be based on both audio and video streams, where the results of traditional audio-based speaker detector and video-based speaker detector could be combined. It is expected that the multimodal processing would provide better results compared to a scenario with only one modality.

7   Conclusions

We have proposed a real-time speaker change detection system based on face recognition. It can be incorporated into an automatic speech recognition system to switch acoustic models, which leads to the reduction of Word Error Rate (WER). Several experiments based on gender or speaker change detection were performed on a test recording simulating on-line transcriptions. Relative WER reduction of 7% was achieved over baseline system with speaker independent acoustic model using only a gender switch detection. Furthermore, by switching off-line prepared adapted acoustic models of speakers, we reduced WER by 11% relatively.

Acknowledgments. This research[1] was supported by the Grant Agency of the Czech Republic, project No. GAČR P103/12/G084.

References

1. Pražák, A., Zajíc, Z., Machlica, L., Psutka, J.V.: Fast Speaker Adaptation in Automatic On-line Subtitling. In: International Conference on Signal Processing and Multimedia Applications, vol.(1), pp. 126–130 (2009)
2. Ajmera, J., McCowan, I., Bourlard, H.: Robust Speaker Change Detection. IEEE Signal Processing Letters 11(8), 649–651 (2004)
3. Pražák, A., Psutka, J.V., Hoidekr, J., Kanis, J., Müller, L., Psutka, J.: Automatic Online Subtitling of the Czech Parliament Meetings. In: Sojka, P., Kopeček, I., Pala, K. (eds.) TSD 2006. LNCS (LNAI), vol. 4188, pp. 501–508. Springer, Heidelberg (2006)
4. Trmal, J., Pražák, A., Loose, Z., Psutka, J.: Online TV Captioning of Czech Parliamentary Sessions. In: Sojka, P., Horák, A., Kopeček, I., Pala, K. (eds.) TSD 2010. LNCS (LNAI), vol. 6231, pp. 416–422. Springer, Heidelberg (2010)

[1] The access to computing and storage facilities owned by parties and projects contributing to the National Grid Infrastructure MetaCentrum, provided under the programme "Projects of Large Infrastructure for Research, Development, and Innovations" (LM2010005) is highly appreciated.

5. Psutka, J.V., Vaněk, J., Psutka, J.: Speaker-clustered Acoustic Models Evaluated on GPU for on-line Subtitling of Parliament Meetings. In: Habernal, I., Matoušek, V. (eds.) TSD 2011. LNCS (LNAI), vol. 6836, pp. 284–290. Springer, Heidelberg (2011)
6. Povey, D.: Discriminative Training for Large Vocabulary Speech Recognition. PhD thesis, Cambridge University, Engineering Department (2003)
7. Zajíc, Z., Machlica, L., Müller, L.: Robust Statistic Estimates for Adaptation in the Task of Speech Recognition. In: Sojka, P., Horák, A., Kopeček, I., Pala, K. (eds.) TSD 2010. LNCS (LNAI), vol. 6231, pp. 464–471. Springer, Heidelberg (2010)
8. Uřičář, M., Franc, V., Hlaváč, V.: Detector of Facial Landmarks Learned by the Structured Output SVM. In: VISAPP 2012: Proceedings of the 7th International Conference on Computer Vision Theory and Applications1, Rome, Italy, pp. 547–556 (2012)
9. Sonnenburg, S., Franc, V.: COFFIN: A Computational Framework for Linear SVMs. Technical Report 1, Center for Machine Perception, Czech Technical University, Prague, Czech Republic (2009)
10. Franc, V., Sonneburg, S.: Optimized Cutting Plane Algorithm for Large-Scale Risk Minimization. Journal of Machine Learning Research 10, 2157–2192 (2009)
11. El Khoury, E., Sénac, C., Joly, P.: Audiovisual diarization of people in video content. In: Multimedia Tools and Applications, pp. 1–29 (2012)
12. Bendris, M., Charlet, D., Chollet, G.: People indexing in TV-content using lip-activity and unsupervised audio-visual identity verification. In: 2011 9th International Workshop on Content-Based Multimedia Indexing (CBMI), pp. 139–144. IEEE (2011)

Ontology of Rhetorical Figures for Serbian

Miljana Mladenović[1] and Jelena Mitrović[2]

[1] Faculty of Mathematics, University of Belgrade, Serbia
{ml.miljana,jmitrovic}@gmail.com
www.matf.bg.ac.rs
[2] Faculty of Philology, University of Belgrade, Serbia
www.fil.bg.ac.rs

Abstract. The paper presents *RetFig*, a formal domain ontology of rhetorical figures for Serbian. This ontology is one of the necessary steps in developing tools
for Natural Language Processing in the Serbian language, especially for tools per-
tinent to discourse analysis, sentiment analysis and opinion mining. The RetFig
ontology was developed taking into account a plethora of rhetorical figures in the
morphologically rich Serbian language, as well as in regard to various classifications of rhetorical figures that exist. We propose a system of linguistic classes and
properties that are best suited for this ontology, as well as some of the possible
usages for this particular ontology of rhetorical figures.

Keywords: domain ontology, rhetorical figures, Semantic web.

1 Introduction

Natural language texts are not always "flat" with unique, ordinary, untwisted literal
meaning. On the contrary, texts written in a natural language almost always have more
than one meaning, due to the usage of various linguistic operations over words, phrases,
sentences, et cetera. Without taking these facts into consideration, we can get incomplete and imprecise results in some NLP tasks. This especially holds true in areas of
opinion mining, sentiment analysis and discourse analysis. For example, if we say "He
is as fast as light", this statement will be marked as a positive opinion statement. On
the other hand, if we say "He is as fast as a turtle", opinion mining techniques will not
show the correct result unless we include the process of detection of rhetorical figures.
Our first task, in this direction, is to create the very first formal and comprehensive domain ontology of rhetorical figures in Serbian that will lead us, primarily, towards an
ontology based semantic tool for annotation of rhetorical figures and implementations
in other NLP tasks.

2 Related Work

Rhetoric is a means of spoken or written communication that we use in order to influence our listeners or our readers in a special way. Rhetorical figures (rhetorical devices,
stylistic figures or figures of speech) have been a subject of research since ancient times
in Aristotles major work, Rhetoric, it was pointed out that: "Rhetorician is someone

I. Habernal and V. Matousek (Eds.): TSD 2013, LNAI 8082, pp. 386–393, 2013.
© Springer-Verlag Berlin Heidelberg 2013

who is always able to see what is persuasive". Furthermore, the first rhetorical classification originated from Latin. It is known as "quadripartita ratio" and it describes four fundamental rhetorical operations on linguistics elements: addition, omission, permutation and transposition. Classical rhetoricians claimed that for any text taken as a literal model, all figures could be obtained with a combination of the four fundamental rhetorical operations by application on different linguistic levels: word forms, phrases, sentences, paragraphs, texts, etc. This kind of classification had been applied by Peacham [1] from 16th century on schemates rhetorical. He used rhetorical operations: repetition, omission, separation and conjuction. Also, it was the basis for later research. In that regard, Morris [2] created a (semio-) syntactic twodimensional classification table made of: linguistic operations and linguistic levels. Similarly, Durand introduced linguistic elements relationships like: identity, similarity, difference, opposition and false homologies [3]. Nowadays, we meet different classification systems. Harris[1] classifies rhetorical figures into three groups: "those involving emphasis; those involving physical organization, transition, and disposition; and those involving decoration." Sutcliffe[2] gives us a classification into six categories: figures of grammar, figures of meaning, figures of comparison, figures of parenthesis, figures of repetition and figures of rhetoric.. Another classification is made by Schwartz[3]: figures of speech, sounds and other rhetorical devices. One of the most comprehensive researches on rhetorical figures can be found in the Inkpot[4] Rhetfig project. According to Kelly et al. [4], the members of Inkpot group, there are four kinds of conceptual classifications of rhetorical figures. The first kind of classification represents the basic classification based on rhetorical characteristics, the classification into: *tropes, schemes* and *chroma*. The second classification is based on linguistic characteristics of the figures. Harris and DiMarco [5] marked it as a *"linguistic domain"* classification where domains are branches of linguistics. The third kind of conceptual classifications of rhetorical figures refers to different linguistic techniques used in rhetorical figures creation: *repetition, omission, series, identity, similarity, symmetry* and *opposition* applied over letters, words, clauses, phrases, sentences, etc. The fourth kind of conceptual classification is based on the generalization-specialization relationship of certain rhetorical figures and groups of figures. Evidently, many different methods for classification of rhetorical figures exist. Some of them are made from the perspective of rhetoric, some are made from the perspective of linguistics, and the others took both of those approaches into account. In order to create an ontology, we must consider all aspects of rhetorical figures research.

3   Building the Ontology of Rhetorical Figures for Serbian

Keeping in mind the complex and modular approach to building an ontology, which by Devedzić [6] includes: gathering and organizing of domain knowledge, defining usage, the range of validity and granularity in ontology, building the taxonomy, defining

---

[1] http://www.virtualsalt.com/rhetoric.htm

[2] http://opundo.com/figures.php

[3] http://cla.calpoly.edu/~dschwart/teaching.html

[4] http://create.uwaterloo.ca/matt/inkpot/projects/
rhetorical_about.html

relations, restrictions and rules over ontology entities, we divided our work into two phases. First, we collected rhetorical figures and their examples in order to create domain knowledge of rhetorical figures and a solid basis for further procedures in the process of building of our ontology. Second, we developed a formal domain ontology of rhetorical figures for Serbian and prepared it for further usage.

3.1 Creating a Collection of Rhetorical Figures in Serbian

In the process of gathering and organizing domain knowledge, the first step was to create a database structure for collecting rhetorical figures [7]. It contains information about: rhetorical figure name in Serbian, name that is referred to corresponding rhetorical figure in English[5], a definition or description, etymology of the name and additional notice. Also, three types of classifications according to rhetorical types, linguistic types and linguistic operations were introduced. We have developed and installed a web application[6] for maintaining and serialization of the database (*RetFig*). In the process of acquiring data about rhetorical figures, we searched novels, poems and journal texts in order to find examples of all of the relevant figures. As the Corpus of contemporary Serbian language mostly consists of daily newspaper articles, we needed to find relevant texts elsewhere. We marked 98 distinct rhetorical figures and manually classified them into 4 rhetorical types: *figures of pronunciation* (figure naglašavanja), *figures of meaning tropes* (figure zamene značenja tropi), *figures of construction* (figure konstrukcije) and *figures of thoughts* (figure širenja i sužavanja misli). Typical representatives of the group of figures of construction are: aphaeresis, apocope, diaresis, ellipsis, etc; representatives of the group of figures of pronunciation are: alliteration, anaphora, paromoiosis, epistrophe, etc; of the group of tropes are metaphor, metonymy, oxymoron, simile, etc.; of the group of figures of thoughts: antitheton, auxesis, climax, paradox, etc. All figures are also divided into five linguistic categories. If the linguistic elements participating in the creation of a rhetorical figure are letters or groups of letters or syllables, we are talking about a group of *phonological rhetorical figures*. If a rhetorical figure is created using Inflectional forms of a word, or a word formation, that figure belongs to the *morphological* group. If a rhetorical figure changes ordinary linguistic order of words in a sentence or if it changes lexical categories of some words, if it adds or omits parts of a sentence, that figure belongs to the *syntax* group. In the case when figures are used to change the literal meaning of a sentence, they belong to the *semantic* group. When a change of literal meanings spreads over the context of more sentences, we are talking about a *pragmatic* group. At last, every rhetorical figure must also be defined by linguistic operations over linguistic elements. We use linguistic operations of *addition, omission, repetition, transposition, joining, separation* and *symmetry*. The RetFig XML file can be downloaded and used locally from the web application address.

3.2 Creation of the RetFig Ontology

Ontology of rhetorical figures in Serbian (The RetFig ontology) is meant to have the following roles: to represent a formal domain ontology that unambiguously describes

---

[5] http://rhetfig.appspot.com/
[6] http://resursi.mmiljana.com/MemberZone/RetFig.aspx/

and defines rhetorical figures in Serbian; to be shared and merged with other linguistic resources and ontologies, such as Serbian WordNet (SWN) [8], Princeton WordNet [9] and Suggested Upper Merged Ontology (SUMO) [10]; to represent the basis upon which a task ontology will be built and used in processes of ontological annotation of rhetorical figures in Serbian. We have decided to use the *top-down* modelling technique. The RetFig ontology was created in Protege 4.2., the free, open source ontology editor and knowledge-based framework, using of OWL 2 Web Ontology Language. RetFig ontology is a domain ontology filled manually. Its growth is not intensive and will depend on the instantiation of new types of figures.

3.3 Building a Taxonomy

As the term "rhetorical figure" is used equally in the fields of rhetoric and linguistics, we have primarily defined two top-concepts: the *RhetoricalEntity* ("RetorickiEntitet") and the *LinguisticEntity* ("LingvistickiEntitet"). The concept *RhetoricalFigure* ("RetorickaFigura") is defined as both a rhetorical and a linguistic concept. On a lower level, the rhetorical concept is represented by concepts: *RhetoricalGroup* ("RetorickaGrupa") and *RhetoricalFigure* ("RetorickaFigura"), while the linguistic concept is represented by concepts: *LinguisticObject* ("LingvistickiObjekat"), *LinguisticRange* ("LingvistickiOpseg"), *LinguisticGroup* ("LingvistickaGrupa"), *LinguisticPosition* ("LingvistickaPozicija"), *LinguisticElement* ("LingvistickiElement") and *RhetoricalFigure* ("RetorickaFigura") (Figure1).

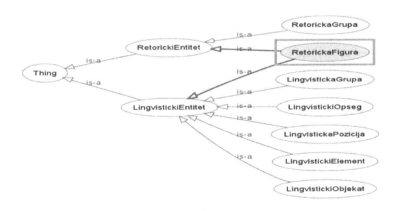

Fig. 1. Taxonomy of linguistic and rhetorical concepts

Each rhetorical figure in a text is characterized by the scope or the range of the text (the context) in which it appears. Looking from the inside out, the scope can be: a word, a phrase, a sentence, a verse, a strophe, a paragraph. Inside of such *linguistic scope*, we defined a *linguistic object* whose transformation via *linguistic operations* leads us to the structure that can be recognized as a certain *rhetorical figure*. *Linguistic object* can be a word, a phrase, a verse or a sentence. Transformation processes are either done over the entire *linguistic object* or over a part of that object. In that regard, we defined *linguistic element* as part of *linguistic object* that is being transformed. If the *linguistic operation* is being performed over the entire object of transformation the *linguistic object* and the *linguistic element* are identical. Otherwise *linguistic element* is smaller than the *linguistic object*. In Figure 2 an example of rhetorical figure Aphaeresis (afereza) detection by *RetFig* ontology is shown. The given example is taken from Shakespeares "King Lear" — "The King hath cause to plain."

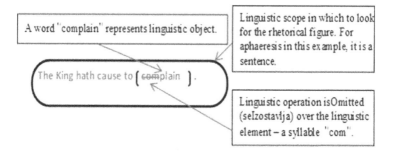

Fig. 2. Mutual relationship of linguistic scope, object and element

Analysis of processes of rhetorical figures creation has shown that mutual relationship between *linguistic objects* and *linguistic elements* differentiates the figures themselves. For example, if a *linguistic object* is a word, *linguistic element* is a letter, and a *linguistic operation* is "letter omission", than we could have: rhetorical figure aphaeresis (afereza), if an omitted letter is the first letter in the word; rhetorical figure apocope (apokopa), if an omitted letter is the last letter in the word; rhetorical figure syncope (sinkopa), if an omitted letter is not in the first or in the last position in the word to which it belongs. Because of that, positional relation between the *linguistic element* and the *linguistic object* is important, therefore, we introduce the concept: *Linguistic position*("LingvistickaPozicija"), in order to define the position in which a *linguistic element* appears inside of a *linguistic object*.

Linguistic operations are defined in the ontology as relations that connect instances of *RhetoricalFigure* ("RetorickaFigura") class (Domain) and instances of *LinguisticElements* ("LingvistickiElement") class (Range). The division of relations in the RetFig ontology is to the following relations: *addition, omission, repetition, trans-position, joining, separation* and *symmetry* at the ObjectProperty level. Also, SubObjectProperty levels are defined.

3.4 Inserting Attributes and Individuals

The most important set of members of the RetFig ontology is the *RhetoricalFigure* ("RetorickaFigura") set of Individuals. They represent rhetorical figures themselves and these Individuals are formally defined to be uniquely identified. For each Individual, the rhetorical and linguistic groups it belongs to are defined, the linguistic scopes, objects, elements and linguistic operations used for the creation of the said rhetorical figure. Conclusively, each rhetorical figure was appointed with its appropriate annotation: *comment* a short definition of the rhetorical figure, *seeAlso* information about the name of the rhetorical figure in English (keeping in mind the goal of mapping to linguistic ontologies [11] in English) and the alternative name of the same rhetorical figure in Serbian. This naming principle has been chosen to allow easier usage of this ontology for Serbian annotators but also to keep the possibility of alignment. Each *Rhetorical-Figure* ("RetorickaFigura") class member is declared as it is shown in the declaration of rhetorical figure Dysphemismus (Disfemizam):

```
<owl:NamedIndividual rdf:about="&ont;DISFEMIZAM">
<rdf:type rdf:resource="&ont;RetorickaFigura"/>
  <ont:naziv rdf:datatype="&xsd;string">DISFEMIZAM
      </ont:naziv>
  <rdfs:comment>Namerno koriscenje ruznijeg, ostrijeg
          izraza umesto normalnog.</rdfs:comment>
  <rdfs:seeAlso xml:lang="en">dysphemismus</rdfs:seeAlso>
  <rdfs:seeAlso xml:lang="sr">KAKOFEMIZAM</rdfs:seeAlso>
  <ont:jeNaPoziciji rdf:resource="&ont;CELINA"/>
  <ont:jeRetorickaGrupa
          rdf:resource="&ont;FIGURE_ZAMENE_ZNACENJA-TROPI"/>
  <ont:seZamenjujeDrugimElementomJacegZnacenja
          rdf:resource="&ont;FRAZA-LELEMENT"/>
  <ont:jeNadObjektom rdf:resource="&ont;FRAZA-LOBJEKAT"/>
  <ont:seZamenjujeDrugimElementomJacegZnacenja
          rdf:resource="&ont;REC-LELEMENT"/>
  <ont:jeNadObjektom rdf:resource="&ont;REC-LOBJEKAT"/>
  <ont:jeNadOpsegom rdf:resource="&ont;RECENICA"/>
  <ont:jeLingvistickaGrupa rdf:resource="&ont;SEMANTIKA"/>
  <ont:jeNadOpsegom rdf:resource="&ont;STIH"/>
</owl:NamedIndividual>
```

From the example given above, we can see that this figure represents usage of an intentionally harsh word or expression instead of an expected, or a polite one. We also find that the name of this figure in English is *Dysphemismus*, and that there is also an alternative name for this figure in Serbian *Kakofemizam* (*Kakophemismus*). *Disfemizam* is a rhetorical figure from the group named tropi (tropes), it is a subject of research in the area of linguistics called Semantika (Semantics). It can be found inside a sentence or a verse (*linguistic scope*) and it is formed by replacing the existing phrase or a word (*linguistic object/element*) in its entirety (*linguistic position*) by a different phrase or a phrase or a word of a stronger meaning (*linguistic operation* – "seZamenjujeDrugimElementomJacegZnacenja").

4   RetFig Ontology Testing

RetFig ontology is meant to give a couple of significant answers. First, it is prepared for ontological annotation of rhetorical figures in Serbian. In this regard, for pre-selected individuals for *linguistic scope* and/or *linguistic object* of observation, this ontology has to give a candidate or candidates for a certain rhetorical figure. For example, if the pre-selected individual is "REC" ("word") for the *linguistic scope*, there is no sense in expecting for the rhetorical figure Dysphemismus (Disfemizam) to be annotated, but it is expected that figures like apheresis (afereza), diaresis (dijareza), protesis (proteza) et cetera will appear. Second, it gives us an insight into the rhetorical figures used in a particular text. For example, if we determine that the analysis of a certain text shows a frequent loss of letters in words, mapping onto the ontological relation *"seIzostavlja"* (isOmmited) gives us a set of rhetorical figures that are formed that way, by omission of letters. Those figures are: aphaeresis (afereza), syncope (sinkopa), apocope (apokopa) and ecthlipsis (elizija). The SPARQL (recursive acronym for SPARQL Protocol and RDF Query Language) queries that give the answers to both of the mentioned tasks are represented in Figure 3. Moreover, the RetFig ontology will also be able to give answers about the statistical data regarding the annotated rhetorical figures. Each rhetorical figure in the RetFig ontology is defined by a finite set of RDF triples that uniquely describe that figure. With SPARQL queries, figures can be selected individually or in groups, which is the purpose of the ontology. The RetFig.owl ontology can be downloaded from the address of the web application.

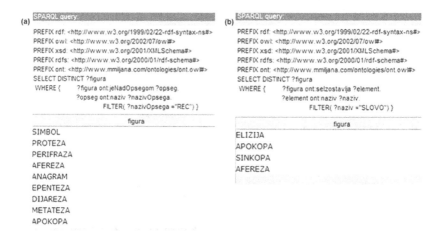

Fig. 3. (a) SPARQL query used to find the rhetorical figures generated over words (b) Query for finding rhetorical figures formed by omission of letters in words

5 Conclusion and Future Work

In this paper we have proposed the *RetFig* ontology that describes and defines linguistic entities and relations used in generation of rhetorical figures in Serbian. We have collected linguistic knowledge about rhetorical figures and their examples in the Serbian language. Furthermore, we have shown that the *RetFig* ontology can be our starting point in the future process of annottion of rhetorical figures in Serbian. Our future work will include connecting RetFig ontology with SUMO, Adimen-SUMO [12] and with the SWN (especially after recent enhancements of SWN which include transformation to RDF) as a part of the process of development of the RetFig application ontology in order to get a usable, ontology based semantic annotation tool for the rhetorical figures in Serbian. Metaphors are the focus of our research in the scope of developing the *SimNet* resource for automatic annotation of metaphors, which will further improve the results of semantic analysis processes aided by the tools we are developing. Implementation of annotation methods for figurative types will also be leaning on extensive research done in the field of MWEs exploration [13] so far.

References

1. Peacham, H.: The Garden of Eloquence. Perseus Digital Library (1593),
   `http://rhetoric.byu.edu/primary%20texts/Peacham.htm`
2. Sloane, T.O.: Encyclopedia of Rhetoric. Oxford University Press, New York (2001)
3. Durand, J.: Rhetorical Figures in the Advertising Image. In: Marketing and Semiotics: New Directions in the Study of Signs for Sale, pp. 295–318. Walter de Gruyter, New York (1987)
4. Kelly, R.A., Abbott, A.N., Harris, A.R., DiMarco, C., Cheriton, D.: Toward an ontology of rhetorical figures. In: SIGDOC 2010 Proceedings of the 28th ACM International Conference on Design of Communication, pp. 123–130 (2010)
5. Harris, R.A., DiMarco, C.: Constructing a rhetorical figuration ontology. Symposium on Persuasive Technology and Digital Behavior Intervention, Convention of the Society for the Study of Artificial Intelligence and Simulation of Behaviour (AISB), Edinburgh (2009)
6. Devedžić, V.: Understanding Ontological Engineering. Communications of the ACM 45(4), 136–144 (2002)
7. Tartalja, I.: Teorija književnosti. Zavod za udžbenike i nastavna sredstva, Beograd (2003)
8. Krstev, C., Lažetić, G.P., Vitas, D., Obradović, I.: Using Textual and Lexical Resources in Developing Serbian Wordnet. Romanian Journal of Information Science and Technology 7(1-2), 147–161 (2004)
9. Fellbaum, C.: WordNet: An Electronic Lexical Database. MIT Press, Cambridge (1998)
10. Niles, I., Pease, A.: Linking lexicons and ontologies: Mapping wordnet to the suggested upper merged ontology. In: Proceedings of the IEEE International Conference on Information and Knowledge Engineering, pp. 412–416. IEEE (2003)
11. Bateman, J.A.: Ontologies of Language and Language Processing. In: Theory and Applications of Ontology: Computer Applications, pp. 393–409. Springer, Netherlands (2010)
12. Àlvez, J., Lucio, P., Rigau, G.: Adimen-SUMO: Reengineering an Ontology for First-Order Reasoning. International Journal on Semantic Web and Information Systems (IJSWIS) 8(4) (2012)
13. Sag, I.A., Baldwin, T., Bond, F., Copestake, A., Flickinger, D.: Multiword Expressions: A Pain in the Neck for NLP. In: Gelbukh, A. (ed.) CICLing 2002. LNCS, vol. 2276, pp. 1–15. Springer, Heidelberg (2002)

Phoneme Set and Pronouncing Dictionary Creation for Large Vocabulary Continuous Speech Recognition of Vietnamese

Thien Chuong Nguyen and Josef Chaloupka

The Institute of Information Technology and Electronics (ITE), Technical University of Liberec,
Studentska 1402/2, 461 17, Liberec, Czech Republic
{chuong.nguyen.thien,josef.chaloupka}@tul.cz

Abstract. This paper describes our study on solving two basic problems of large vocabulary continuous speech recognition (LVCSR) of Vietnamese, which can be used as a standard reference for Vietnamese researchers and other researchers who are interested in Vietnamese language. First, a standard phoneme set is proposed with its corresponding grapheme-to-phoneme map. This phoneme set is the core to solve other problems related to LVCSR of Vietnamese. Then the creation of standard pronouncing dictionary based on the grapheme-to-phoneme map and the analysis of Vietnamese syllable is also described. Finally, we present the results on LVCSR using different types of pronouncing dictionary, which show some interesting aspects of Vietnamese language such as the structure of Vietnamese syllable and the effect of tone in the relationship with syllable.

Keywords: Automatic speech recognition, LVCSR of Vietnamese, phoneme set, pronouncing dictionary, tonal language.

1   Introduction

Vietnamese is officially used in speaking and writing system of Vietnam. It is an isolating language in which the words are invariable, and grammatical relationships are determined by word order and function words. Vietnamese is also a tonal language with six tones [1]. In Vietnam, the research on speech recognition began about twenty years ago, and researchers when doing experiments on LVCSR of Vietnamese have to deal with two basic problems: the selection of phonetic units (phonemes or the combination of phonemes and tones) to build acoustic models and the creation of pronouncing dictionary.

The selection of phonetic units is not straightforward as in English case where each phoneme is a phonetic unit, but will very much depend on the analysis of Vietnamese syllable and the knowledge of Vietnamese phonology. Base on the structure of Vietnamese syllable (Fig. 1) there are two main phonetic unit types: phoneme-based and rhyme-based phonetic unit. In the first type [2,3,4,5,6,7], the smallest components of syllable (Onset, Medial, Nucleus and Coda) are used as phonetic units and in the second type [2,5], the phonetic units are Onset (Initial) and Rhyme (Final) of the syllable. These two types are different in the properties of phonetic units obtained from each syllable and in the total number of distinct phonetic unit. In addition, using the knowledge

I. Habernal and V. Matousek (Eds.): TSD 2013, LNAI 8082, pp. 394–401, 2013.
© Springer-Verlag Berlin Heidelberg 2013

of Vietnamese phonology, other types of phonetic unit are also proposed. In [2], the phonetic units are the mixture of units from the two types above in which some units (considered as type 2) are the combination of two short neighbor phonetic units and the others are used as phonetic units of type 1. In [7,8], the syllable is analyzed into consonant (Onset, Coda), and monophthong (Nucleus) or diphthong, triphthong (combination of Medial, Nucleus and/or Coda) which are used as basic phonetic units. Note that in these methods, the way in which tone is integrated into the syllable will affect the number of distinct phonetic unit as well as the performance of acoustic model.

Another difficulty when doing experiments on LVCSR of Vietnamese is the lack of a standard pronouncing dictionary. In most of the works [3,5,7,8,9,10], the pronouncing dictionary is created by simply splitting the dictionary entries (syllables, words) into their corresponding graphemes. In other words, they apply a one to one grapheme-to-phoneme mapping of the dictionary entries, and so we call it the grapheme-based pronouncing dictionary. In some other works [11,12], phoneme-based pronouncing dictionary is used but a standard phoneme set is yet to be proposed.

To provide a baseline approach for dealing with all of the problems mention above, in this paper we first propose a standard phoneme set for Vietnamese which contains 39 basic phonemes. This phoneme set is the core to solve all other problems of LVCSR of Vietnamese. Using this phoneme set, researchers can methodically deal with many interesting aspects of LVCSR of Vietnamese such as phonetic unit creation, the effect of tone on Vietnamese syllable in the form of tone's role and position, or dialect in Vietnamese. We also propose a grapheme-to-phoneme mapping table and use it to create different types of pronouncing dictionary for Vietnamese. In our experiments, five different types of pronouncing dictionary are created and examined using the proposed mapping table.

2   Phoneme Set Proposal

For experiments on LVCSR of Vietnamese, the need for a set of standard phonemes is obvious but the problem is more complicated. In most of the previous works, graphemes are used in place of phonemes which are not always true. For example, the grapheme pairs "g, gh" and "ng, ngh" both present the same phonemes /ŋ/ and /ɣ/ respectively but are used separately as phonemes. In some other works [9,10], phonemes are used but the phoneme set is not given. To solve this problem, in this paper we propose a standard phoneme set and a many to one grapheme-to-phoneme mapping table. The Vietnamese phoneme set is composed of 23 consonants, 11 monophthongs, 3 diphthongs, one medial phoneme and one semivowel phoneme. Table 1 shows the complete phoneme set with its corresponding graphemes, International Phonetic Alphabet (IPA) representations and properties. Note that IPA representations and properties of some phonemes can vary depending on dialect.

3   Pronouncing Dictionary Creation

To create pronouncing dictionary, the grapheme-to-phoneme mapping table and the methods of analyzing of Vietnamese syllable are utilized. Syllable is the core of all

Table 1. Grapheme-to-phoneme mapping table

Type	IPA	Grapheme	Phoneme	Properties
Consonant	ɓ	b	B	voice bilabial implosive
	c	ch	CH	voiceless palatal stop
	ɗ	đ	D	stop, voiced alveolar implosive
	f	ph	F	voiceless labiodental fricative
	h	h	H	voiceless glottal fricative
	j	d	Y	palatal approximant
	k	k, q, c	K	voiceless velar stop
	l	l	L	alveolar lateral approximant
	m	m	M	bilabial nasal
	n	n	N	alveolar nasal
	ɲ	nh	NH	palatal nasal
	ŋ	ng, ngh	NG	velar nasal
	p	p	P	voiceless bilabial stop
	s	x	S	voiceless alveolar sibilant
	ʂ	s	SH	voiceless retroflex sibilant
	t	t	T	voiceless alveolar stop
	tʰ	th	TH	stop, aspirated, alveolar
	ʈʂ	tr	TR	voiceless retroflex affricate
	v	v	V	voiced labiodentals fricative
	x	kh	KH	voiceless velar fricative
	ɣ	g, gh	G	voiced velar fricative
	z	gi	ZH	voiced alveolar sibilant
	ʐ	r	R	voiced retroflex sibilant
Medial	w	u, o	W	velar glide
Semi-vowel	j	y, i	IH	palatal glide
Monophthong	a	a, ă	AU	open front unrounded
	e	ê	EE	close-mid front unrounded
	ɛ	e	EH	open-mid front unrounded
	ə	â	AH	mid-central
	əː	ơ	AX	lower mid, central
	i	y, i	IY	close front unrounded
	ɨ	ư	UH	close central unrounded
	o	ô	AO	close-mid back rounded
	u	u	UW	close back rounded
	ɔ	oo, o	OO	open-mid back rounded
	aː	a	AA	open unrounded
Diphthong	iə	yê, iê, ya, ia	IE	
	ɨə	ươ, ưa	UA	
	uə	uô, ua	UO	

Vietnamese words in which a meaning word can be constructed from one, two, three or more syllables. In Vietnamese modern language, there are about 8000 syllables [14]. Each syllable is modeled as in Fig. 1 and contains five main components: initial consonant (C1), medial (w), vowel (V), final consonant/semi-vowel (C2) and tone (T). Not all syllables have their own complete form, some syllables can appear without one of the following components: C1, w, or C2. It means vowel and tone are always presented in a syllable and are the main components of a syllable.

The type of created pronouncing dictionary is depended on the analysis of Vietnamese syllable, namely the role and position of tone in a syllable. In this paper, we propose five different pronouncing dictionary types corresponding to five different methods of analyzing of the Vietnamese syllable. Each analyzing method will differ from the others in type, properties and total number of phonetic units. Table 2 shows how to analyze a Vietnamese syllable into basic phonetic units and Table 3 shows their corresponding number of possible distinct phonetic units. As you can see, although the number of phonemes is invariable (39 phonemes), the total number of phonetic units will vary depended on the way in which tones are integrated into the syllable. In methods M1 and M2, the phonetic units are very different from phonemes and the number of distinct units is also larger. In methods M3 and M4, the types and the number of phonetic units are the same to one another, the difference is only from the tone's position. Method M5 does not contain tone's information and has the same number of phonetic units as phonemes.

By applying the grapheme-to-phoneme mapping table on the output of the syllable analyzing methods, a standard syllable-based or word-based pronouncing dictionary can be constructed. Each entry of the dictionary is formed using the procedure described below:

Procedure 1: Dictionary Entry Creation.

1. Separate dictionary entry into syllables.

2. Select the first syllable and analyze it into five components: C1, w, V, C2 and T

3. Using mapping table to convert four components C1, w, V and C2 into their corresponding phonemes.

4. Integrate tone T into phonemes to create appropriate phonetic units based on hypotheses about role and position of tone.

5. Concatenate these phonetic units to those of the previous syllables (if any) to create the pronouncing representation of the entry.

6. Select the next syllable (if any), analyze it into five components: C1, w, V, C2, T and go to step 3.

Table 2. Methods to analyze Vietnamese syllable

Phonetic Unit Type		Tone position	
		After main vowel	*At the end of syllable*
Tone role	*Dependent*	**M1**: C1 w VT C2	**M2**: C1 w V C2T or C1 w VT
	Independent	**M3**: C1 w V T C2	**M4**: C1 w V C2 T
No Tone		**M5**: C1 w V C2	

Table 3. Number of distinct phonetic unit

Method	Number of phonetic unit
M1	109
M2	153
M3	45
M4	45
M5	39

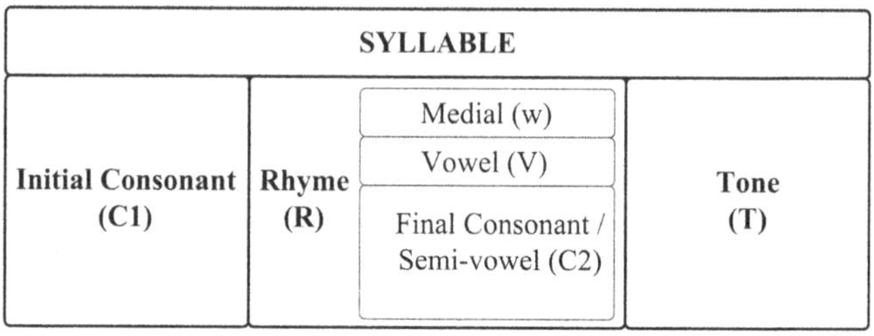

SYLLABLE			
Initial Consonant (C1)	Rhyme (R)	Medial (w)	Tone (T)
		Vowel (V)	
		Final Consonant / Semi-vowel (C2)	

Fig. 1. Structure of Vietnamese syllable

4 Text and Speech Corpus Acquisition

4.1 Text Corpus

For our study on LVCSR, we collect two types of text corpora. First, a general purpose text corpus is obtained by using a two-step procedure. In step 1, we collect a small text corpus from a well-known web's resource: the Wikipedia. Because the corpus resulted from this resource is not large and diverse enough, we just use this corpus to extract seed words which have to meet some requirements such as the frequency of occurrence of these words has to be not too high and they have to contain at least one of the specific Vietnamese letters and/or six diacritics to mark tones. In step 2, from these seed words, we generate a list of web queries and feed them to the search engine to collect only links available for Vietnamese language. After downloading and extracting all the useful text from these links, we obtain the final text corpus that good enough for multi-purpose research. Also, a specific purpose text corpus is collected using some websites which are very rich of text data in the field of news and literature. These two corpora are then combined and filtered to obtain the final text corpus. In this corpus, the VN set contains only sentences with 100% Vietnamese syllable and the VNF set contains all sentences in the filtered corpus (Table 4).

The resulted text corpus will be used to construct syllable-based language model (LM) for LVCSR of Vietnamese. In our experiments, we train the LM using the system's vocabulary with the size of 5000 syllables. Those syllables are selected as follows: create the count table of all the syllables in the text corpus and sort them in decreasing order. Then select the first 5000 syllables with the highest frequency of occurrence in the corpus and store them in the system dictionary. Table 5 shows the goodness of bi-gram and tri-gram LM trained on two data set of the text corpus.

Table 4. Statistics of the text corpus

	Train		Test
	VN	VNF	
Text size (Mb)	438.7	738.7	37.3
No. of sentence	5,907,440	8,681,869	426,480
No. of Vietnamese syllable	75,908,552	124, 006, 866	6,468,853
No. of foreign word	0	4,637,163	0

Table 5. Perplexity of LM

Vocabulary size	Perplexity (VN)		Perplexity (VNF)	
	bi-gram	tri-gram	bi-gram	tri-gram
5000	218.6337	146.7583	229.4118	146.5898

4.2 Speech Corpus

The speech corpus in our experiments is collected from the internet resource which contains speech mainly of types: story reading, news report, weather forecast, conversation and has the following properties: it covers three main dialects of Vietnamese language, has many ranges of speaking rate and contains varying types of background noise (room, studio, outdoor...). First, sound files of the above types are downloaded from some main websites which are rich of speech data. Then they are converted into 16 bits wave format with sampling rate of 11025 Hz. In the next step, we select only good utterances which contain only Vietnamese syllables and manually transcribe them resulted in a total of 16855 utterances with the length of 31 hours 58 minutes. The detail information about the speech corpus is shown in Table 6.

Table 6. Statistics of Speech Corpus

Data set	Number of speaker		Number of utterance	Duration
	Male	Female		
Train	52	102	13600	27h7m
Test	4	8	3255	4h51m
Total	56	110	16855	31h58m

5 Experiments

In this paper, the HTK Toolkit [15] is used to train and test the recognizers for LVCSR of Vietnamese. For each method (Section 3), the phonetic units are trained with 3 states Hidden Markov Model (HMM) using flat-start procedure. Then a set of context-dependent word internal triphone acoustic models are trained using the training speech corpus described in section 4. Similar acoustic states of these triphones are tied using tree-based clustering method. Each state of the phonetic unit HMM consists of 8 Gaussian mixtures. In all experiments, we extract feature vector every 10ms with window size of 25ms. Each feature vector has 39 dimensions containing 13 MFCC with their first and second derivatives. Because in the speech corpus, we only transcribe sentences which contain 100% of Vietnamese syllable, we will use the bi-gram LM trained from the VN set of the text corpus for all of our experiments.

In the first experiment, we will examine the performance of LVCSR recognizer using five different pronouncing dictionary types. Table 7 shows that the best word error rate (WER) is obtained when using method M4 (32.55%) in which tone is considered as an independent component locating at the end of syllable, and the worst WER is for method M5 (36.63%) when tone is not modeled in the recognizer.

In the second experiment, we optimize our system by applying the gender dependent recognizer on the best method M4 (GDM4). Our gender recognizer is trained using Gaussian Mixture Model (GMM) on MFCCs feature vectors. Experiment on our gender database with 200 speakers for training (100 male and 100 female speakers) and 40 speakers for testing (20 male and 20 female speakers) shows that GMM model with seven mixtures for each gender's model obtain the recognition rate of 100%. Using this gender recognizer on our LVCSR recognizer, we can improve the WER to 31.18% for the best method (Table 7).

Table 7. WER for different pronouncing dictionary types

Method	M1	M2	M3	M4	M5	GDM4
WER [%]	34.42	34.06	34.23	32.55	36.63	31.18

6   Conclusions

This paper proposes a standard phoneme set which is used to create five different types of pronouncing dictionary for our LVCSR of Vietnamese experiments. The best WER (32.55 %) achieved when using the method M4. This result indicates that tone tend to locate at the end of syllable as an independent phonetic unit. Also, we can see that, for other methods (M1, M3) where tone is located after main vowel, the WER is not different much compare to the best method. This means that, part of the tone is also present on or right after the main vowel. In the case in which tone's information is not integrated into syllable, the WER is not as good as the other methods and so it means tone is an important part when dealing with Vietnamese language.

In our works, only gender dependent optimization method is applied. We believed that the WER can be improved dramatically when applying other optimization methods on two aspects: Vietnamese specific optimization strategies (using word-based LM, extract pitch information, more tone processing methods, dealing with dialect) and general optimization strategies (signal adaptation, dealing with noise, LM tuning...). So in our next work, the recognizer will be trained and optimized based on the total examination of these aspects.

Acknowledgments. This research was supported by the Student Grant Scheme (SGS) at the Technical University of Liberec.

References

1. Chuong, N.T.: Selection of Sentence Set for Vietnamese Audio-Visual Corpus Design. In: IDAACS 2011, Praha, Czech Republic, pp. 492–495 (2011)
2. Vu, T.T., Nguyen, D.T., Luong, C.M., Hosom, J.P.: Vietnamese large vocabulary continuous speech recognition. In: INTERSPEECH 2005, pp. 1689–1692 (2005)
3. Vu, Q., Demuynck, K., Van Compernolle, D.: Vietnamese Automatic Speech Recognition: The FLaVoR Approach. In: Huo, Q., Ma, B., Chng, E.-S., Li, H. (eds.) ISCSLP 2006. LNCS (LNAI), vol. 4274, pp. 464–474. Springer, Heidelberg (2006)
4. Nguyen, H.Q., Nocera, P., Castelli, E., Trinh, V.L.: Using tone information for Vietnamese continuous speech recognition. In: RIVF 2008, pp. 103–106 (2008)
5. Nguyen, H.Q., Trinh, V.L., Le, T.D.: Automatic Speech Recognition for Vietnamese Using HTK System. In: RIVF 2010, pp. 1–4 (2010)
6. Vu, Q., et al.: A Robust Transcription System for Soccer Video Database. In: ICALIP, Shanghai (2010)
7. Nguyen, T., Vu, Q.: Advances in Acoustic Modeling for Vietnamese LVCSR. In: IALP 2009, pp. 280–284 (2009)
8. Vu, N.T., Schultz, T.: Vietnamese Large Vocabulary Continuous Speech Recognition. In: ASRU IEEE, Italy, pp. 333–338 (2009)
9. Vu, N.T., Schultz, T.: Optimization On Vietnamese Large Vocabulary Speech Recognition. In: Workshop on Spoken Languages Technologies for Under-Resourced Languages, SLTU 2010, Penang, Malaysia (May 03, 2010)
10. Le, V.B., Besacier, L.: Comparison of Acoustic Modeling Techniques for Vietnamese and Khmer ASR. In: ICSLP 2006, Pittsburgh, PA (September 2006)
11. Nguyen, H.Q., Nocera, P., Castelli, E., Trinh, V.L.: Large vocabulary continuous speech recognition for Vietnamese, an under-resourced language. In: SLTU 2008, Ha Noi, Vietnam, May 5-7 (2008)
12. Le, V.B., Tran, D.D., Besacier, L., Castelli, E., Serignat, J.-F.: First steps in building a large vocabulary continuous speech recognition system for Vietnamese. In: RIVF 2005, Can Tho, Vietnam (February 2005)
13. Le, T., Nguyen, H., Vu, Q.: Progress in Transcription of Vietnamese Broadcast News. In: Proc. International Conference on Communications and Electronics, ICCE 2006 (October 2006)
14. Hoang, P.: Syllable Dictionary. Danang Publisher, Vietnam (1996)
15. Steve, Y., Odel, J., Ollason, D., Valtchev, V., Woodland, P.: The HTK Book, version 3.2. Cambr. Univ., UK (2002)

Phonetic Spoken Term Detection in Large Audio Archive Using the WFST Framework*

Jan Vavruška, Jan Švec, and Pavel Ircing

Department of Cybernetics, University of West Bohemia, Plzeň, Czech Republic
{vavruska,honzas,ircing}@kky.zcu.cz

Abstract. The paper presents a technique for phonetic spoken term detection in large audio archive. It is designed within the framework of weighted finite-state transducers and utilizes the rather recently developed notion of factor automata, which we have enhanced with a score normalization and a technique for systematic query expansion which allows for phone deletions and substitutions and consequently compensates for frequent pronunciation imperfections and systematic phoneme interchanges occurring during the ASR decoding process. The experiments presented in the paper show that the new WFST-based method outperforms the baseline system both in terms of search performance and speed. Finally, the paper discusses the issues of the proposed techniques that need to be addressed before the application in real-life tasks.

Keywords: spoken term detection, finite state automata.

1 Introduction

Systems for effective searching in vast and rapidly growing audio(visual) archives are becoming one of the most useful applications of automatic speech recognition (ASR) technology nowadays. There are actually several different approaches to this task. One of them for example uses the ASR engine just to convert the speech into plain text and then employs standard information retrieval (IR) techniques designed for text documents. However, the methods presented in this paper address the search problem using the techniques of so-called spoken term detection (STD) which does not care about the somehow abstract "topic" of the documents (as does the traditional IR) but rather attempts to find only the occurrences of query terms. In this way, it is similar to keyword spotting but the difference lies in the fact that in the STD task, the keywords are not known in advance. Therefore one of the main challenges of the STD is the construction of an efficient index — with regard to both the storage requirements and the search speed.

It is evident that not only in comparison with textual "term detection" (which is just a fancy name for a trivial full text search) but also in contrast to more sophisticated textual IR tasks, the STD has to deal with two substantial issues. Firstly, the output of even the state-of-the-art speech recognizers is far from being perfect; the word error

---

* This research was supported by the Ministry of Culture Czech Republic, project No. DF12P01OVV022.

I. Habernal and V. Matoušek (Eds.): TSD 2013, LNAI 8082, pp. 402–409, 2013.

rate (WER) for the spontaneous speech is often around 30% or even higher. Secondly, the ASR system always works with a limited recognition vocabulary and thus the out-of-vocabulary (OOV) words cannot be recognized and consequently found by the STD system.

The first issue can be partially alleviated by using the ASR lattices instead of the one-best hypothesis only. Each path in the ASR lattice (technically, a directed acyclic graph) represents a recognition hypothesis described as a string of recognition units (typically words) and the associated probability of the hypothesis. That way we can search the alternative recognition hypotheses (which might in fact be better transcriptions of the actual utterance than the most probable one) and also utilize the information about their likelihood, which is an important information that is lost when using the plain text transcription only. However, the problem of OOVs cannot be solved at the word level. The common practice for addressing this issue is to use sub-word units (such as phonemes) in the recognition process and construct the index from those smaller units.

Consequently, this work mainly focuses on the methods for effective indexing of the phoneme lattices. It is clear that the processing requirements for phoneme lattices are much higher than for the word ones because the number of plausible alternative hypotheses substantially increases at the phoneme level.

In [1], we have presented an STD system designed for the MALACH audiovisual archive [2]. The system decomposes each phoneme lattice into triplets and stores them in the index implemented as an SQL database. That way the important structural information contained in the original lattice is lost. In this paper, we will focus on the progressive indexation technique presented in [3] that is based on the weighted finite-state transducers (WFSTs). This approach allows to effectively represent the index of lattices in an optimal way based on the idea of factor transducer [4]. The optimal search time is reached by using the WFST optimization algorithms such as determinization and minimization. The resulting index can be queried in the time which is linear with respect to the length of the query and to the number of query results.

Our paper presents a novel application of this technique to phoneme indexing of data from the MALACH project. It introduces a method for length-based posterior probability normalization and also proposes further modification of the search process in order to allow for approximated match of the query and the occurrence. The performance of the described method is compared with the results obtained with our previous (baseline) system.

2   WFST-Based Spoken Term Detection

2.1   Preliminaries

In this section, we introduce the basic terminology for the finite-state automata and strings from [3] that we will use through this paper.

Weighted finite-state transducer T over a set $\mathbb{K}$ is an 8-tuple $T = (\Sigma, \Delta, Q, I, F, E, \lambda, \rho)$ where Σ is the finite input alphabet, Δ is the finite output alphabet, Q is the finite set of states, $I \subseteq Q$ is the set of initial states, $F \subseteq Q$ is the set of final states, E is the finite set of arcs: $E \subseteq Q \times (\Sigma \cup \epsilon) \times (\Delta \cup \epsilon) \times \mathbb{K} \times Q$, $\lambda : I \mapsto \mathbb{K}$ is the initial weight function and $\rho : F \mapsto \mathbb{K}$ is the final weight function.

Weighted finite-state acceptor A over the set $\mathbb{K}$ is an 7-tuple $A = (\Sigma, Q, I, F, E, \lambda, \rho)$ defined in a similar way as a transducer by simply omitting the output alphabet Δ.

An automaton A is *deterministic* if no state has more than one output transition with the same input label. *Minimal* automaton B from the automaton A is the automaton with a minimal number of states that is equivalent to A.

A *path* $\pi = e_1 \ldots e_k$ is an element of E^* with consecutive arcs satisfying the condition $next\_state[e_{i-1}] = previous\_state[e_i]$, $i = 2, \ldots, k$. A symbol sequence $x \in \Sigma^*$ is recognized by the automaton if there exists a path from an initial state to a final state, labeled with x on input side.

If two strings $u, v \in \Sigma^*$ are given, v is a *factor* (substring) of u if $u = xvy$ for some $x, y \in \Sigma^*$.

Factor automaton $F(u)$ of a string u is the minimal deterministic finite-state acceptor recognizing exactly the set of factors of u. Similarly, we denote by $F(A)$ the minimal deterministic acceptor recognizing the set of factors of a finite acceptor A, that is the set of factors of the strings recognized by A.

2.2   Indexation

As was already mentioned, the main distinction between the baseline system described in [1] and the technique employed in this work is that now not only the selected arcs, but the whole lattices are indexed together and the algorithms take into account the probability of the indexed string in the lattice. The following paragraph briefly outlines the principle of the indexing technique (see the original work [3] for the details).

Assume that the spoken archive consists of utterances u_i, $i = 1, \ldots, n$. The corresponding lattices A_i are generated using the ASR system. The lattices A_i are represented as WFSTs over an input alphabet Σ which represents the phoneme set. The indexing algorithm consists of three steps: (1) lattice preprocessing (2) construction of factor automaton and (3) factor automaton optimization.

Preprocessing. The algorithm starts with preprocessing of each input lattice A_i. The overlapping arcs labeled with same phoneme are clustered and merged. Then the low probability arcs are pruned using a fixed threshold. Finally, the arc weights are normalized using the *weight-pushing* algorithm. The result of the preprocessing step are the lattices B_i. To determine the overlapping arcs in A_i, the time alignment of lattice states has to be supplied.

Construction of factor automaton. The use of a factor automaton in lattice indexing task was proposed in [5]. The approach described in [3] differs in the use of $\mathcal{T} \times \mathcal{T} \times \mathcal{T}$ semiring (Cartesian product of three tropical semirings) for the factor automaton construction. Each weight assigned to an arc is represented as a triplet (p, t_s, t_e) where t_s, t_e specify the time interval and p is the posterior probability of a factor x occurring in the utterance u_i in the specified interval.

The factor automaton is constructed from the lattice B_i where the transitions leading to final state are labeled with output symbol i serving as the utterance identifier in the search results. Then for each state q the shortest distance $\alpha[q]$ from the origin state to q and the shortest distance $\beta[q]$ from q to the final state are computed. The factor automaton $F(B_i)$ is constructed by starting with a copy of B_i and by adding two new states r

and s which are marked as a new origin and final state, respectively. For each state q, two new transitions are added. The transitions from r to q are created as ϵ-transitions with weights consisting of triplets $(\alpha[q], t_s, 0)$ denoting the forward probability of factors starting in q and the time offset of the beginnings of these factors. The transitions from q to r are again ϵ-transitions with weights $(\beta[q], 0, t_e)$ representing the backward probability of factors ending in q and the time offset of the endings of these factors.

Finally, the factor automaton U for the whole set of utterances u_i is constructed as an union of partial factor automata $F(B_i)$ (remember that the utterance id is encoded in the output symbols of B_i):

$$F(B) = \bigoplus_{i=1}^{n} F(B_i) \tag{1}$$

Factor automaton optimization. Since $F(B)$ is a non-functional transducer (it can generally assign more output sequences to a single input sequence), its input and output symbols must be first encoded for optimization. That is, the transducer U is converted to a finite state acceptor whose transitions are labeled with symbols denoting pairs *(input symbol, output symbol)* in the original transducer. This acceptor can be directly optimized using ϵ-removal, determinization and minimization algorithms. The resulting acceptor is converted back to the corresponding transducer U which represents the optimal inverted index of the lattice set $A_i, i = 1, \ldots n$.

2.3   Search

The query is again represented as an arbitrary automaton X, e.g. unweighted string or regular expression compiled into such representation. The search algorithm can be described using the WFST operations: Transducer X is composed with U and the result is projected onto output symbols. The result of the projection is an acceptor R. Finally the ϵ-removal algorithm is performed over R. Each successful path π in R represents one occurrence of a query X in the lattice set represented by the factor transducer U. The symbols of π represent the identifiers i of original utterances u_i and the weights correspond with the triplet (-log posterior probability, start time, end time) assuming that the weights are defined over the $\mathcal{T} \times \mathcal{T} \times \mathcal{T}$ semiring.

3   Proposed Modification

Even though our system uses the phonetic representation of the words for searching, it would be very inconvenient for users if they had to type the queries directly in phonemes. Moreover, we often want to combine word-based and phoneme-based search and, most importantly, shield the users from those technical details. Thus we let the user to enter the query in the standard word form and then obtain the phonetic representation either by looking up the word in the lexicon or, in case of OOV words, by using the rule-based transcription system that generates all the pronunciation variants. These are then compiled into WFST (a small example is shown on Fig 1).

But we have observed that the queries generated in this manner yielded high error rate and negligible detection rate due to very low posterior score of returned correct

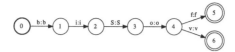

Fig. 1. WFST representing two different pronunciations of the word "Bischof" in Czech

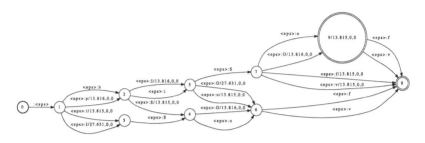

Fig. 2. Final phonetic query produced by a composition of the query from Fig. 1 with a confusion table

results. The problem was that in many cases the correct path returned from the index contained one or more phonemes with very low score, causing the overall score of the whole phone sequence to be low as well. So we have normalized the overall score by dividing it by the number of phonemes belonging to the given path:

$$\mathcal{W}_W(w_1, \ldots, w_n) = \frac{1}{n} \cdot [\log(w_1) + \ldots + \log(w_n)] = \log \sqrt[n]{w_1 \cdots w_n},$$

where w_i is the score of the i-th phoneme and n is the number of phonemes in the found word W. As can be seen from the equation, this normalization in fact corresponds to using the geometric mean of individual phoneme scores.

However, even with such normalization, the algorithm was still finding rather few paths matching the queries in the index. After closer inspection, we have discovered that there are many phones missing in the phoneme lattices due to the sloppy pronunciation of the speakers. So we expanded the query by composing it with the transducer representing a confusion table that contains all possible variants of individual phoneme deletions. The penalty $\alpha = -\log(1 \cdot 10^{-6}) = 13,815$ was determined heuristically and assigned to each such deletion. The search performance has then significantly increased. Further improvement was obtained by allowing the interchange between specific phonemes — such substitutions might appear as a result of the systematic recognition errors of the phoneme ASR decoder. So we supplemented the "confusion" transducer with arcs representing the following interchanges: voiced/unvoiced consonant, short/long vowel, all with the same penalty α as mentioned above. The example of resulting query term transducer that allows those edits (after determinization and minimization) is shown on Fig. 2.

4 Experiments

4.1 Experimental Data

The data used for our experiments were selected from the Czech portion of the USC-SFI archive (http://sfi.usc.edu/) processed within the MALACH project [2]. The whole Czech part contained more than 550 testimonies (almost 1,000 hours of video) of Holocaust survivors. Selected 100 hours from this data was transcribed and used as a training corpus for the ASR system development — in order to facilitate the processing, we have used only 47 hours from this corpora as our experimental data.

Our ASR system, employing state-of-the-art acoustic and language modeling techniques (see [1] for details), processes input utterances in two passes — outputs from the first pass are used for adaptation of the acoustic models and the adapted models are then used for lattice generation. The only distinction from the procedure described in [1] is that also the phoneme lattices using 5-gram phoneme language models were generated for the experiments reported in this paper. Lattices were converted into the standard OpenFST format [6].

4.2 Results

We performed a set of experiments using different combinations of various configuration parameters and tried to find an optimal configuration settings w.r.t. the detection rate (DR), false alarm (FA), and equal error rate (EER) values. Results of the WFST-based STD system are influenced mainly by the depth of lattice pruning. There are actually two levels of pruning — first in the ASR decoder where we can restrict the maximum number of lattice arcs outgoing from each state (let us denote it as I), and then the standard OpenFST pruning method with path weight threshold p, used while preprocessing the lattices before indexation. We performed experiments with combination of both values ranging from 1 to 5. These experiments were done on two phoneme lattices datasets, generated with 0-gram and 5-gram phoneme language model, respectively. Thus the total number of experiments was 50.

We found that 0-gram dataset yields unsatisfying results, so we further experimented with the 5-gram phoneme lattices only. The best values (also with regard to the index size) were observed for $I = 5$ and $p = 5$. Fig. 3 and Table 1 show the results for this setting in comparison with our baseline system [1] — the steeper WFST curve indicates better DR values at individual FA levels. The search with the WFST system is also faster, as shown in the last column of the table.

However, there is an important issue of the WFST indexation approach that still remains to be solved — huge storage requirements of the phonetic index. During our experiments, we have found out that the index size grows exponentially with the size of the indexed data. This is caused by the presence of many low-weight paths in indexed lattices; those paths fall under the detection threshold θ, but still are able to pass through the pruning and remain in the lattices. So our future work will be aimed mainly in this direction.

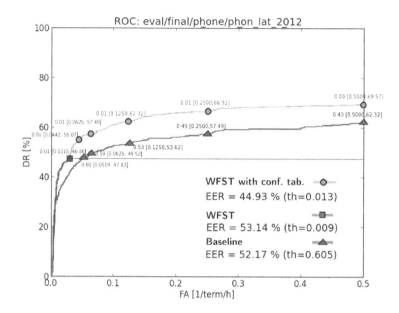

Fig. 3. The ROC curve for the WFST and baseline STD systems

Table 1. Performance characteristics of both systems

	$DR_{0.125}$	$DR_{0.25}$	$DR_{0.5}$ [%]	EER [%]	Avg. search time [s]
Baseline	53.62	57.49	62.32	52.17	2.13
WFST	62.32	66.91	69.57	44.93	1.99

5   Conclusions and Future Work

The paper has presented an innovated approach to phonetic STD using the WFST framework. We have employed the recently developed technique utilizing the principle of factor automata, supplemented it with score normalization and also implemented a technique for systematic query expansion which allows for phone deletions and substitutions and consequently compensates for frequent pronunciation imperfections and systematic phoneme interchanges occurring during the ASR decoding process. However, this expansion technique still leaves some room for improvement, as we have set the deletion and substitution penalties rather heuristically.

Using this enhanced WFST method, we have managed to outperform our baseline system [1] both in terms of search performance and speed. The more complex and structurally rich WFST index however leads to huge index sizes, growing exponentially with the size of the data. So before the application of the system in real tasks we have to deal with this fact and apply some more sophisticated pruning strategies. One of the natural choices would be to remove unuseful parts of index like stop words or in-vocabulary terms (which could be searched in the word-based index). Further choice might include removing automata paths that fall under the detection threshold, as described in Sec. 4.2.

Thinking about the practical applications of the search system brings up yet another problem. When we add a new item to the archive, we first need to process it by the ASR system to produce a corresponding lattice. When using the presented techniques without further modification, adding a new lattice to the index would require the reconstruction of the whole index and although it is an off-line process, it becomes noneffective due to large computational requirements. So we would like to devise a technique for dividing the archive into several smaller subindexes, such that adding of a new item to the archive would require just the rearrangement of a particular subpart. The whole index then would be represented for example as an union of the partial index transducers via initial ϵ-transitions. It is clear that index constructed in such a way will not longer be a deterministic machine, so we expect some search time increase. As a compensation of this disadvantage, we could be able to provide searching as a parallel process, distributed to several nodes of a distributed cluster.

References

1. Psutka, J., Švec, J., Psutka, J., Vaněk, J., Pražák, A., Šmídl, L., Ircing, P.: System for fast lexical and phonetic spoken term detection in a Czech cultural heritage archive. EURASIP Journal on Audio, Speech, and Music Processing 2011(1), 10 (2011)
2. Byrne, W., Doermann, D., Franz, M., Gustman, S., Hajič, J., Oard, D., Picheny, M., Psutka, J., Ramabhadran, B., Soergel, D., Ward, T., Zhu, W.J.: Automatic Recognition of Spontaneous Speech for Access to Multilingual Oral History Archives. IEEE Transactions on Speech and Audio Processing 12(4), 420–435 (2004)
3. Can, D., Saraclar, M.: Lattice indexing for spoken term detection. IEEE Transactions on Audio, Speech, and Language Processing 19(8), 2338–2347 (2011)
4. Mohri, M., Moreno, P., Weinstein, E.: Factor automata of automata and applications. In: Holub, J., Žďárek, J. (eds.) CIAA 2007. LNCS, vol. 4783, pp. 168–179. Springer, Heidelberg (2007)
5. Allauzen, C., Mohri, M., Saraclar, M.: General indexation of weighted automata - application to spoken utterance retrieval. In: Ramabhadran, B., Douglas, O. (eds.) HLT-NAACL 2004 Workshop: Interdisciplinary Approaches to Speech Indexing and Retrieval, pp. 33–40. Association for Computational Linguistics, Boston (2004)
6. Allauzen, C., Riley, M.D., Schalkwyk, J., Skut, W., Mohri, M.: OpenFst: A general and efficient weighted finite-state transducer library. In: Holub, J., Žďárek, J. (eds.) CIAA 2007. LNCS, vol. 4783, pp. 11–23. Springer, Heidelberg (2007)

Ranking Machine Translation Systems via Post-editing

Wilker Aziz[1], Ruslan Mitkov[1], and Lucia Specia[2]

[1] Research Group in Computational Linguistics
University of Wolverhampton, UK
{W.Aziz,R.Mitkov}@wlv.ac.uk
[2] Department of Computer Science
University of Sheffield, UK
l.specia@sheffield.ac.uk

Abstract. In this paper we investigate ways in which information from the post-editing of machine translations can be used to rank translation systems for quality. In addition to the commonly used edit distance between the raw translation and its edited version, we consider post-editing time and keystroke logging, since these can account not only for technical effort, but also cognitive effort. In this system ranking scenario, post-editing poses some important challenges: i) multiple post-editors are required since having the same annotator fixing alternative translations of a given input segment can bias their post-editing; ii) achieving high enough inter-annotator agreement requires extensive training, which is not always feasible; iii) there exists a natural variation among post-editors, particularly w.r.t. editing time and keystrokes, which makes their measurements less directly comparable. Our experiments involve untrained human annotators, but we propose ways to normalise their post-editing effort indicators to make them comparable. We test these methods using a standard dataset from a machine translation evaluation campaign and show that they yield reliable rankings of systems.

Keywords: machine translation evaluation.

1 Introduction

Machine Translation (MT) evaluation is the problem of assessing the quality of machine translated text. It is used both to measure the quality of an individual system and to rank a set of systems with respect to the quality of the translations they produce. MT evaluation can be automatic, semi-automatic and manual. In automatic evaluation, MT output is automatically and systematically compared to an independently obtained set of human-produced gold-standard (reference) translations. Popular metrics are based on shallow comparisons such as edit distance (e.g. TER), exact matching of n-grams (e.g. BLEU), or matching of synonyms and short paraphrases (e.g. METEOR, TERp). Overall, these metrics compute a shallow notion of overlap between the MT output and the available reference translation(s). Due to their limited level of linguistic analysis, they only cover a few aspects of quality.

Automatic metrics may easily penalise a valid translation because it has been phrased differently from the available references. Increasing the number of references per input segment is known to be helpful in that it increases the chance of finding a close match

I. Habernal and V. Matousek (Eds.): TSD 2013, LNAI 8082, pp. 410–418, 2013.

to the MT output. However, the space of possible translations is very large [1], and except for very small scale datasets, it is impractical to prepare gold-standard sets to account for such variability. To overcome this problem, in a typical scenario of semi-automatic evaluation, human annotators are asked to post-edit the output of MT systems producing a set of targeted reference translations. Post-editing guidelines may be task dependent, but generally it is expected that meaning and grammar issues (and less often, style issues) in the original MT output are fixed. In settings where the goal is to perform cheap and fast evaluation of MT quality, a common strategy is that of minimally post-editing the MT output, where only major issues are fixed. Once the targeted references are obtained, evaluation can be performed using standard automatic metrics. A popular metric is the Human-mediated Translation Edit Rate([2]), or HTER. HTER estimates the minimum number of edit operations necessary to fix the MT output. In this setting, MT quality is quantified by the amount of post-editing effort necessary to fix it. HTER was used as the official metric to compare MT systems as part of the recent DARPA GALE program [3]. Provided with extensive and comprehensive guidelines and training, professional translators involved in the evaluation were reported to have achieved high inter-annotator agreement, making the HTER assessment reliable. However, most evaluation settings are not able to afford the preparation, training, time and costs required to achieve similar results.

Lastly, manual evaluation is usually done via human scoring/error counts or ranking, that is, annotators are given guidelines on how to score or rank machine translated text. Manual evaluation is known to suffer from low inter-annotator agreement due to the subjectivity of the task and it is also cognitively demanding. It typically involves mentally constructing a reference translation which is then used to penalise/reward the MT at a sub-sentence level. Particularly for MT system ranking, because of the high cost of this type of evaluation, it is often not possible to compare all systems against all other systems for the complete dataset. Instead, samples are considered for partial rankings and global rankings are extrapolated from these using heuristics. For example, the evaluation campaigns organised yearly by the Workshop on Machine Translation (WMT) [4] use human ranking for both MT evaluation and metrics evaluation (i.e., assessment of automatic metrics of MT evaluation). In addition, WMT rely on volunteers or mechanical turkers to produce annotations, who are not professionals and would not be willing to undergo many hours of training activities. Therefore, the reliability of the global rankings provided have been frequently questioned, e.g., [5,6].

In this paper we focus on post-editing as a more natural and objective way of producing gold-standard annotations for MT evaluation. However, our goal is to move away from standard metrics like HTER. A limitation of this and other edit distance metrics is that they cannot fully capture the effort resulting from post-editing. Certain operations can be more difficult than others, based not only on the type of edit (deletion, insertion, substitution), but also on the words being edited. We use information about the post-editing process to attempt to quantify this effort: post-editing time and keystrokes (counts, groups, etc.). Previous work has proposed post-editing time as a practical way of capturing both technical and cognitive aspects of post-editing effort [7]. However, so far post-editing time has only been used as a measure of translation quality in very controlled environments [8], mostly comparing post-editing against translation from scratch, based on annotations from the same translator performing both tasks. A challenge that arises from the use of post-editing for the ranking of MT systems is the fact that multiple annotators are necessary, since having the same annotator fixing alternative translations for the same source text can bias their post-editing. For example, an annotator may tend to produce a translation that is similar to the one previously post-edited for another system, possibly making fewer edits than if the source sentence has been seen for the first time. This becomes particularly a problem when post-editing time and keystrokes are measured, since different annotators work at different paces and may use different strategies to perform the same post-editing. While on average certain post-editors are simply consistently slower/faster than others, other factors impact post-editing speed and strategies for particular segments (e.g. the length of the segment), making normalisations across annotators far from trivial.

We investigate the use of post-editing to rank MT systems using very simple guidelines and multiple un-trained, non-professional post-editors, followed by strategies to normalise their post-editing effort indicators so that they become comparable across editors. Using a subset of the WMT12's dataset of English-Spanish translations, we show that metrics based on post-editing time and edit distance are subject to large variability across editors and should therefore be normalised before used for comparison.

This paper is organized as follows: Sect. 2 describes the normalization techniques, Sect. 3 contains experimental results and Sect. 4 summarises our findings.

2   Method

The difficulty in using post-editing effort indicators to rank MT systems arises from the "human in the loop" aspect of post-editing. People work at different paces, they have different skills, and they react differently to different type of errors. This variation becomes more evident when non-professional annotation is used.

Consider a set of S input sentences and their alternative translations produced by M systems. In this dataset a task is a pair made of an input sentence and one possible translation produced by a specific system. Each of the $S \times M$ tasks is identified by $t = (i, m)$ where i refers to the input sentence, and m refers to the system that produced the translation. Tasks are assigned to a set of A annotators respecting the constraint that annotators should never be presented with a task that contains an input sentence that they have already seen (let us call this constraint **one-only**). This means that to gather annotation for the whole dataset it is necessary to have as many annotators as there are systems in the comparison (i.e. $A \geq M$). In general terms, to have k annotations for each task, it is necessary to have $k \times M$ annotators. However if it is not possible to have as many annotators, one can randomly skip tasks (this relaxation is similar to observing partial ordering in human ranking).

Assume that annotators post-edit their tasks following simple guidelines producing annotations that include post-editing effort indicators such as the targeted reference, post-editing time, keystrokes count and edit distance. The challenge at hand consists in ranking translations of the same input sentence, therefore post-edited by different annotators, using the effort indicators available without assuming they are directly comparable. For instance, one could be interested in using post-editing time (or the HTER score) as observed from different annotators to rank alternative translations of the same input sentence according to how time consuming they are (or how many edit operations they require), but we know these are not comparable across annotators.

2.1 Mean Normalisation

Consider as variables all measurements from the post-editing process, e.g., post-editing time. Mean normalisation is a fairly standard technique broadly used in machine learning to make a variable assume mean value close to 0 and variance close to 1. This technique adjusts the values of a variable to $(x - \mu)/\sigma$, where x is the original value of the variable and μ and σ are the variable's mean and standard deviation, as observed in a large dataset. This normalisation is done independently for each annotator as well as independently for each variable. We then proceed by sorting the tasks in terms of the normalised feature.

2.2 Regression

Mean normalisation is simply a mathematical trick to adjust a variable as to make it have standard parameters. It oversimplifies the complex problem of comparing indicators from different annotators. Furthermore, it can only be used with one variable at a time, preventing the combination of different types of effort indicators.

To cope with these limitations, we propose to learn how to compare effort indicators across annotators based on a training set in which we can observe their variance in performing the same tasks. In this training set, each input sentence is arbitrarily assigned one - and only one - translation, and all annotators post-edit it. Figure 1 exemplifies how tasks are assigned differently in the two cases. Treating these indicators as features in machine learning scenario is quite appealing, but there is no gold-standard annotation for post-editing effort.[1]

This problem can be seen as the problem of learning a latent scale of post-editing effort onto which all the annotators' effort indicators can be mapped. If we had at our disposal an infinite number of annotators performing the same tasks, we would have access to each task's expected post-editing time (the same for HTER and keystrokes). It would not be absurd to claim that the expected post-editing time can be seen as a gold-standard annotation onto which the features of an individual annotator could be mapped. In practice, because of annotation costs, no more than a few annotators can be used. Therefore we use the sample mean in the training set instead of the unknown expectation. Section 3 presents empirical results to support that such approximation is reasonable.

In a nutshell, we use regression techniques to learn from the training set how to map input features in the form of post-editing effort indicators onto the approximated gold-standard. We learn one such a function for each annotator. At test time, we use them to convert effort indicators of each annotator individually into the gold-standard scale

---

[1] Coming up with gold-standard annotation for post-editing effort is a non-trivial task. In Sect. 3 we show that an attempt using human scoring on post-editing effort resulted in a dataset with extremely low inter-annotator agreement.

Input System		A1	A2	A3	Effort	Approximation
			Training			
50	4	$\mathbf{F}^1_{(50,4)}$	$\mathbf{F}^2_{(50,4)}$	$\mathbf{F}^3_{(50,4)}$	?	$\mu^f_{(50,4)}$
51	5	$\mathbf{F}^1_{(51,5)}$	$\mathbf{F}^2_{(51,5)}$	$\mathbf{F}^3_{(51,5)}$	?	$\mu^f_{(51,5)}$
52	6	$\mathbf{F}^1_{(52,6)}$	$\mathbf{F}^2_{(52,6)}$	$\mathbf{F}^3_{(52,6)}$	?	$\mu^f_{(52,6)}$
			Test			
	4	$\mathbf{F}^1_{(100,4)}$			?	$y^1_{(100,4)}$
100	5		$\mathbf{F}^2_{(100,5)}$		?	$y^2_{(100,5)}$
	6			$\mathbf{F}^3_{(100,6)}$	?	$y^3_{(100,6)}$

Fig. 1. Example of how tasks are assigned differently for training and test. $\mathbf{F}^a_{(i,m)}$ denotes the vector of post-editing indicators from annotator a for task (i, m). At training time annotators perform the same tasks, the gold-standard effort label is unknown and approximated as $\mu^f_{(i,m)}$, i.e., the mean value of feature f observed for task (i, m). At test time each annotator is assigned an alternative translation of the same input. We use $y^a_{(i,m)}$, the predicted value of feature f given annotator's a effort indicators for task (i, m), to rank the alternative translations.

where they can be compared (column "Approximation" in Fig. 1). We then rank tasks w.r.t. the predicted mean.

3 Experiments

Our task is to rank alternative translations of an input sentence on the basis of post-editing effort indicators. This section presents the collection of the dataset and experiments with different normalisation and regression techniques.

3.1 Data Collection

We selected a subset of the WMT12's English-Spanish translation evaluation test set. For the training set we selected 200 input sentences. Each of these sentences was then assigned one translation by a randomly selected system (resulting 200 tasks). The test set, on the other hand, is made of 100 input sentences along with their alternative translations by 10 systems[2] (resulting 1,000 tasks). We had 10 volunteer English-Spanish bilingual speakers (native Spanish speakers) performing the post-editing of these machine translations. In summary, there were i) 200 tasks in the training set and all annotators performed each one of them, and ii) 1,000 tasks in the test set shared between 10 annotators observing the **one-only** constraint.

Post-editing was performed using the post-editing tool PET [9], which collects information such as post-editing time and keystrokes at the segment-level in the background, while post-editing is done. The participants were provided with simple guidelines (perform minimum post-editing) and a 10-minute video demonstrating how the post-editing tool should be used. After they post-edited each translation, they were asked to assign a score from 1 to 4 to quantify the effort spent on post-editing.[3] To illustrate how

---

[2] Note that there were 11 systems participating in this task, we left **UK** out.

[3] 1 - complete re-translation needed; 2 - a lot of post-editing needed, but quicker than re-translation; 3 - a little post-editing needed; and 4 - no modification needed.

Table 1. System-level rank correlation

Target	AI	MN	Regression 1	Regression 4
time	0.3696	**0.7333****	**0.6969** (R)	**0.5757** (B)
keystrokes	0.4787	0.6121*	0.5878 (R)	0.5636 (R)
HTER	**0.5393**	0.3939	0.4181 (S)	0.5636* (R)

post-editing information can be more reliable than human scoring, we computed Co-ehn's κ inter-annotator agreement [10] for these scores on post-editing effort. We observed very low values of κ, ranging from 0.12 to 0.44 with average 0.269 ± 0.082.

3.2 Ranking with Post-Editing

The most obvious baseline strategy for ranking translations on the basis of post-editing annotation is to chose an effort indicator, assume it is comparable across annotators, and rank translations according to its observed values. We refer to this baseline as **AI** (as in "as is"). The second strategy is mean normalisation, referred to as **MN** (Sect. 2.1), where we extracted the parameters μ and σ from the training set for each annotator and for each of the target features. Finally, we tested regression algorithms (**R** - regularised linear regression, **B** - Bayesian regression, and **S** - SVR with *rbf* kernel)[4] using different feature sets to predict the expected post-editing time.

WMT assigns an overall quality score to each system by drawing pairwise comparisons from the human rankings and computing the proportion $\text{score}(s) = \frac{\text{win}(s)}{\text{win}(s)+\text{loss}(s)}$ [5], that is, how many times a system wins a pairwise comparison out of the comparisons it won or lost. We use the same equation, however rather than using human rankings to decide who wins or loses a pairwise comparison, we use the value of the target feature: post-editing time in **AI**, normalised post-edited time in **MN**, or the expected post-editing time as predicted from effort indicators in **R**, **B** and **S**.

For each alternative translation of an input segment, we observe post-editing features from different annotators. Using one of the aforementioned methods we predict their expected values for the target feature, making them comparable. In this space of comparable feature values we perform ranking of alternative translations. We do that for three different target features (time, keystrokes and HTER) and report the results in terms of rank correlation to WMT's official ranking. Table 1 shows the system-level rank correlation in terms of Spearman's ρ coefficient. We also investigate the agreement between our strategies and WMT's human rankings w.r.t the pairwise comparisons themselves. Table 2 shows the segment-level rank correlation in term of Kendall's τ coefficient.

In *Regression 1*, the annotator's time is used alone to predict the task's expected time (same for keystrokes and HTER). In *Regression 4*, the length of the input (in number of tokens), the annotator's time, keystrokes and HTER score are used as features to predict the task's expected value for time, keystrokes or HTER. We boldface the most successful target in each column and star the best strategy in each row. A double star shows the best among all.

---

[4] As available in the scikit learn toolkit (http://scikit-learn.org/), with hyper-parameters tuned using the toolkit's randomised search in 5-fold cross-validation.

Table 2. Segment-level rank correlation

Target	AI	MN	Regression 1	Regression 4
time	0.0975	0.1555	0.2054 (R)	0.2451* (S)
keystrokes	**0.1941**	0.2189	**0.3065*** (S)	0.2870 (S)
HTER	0.1794	**0.2637**	0.2693 (R)	**0.3559**** (S)

First of all, both tables show that using the post-editing indicators "as is" is the worst strategy. Note that even HTER is improved by the proposed normalisations: segment level rank correlation jumps from 0.17 (AI) to 0.26 without additional features (*Regression 1*) and to 0.35 with additional features (*Regression 4*). Moreover the segment-level correlation for post-editing time improves from 0.09 to 0.245. Table 2 shows that post-editing time is less efficient than HTER, possibly because HTER ranges over a smaller interval than post-editing time, making the regression problem easier (more predictable scores).

It is interesting to notice the mismatch between segment- and system-level correlation w.r.t. the best performing strategies. The best performing system-level metric is the mean normalised time, while HTER using SVR performs the best at the segment-level. Mean normalisation relies on accumulated statistics over the entire training set, perhaps this explains why it is better at reflecting the overall tendency of a system to win comparisons against others. On the other hand, for the harder and finer-grained problem of ranking alternative translations of an input segment, the regression techniques are superior. We note that regression task attempts to predict the expected value of the target feature at the segment-level.

The last row in Tab. 2 shows how the performance of HTER improves with the addition of the other effort indicators as features. We further investigated this direction by removing features from the starred system and noticed that the correlation drops to 0.3458 (no time) and to 0.3213 (neither time, nor keystrokes). These observations show another interesting outcome of using machine learning to normalise the post-editing annotation. We can benefit from effort indicators of different nature despite their low inter-annotator agreement. To further investigate this observation we used the human score on post-editing effort (from 1-4) as one more feature in regression. Segment-level time and HTER with SVR improved to 0.2754 and 0.3633, respectively, in the presence of that feature, even though in isolation the feature had shown low agreement (Sect. 3.1).

Table 3 shows the overall ranking of systems (and their rank correlation ρ) according to the WMT (based on human rankings), and the best performing strategy for each target feature (starred systems in Tab. 1). There are some interesting differences between the overall rankings obtained via post-editing and the one obtained via human ranking. Note how RBMT-3 is consistently ranked worse by all methods based on post-editing. On the other hand, JHU (officially at the bottom of the ranking) and UEDIN (officially in the middle) both gained a few positions. These results indicate that using post-editing to gather annotation for ranking MT systems could be a promising direction. The mismatches between our ranking and the official one should not be seen as flaws, but rather as the result of a different, perhaps more objective way of addressing the problem as compared to the one used currently, where humans explicatively perform the ranking.

Table 3. Overall ranking using different annotation

Human ranking	Time	HTER	Keystrokes
0.65: Online-B	0.66: Online-B	0.67: Online-B	0.63: Online-B
0.58: RBMT-3	0.54: Online-A	0.56: UEDIN	0.55: Online-A
0.56: Online-A	0.53: UEDIN	0.55: Online-A	0.53: UEDIN
0.55: PROMT	0.51: PROMT	0.53: UPC	0.52: RBMT-1
0.52: UPC	0.50: UPC	0.49: PROMT	0.51: UPC
0.52: UEDIN	0.48: RBMT-3	0.48: RBMT-1	0.49: PROMT
0.46: RBMT-4	0.46: Online-C	0.46: JHU	0.45: RBMT-3
0.45: RBMT-1	0.45: JHU	0.42: RBMT-3	0.45: JHU
0.43: ONLINE-C	0.43: RBMT-1	0.41: RBMT-4	0.43: RBMT-4
0.36: JHU	0.43: RBMT-4	0.41: Online-C	0.42: Online-C

4   Conclusions

In this paper we have argued for using post-editing as a more natural and objective way to produce gold-standard annotation for ranking MT systems for quality. We have proposed normalisation techniques to cope with important challenges from using post-editing for this task, such as low inter-annotator agreement. Our technique learns how to compare effort indicators produced by different annotators. By making post-editing effort indicators comparable, we were able to rank translations from different MT systems without having to directly collect a dataset of explicit human rankings. Extrapolating system rankings from a more natural annotation relieves people from the burden of having to do it themselves – which seems to be very cognitively demanding. Furthermore it allows for different notions of effort to be used altogether, alleviating low inter-annotator agreement issues and strengthening our normalisation techniques. Finally, the data required for our approach, i.e. post-edited translations with implicitly collected effort indicators, is a by-product of a process that is becoming increasingly popular, particularly in the translation industry. Therefore, this can also be considered a more cost-effective way of collecting data.

Future directions for this research include using latent variable models to address a task's expected effort. Related work on post-editing as a way to capture cognitive effort suggests that linguistic features of the input text play an important role at classifying edit operations according to their complexity. Adding such features to our models should increase their prediction power, leading to better normalisation.

References

1. Dreyer, M., Marcu, D.: HyTER: Meaning-Equivalent Semantics for Translation Evaluation. In: Proceedings of the 2012 Conference of the North American Chapter of the Association for Computational Linguistics: Human Language Technologies, pp. 162–171. Association for Computational Linguistics, Montréal (2012)
2. Snover, M., Dorr, B., Schwartz, R., Micciulla, L., Makhoul, J.: A Study of Translation Edit Rate with Targeted Human Annotation. In: Proceedings of the 7th Conference of the Association for MT in the Americas, Cambridge, Massachusetts, pp. 223–231 (2006)

3. Olive, J., Christianson, C., McCary, J.: Handbook of Natural Language Processing and Machine Translation: DARPA Global Autonomous Language Exploitation. Springer (2011)
4. Callison-Burch, C., Koehn, P., Monz, C., Post, M., Soricut, R., Specia, L.: Findings of the 2012 Workshop on Statistical Machine Translation. In: Proceedings of the 7th WMT, Montréal, pp. 10–51 (2012)
5. Bojar, O., Ercegovčević, M., Popel, M., Zaidan, O.: A Grain of Salt for the WMT Manual Evaluation. In: Proceedings of the 6th WMT, Edinburgh, pp. 1–11 (2011)
6. Lopez, A.: Putting human assessments of machine translation systems in order. In: Proceedings of the 7th WMT, Montréal, pp. 1–9 (2012)
7. Koponen, M., Aziz, W., Ramos, L., Specia, L.: Post-editing time as a measure of cognitive effort. In: Proceedings of the AMTA 2012 Workshop on Post-editing Technology and Practice, San Diego (2012)
8. Plitt, M., Masselot, F.: A Productivity Test of Statistical Machine Translation Post-Editing in a Typical Localisation Context. The Prague Bulletin of Mathematical Linguistics 93, 7–16 (2010)
9. Aziz, W., de Sousa, S.C.M., Specia, L.: PET: A tool for post-editing and assessing machine translation. In: Proceedings of the 8th Conference on Language Resources and Evaluation, Istanbul (2012)
10. Cohen, J.: A coefficient of agreement for nominal scales. Educational and Psychological Measurement 20, 37–46 (1960)

Recursive Part-of-Speech Tagging Using Word Structures

Samuel W.K. Chan and Mickey W.C. Chong

Dept. of Decision Sciences
The Chinese University of Hong Kong
{swkchan,mickey_chong}@cuhk.edu.hk

Abstract. This research takes advantage of word structures and produces a good estimate of part-of-speech tags of Chinese compound words before they are fed into a tagger. The approach relies on a set of features from Chinese morphemes as well as a set of collocation markers which provide hints on the syntactic categories of compound words. A recursive inferential mechanism is devised to alleviate the riffle effect from changes made at its neighbors during tagging. The approach is justified with a compound words database with more than 53,500 words. Experimental results with 500,000 words show the approach outperforms its counterparts.

Keywords: Part-of-speech tagging, Tree-based classifier, Chinese morphemes.

1 Introduction

Using lexical resources in all domains of text processing, ranging from word segmentation, part-of-speech (POS) tagging, sentence parsing, sense tagging, speech recognition and text classification, have reached a common consensus. However, even large lexical databases, such as WordNet [3], do not include all the words encountered in broad-coverage text applications. The quality of these resources depends certainly, to a large degree, on the considerable efforts of lexicographers, who must keep pace with both language change and knowledge development in all relevant domains. In Chinese, the situation is even more taxing, new words can be simply constructed by the concatenation of morphemes, and there is no delimiter between words. As a result, the number of out-of-vocabulary (OOV) words in Chinese is huge. Even narrow coverage of OOV words produces explosive ambiguity in the text processing. At the same time, a completely unsupervised and refined POS tagging is impossible without any help from lexicographers. In this research, we have designed and implemented a recursive means of predicting the POS tags of Chinese words based on two important features: *word structure* and *word sequence in raw text*. It is based on an observation that a non-native reader, who understands the right-headedness of nouns and their relevant word structures, knows that 綠葉 /lv4 ye4/(green leaf: *leaf*) is some kind of leaf, even though the reader never comes across the word. Given more than half a million of words could be found in most lexicons, there are at most 6,000 morphemes that are commonly observed in modern Chinese. It is imperative to devise an objective means from the morphemes

I. Habernal and V. Matousek (Eds.): TSD 2013, LNAI 8082, pp. 419–425, 2013.

so as to make a good approximation of POS tags in face of the numerous OOV words in Chinese. In this research, a tree-based classifier is devised and applied recursively to predict the POS tags of the word using its orthographic form and its contextual neighbors without becoming trapped in a subjective linguistic quagmire. The paper is organized as follows. In Section 2, we first provide a review of the related work. We then describe, in Section 3, a technique in predicting the POS tag of a word. The technique relies on a supervised ensemble machine learning technique which makes use of a set of linguistic features for the predictions. It shortlists the potential POS tags by imposing necessary, even not sufficient, constraints on their features. At the same time, as in other tagging problems which heavily rely on a fixed window size, the POS tag of the target words will be affected by the riffle effect from changes made at its neighbors. A recursive inferential mechanism is devised until the prediction is steady. The detailed discussion of the recursive inferential mechanism will also be delineated in this section. In order to demonstrate the capability of the technique, a POS engine is devised to tag a test corpus used in an open evaluation. The detailed results are given in Section 4, followed by a conclusion.

2   Related Work

One of the major strategies in tagging out-of-vocabulary (OOV) words is based on Freges Principle of Compositionality, which states the sense of a complex can be compounded out of the senses of the constituents [4]. The meaning of a complex, such as an OOV word, can be identified by combining or concatenating the meanings of the morphemes that make up the word. Tseng & Chen [10] make use of the k-NN classifier to devise a morphological analyzer that can segment a word into a sequence of morphemes. The analyzer can also predict the morpho-syntactic relationships between morphemes and is based on the assumption that the morpho-syntactic relation of an OOV word reflects its sense. Ng & Low [8] suggest that a character-based approach is better than a word-based approach for POS tagging in Chinese, simply because Chinese morphemes have well defined meanings. Gao et al. [5] use a support vector machine (SVM) to estimate the likelihood that two adjacent morphemes will form a new word. Although they do not target POS tagging, they find that their SVM classifier fits well with OOV identification. Chung & Chen [1] analyze the morpho-syntactic behaviors of about 4,025 morphemes and classify them into 4 semantic types. They also propose constraint-based resolutions and a set of composition rules to predict the POS tags. Inspired by the work above, we take one step further to propose and implement a mechanism which is based on a distributional hypothesis that words that occur within similar neighbors are semantically similar [2,6]. The detailed discussion of the mechanism is as follows.

3   System Architecture of the POS Engine

The basic idea of the engine in predicting the POS tags is based on two important types of features: morpheme properties and their word neighbors in raw text. The system architecture of the engine is shown in Figure 1.

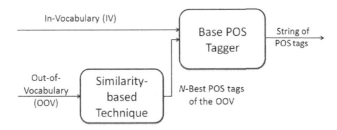

Fig. 1. System architecture of the POS Engine

Whenever there is an OOV word in the input sentence, the word is first subject to a similarity-based technique to unveil its potential POS tags before it is further processed in a base POS tagger. The fundamental rationale of our similarity-based technique is that whenever we have zero evidence for a higher-order, we "back off" to a lower-order. We approximate the POS of the OOV words by their constituent morphemes information as shown in Table 1.

Table 1. Categories of features used in the similarity-based technique

	Feature Template
(a)	Morpheme features, $M_n(n = 0, 1)$
(b)	Radical features, $R_n(n = 0, 1)$
(c)	POS of the morphemes, $MPOS_n(n = 0, 1)$
(d)	Phonetic components
(e)	Collocation marker features

In the feature template, M, R, and $MPOS$ refer to the morpheme, the radical of the morpheme, and its morpheme POS respectively. We denote $n = 0$ for the left morpheme, 1 for the right one. In addition, collocation marker features are used to uncover the co-occurrence between the word W and some key linguistic clues in a corpus under the distributional hypothesis. The major categories of the collocation markers include (i) tense; (ii) function words; (iii) comparable; (iv) negative adverb; (v) complement. These collocation markers reveal the syntactic context of the words and provide the useful hints on their POS tags. On the other hand, the base POS tagger in the engine, as shown in Figure 1, is used to predict the POS tag of each word, both IV and OOV, from word sequence S. The shortlisted POS tag of the OOV words from the similarity-based technique will be further testified with the presence of their neighbors. A set of features is elicited from the training data. That includes the additional POS tags of its preceding and subsequent words in the training data. The assumption is that the POS tag of the current word in the sequence S is not independent but rather mutually related. One can model this dependence implicitly by including information about the preceding and subsequent words. In this base POS tagger, we consider the following feature set.

– Word structure features as shown in (a)-(d) of Table 1;
– POS_{i-1} and POS_{i-2}, POS tags of two preceding words, W_{i-1} and W_{i-2};

- POS_{i+1} and POS_{i+2}, POS tags of two subsequent words, W_{i+1} and W_{i+2};
- Relative position of the word W_i found in S;
- Collocation of three pairs of POS tags, i.e., (POS_{i-1}, POS_{i-2}), (POS_{i+1}, POS_{i+2}), (POS_{i-1}, POS_{i+1});
- Two major information-theoretic functions, mutual information and the likelihood ratio, in quantifying the collocation of the POS tags;
- Learning target is the POS tag of the current word W_i.

To train the classifier to learn the POS tag of the current word W_i, we provide the POS tags of its neighbors in the word sequence S during the training. The relative position of the word W_i in the sequence S is calculated by dividing the absolute position in the sequence by its length. Two kinds of information-theoretic functions, namely, pointwise mutual information (PMI) and the likelihood ratio (LR), are applied in quantifying the collocation of tags. They both quantify the collocation disparity between the POS tags around the target word. The training of the base POS tagger proceeds as follows. Given a word sequence S, the above feature set is computed with respect to each learning target, thereby producing an n-tuple depicting the constraints laid down from its word structure as well as from its neighbors. All the tuples are then presented to a tree-based learning. As the result, a tree-based classifier is produced from the tuples, which becomes the major component in our engine. During the prediction phase, this process is repeated as in the training, producing an n-tuple for each unknown POS tag. The n-tuple is then subject to the verdict of the tree-based classifier. However, one major problem in the prediction is the noise given rise from the feature set. The accuracy of the current prediction relies completely on the n-tuple that involves the POS tags of its preceding and subsequent words. Obviously, the correctness of the n-tuple is in doubt since the tuple also relies in turn on the current POS tag. That is, the adjacent POS tags will surely be affected if the current POS tag is being modified during the sequential POS tagging process. In other words, due to this cyclic dependence between the current POS tag with the POS tags of its neighbors, the tree-based classifier receives a noisy set of input features in its current prediction. This situation is remedied in two phases in our engine. In the first phase, a default POS is first assigned to the word. The default POS is the tag which is most frequently found if it is an IV word, or it is suggested by the similarity-based technique for an OOV word. Although this default POS is clearly far from perfect, it provides a good approximation, associated with an uncertainty, for the initial guess. Each prediction is associated with a certainty factor (CF). The factor indicates the confidence level of the prediction.

In the second phase, instead of predicting the POS tags consecutively from the start to the end of the sequence S, our prediction begins from the POS tag with least certainty, in the hope that its adjacent words with higher certainty, will shed some light on the current prediction. In other words, all POS tags in the sequence will be evaluated, from the highly unresolved to the most promising, in one complete cycle. The intertwined cyclic dependence will be eased by allowing the whole prediction mechanism to repeat iteratively for several cycles until the final POS sequence becomes steady. A POS sequence is defined as steady if there is no further, or little, modification of POS tags during the prediction cycle. Similarly, the certainty factor of POS tags predicted by the classifier in the i-th cycle defines the precedence order of the predictions in the

Table 2. Recursive inferential mechanism in the POS engine

COMPUTE the certainty factor for all POS tags $CF(POS_j)$ for $j = 1 \ldots k$ in the word sequence S as described in the first phase, where k is the number of words in S.
WHILE the prediction is not steady
    NORMALIZE all of the certainty factors $CF(POS_j)$ for $j = 1 \ldots k$
    PUSH the sorted POS_j into a stack with the smallest $CF(POS_j)$ at the top
    WHILE the stack is not empty
        POP POS_j from the stack
        READ the features set of its neighbors W_{j-2}, W_{j-1}, W_{j+1}, and W_{j+2}
        GENERATE the POS features for the word W_j
        ACTIVATE the tree-based classifier for the prediction
        FOR each fired rule in the classifier
            COMPUTE the modified Laplace estimate L for each POS tag n
            $L_n \leftarrow L_n \times \max[CF(POS_l)]$ for all adjacent words
                where $l = j - 2, j - 1, j + 1, j + 2$
            $v_n \leftarrow v_n + L_n$, where v_n is the vote of the POS tag n
        ENDFOR
        RETRIEVE the POS tag with the largest average vote
        NORMALIZE all of the certainty factors $CF(POS_j)$ for $j = 1 \ldots k$
    ENDWHILE
ENDWHILE

$(i + 1)$-th cycle. Table 2 describes the main rationale behind the POS engine in more details.

4 Experiments and Results

We train and test the above feature templates in a tree-based supervised machine learning model [9]. The feature spaces partition is fully described by a single tree. Usually, if the training data exhibit regular patterns and are not random, a classifier will be constructed after training. In the preparation of linguistic data as shown in Table 1, we take advantage of a Chinese compound words database which is developed at Peking University (PKU). The database originally contains more than 53,000 Chinese compound words with their word POS tags, morpheme POS tags as well as pinyin [7]. It involves 21 different types of word POS tags and 23 types of morpheme POS tags. In addition, more than 60 collocation markers are extracted from a Chinese corpus that is segmented by a latest developed Chinese word segmenter[1]. As a result, more than 80 different features, as shown in Table 1, are then subject to the training in the similarity-based technique. At the same time, while it is unreasonable to assume the POS of the morphemes for every words in a real corpus, we take a further approximation by multiplying the conditional probability of the POS s, given the morpheme m at position l, to the certainty factor CF_i^j of the rules r_i generated from the classifiers c_j learned under the tree-based model.

---

[1] http://HanMosaic.baf.cuhk.edu.hk

$$CF_i^j = \begin{cases} CF_i^j \times \prod_{k=0,1} P(s_k|m_k, l_k) & \text{if } s_k \text{ is in } r_i \\ CF_i^j & \text{otherwise} \end{cases} \quad (1)$$

To evaluate the overall performance of our POS engine, we test our tagger using a corpus that was provided in a POS tagging track of an open evaluation CIPS-ParsEval-2009, held in Beijing. The corpus provides 389,170 and 92,687 words for the training and test data respectively. In addition, we adopt the accuracy of POS tagging is equal to the ratio of sum of words with correct POS tags to sum of words in gold-standard sentences. Table 3 demonstrates the performance of our engine with and without the information from word structures. During the experiment, 3-best POS tags of the OOV, which are deduced by the technique, are fed into the base tagger as shown in Figure 1. As a result, the final accuracy of tagging the OOV surges more than 24%, with a ripple effect of 1.1% increase for the IV. This raises the final accuracy to 95.3% while the state-of-the-art in the competition for the closed and open tracks are 93.41% and 93.40% respectively. In addition, while the overall accuracy of POS tagging may vary from taggers, corpus, or more importantly, the percentage of OOV words in the test corpus, the overall tagging accuracy of 91.9% was reported in the Ng & Low tagger [8]. Our POS engine has improved significantly over the previous taggers.

Table 3. Performance of our tagger with and (*without*) the information from word structures, where IV and OOV represent the in-vocabulary and out-of-vocabulary words respectively

	Total	IV	OOV
Correct	88,290 (*85,684*)	82,881 (*81,891*)	5,409 (*3,793*)
Incorrect	4,397 (*7,003*)	3,255 (*4,245*)	1,142 (*2,758*)
Accuracy	95.26% (*92.44%*)	96.22% (*95.07%*)	82.57% (*57.90%*)

5 Conclusions

While previous research on Chinese OOV words mostly focuses on the identification of proper nouns, in this paper, we take advantage of the features from word structures, and the linguistic hints from the collocation markers. A similarity-based technique to determine the POS tags, into which an OOV word fits mostly, is devised. Results show the morpho-syntactic relations of the words are highly indicative for their POS tags. Whereas it is a common approach to incorporating the contextual information into the POS tagging, a recursive inferential technique is also delineated and implemented to ease the ripple effects from any changes made in its neighbors. We apply the technique into a real POS tagger. Experiments show our technique renders a relatively good POS tagging accuracy.

Acknowledgement. The valuable comments from the anonymous reviewers are highly appreciated. The work described in this paper was partially supported by a grant from the Research Grants Council of the Hong Kong Special Administrative Region, China (Project No. CUHK440609).

References

1. Chung, Y.-S., Chen, K.-J.: Analysis of Chinese morphemes and its application to sense and part-of-speech prediction for Chinese compounds. In: Proceedings of the Joint Conference of 23rd International Conference on the Computer Processing of Oriental Languages (2010)
2. Dagan, I., Lee, L., Pereira, F.: Similarity-based models of word co-occurrence probabilities. Machine Learning Journal 34(1-3), 43–69 (1999)
3. Fellbaum, C.: WordNet: An Electronic Lexical Database. MIT Press, Cambridge (1998)
4. Frege, G.: On sense and reference. The Philosophical Review 57, 207–230 (1948)
5. Gao, J., Li, M., Wu, A., Huang, C.-N.: Chinese word segmentation and named entity recognition: A pragmatic approach. Computational Linguistics 31(4), 531–574 (2006)
6. Lin, D., Zhou, S., Qin, L., Zhou, M.: Identifying synonyms among distributionally similar words. In: Proceedings of the 18th International Joint Conference on Artificial Intelligence, pp. 1492–1493 (2003)
7. Liu, Y., Yu, S., Zhu, X.: Construction of the contemporary Chinese compound words database and its application. In: Zhang, P. (ed.) The Contemporary Educational Techniques and Teaching Chinese as a Foreign Language, pp. 273–278. Guangxi Normal University Press (2000)
8. Ng, H.T., Low, J.K.: Chinese part-of-speech tagging: One-at-a-time or all-at-once? Word-based or character-based? In: Proceedings of EMNLP, Barcelona, Spain (2004)
9. Quinlan, J.R.: C4.5: Programs for Machine Learning. Morgan Kaufmann (1993)
10. Tseng, H., Chen, K.-J.: Design of Chinese morphological analyzer. In: Proceedings of the First SIGHAN Workshops on Chinese Language Processing (2002)

Resolving Ambiguities in Sentence Boundary Detection in Russian Spontaneous Speech

Anton Stepikhov

Department of Russian, St. Petersburg State University,
11 Universitetskaya emb., 199034 St. Petersburg, Russia
a.stepikhov@spbu.ru

Abstract. The paper analyses inter-labeller agreement within manual annotations of transcribed spontaneous speech and suggests a way to resolve ambiguities in expert labelling. It argues that the number of controversial sentence boundaries may be reduced if some of them are regarded as "zones". We describe a technique of detecting these zones and analyse which syntactic structures are the most likely to appear in them. Though the approach is based on Russian language material, it may be applied to oral texts in other languages.

Keywords: sentence boundary detection, manual annotation, segmentation, spontaneous speech, oral text, monologue, Russian language resources.

1 Introduction

The problem of sentence boundary detection in spontaneous speech has been known for decades. As [1] points out, though any text can be segmented into sentences, not every text can be segmented unambiguously. At the same time, obtaining information about sentence boundaries is a key issue for natural language processing. This data improves linguistic analysis for text mining systems, information retrieval, and text processing techniques such as parts of speech tagging, parsing and summarisation [2]. The presence of sentence boundaries in speech recognition output enhances its human readability [3].

Information about sentence boundaries in unscripted speech is acquired through sentence boundary labelling which can be performed manually or automatically. There are various models of automatic sentence boundary detection ([2,3,4,5,7]), and most of them are focused on reproducing expert manual annotation. It is expert annotation that is the subject of this paper.

It is known that the extent of inter-labeller agreement varies (e.g. [3]). This paper is an in-depth exploration of manual annotation ambiguity. Understanding of its nature may suggest ways of resolving ambiguities and, consequently, enhance human standard of automatic sentence boundary detection models.

2 Data and Method Description

2.1 Corpus

The study is based on a corpus of spontaneous monologues. The corpus is balanced with respect to speakers' age, gender and profession (linguists and non-linguists). It consists

I. Habernal and V. Matousek (Eds.): TSD 2013, LNAI 8082, pp. 426–433, 2013.

of 160 texts obtained from 32 speakers. They were well acquainted with the person making the recording, which made their speech natural to a maximum extent. Each speaker engaged in 5 different tasks: two story retellings, two types of picture description and a topic-based story. For the retelling, the stories were read and subsequently retold from memory. For picture description, the speakers examined and described pictures simultaneously. The stories and pictures were the same for each speaker. For the topic-based story, they commented on one of two themes: "My leisure time" or "My way of life". The recordings were done either in a soundproof room at the Department of Phonetics of St. Petersburg University or in a quiet space in field conditions. Overall duration of the recorded texts is about 9 hours.

The corpus was annotated according to the technique described below. The corpus and annotation were released as part of the Russian National Corpus [9] (spoken corpora).

2.2   Method

Recordings of unscripted speech were transcribed orthographically by the author. The transcription did not contain any punctuation. To make text reading and perception easier, graphic symbols of hesitation (like *eh, uhm*) and other comments (e.g. [sigh], [laughter]) were also excluded.

The transcripts were then manually segmented into sentences by a group of experts consisting of 20 professors and students of the Philological Faculty of St. Petersburg University (Russia) and of Faculty of Philosophy of the University of Tartu (Estonia). All were Russian native speakers. The experts were asked to mark sentence boundaries using conventional full stops or any other symbol of their choice (e.g. a slash).[1] The experts were presumed to have a native intuition of what a sentence is and, thus, it was left undefined. They were not time-constrained while performing the annotation.

Sentence boundary identification can be based on textual and prosodic information. The interaction between the two is not yet fully understood. For example, [6] showed that the influence of the semantic factor on segmentation outweighs that of the tone factor. In addition, the analysis of sentence boundary detection in a Russian ASR system reveals that in Russian spontaneous speech it is difficult to detect boundaries based on prosodic clues alone [7]. In our experiment, the experts had no access to the actual recordings. This approach allows us to focus on semantic and syntactic factors only and separate them from prosodic factors. At the same time, text reading suggests text reproducing and, hence, segmentation in inner speech. Thus, the lack of information about a speaker's intonation is to some extent compensated by the reader's prosodic competence, allowing him or her to feel the rhythm and melodic texture of sentences without their physical conversion into sound [8].

As a result of this experiment, 20 versions of syntactic segmentation of the proposed texts into sentences were obtained (3,200 marked texts in total). They were then subjected to further statistical analysis.

---

[1] A similar approach to the analysis of Russian spontaneous speech has been used by some researchers earlier (e.g. [6]). The suggested method, however, allows predicting and resolving ambiguities in sentence boundary detection as will be shown further.

3   Data Analysis

3.1   Analysis of Inter-Labeller Agreement

The analysis of the inter-labeller agreement shows that experts disagree in their marking of sentence boundaries.

For each position in the text the number of experts who had marked the boundary at this position was computed. This number can be interpreted as a "boundary confidence score" (BCS) which ranges from 0 (no boundary marked by any of the experts) to 20 (boundary marked by all experts = 100% confidence). The distribution of BCS in the corpus is illustrated in Fig. 1.

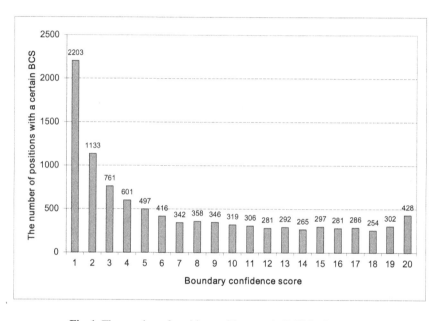

Fig. 1. The number of positions with a certain BCS in the corpus

A reasonable approach would be to accept as a sentence end a position with a BCS no less than 12 (60% of the experts). With this approach, we found that almost 70% of positions marked as sentence boundaries should not be taken into consideration for sentence boundary detection because the BCS failed to reach the threshold. It is worth mentioning that the largest number of positions with BCS > 0 have BCS = 1 (22.1% of all mentioned positions). This fact reveals the high extent of variability in detecting possible boundaries.

3.2   Resolving Ambiguities

Using the threshold approach is, however, just the first step in detecting sentence boundaries. The data examination reveals that there are labelled positions which require more careful consideration. These places are underlined in the example below, figures in bold mark the BCS:

*Пошел в модный салон **10** вот **6** примерил шляпу **13***
*'He went to a boutique **10** so **6** tried on a hat **13**'.*

Based on the threshold defined above (BCS = 12), the boundary marks in the underlined fragment should be ignored. On the other hand, some experts could have detected the boundary before the underlined word and others after it. In this case 16 (10+6) experts might have associated the sentence boundary with the word *вот* (*'so'*). If experts' estimation is ignored, two sentences are combined into one, which results in the distortion of the text's linguistic analysis. Thus, the simple threshold approach in similar situations routinely underestimates the number of sentences in the data and misses possible boundaries.

Another situation is, however, possible. The same experts might have marked the boundary twice – on both sides of the word. It means that the underlined words might be treated as a sentence boundary by only 10 experts rather than 16; that is, the final confidence score for these boundaries is less than the threshold amount.

To resolve this ambiguity, the following method was developed. We identified single words and short segments that had the following conditions: 1) they were marked both before and after, 2) each position had a confidence limit below the established threshold, but: 3) both could reach it in sum. For example: *7 вот 9* (*'7 so 9'*). The next step was to ascertain information about the number of experts who had marked the sentence boundary in a certain zone. In this regard, only one sentence boundary marker from an expert was considered relevant in this zone. If a given segment had been labelled twice by the same expert – before and after it – one of the markers was ignored. As a result, the cases of the "double boundary" in a certain zone were excluded and the information about the exact number of experts who had associated the boundary between sentences with a certain zone was obtained. We propose that if the final confidence score passes the threshold, the word or a segment should be treated as a special "boundary zone". Such zones indicate a boundary between two sentences without specifying its exact location.

After the described processing the above given example looks like this[2]:

Пошел в модный салон [вот] (12). Примерил шляпу (13).
'He went to a boutique [so] (12). Tried on a hat (13).'

The final confidence score 12 rather than 16 means that 4 annotators had marked the boundary twice and in these cases one of the labels was ignored. The other 8 experts had labelled the boundary before or after the word *вот* (*'so'*).

On the base of the analysis of manual annotation, the transcribed texts were segmented into sentences. With this approach, we identified 284 such zones in the corpus. This increased the total number of sentences from 2,690 to 2,974, with zone boundaries accounting almost 10% of all boundaries.

---

[2] Sentence boundary zones are enclosed in square brackets. In round brackets is shown the final boundary confidence score. In further examples the boundary zones are given without underlining.

3.3 Syntactic Content of Sentence Boundary Zones

To understand the variability in sentence segmentation, the syntactic content of these zones was examined.

Table 1. Syntactic content of sentence boundary zones

	Syntactic Content of the Boundary Zone	Count	Percentage
1	**Discourse markers / filler words, parenthesis, function words (particles, conjunctions), interjections, onomatopoeia**	156	54.9%
1a	*Discourse marker / filler word*	*111*	*39.1%*
1b	*Parenthesis*	*14*	*4.9%*
1c	*Combination of two or three discourse markers and / or function words, parenthetic words, interjection*	*13*	*4.6%*
1d	*Particle, conjunction*	*12*	*4.2%*
1e	*Interjection, onomatopoeia*	*6*	*2.1%*
2	**Subordinate parts of the sentence**	39	13.7%
3	**Disfluencies**	38	13.4%
4	**Syntactically ambiguous content**	19	6.7%
5	**Principal parts of the sentence**	16	5.6%
6	**Other cases**	16	5.6%
	TOTAL AMBIGUOUS CASES	**284**	**100%**

Table 1 reveals that most of the boundary zones are filled with semantically and syntactically weakened elements: discourse markers, parenthesis, function words, interjections, onomatopoeia and disfluencies such as breaks, repetitions and revisions. Thus, types 1 and 3 constitute almost 70% of all cases. The most considerable number of sentence boundary zones (39.1%) contains discourse markers / filler words (see 3.2 and Appendix for examples).

Rather often boundary zones contain subordinate parts of the sentence (type 2, example see in Appendix) – mostly adverbial modifiers of different types (10.9%). The appearance of these zones may be explained by the fact that these boundary units can be associated with both previous and subsequent segments. The other reason may be language-specific features such as rich inflectional morphology of Russian and, hence, relatively free word order.

Type 4 is functionally close to type 2. The distinction between them is that the content of sentence boundary zones of type 4 can perform various syntactic functions depending on their association with the previous or the following sentence. In example 4 (see Appendix) the content of the boundary zone may be either a predicate (within the left context) or parenthesis (within the right one).

Besides the common explanation, types 5 and 6 are also determined by the possibility of asyndetic connection between homogeneous parts of the sentence and clauses.

4   Discussion and Conclusions

We have presented a corpus of Russian spontaneous monologues and a possible approach to sentence boundary detection in oral speech. The approach is based on the analysis of the expert manual annotation of transcriptions.

The proposed method involves a linguistic experiment and can be used without drawing on a recorded speech. The enlistment of sufficient number of experts allows acquiring data without explicit information about text prosody. It should also be mentioned that manual annotation of sound recordings with a considerable number of experts is in comparison a rather demanding if not impossible task.

Our approach provides a confidence score for each position which was interpreted as a sentence boundary by the experts (boundary confidence score, BCS). The BCS allows for obtaining information about the unequal status of annotated boundaries and is therefore a valuable resource for development and / or adjustment models of automatic sentence boundary detection in unscripted speech.

The analysis of inter-labeller agreement reveals the problem of spontaneous text segmentation into sentences. It is only about 30% of all labelled positions in the corpus that reached the threshold and, thus, proved relevant for sentence boundary detection. This figure indicates the high extent of variability in the detection of possible boundaries, which may be explained by the widespread asyndetic and paratactic connection between clauses in spontaneous speech. The possibility for subordinating conjunctions to begin a sentence may as well be a contributing factor.

We demonstrate that the ambiguity of expert annotation may be resolved by considering the sentence boundary in spontaneous speech not only as a certain point between two sentences but also some zone of relatively small length (1–3 words as a rule). The phenomenon of sentence boundary zones may be explained by language-specific features such as the rich inflectional morphology of the Russian language and its relatively free word order. Or they may be features general to all languages that merely await identification. Detecting these zones allows acquiring more precise data for the syntactic analysis of unscripted speech. In our corpus almost 10% of sentence boundaries were detected as zones, with about 70% of them filled with semantically weakened elements.

It is worth mentioning that some discourse markers were interpreted by the experts as a separate sentences, with boundary marks on both sides of the words. This segmentation reflects the real prosodic characteristics of such utterances, as in oral speech, their tone contour corresponds with that of a declarative sentence. This fact reveals the reality of human prosodic competence and shows that information about sentence boundaries may be obtained without explicit information about text intonation.

In some cases the exact place of a sentence boundary can be disambiguated by the prosodic characteristics of recorded speech. Therefore in future we plan to collect expert annotations of recorded speech and to compare the results of the two types of labelling (based on prosodic and textual information). This study is also considered to be groundwork for developing a model to predict sentence boundaries in oral speech.

5 Appendix

The examples below represent some of the syntactic content of sentence boundary zones. Examples are limited to the most illustrative in terms of translation. Numbering corresponds with the type of syntactic content given in Tab. 1.

1b. В общем-то изображено на нее на ней изображен какой-то крестьянский хутор ну такой довольно основательный [как я понимаю] (18). Это каменные здания каменный забор какие-то люди ходят во дворе (12).

'*In fact it depicts a pretty well-built peasant farm [as I understand] (18). These are stone buildings stone wall some people are walking around the yard (12).*'

1e. Из-за этого меня иногда путали с пугалом думали что это я в огороде стою окликали [эй] (16). Нет я не откликался вернее я хэ я вообще не слышал [пуг...] (13).

'*Because of this, I was sometimes mistaken for a scarecrow they thought that it was me who was standing in the garden and called [hey] (16). No, I didn't reply that is I ha I didn't hear at all [sca...] (13).*'

2. Ну вообще люблю например Зоологический музей кроме того ну и чего-то почитать на эту тему тоже не прочь съездить на природу [зимой] (13). Катаюсь на горных лыжах <...>

'*Well in general I like for example The Museum of Zoology too and to read something on the topic as well I would not mind a trip to the country [in winter] (13). I go downhill skiing <...>.*'

3. Значит речь идет это самое дело происходит в селе [в селе] (19). Это осень и это время когда поспели яблоки (17).

'*Well it is about well it happens in the village [in the village] (19). It is autumn and it's the time when the apples have ripened (17).*'

Рассказчик вспоминает раннюю осень когда собирают урожай яблок [и ему] (15). Урожай очень большой (16).

'*The narrator recalls early autumn when people are harvesting apples [and he] (15). The harvest is very large (16).*'

Такой ненастный день видимо какой-то [хотя вроде бы нет] (13). Ну нет дождя нет конечно но ветер там не знаю дым дым из трубы идет <...>

'*Nasty day, it looks like [though maybe not] (13). Well there's no rain of course but wind there I don't know smoke smoke is coming out of the chimney <...>*'

4. Нашли человека и поняли человек не мог пешком далеко-то уйти значит где-то не так далеко и жилье [может быть] (16). Это и спасло новоселковских мужиков <...>

'*They found a man and realised the man couldn't walk a long distance on foot it means that somewhere not far away a house [maybe] (16). It saved the men from Novosyolki <...>*'

5. Сделал себе покушать [поел] (12). Вот пошел естественно к компьютеру проверил почту посмотрел кто тебе позвонил кто не позвонил (14).

'You cook some food [eat it] (12). You go of course to the computer check the mail look who has given you a call who hasn't (14).'

6. Возможно это сумерки [тени ложатся] (12). И вот хорошо тепло <...>
'It may be twilight [shadows are falling] (12). It's pleasant warm <...>'

Acknowledgments. The paper has benefited greatly from the valuable comments and questions of Dr. Anastassia Loukina and Dr. Walker Trimble. The author also thanks all the speakers and experts who took part in the experiment.

References

1. Skrebnev, Y.M.: Vvedenie v kollokvialistiku. Izdatel'stvo Saratovskogo universiteta, Saratov (1985) (in Russian)
2. Nasukawa, T., Punjani, D., Roy, S., Subramaniam, L.V., Takeuchi, H.: Adding Sentence Boundaries to Conversational Speech Transcriptions using Noisily Labelled Examples. In: AND 2007, pp. 71–78 (2007)
3. Liu, Y., Chawla, V.N., Harper, M.P., Shriberg, E., Stolcke, A.: A study in machine learning from imbalanced data for sentence boundary detection in speech. Computer Speech and Language 20(4), 468–494 (2006)
4. Gotoh, Y., Renals, S.: Sentence Boundary Detection in Broadcast Speech Transcripts. In: Automatic Speech Recognition: Challenges for the new Millenium, ISCA Tutorial and Research Workshop (ITRW), Paris, France, September 18-20, pp. 228–235 (2000)
5. Kolář, J., Liu, Y.: Automatic sentence boundary detection in conversational speech: A cross-lingual evaluation on English and Czech. In: Proceedings of the IEEE International Conference on Acoustics, Speech, and Signal Processing, ICASSP 2010, Sheraton Dallas Hotel, Dallas, Texas, USA, March 14-19, pp. 5258–5261 (2010)
6. Vannikov, Y., Abdalyan, I.: Eksperimental'noe issledovanie chleneniya razgovornoj rechi na diskretnye intonacionno-smyslovye edinicy (frazy). In: Sirotinina, O.B., Barannikova, L.I., Serdobintsev, L.J. (eds.) Russkaya Razgovornaya Rech, Saratov, pp. 40–46 (1973) (in Russian)
7. Chistikov, P., Khomitsevich, O.: Online Automatic Sentence Boundary Detection in a Russian ASR System. In: Potapova, R.K. (ed.) The 14th International Conference "Speech and Computer", SPECOM 2011, Kazan, Russia, September 27-30, pp. 112–117 (2011)
8. Gasparov, B.M.: Yazyk, pamyat', obraz. Lingvistika yazykovogo sushchestvovaniya. Novoe literaturnoe obozrenie, Moscow (1996) (in Russian)
9. Russian National Corpus, http://www.ruscorpora.ru/en/index.html

Revealing Prevailing Semantic Contents of Clusters Generated from Untagged Freely Written Text Documents in Natural Languages

Jan Žižka and František Dařena

Department of Informatics, FBE, Mendel University in Brno
Zemědělská 1, 613 00 Brno, Czech Republic
{zizka,darena}@mendelu.cz

Abstract. The presented work deals with automatic detection of semantic contents of groups of textual documents, which are freely written in various natural languages. The large original set of untagged documents is split between a requested number of clusters according to a user's needs. Each cluster is taken as a class and a classifier (decision tree) is induced. The words used by the tree represent significant terms that define semantics of individual clusters. The importance (weights) of the terms combined in individual tree branches are computed according to their particular meaning from the correct classification viewpoint – a certain word combined with other words may lead to different classes but a specific class can strongly prevail. The results are demonstrated using large data sets composed from many hotel-service customers' reviews written in six different natural languages.

Keywords: textual documents, semantics, natural language, clustering, tagging, classification, automatic disclosure of meaning, Cluto, c5/See5.

1 Introduction

Having a very large collection of untagged and unformatted textual documents freely written in a natural language, a question suggesting itself can be: Would it be possible to automatically find some document groups characterized by the same or very similar semantic contents? For thousands or millions of documents, it would be quite impractical or impossible to do it manually within an acceptable time and cost range. Computers are able to find such groups using appropriate clustering algorithms and procedures, however, a typical problem often is whether the individual clusters represent any *reasonable semantic meanings*.

The following sections describe a method that investigates a given set of document clusters from the semantic point of view. Inspired by [11], where the authors applied the significance of entropy-lowering words to classification of textual documents, the research presented here demonstrates how it is possible to specify the main sense of a short textual document, which does not have a particular structure and is freely written in a natural language to represent a certain opinion. Such documents are typical for various web-based applications as blogs, mind expression, opinion formulation, etc.

I. Habernal and V. Matousek (Eds.): TSD 2013, LNAI 8082, pp. 434–441, 2013.

The accent was put on processing large number of documents without tags that could provide bias concerning their sense from the semantic point of view. As the data, the research used reviews of customers of hotel services. The reviews can be easily written using a common Internet web-browser. Providers of various services today collect meaning of customers as a certain feedback. Many reviews give typically a lot of information referring to a certain matter – here, the quality of hotel services and customers' (dis)satisfaction. Such reviews are usually ranked by their authors using a given scale like from five stars (complete satisfaction) to one star (complete dissatisfaction). This labeling is often too rough because the service provider may be interested in, e.g, what is very important, what can be ignored, or what is typical. In this case, the set of reviews should be additionally categorized. A typical method is an application of clustering, which generates groups of instances having similar characteristics representing certain contents.

Here, it is necessary to emphasize that 1) the research goal was not creating classes for classification, and 2) the text-mining problem belongs among the strongly *data-driven* tasks [2]. The primary aim was to find a method how to *reveal the main semantic contents of clusters* generated in different numbers. As almost each review described several aspects of evaluating the hotel service, the individual clusters expectedly did not represent just one topic. However, there typically was one prevailing theme accompanied by other minor ones. In some cases, certain clusters represented just a mixture of various aspects without any particular preference of some of them. Anyway, it corresponds to the different way how individual users look at the reviewed service: different people prefer different things (breakfast, transportation, price, etc.) while some service qualities are commonly shared (cleanness, food, quiet hours, staff friendliness, etc.).

2 Generating Clusters Using the Cluto System

Various clustering methods can be applied to untagged data sets. A report concerning the unsupervised approach can be found in [10]. Here, the research was based on a developed clustering system known as *Cluto* [6], which is very suitable for clustering high-dimensional data-sets like textual documents. A user has to specify a set of parameters as the number of clusters plus the method of particular clustering (details can be found in [6]), and Cluto divides the data into groups with minimized inter- and maximized intra-group group similarity. The essential question was how many clusters should be generated. It is necessary to avoid too high generality (just one/two cluster/s) or concreteness (as many clusters as instances). The generality is important as it can say what is common in a certain group of reviews. However, larger number of smaller and more concrete clusters may provide more detailed information. Thus, a user interested in what individual clusters represent can opt for different number of clusters and then investigate what is typically general and what are the details – in such a case, there is no optimal cluster number as it depends on a particular application.

2.1 Data Preparation

After the preliminary phase, the authors decided to experiment with three numbers of clusters: two (the highest generality), five (medium generality), and ten (higher specificity). These numbers depended on the total number of available reviews (each cluster should contain sufficient amount of documents to avoid too low number of reviews per cluster). Another goal was finding out how the suggested method worked for different languages: DE (German), EN (English), ES (Spanish), FR (French), IT (Italian), and CS (Czech). The first five languages represented the 'big' ones as there were hundreds of thousands reviews available (for EN almost two millions) – due to too high computational complexity, each of those languages was finally represented by a repeated random selection of 50,000 reviews. The only exception was the CS-set because it contained only 17,103 reviews – it was used in one piece as a representative of 'smaller' languages.

The data source was prepared using the method *bag-of-words* [1]: each review was transformed into a vector where a word was replaced by its weight using the *tf-idf* formula ($term\ frequency \times inverted\ document\ frequency$) [9]. Alternative word representations like *n*-grams provided worse results due to increasing the original very high sparsity (av. 99.85%), which is one of difficult problems [7]. Because the experiments used data from six languages and there was no available *unified* tool for *uniform* stemming, stop-words removing, etc., only digits were removed (various tested tools were giving different results for the same data). The rest was left as it was because the intention was also to compare mutually the results for the six tested languages under the as same conditions as possible. However, when not thoroughly reducing irrelevant or grammatically incorrect terms, the result contains a lot of redundant terms that wrongly increase the dimensionality and introduce noise – for example, a certain word can be mistyped in several ways, which leads to an artificial increase of the dictionary size.

2.2 Clustering

The clustering procedure used the tool Cluto [6]. Except its many parameters that allow various experiments and looking for the best parameter combination, Cluto implementation is also very fast and uses the computer memory (RAM) very efficiently with relation to the possible sparsity of vectors. The primary parameter was the requested number of clusters, which was from 1 to 20. Because of the limited space, only the results for 10 are here demonstrated but they are quite representative and illustrative. As the method of clustering, the experiments employed the so-called *direct* one, which is Cluto's implementation of k-means [5]. The similarity between reviews was measured using the cosine of an angle between vectors (often used in text mining), and the criterion function for evaluating the clusters' quality was hybrid $H2$ based on combination of the internal criterion $I2$ (the intra-cluster similarity maximization based on a cluster's centroid vector) and external $E1$ (the inter-cluster similarity minimization); all the parameter details can be found in mentioned [6].

3   Searching for the Prevailing Semantic Contents

The semantic content of the generated clusters is – due to the applied *bag-of-words* representation – given by words (terms) that are significant for expressing the meanings revealed by the used data-mining techniques. Certain important terms relate to a specific topics while other significant words to different ones. Now the question is how to find those significant words that would express the particular semantic meanings. In [11] and [4], the authors applied a generator of decision trees [8] that provided a rank of attributes the values of which were decisive for minimizing the entropy. The heterogeneous set of instances mixed from different classes was split between more homogeneous subsets representing instances belonging to more specific classes.

3.1 Looking for Significant Words

The main idea is to find such document elements that can say what a document is talking about – here, the *significant words*. The most significant word is in the root of the decision tree because the tree asks each time for the value representing the word. Other words in the rank get gradually lower significance according to their importance for decreasing entropy with respect to the classification accuracy. As it was shown in [12], those significant words (and phrases composed from them) corresponded very well to a reader's point of view. For the presented data type, such words represent the semantic contents of the clusters, see [12]. The most significant words (from the root and levels below the root) are the leading exponents, which provide the main meaning of the document.

The search for the prevailing semantic contents starts from creating a given number of clusters according to the need of generality or the level of details. Each of the clusters is taken as a class. Then, using the clusters, a decision tree is constructed (c5/See5 [3]) and the byproduct of the tree is the rank of significant words. The top-level words in the rank give the semantic meaning of the cluster: a group of reviews dealing with the same matter. The words appearing in the decision tree create a dictionary composed only from the significant words – their number is a small fraction of the all words used in the reviews, typically a couple of hundreds from tens of thousands (for EN it was just 198 significant words from 26,092, for CS 287 from 29,023, etc.). The classification using generated clusters as tags of the reviews worked with a relatively small accuracy error that was (applying 10-fold cross-validation testing) 8-13% and slightly higher for CS due to the smaller number (17,103 vs. 50,000) of review samples – 16.4%.

3.2 Weighting the Importance of Different Significant Words

The words contained in each branch of the classification tree present combinations of terms leading to a certain class. If a branch ending in a leaf (which represents a class) contains words that lead exclusively to that class, such words are typical just for that class. Branches leading to different classes may contain some identical words, for example, *always-bad-breakfasts*, or *always-almost-not-bad-breakfasts*, where only the word *bad* makes a difference while other words are the same. Another branch can contain *almost-not-friendly-personnel*, where the semantic meaning does not deal with breakfast even if there are also some identical words – in spite of certain identical words, the three branches lead to different classes (*good breakfast, bad breakfast, unfriendly personnel*). Anyway, the most significant word, which is in the tree root, is part of *every* branch.

Thus, there is a problem how to assign a degree of strength to words pointing correctly (in combinations with other words) via different branches to different classes, that is, to different semantic meanings. For example, a word W_1, *bad*, can be used 30 times for correct and 0 times for incorrect classification to a class B (bad breakfasts), and another word W_2, *always*, can be used 30 times for correct and 20 times for incorrect classification to that class B (50 times in total). Which of these two words contribute more to B? The word W_1 was used less times but in 100% correctly, while the word W_2 was used 5 times more but with only 60% correctness.

According to the method described in [4], the frequencies of the correct and incorrect directing by a given word in the tree were represented using a two-dimensional vector space to introduce a weight that balanced those two frequencies. The word weight w_w was given by the following formula:

$$w_w = \frac{N_{correct}}{N_{all}} \cdot \frac{ln\sqrt{N_{correct}^2 + N_{all}^2}}{ln(N_{max})}, \tag{1}$$

where N_{max} is the maximum of $N_{correct}$ (the number of a word usages for correct classifications) and N_{all} (the sum of a word usages for all classifications). The calculated weight then determines the importance of a word in relation to a given class – higher numbers mean greater relevancy.

4   Results of Experiments

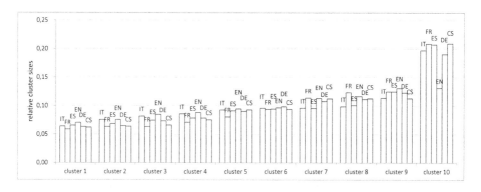

Fig. 1. The distribution of relative sizes of ten individual clusters for the six investigated natural languages. The last cluster 10 represents a mixture of several different topics where none of them prevails semantically others.

As described above, the first step included the clustering, which separated certain numbers of reviews between groups having high similarity inside and low outside. Fig. 1 illustrates how the original sets of reviews for individual languages were relatively split among requested 10 cluster groups (10 clusters for each investigated language). After looking at the generated significant words, the cluster 10 was semantically indifferent because it contained various aspects while none of them prevailed. Other clusters represented prevailing topics. For other numbers of clusters, the situation was very similar; only the lower numbers (below 5) did not show big differences in the number of review distributions between clusters, especially for the minimum given by two clusters. The semantically relevant clusters had very close sizes within each group (cluster 1 to cluster 9).

The results are summarized in Table 1. The semantic contents (main topics) was assigned to individual clusters manually after looking at a list of significant words weighted in combinations with other ones by w_w's. It is interesting to see that for different languages (various nationality of hotel customers), there are often the identical topics and meanings, for example, *location, staff, breakfast*, and so like. As a brief illustration, the top terms representing English for *room negatives* according to their (gradually decreasing) weights: *too, small, noisy, little, old, dated, ...*; French: *peu, trop, manque, absence, odeur, insonorisation, bruit, ...*; Spanish: *poco, falta, caro, algo, antiguo, escaso, ...*; Czech: *hluk, malý, klimatizace, chybí, ...* Similarly, *room positives* for English: *comfortable, spacious, clean, modern, quiet, large, lovely, well, ...*; German: *schöne, schöner, wunderschöne, saubers, grosse, ...*; Italian: *camere, pulite, stanze, confortevoli, ...*, and so like. An example of an English *N/A* cluster: *I, nothing, my, have, like, didn't, that, would, they, ...*, and very similarly for other used languages.

5   Conclusions

The presented method of disclosing semantic contents from a very large volume of untagged textual documents written in natural languages demonstrates that it is possible to carry out it with a useful machine support. A user interested in the possible contents of textual documents has to decide how many groups the collected data set should be separated in. Such a separation can be realized by clustering, after which a user gets potential classes. Then, a classifier of the decision tree type is induced, which generates a small set of words significant for expressing individual semantic contents. Because the significant words are combined in each tree branch, it is necessary to give them weights that represent the words' importance for each possible semantic contents hidden in each cluster.

The experiments were carried out with large real-world data created in six different natural languages by customers reviewing used hotel services. The results demonstrated that – depending on the generality or specificity given by the requested number of created clusters – computers are able to reveal meaning of groups of textual documents and that such meanings are very often identical or similar between various languages. At the same time, the experiments shown that a user (which can be a hotel manager) can also reveal groups of meanings that are not very specific and, according to her or his needs, it is possible to more deeply study reduced number of reviews, which is without such a support impossible due to the extremely large volume of data.

Unfortunately, comparing and evaluating different similar systems is extremely difficult because of the different used data sets, sense inventories, and knowledge resources adopted. Text-mining belongs among strongly data-driven areas from the machine-learning viewpoint and the inductively obtained results often depend on particular data

Table 1. The revealed prevailing semantic contents (for 10 generated clusters) based on significant words in individual reviews for the six tested languages. N/A means that no specific topic could be derived from significant words and the cluster represented a mixture of several topics approximately balanced.

language	main topic of the clusters				
	1	2	3	4	5
CS	general positives	breakfast	N/A	positives, no diacritic	staff
DE	N/A	general positives	location	N/A	breakfast
EN	N/A	value	N/A	hotel facilities	room positives
ES	rooms	environment	location (no diacritic)	N/A	location (with diacritic)
FR	N/A	breakfast, facilities	environment	room negatives	location
IT	location	staff, facilities	rooms	N/A	convenience

language	main topic of the clusters				
	6	7	8	9	10
CS	room negatives	staff, cleanliness	location	surroundings	location
DE	N/A	general positives	room positives	quality/price	atmosphere
EN	room negatives	N/A	staff	room facilities	location
ES	general negatives	N/A	rooms	quality/price	location
FR	N/A	location	price, quality	N/A	comfort
IT	room facilities	location	N/A	good quality	room positives

[9]. In addition, comparing methods even on the same corpus is not eligible if there is different sense inventories. Primarily, the presented research aimed at particular large real-world data-sets with very sparse vectors, looking for an uncomplicated method applicable to not only one specific language.

The following research work aims at deeper analysis of such clusters, including more languages as well as more sophisticated data preparation (at least, removing stop-words and applying a kind of stemming). A big problem is subsequently (in bulk, after writing reviews) correcting mistyping of very large data volumes – it would be much better to apply this function simply during writing the reviews to get not so noisy data. It should be also investigated how (and in which) the various number of requested clusters differs and for what number of clusters the suggested method begins to be useless due to the loss of generality.

Acknowledgments. This work was supported by the research grants of the Czech Ministry of Education VZ MSM No. 6215648904 and IGA of the Mendel University in Brno No. 4/2013.

References

1. Berry, M.W., Kogan, J. (eds.): Text Mining: Applications and Theory. John Wiley & Sons, Chichester (2010)
2. Bloedhorn, S., Blohm, S., Cimiano, P., Giesbrecht, E., Hotho, A., Lösch, U., Mödche, A., Mönch, E., Sorg, P., Staab, S., Völker, J.: Combining Data-Driven and Semantic Approaches for Text Mining. In: Foundations for the Web of Information and Services: A Review of 20 Years of Semantic Web Research, pp. 115–142. Springer, Heidelberg (2011)
3. c5/See5 (June 2013), http://www.rulequest.com/see5-info.html
4. Dařena, F., Žižka, J.: Text Mining-Based Formation of Dictionaries Expressing Opinions in Natural Languages. In: Proceedings of the 17th International Conference on Soft Computing Mendel 2011, Brno, June 15-17, pp. 374–381 (2011)
5. Karypis, G., Zhao, Y.: Criterion Functions for Document Clustering: Experiments and Analysis. Technical Report 01-40, University of Minnesota, USA (2001)
6. Karypis, G.: Cluto: A Clustering Toolkit. Technical report 02-017, University of Minnesota, USA (2003)
7. Qu, L., Ifrim, G., Weikum, G.: The Bag-of-Opinions Method for Review Rating Prediction from Sparse Text Patterns. In: Proceedings of the 23rd Intl. Conference on Computational Linguistics, COLING 2010, Beijing, China, August 23-27, pp. 913–921 (2010)
8. Quinlan, J.R.: C4.5: Programs for Machine Learning. Morgan Kaufmann, San Francisco (1993)
9. Sebastiani, F.: Machine Learning in Automated Text Categorization. ACM Computing Surveys 1, 1–47 (2002)
10. Traupman, J., Wilensky, R.: Experiments in Improving Unsupervised Word Sense Disambiguation. Technical Report UCB/CSD-03-1227, February 2003, Computer Science Division (EECS), University of California, Berkeley (2003)
11. Žižka, J., Dařena, F.: Mining Significant Words from Customer Opinions Written in Different Natural Languages. In: Habernal, I., Matoušek, V. (eds.) TSD 2011. LNCS, vol. 6836, pp. 211–218. Springer, Heidelberg (2011)
12. Žižka, J., Dařena, F.: Mining Textual Significant Expressions Reflecting Opinions in Natural Languages. In: Proc. of the 11th Intl. Conf. on Intelligent Systems Design and Applications, ISDA 2011, Córdoba, Spain, November 22-24, pp. 136–141 (2011)

Robust Methodology for TTS Enhancement Evaluation*

Daniel Tihelka, Martin Grůber, and Zdeněk Hanzlíček

University of West Bohemia, Faculty of Applied Sciences, Dept. of Cybernetics
Univerzitni 8, 306 14 Plzeň, Czech Republic
{dtihelka,gruber,zhanzlic}@kky.zcu.cz

Abstract. The paper points to problematic and usually neglected aspects of using listening tests for TTS evaluation. It shows that simple random selection of
phrases to be listened to may not cover those cases which are relevant to the
evaluated TTS system. Also, it shows that a reliable phrase set cannot be chosen
without a deeper knowledge of the distribution of differences in synthetic speech,
which are obtained by comparing the output generated by an evaluated TTS system to what stands as a baseline system. Having such knowledge, the method
able to evaluate the reliability of listening tests, as related to the estimation of
possible invalidity of listening results-derived conclusion, is proposed here and
demonstrated on real examples.

Keywords: speech synthesis, TTS evaluation, listening tests, statistical reliability.

1   Introduction

During the development, enhancement, or experiments with text-to-speech (TTS) system, researchers usually face the problem of reliable evaluation of the new or enhanced
system. Contrary to speech recognition (ASR, LVCSR) [1,2], which can be evaluated
by using an input signal to be recognised accompanied by the text transcript of the signal expected to be recognised, and the use of mathematical methods of text difference
evaluation, the evaluation of a TTS system must still rely on (rather a larger number of)
subjective responses of listeners evaluating the naturalness, or comparing two versions
of a TTS system. Unfortunately, the comparison of TTS outputs on the signal level, although being mathematically rigorous, does not correspond very much to the listeners'
perception [3,4].

In addition, there is a problem with the absence of a unified methodology of comparing various TTS systems. With the exception of the traditional Blizzard Challenge
evaluation event [5,6], there is no common public database (or a set of language-specific
databases) from which synthetic speech could be created and compared to speech generated by other synthesizers. The existence of such, within TTS community generally
agreed, database would allow the robust and reliable comparison of TTS systems with
results mutually comparable across individual systems, whether evaluating the enhancement of a particular TTS engine, or comparing one engine to another. Without this, the

---

* The research has been supported by the Technology Agency of the Czech Republic, project No.
TA01030476 and by the European Regional Development Fund (ERDF), project "New Technologies for Information Society" (NTIS), European Centre of Excellence, ED1.1.00/02.0090.

I. Habernal and V. Matousek (Eds.): TSD 2013, LNAI 8082, pp. 442–449, 2013.

TTS researchers will continue to face the problem of how to validly interpret non-self-repeatable private-data-collected evaluations like "on our data, the performance was increased significantly", which we were faced with e.g. in [7].

However, the situation in the field of TTS evaluation is even worse. The usual procedure is to randomly select [8,9,10,11,12] a number of sentences, synthesize them by various synthesizers (or by various versions of the same synthesizer) being compared, and let the listeners evaluate individual outputs independently (using the MOS test), or confront one with the other (using the CCR test). The overlooked side of such an approach is that it may not reveal very much about the enhancement or the lowering of the quality, since the significant cases (see the definitions of δ for what they mean) that may show different results have not even been tested.

Therefore, the present paper aims to offer a rigorous methodology which enables lining up and quantifying the possibility that the results drawn from listeners' evaluation may not be entirely trustworthy. We are convinced that this information should be included in each listening test-based TTS evaluation.

2   Reliability of Listening Tests

When experimenting with enhancements of a TTS system, it is usually a rather small part of the whole system which is actually changed. Therefore, when synthesizing a set of test stimuli by the baseline and the enhanced version, some parts of the generated speech and even whole phrases will remain identical for both versions.

However, to be able to evaluate results, only a reasonable number of stimuli is generated for the listening test, usually ranging from 10 to 30. Otherwise, i.e. when having a larger number of listening stimuli, there is a risk of not finding a sufficient number of listeners willing to participate in the test, or they will not carry out the test as carefully as they should.

2.1   Suitability of Phrases to Be Evaluated

Let us define a difference function $\delta(a, b)$ determining how much two variants a and b of the same phrase (generated by different synthesizers or synthesizer versions) differ from the point of view of the material used to build the phrases:

$$\delta(a, b) \in \langle 0, 1 \rangle$$

with boundary value $\delta(a,b) = 0 \Leftrightarrow a = b$ and $\delta(a,b) = 1 \Leftrightarrow a \cap b = \emptyset$. For example, in the case of unit selection TTS, the $\delta(a, b) = 1$ is the case where no unit candidate occurring in a is also used in the same position in b. Nevertheless, it is not strictly necessary to limit the range in any way; we have simply used it for higher readability.

Having the difference function, we can build its probability mass function $P(\delta)$ (shown in Figure 1) and distribution function $F(\delta)$. Note that we need to work with probability space, since the set of phrases to be synthesized, and to be possibly listened to, is countable but not finite[1]. And now it is simple to compute

$$P(X \geq \delta) = 1 - P(X < \delta)$$

---

[1] In general case. For pragmatic reasons, however, we limit the length of phrases for listening tests to a certain length and to natural sentences only.

which represents the probability of synthesizing a phrase with the value δ higher than a given value.

The interpretation of this number is now rather straightforward and provides important information usually neglected in the "classic" approach — how large is the expected probability of the occurrence of a synthetic phrase not covered by the listening tests, i.e. the probability of the occurrence of a phrase for which the result of the listening test is not valid. Or alternatively, what is the probability of the occurrence of the worst possible case covered by the tests (the δ of phrases used for listening), while one can expect (or test) that all of the better cases (lower δ) will not sound worse, simply due to their higher similarity to the version of TTS system being compared to[2].

2.2   Real Example Demonstration

To illustrate this on some real examples, let us use the statistics obtained during the research into our TTS system optimization (described in [13]) as well as a "dummy" experiment in which we changed a small part of concatenation cost computation:

data size reduction, experiment 1. We have removed some phrases from the full corpus to reduce it to approximately 66% of the original size, which is a size acceptable for the joining with screen-reader programs used by blind or semi-blind people on their home PCs. In this experiment, the synthesis is actually forced to use different unit candidates than those used by the baseline, since they are absent from the reduced dataset.

data size reduction, experiment 2. It is basically the same as the previous experiment, except that the number of phrases removed was much larger – the reduced data size is approximately 17% of the original data. Thus, the ratio of different candidates is much higher here.

unit selection feature change. It is a kind of artificial example of unit selection algorithm tuning, where we slightly changed concatenation cost computation – instead of Euclidean distance between 12 MFCC vectors, the average absolute difference of first 2 MFCC is computed. Contrary to the previous experiments, the number of units remain the same and the differences in the output (if there are any) are caused by the different behaviour of the unit selection algorithm.

The outputs of the modified system were always compared to the baseline of our TTS. In the illustration, the following two schemes of δ computation were used:

1. the difference of unit candidates. For a and b variants of a phrase consisting of N units with candidates in sequence $a_1, a_2, \ldots, a_N$ and $b_1, b_2, \ldots, b_N$ respectively, the difference is computes as

$$\delta^k(a, b) = \frac{\sum\limits_{i=1}^{N} \|a_i, b_i\|}{N}$$

---

[2] Of course, even a small change in a single place may decrease the quality of the TTS speech, but it is still more probable that if this single change makes the speech worse, more changes will make it even worse. Or to put it differently, it is very unlikely that when synthetic speech with higher δ is evaluated consistently better, synthetic speech with lower δ will sound consistently worse.

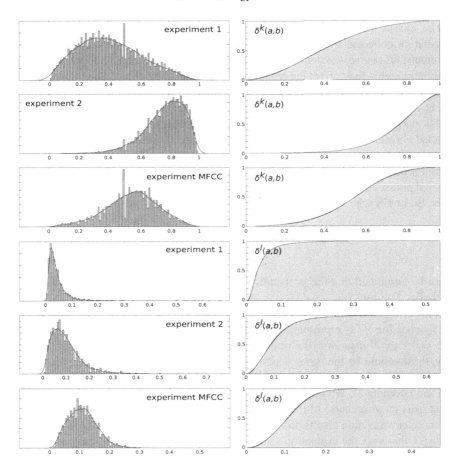

Fig. 1. The visualization of $\delta(a, b)$ probability mass function (sampled to 100 equidistant intervals) and the corresponding distribution function for *speaker 1*. The solid line represents the estimated probability density function (see Section 2.3).

where $\|a_i, b_i\| = 0 \Leftrightarrow a_i = b_i$ and $\|a_i, b_i\| = 1$ otherwise.

2. the difference in the number of concatenation points being defined for the same sequences a and b as

$$\delta^l(a, b) = \frac{|\sum_{i=1}^{N-1} \mathcal{D}(a_i, a_{i+1}) - \mathcal{D}(b_i, b_{i+1})|}{N - 1}$$

where $\mathcal{D}(a_i, a_{i+1}) = 0 \Leftrightarrow$ there is no discontinuous concatenation point between candidates a_i and a_{i+1}, meaning that a_i is the natural and immediate predecessor of a_{i+1} in natural speech, i.e. $next(a_i) = a_{i+1}$; $\mathcal{D}(a_i, a_{i+1}) = 1$ *otherwise*.

The charts in Figure 1 represent the probability mass function $P(\delta)$ and its distribution collected by the synthesis of more than 1 million phrases for all the experiments. Of course, the definition of δ depends on the form of TTS tested and on the expectation what may affect the resulting speech; e.g. there is no straightforward way of unit candidates comparison in the case of HMM-based synthesizer.

Let us now assume the "classic" evaluation test procedure without the knowledge of this statistics. For the purpose of the listening test, let us imagine that 30 randomly selected phrases would be synthesized, which is quite an unusually large number. And naturally, we hope (or claim in the results) that the evaluated phrases represent an typical overall behaviour pattern of the TTS work, or rather that the synthesized versions of the selected phrases differ from the baseline preferably more than less. However, according to the Bernoulli schema

$$P_{<x,y>} = \sum_{i=x}^{y} \binom{y}{i} P^i (1-P)^{y-i}$$

it can be computed that the probability $P = P(X >= \delta)$ for at least 15 of the 30 synthesized phrases (i.e. $x = 16, y = 30$) is not very high, especially for experiments with lower amount of δ changes in the synthesized speech. In Table 1, we have chosen $X >= 0.6$ for δ^k as representing the case with 60% or more of changed units, and $X >= 0.1$ for δ^l meaning 10% or higher change of concatenation points. And, preferably, those would be the cases which the listening tests should focus on, since those cases are more likely to manifest benefits or harms of the evaluated TTS (for vindication see Footnote 2). We hope that this example clearly shows the need of knowledge (or at least an awareness) of the synthesized data behaviour, as well as the estimation of incorrect evaluation results probability, analogous to significance level in hypothesis testing. Otherwise, the evaluation results (and thus conclusion claims) may not be very representative.

2.3   Use of the Probability with Smaller Dataset

For the computation, we have directly used the relative occurrence values of δ collected from the 1 million synthesis output. Let us note that the synthesis ran more than 17 hours on MPI-parallelized 32 Intel Xeon X5680 3.33GHz cores, which would take approximately 23 days on the single core. Of course, it is not usually necessary nor meaningful to collect such an extensive amount of data, since standard kernel density estimation technique [14] can be used to model the probability distribution function from a much lower δ dataset. To prove it, we have randomly selected $5,000$ δ values, for which we have done this estimate. This case is illustrated in Figure 1 by the solid line and the values of $P_{<16,30>}$ are shown in Table 2.

Simple comparison of the estimated values in Table 2 with data-computed values in Table 1 shows that the estimate is sufficiently precise.

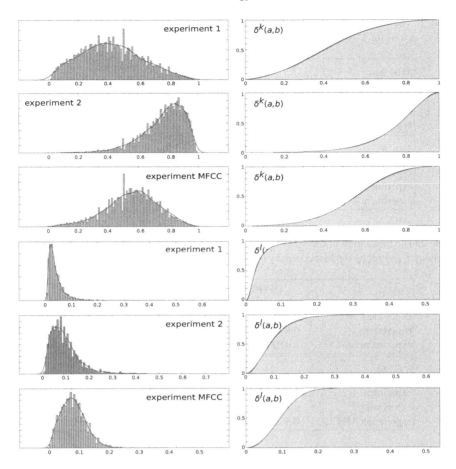

Fig. 2. The visualization of $\delta(a,b)$ probability mass function and the corresponding distribution function for *speaker 2*. The structure is the same as in Figure 1.

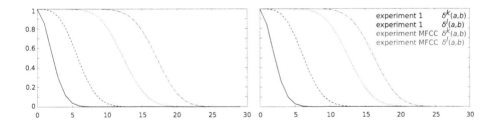

Fig. 3. The visualization P_i for all the cases where at least $i = 1, 2 \ldots, 30$ out of 30 phrases match the required $P(X)$; left for *speaker 1*, right for *speaker 2*.

Table 1. The illustration of probability of the selection of at least 15 out of 30 phrases with $P(X >= 0.6)$ for $\delta^k(a, b)$ and $P(X >= 0.1)$ for $\delta^l(a, b)$ for all the experiments

$\delta^k(a, b)$	speaker 1		speaker 2	
	$P(X >= 0.6)$	at least 15×	$P(X >= 0.6)$	at least 15×
experiment 1	0.194	0.00	0.201	0.00
experiment 2	0.887	1.00	0.882	1.00
experiment MFCC	0.409	0.09	0.421	0.11
$\delta^l(a, b)$				
	$P(X >= 0.1)$	at least 15×	$P(X >= 0.1)$	at least 15×
experiment 1	0.075	0.00	0.061	0.00
experiment 2	0.408	0.08	0.386	0.05
experiment MFCC	0.572	0.66	0.545	0.55

Table 2. The illustration of the same values as in Table 1, but here computed from distribution functions $F(x)$ get by the kernel density estimation technique

$\delta^k(a, b)$	speaker 1		speaker 2	
	$P(X >= 0.6)$	at least 15×	$P(X >= 0.6)$	at least 15×
experiment 1	0.192	0.00	0.204	0.00
experiment 2	0.890	0.99	0.882	1.00
experiment MFCC	0.410	0.09	0.450	0.18
$\delta^l(a, b)$				
	$P(X >= 0.1)$	at least 15×	$P(X >= 0.1)$	at least 15×
experiment 1	0.073	0.00	0.061	0.00
experiment 2	0.431	0.13	0.418	0.10
experiment MFCC	0.590	0.68	0.568	0.65

3   Conclusion

We aimed to show that there is a non-negligible (and even computable) possibility of drawing unreliable results from listening test evaluation, when phrases to be evaluated are chosen at random, which is, however, the usual manner of their selection. Moreover, without some knowledge of differences between the outputs from the baseline and the tested TTS, no-one can be sure which of the cases is actually evaluated by the listening tests — frequent cases with slight differences (i.e. $X < \delta$ for smaller δ), or cases with larger changes and thus with higher informative capability (see Footnote 2)?

The point is that the knowledge of $F(X)$ is important for the estimation of the expected listening tests validity. Having the selected set of phrases to be listened to, the minimum, maximum and average δ can be obtained for them and used to compute $P(X \geq \delta)$. This value estimates, among others, the probability of the occurrence of a phrase for which the result of the listening test is not valid (see the interpretation in Section 2.1).

With all this information, the results of various TTS comparisons will gain higher level of reliability and will become more trustworthy for their readers.

The access to computing and storage facilities belonging to the National Grid Infrastructure MetaCentrum was provided under the program LM2010005 "Projects of Large Infrastructure for Research, Development, and Innovations".

References

1. Ircing, P., Psutka, J., Psutka, J.V.: Using morphological information for robust language modeling in Czech ASR system. IEEE Transactions on Audio Speech and Language Processing 17, 840–847 (2009)
2. Pražák, A., Psutka, J.V., Hoidekr, J., Kanis, J., Müller, L., Psutka, J.: Automatic online subtitling of the Czech parliament meetings. In: Sojka, P., Kopeček, I., Pala, K. (eds.) TSD 2006. LNCS (LNAI), vol. 4188, pp. 501–508. Springer, Heidelberg (2006)
3. Hanzlíček, Z., Matoušek, J.: Voice conversion based on probabilistic parameter transformation and extended inter-speaker residual prediction. In: Matoušek, V., Mautner, P. (eds.) TSD 2007. LNCS (LNAI), vol. 4629, pp. 480–487. Springer, Heidelberg (2007)
4. Tihelka, D., Romportl, J.: Exploring automatic similarity measures for unit selection tuning. In: INTERSPEECH 2009, Proceedings of the 10th Annual Conference of International Speech Communication Association, pp. 736–739. ISCA, Brighton (2009)
5. Black, A.W., Tokuda, K.: The Blizzard Challenge – 2005: Evaluating corpus-based speech synthesis on common datasets. In: Interspeech, Lisbon, Portugal, pp. 77–80 (2005)
6. Yamagishi, J., Zen, H., Jian Wu, Y., Toda, T., Tokuda, K.: The HTS-2008 system: Yet another evaluation of the speaker-adaptive hmm-based speech synthesis system in The 2008 Blizzard Challenge. In: Proc. Blizzard Challenge 2008 (2008)
7. Tihelka, D., Kala, J., Matoušek, J.: Enhancements of viterbi search for fast unit selection synthesis. In: Interspeech, pp. 174–177 (2010)
8. Eide, E., Aaron, A., Bakis, R., Cohen, P., Donovan, R., Hamza, W., Mathes, T., Picheny, M., Polkosky, M., Smith, M., Viswanathan, M.: Recent improvements to the IBM trainable speech synthesis system. In: Proceedings of the IEEE International Conference on Acoustics, Speech, and Signal Processing, ICASSP 2003, vol. 1, pp. 708–711 (2003)
9. Vepa, J., King, S.: Subjective evaluation of join cost functions used in unit selection speech synthesis. In: Interspeech, Jeju Island, Korea, pp. 1181–1184 (2004)
10. Grůber, M., Tihelka, D.: Expressive speech synthesis for Czech limited domain dialogue system – basic experiments. In: IEEE 10th International Conference on Signal Processing (ICSP), Beijing, China, vol. 1, pp. 561–564 (2010)
11. Grůber, M.: Enumerating differences between various communicative functions for purposes of Czech expressive speech synthesis in limited domain. In: Interspeech, pp. 650–653. Curran Associates, Inc., Morehouse Lane (2012)
12. Přibil, J., Přibilová, A.: Distributed listening test program for synthetic speech evaluation, pp. 241–242 (2008)
13. Hanzlíček, Z., Matoušek, J., Tihelka, D.: Experiments on reducing footprint of unit selection TTS system (2013)
14. Bowman, A.W., Azzalini, A.: Applied Smoothing Techniques for Data Analysis. Oxford University Press (1997)

Romanian Syllabication Using Machine Learning

Liviu P. Dinu[1,2], Vlad Niculae[1,3], and Octavia-Maria Şulea[1,2]

[1] Center for Computational Linguistics, University of Bucharest
[2] Faculty of Mathematics and Computer Science, University of Bucharest
[3] University of Wolverhampton
ldinu@fmi.unibuc.ro, vlad@vene.ro, mary.octavia@gmail.com

Abstract. The task of finding syllable boundaries can be straightforward or challenging, depending on the language. Text-to-speech applications have been shown to perform considerably better when syllabication, whether orthographic or phonetic, is employed as a means of breaking down the text into units bellow word level. Romanian syllabication is non-trivial mainly but not exclusively due to its hiatus-diphthong ambiguity. This phenomenon affects both phonetic and orthographic syllabication. In this paper, we focus on orthographic syllabication for Romanian and show that the task can be carried out with a high degree of accuracy by using sequence tagging. We compare this approach to support vector machines and rule-based methods. The features we used are simply character n-grams with end-of-word marking.

1 Introduction

In this paper we describe systems for solving end-of-the-line hyphenation in Romanian. The most challenging aspect is that of distinguishing between hiatus and diphthongs, an ambiguity exemplified in Table 1, as well as between the letter *i* which can surface either as a non-vocalic element, or as a proper vowel, affecting thus the syllable boundary. Although the words in the dataset we used for this task were marked for main stress, we ignored this information in order to build a system that works on plain text.

Table 1. Romanian hiatus/ diphthong ambiguity

sequence	hiatus		diphthong	
ai	*ha-i-nă*	(heinous)	*hai-nă*	(coat)
iu	*pi-u-li-ţă*	(screw nut)	*piu-re*	(purée)
oa	*bo-ar*	(bull herder)	*oa-meni*	(humans)

2 Syllabication and End-of-the-line Hyphenation

Syllabication or syllabification is the task of detecting syllable boundaries within words. The boundaries depend on the phonetic structure of those words and on phonotactic constraints of the respective language. End-of-the-line hyphenation, on the other hand,

I. Habernal and V. Matousek (Eds.): TSD 2013, LNAI 8082, pp. 450–456, 2013.

is the task of determining where to break up a word when it doesn't fit at the end of a line in a written text and this separation may or may not overlap with a syllable boundary, depending on how close its rules are to those of proper syllabication in a particular language. This is why some have dubbed the product of applying these end-of-the-line hyphenation rules to a whole word, and not just in one place to accommodate line length, as *orthographic syllabication* [1].

In Romanian, the rules for end-of-the-line hyphenation, as discussed in the orthographic dictionary [2], are not based on phonetics and, if applied to the whole word, may not render a proper phonetic syllabication, so a distinction can and should be made between phonetic and orthographic syllabication. When analyzing the orthographical rules listed in the dictionary, we see that they diverge from the rules of phonetic syllabication by often times going against the maximal onset principle and, thus, rendering different codas and onsets, but not affecting the nucleus. This means that the diphthong-hiatus ambiguity is maintained in and remains a source of problem for both types of syllabication. The Romanian orthographic syllabication rules also diverge from the phonetic ones in that they are mostly independent from stress placement since Romanian, unlike French, doesn't normally mark stress orthographically. In what follows we will take into account these aspects.

The main application for syllabication, whether phonetic or orthographic, has been shown to be in text-to-speech systems, where a need for letter-to-phoneme conversion is great. Specifically, the performance of the conversion has been shown [1] to increase for English when done on segments smaller than words, namely syllables, with the phonetic syllable increasing performance a bit more. For orthographic syllabication, structured SVMs [1] and CRFs [3] have been used. We investigated unstructured SVMs and CRFs when attempting to solve only the task of orthographical syllabification, due to it being much better resourced for Romanian. Attempts at solving the task of syllabication in Romanian have used rule-based algorithms [4] as well as contextual information (previous and following segments) with contextual grammars [5].

3 Approach

As mentioned previously, one difficulty of the syllabication task for Romanian, of either type, lies in the Romanian diphthong-hiatus contrast, which, although predictable for native speakers from diachronic information [6], is difficult to discriminate based only on synchronic information. This is where statistical machine learning techniques will come into play.

Table 2. Status difference in word-final *i*

part of speech	nucleic		in the coda	
verb	*ci-ti*	(to read)	*ci-teşti*	(you read)
noun	*ti-gri*	(tigers)	*e-levi*	(pupils)
adjective	*ne-gri*	(black)	*albi*	(white)

3.1   Classifiers

Linear Support Vector Machine (SVM). A linear support vector machine is a simple, unstructured learner that predicts one label for each input sample. SVMs are discriminative max-margin models optimizing the hinge loss.

$$L_{\text{hinge}}(y, \hat{y}) = \max(0, 1 - y \cdot \hat{y})$$

The optimization problem can be approached using the *scikit-learn* stochastic gradient descent (SGD) SVM solver. The ℓ_2 regularization is parametrized in terms of α.

The objective function used by the SGD solver is:

$$E(w, b) = \sum_{i=1}^{M} L_{\text{hinge}}(y_i, wx_i + b) + \frac{\alpha}{2} \sum_{i=1}^{d} w_i^2$$

For SGD, the learning rate schedule used for the updates is given by:

$$\eta^{(t)} = \frac{1}{\alpha(t_0 + t)}$$

In the above, t is the time step, going from 1 to $M * N$ where M is the number of samples and N is the number of passes over the training set (epochs). In our case we used the default value of $N = 5$. t_0 is determined based on a heuristic proposed by Léon Bottou in the SvmSgd package[1], such that the expected initial updates are comparable with the expected size of the weights. This schedule is the default one in the *scikit-learn* stochastic gradient descent classifier and it performed well.

Sequence Tagging. Some problems can be structured as related sequential prediction problems, where the classification of one point is related to the classification of the ones right before or after it in the sequence. Such models are common in NLP because of the sequential nature of text, and are used for tasks like POS tagging. The Conditional Random Field (CRF) model has a learning algorithm [7] that minimizes the regularized negative conditional log likelihood of the data. We used L-BFGS optimization. The regularization parameter c is not halved like it is in the SVM formulation:

$$E(\theta) = -\sum_{i=1}^{M} \log p(y_i | x_i; \theta) + c \sum_{i=1}^{K} \theta^2$$

We implemented three systems for orthographic syllabication of Romanian words. The first is a simple rule-based approach which aims to implement as many of the rules [2] for orthographic syllabication given in [2] as possible. The second uses a linear SVM to decide what to do with every possible candidate split in the word. The third approach uses a CRF model. The features used for the two machine learning approaches are character n-grams.

---

[1] http://leon.bottou.org/projects/sgd
[2] Also available, in Romanian, at http://ilr.ro/silabisitor/reguli.php

The advantage of a sequence model over the SVM is that it has the modeling power to jointly analyze consecutive splits, whereas the SVM is limited to local knowledge. It is not aware of interaction between consecutive candidate splits within a same word (e.g. the CRF can learn that it should be rare to split on both sides of a single consonant).

3.2 Software

The software we use is the *scikit-learn* machine learning library for the Python scientific computing environment version 0.12.1 [8]. The library provides efficient text n-gram feature extraction using the sparse matrix implementation in SciPy[3]. We use the SVM implementation by stochastic gradient descent.

We also used *CRFsuite* version 0.12 [9] for its implementation of CRF inference and training.

3.3 Dataset and Preprocessing

The dataset used is based on the morphological dictionary *RoMorphoDict* [10], which is available from the author and has two versions. The resource relevant to our task provides a long list of word forms along with their hyphenated form with accent marking. An online version of this second data resource is available for querying at http://ilr.ro/silabisitor/.

The dataset is annotated for accent by using Unicode accented characters. Unlike, for instance, French, in Romanian the accent is usually not written and therefore not usually available as input, so we strip it completely. We do not strip down to the ASCII character set though, because Romanian has the characters ă, â, î, ș, ț – these are not accented characters, but separate letters.

3.4 Features

The feature extraction consists of extracting character n-grams that form strings. This is parametrized by the n-gram size n. For example, we will consider $n = 3$, the word *dinosaur* and the split between *o* and *s*. The position induces two strings, *dino* and *saur* but we are only interested in the window of radius n around the split, so we are left with *ino* and *sau*. Since the bag-of-n-grams features we use for the SVM loses the order, we consider adding a special marker, obtaining *ino$* and *$sau*. The n-grams of length up to 3 are: *i, n, o, $, in, no, o$, ino, no$* and the analogous for the right hand side. This is then vectorized into counts, with the observation that features that occur with the frequency of 1 across all instances are obviously irrelevant and are ignored.

The vectorization produces one matrix for the left sides of the splits and one for the right side of the splits. These are stacked horizontally to give our data matrix. For the CRF, the feature extraction is the same, but the sparse vectorized representation is replaced with an input like:

```
1 c[-3]=i c[-2]=n c[-1]=o c[-3-2]=in c[-2-1]=no c[-3-2-1]=ino
  c[1]=s c[2]=a c[3]=u c[12]=sa c[23]=au c[123]=sau
```

---

[3] http://docs.scipy.org/doc/scipy/reference/sparse.html

The format above is the one accepted as input by *CRFsuite*. Because the feature names include the offset, the dollar marker would provide no useful information. The names could just as well be arbitrary: *CRFsuite* cannot understand that c[-2-1] means the bigram just before the split, but the values that a certain feature tends to take carry the discriminative information.

Both the SVM and the CRF were trained using ℓ_2 regularization, so we would have another system parameter that controls the amount of regularization applied, α. In the case of the SVM, we also investigate the difference in the results if the boundary marker $ is used or not. These parameters are optimized using grid search to maximize the classification accuracy of individual splits using 3-fold cross-validation over the training set.

3.5 Generating Training Samples

The average word in our dictionary has 9.96 characters and 4.24 syllables. This means that each word generates around 9 training instances (possible splits), out of which we expect around 3 to be labeled as true, and the rest as false. Prior to generating training instances, we split the words into a training and test set, each consisting of 262, 764 words.

For each word of length n we generate $n-1$ instances, corresponding to each position between two letters of the word. Instances are labeled as positive if a hyphen can be inserted there, or negative if not.

This tagging method is called NB labelling [1], because we label each split as boundary (B or 1) or no boundary (N or 0). For example, the word *di-a-mant* (diamond) would be encoded as:

$$
\begin{array}{cccccc}
\text{d} & \text{i} & \text{a} & \text{m} & \text{a} & \text{n} & \text{t} \\
0 & 1 & 1 & 0 & 0 & 0
\end{array}
$$

A slightly more informative way of assigning labels, introduced also in [1], is to use numbered NB (#NB) tags: each split is labeled with the distance from the last hyphen:

$$
\begin{array}{cccccc}
\text{d} & \text{i} & \text{a} & \text{m} & \text{a} & \text{n} & \text{t} \\
1 & 0 & 0 & 1 & 2 & 3
\end{array}
$$

This way the class corresponding to boundaries is class 0, and classes $1, 2, \ldots$ are all negative. We are still dealing with a classification problem, as the classes are discrete. On our dataset the maximum class is 7. This helps if we have a structured sequence model, because from class k the Markov chain can only transition to class 0 or $k + 1$. An unstructured SVM cannot take advantage of this.

3.6 Results

Apart from the accuracy on the test data, we looked at precision and recall scores. All measures are relative to training instances, which correspond to splits within words. A more tough evaluation metric is at the word level: a word is considered correctly hyphenated if all of the splits within it were correctly classified. This score should show the advantage of using a sequence model.

Table 3. Results for hyphenation models. The scores are evaluated on the test set. In the numbered NB case, we are only evaluating with respect to class 0, the one marking the position of the hyphens. Correctly identifying all tags $0 - 7$ is not directly relevant for this application, but their presence helps.

model	n	marker	regularization	Hyphen acc.	Hyphen F_1	Word acc.
RULE -	-		-	94.31%	92.12%	60.67%
SVM NB	4	yes	$\alpha = 10^{-5}$	98.72%	98.24%	90.96%
SVM #NB	4	yes	$\alpha = 10^{-6}$	98.82%	98.37%	91.46%
CRF NB	4	-	$c = 0.1$	99.15%	98.83%	94.67%
CRF #NB	4	-	$c = 1.0$	99.23%	98.94%	95.25%

The rule-based algorithm obtained 60.67% splitting accuracy at word level and 94.31% accuracy, within the word, at split level. The small accuracy at word level we found to not only be influenced by the ambiguity between hiatus and diphthongs or glide+vowel sequences, which cannot be distinguished based on the orthographic rules, but also by the difference in vocalic status of the word-final i, which can form a new nucleus or stay in the coda. Table 2 shows examples of word-final i either determining a new syllable or staying in the coda. Looking at the data relating to word-final i we see that it is not only dependant on phonetic constraints (i.e. it is vocalic after stop+liquid) but also on morphological ones (i.e. it is vocalic in infinitive-based forms). However, since the system does not have access to part of speech tags, we could not refine the rules of the algorithm any further. Because it is not a trained system, we could evaluate the rule-based system on the entire corpus (training and test data). We found it to be consistent, scoring 60.66% word accuracy and 94.32% split accuracy.

We went on to evaluate the SVM system. The hyperparameters found to optimize it are $n = 4, \alpha = 10^{-5}$, marker=true. The grid we investigated ranged over $n \in \{1, 2, 3, 4\}$ and $\alpha \in \{10^{-8}, 10^{-7}, 10^{-6}, 10^{-5}, 10^{-4}\}$. For the CRF system, we did not use marking, because the feature representation includes feature names that encode the offsets. The space over which we looked for n is the same, while for c we searched $\{0.01, 0.1, 1.0, 10.0\}$.

The best results and the associated parameter values are presented in Table 3. The trained models and scripts needed to apply on new data are made available at https://github.com/nlp-unibuc/ro-hyphen.

4   Conclusions

In this paper, we have looked at ways of detecting orthographic syllable boundaries, with the intent to rule over the diphthong-hiatus ambiguity which arises when attempting to implement a rule-based algorithm for either types of syllabication in Romanian. We've seen that this ambiguity along with status ambiguity of word final i causes great drop in performance for a rule-based algorithm (60.66% word level split accuracy vs. 94.32% split accuracy at syllable level), but is handled very well by an SVM or CRF system, with the CRF having the best accuracy at word level (95.25%).

The systems we used can be viewed with respect to a trade-off spectrum. The rules in the rule-based system can take any form and they can model very complex interactions between features. This model has the largest predictive power, but the rules are written by hand, therefore limiting its practicality and its performance. At the opposite end of the spectrum is the SVM classifier, which applies a simple linear decision rule at each point within a word, looking only at its direct context. This simple approach outperforms the rule-based system by being trained on large amounts of data. The sequence tagger is more successful because it exploits the data-driven advantage of the SVM, while having more modeling power. This comes at a cost in model complexity, which influences training and test times.

Acknowledgements. The contribution of the authors to this paper is equal. The work of Liviu P. Dinu was supported by a grant of the Romanian National Authority for Scientific Research, CNCS – UEFISCDI, project number PN-II-ID-PCE-2011-3-0959.

References

1. Bartlett, S., Kondrak, G., Cherry, C.: Automatic syllabification with structured svms for letter to phoneme conversion. In: 46th Annual Meeting of the Association for Computational Linguistics: Human Language Technologies (ACL 2008: HLT), pp. 568–576. Association for Computational Linguistics, Columbus (2008)
2. Collective: Collective: Dictionarul ortografic, ortoepic si morfologic al limbii române., 2nd edn., revised. Romanian Academy, Bucharest (2010) (in Romanian)
3. Trogkanis, N., Elkan, C.: Conditional Random Fields for word hyphenation. In: Proceedings of the 48th Annual Meeting of the Association for Computational Linguistics, pp. 366–374. Association for Computational Linguistics, Uppsala (2010)
4. Toma, S.A., Oancea, E., Munteanu, D.: Automatic rule-based syllabication for Romanian. In: Proceedings of the 5th Conference on Speech Technology and Human-Computer Dialogue (2009)
5. Dinu, A., Dinu, L.P.: A parallel approach to syllabification. In: Gelbukh, A. (ed.) CICLing 2005. LNCS, vol. 3406, pp. 83–87. Springer, Heidelberg (2005)
6. Chitoran, I., Hualde, J.I.: From hiatus to diphthong: the evolution of vowel sequences in romance. Phonology 24, 37–75 (2007)
7. Lafferty, J.D., McCallum, A., Pereira, F.C.N.: Conditional random fields: Probabilistic models for segmenting and labeling sequence data. In: Proceedings of the Eighteenth International Conference on Machine Learning. ICML 2001, pp. 282–289. Morgan Kaufmann Publishers Inc., San Francisco (2001)
8. Pedregosa, F., Varoquaux, G., Gramfort, A., Michel, V., Thirion, B., Grisel, O., Blondel, M., Prettenhofer, P., Weiss, R., Dubourg, V., Vanderplas, J., Passos, A., Cournapeau, D., Brucher, M., Perrot, M., Duchesnay, E.: Scikit-learn: Machine learning in Python. Journal of Machine Learning Research 12, 2825–2830 (2011)
9. Okazaki, N.: CRFsuite: a fast implementation of Conditional Random Fields (CRFs) (2007)
10. Barbu, A.M.: Romanian lexical databases: Inflected and syllabic forms dictionaries. In: Sixth International Language Resources and Evaluation (LREC 2008) (2008)

SVM-Based Detection
of Misannotated Words in Read Speech Corpora*

Jindřich Matoušek and Daniel Tihelka

University of West Bohemia, Faculty of Applied Sciences, Dept. of Cybernetics,
Univerzitní 8, 306 14 Plzeň, Czech Republic
{jmatouse,dtihelka}@kky.zcu.cz

Abstract. Automatic detection of misannotated words in single-speaker read-speech corpora is investigated in this paper. Support vector machine (SVM) classifier was proposed to detect the misannotated words. Its performance was evaluated with respect to various word-level feature sets. The SVM classifier was shown to perform very well with both high precision and recall scores and with $F1$ measure being almost 88%. This is a statistically significant improvement over a traditionally used outlier-based detection method.

Keywords: annotation error detection, classification, support vector machine, read speech corpora.

1 Introduction

Nowadays, the most successful speech processing methods utilize very large speech corpora. Besides speech recordings themselves, the corpora also contain some annotation data which describe what was actually pronounced in each recording. Though many annotation levels may exist (such as phonetic, syntactic, morphological, etc.), a word-level annotation usually constitutes a base, which the other levels are derived from. It is obvious that there is a need to have the word-level annotation as accurate as possible because any discrepancy between speech signal and its linguistic representation is critical, especially in methods which use the speech signal directly. For instance, this is the case of concatenative speech synthesis that works with speech units which are believed to have linguistic properties derived from the corresponding word-level annotation. When the annotation does not match the speech signal, serious speech synthesis errors occur—synthesized speech could be less intelligible, or even other speech than expected may be synthesized [1].

The advantage of corpus-based methods is that large corpora can capture the inherent variability of human speech. On the other hand, the problem with large corpora is that manual annotation is a time-consuming and costly process. However, despite careful

---

* The work has been supported by the Technology Agency of the Czech Republic, project No. TA01030476, and by the European Regional Development Fund (ERDF), project "New Technologies for Information Society" (NTIS), European Centre of Excellence, ED1.1.00/02.0090. The access to the MetaCentrum clusters provided under the programme LM2010005 is highly appreciated.

I. Habernal and V. Matousek (Eds.): TSD 2013, LNAI 8082, pp. 457–464, 2013.

manual annotation, even human annotators do make errors [2]. The typical annotation errors (like missing or extra words, swapped, mispronounced or in other way misannotated words), their frequency in Czech speech synthesis corpus [2], and their impact on the quality of synthetic speech were presented in [1].

In this paper, a procedure for automatic detection of misannotated words is proposed. Unlike other studies [3,4,5,6] which focus rather on *revealing bad phone-like segments*, the proposed method aims mainly at *revealing word-level errors* (i.e., misannotated words). Since phone-level segments are very varied, phone-level error detection usually results in many "false positive" detections. In other words, due to low precision of these methods, many good speech segments had to be unnecessarily checked or removed from speech corpora. Assuming that the location of bad speech segments can be generalized to a word level (in the case of a misannotated word, a sequence of bad segments is often observed), simple *word-level* features can be collected. Then, the whole word is a subject of an automatic classification whether it is good or bad. The aim of this study is to find out whether such word-level error detection could reveal annotation errors both with high recall and precision measures. Lessons learned can be also used for the automatic error detection in synthetic speech [7,8].

In Section 2 we present our data set. Section 3 describes metrics and statistical testing used for evaluation of word-level error detection results. Section 4 presents a baseline detection system used in our previous work. In Section 5 SVM-based detection is described. Word-level error detection results are discussed in Section 6. Conclusions are drawn in Section 7.

2   Experimental Data

We used a Czech read speech corpus of a single-speaker male voice, recorded for the purposes of unit-selection speech synthesis in the state-of-the-art text-to-speech system ARTIC [9]. The voice talent was instructed to read a set of specially prepared text prompts [2] in a "news-broadcasting style" and to avoid any spontaneous expressions. The full corpus consisted of 12,242 utterances (approx. 18.5 hours of speech) segmented to phone-like units using HMM-based forced alignment (carried out by the HTK toolkit [10]) with acoustic models trained on the speaker's data [11,12]. From this corpus we selected 1,335 words in 88 utterances collected during ARTIC system tuning and evaluation, and used them as data for our experiments; 267 words contained some annotation error (207 of them being different) and the rest of 1,068 words were annotated correctly. The decision whether the annotation was correct or not was made by a human expert who analyzed the phonetic alignment.

In order to get more robust results (in the sense of being less dependent on a concrete split of data into training/evaluation partitions and to compensate for a relatively small corpus size), 10 random training/evaluation data splits with 80% of words being used for training and 20% of words being used for evaluation in each split were conducted in each experiment. The splits were made to preserve the ratio of correctly annotated and misannotated words for each class.

3 Evaluation Metrics

3.1 Detection Metrics

Standard metrics like *recall* (R), interpreted as the ability of a classifier to find all misannotated words, *precision* (P), the ability of a classifier not to label as misannotated a word that is annotated correctly, $F1$, a combined measure that results in high value if, and only if, both precision and recall result in high values, and *accuracy* (A), a proportion of correct detections in all detections,

$$R = \frac{t_p}{t_p + f_n}, \qquad\qquad F1 = \frac{2 * P * R}{P + R},$$

$$P = \frac{t_p}{t_p + f_p}, \qquad\qquad A = \frac{t_p + t_n}{t_p + f_p + f_n + t_n}$$

were used to evaluate the performance of the word-level detection. The symbols stand for: t_n, number of correctly annotated words ("true negatives"), t_p, number of words correctly detected as misannotated ("true positives"), f_n, number of misannotated words that were not detected ("false negatives") and f_p, number of words falsely detected as misannotated ("false positives").

3.2 Statistical Significance

Since the used data sets are relatively small, statistical significance tests were performed to see whether the achieved results are significantly better when comparing the proposed SVM classifier to a baseline detection method. We applied McNemar's test because it was found to be simple but powerful [13]. In this test, two classifiers A and B are tested on a test set, and for all testing examples the following four numbers are recorded: number of examples misclassified by both A and B (n_{00}), number of examples misclassified by A but not by B (n_{01}), number of examples misclassified by B but not by A (n_{10}), and number of examples correctly classified by both A and B (n_{11}). Under the null hypothesis, the two classifiers should have the same error rate, i.e. $n_{01} = n_{10}$. McNemar's test is based on a χ^2 test for goodness of fit that compares the distribution of counts expected under the null hypothesis to the observed counts:

$$\chi^2 = \frac{(|n_{01} - n_{10}| - 1)^2}{n_{01} + n_{10}} \tag{1}$$

where a "continuity correction" term (of -1 in the numerator) is incorporated to account for the fact that the statistic is discrete while the χ^2 distribution with 1 degree of freedom is continuous. If the null hypothesis is correct, then the probability that this quantity is greater than $\chi^2_{1,0.95} = 3.841$ is less than 0.05 (the significance level $\alpha = 0.05$). So we may reject the null hypothesis in favor of the hypothesis that the two classifiers have different performance when $\chi^2_{1,0.95} > 3.841$.

In our work, McMenar's test was also used to interpret detection results across different feature sets (but still based on the same training data).

4   Baseline Detection Method

In this study, an outlier-detection-based system inspired by the work described in [6] was taken as a baseline for the detection of misannotated words. For the outliers detection, the most prevalent and intuitive outcome of HMM forced alignment—phone durations and acoustic likelihood of each phone model were used. The duration can help in revealing unusually long or unusually short phone segments that tend to accompany annotation errors. The acoustic likelihood of a phone segment indicates acoustic reliability of the aligned phone. Phone-dependent outlying values, in the case of phone durations both extremely long and short values and in the case of acoustic likelihoods very small log likelihood values, were then used to indicate a potentially badly segmented phone segment. More specifically, for PERCp in Table 1, segments of phone i with duration d_i outside the range $d_i(100 - p) < d_i < d_i(p)$ or log likelihood $l_i < l_i(100 - p)$ were denoted as outliers. Values $d_i(100 - p)$, $d_i(p)$, and $l_i(100 - p)$ denote duration at $(100 - p)$th and pth percentile and log likelihood at $(100 - p)$th percentile of phone i. Words containing at least one phone segment with an outlying value were then marked as misannotated. Detection results for different percentile-based outlying values are shown in Table 1.

Table 1. Evaluation of outlier-based annotation errors detection. The numbers denote mean values $\pm$ standard deviation over the 10 evaluation data sets.

Percentile	Acc [%]	P [%]	R [%]	$F1$ [%]
PERC99.00	89.0 ± 2.0	67.0 ± 4.8	88.5 ± 3.1	76.2 ± 3.6
PERC99.25	90.1 ± 2.2	71.5 ± 6.2	84.7 ± 4.5	77.4 ± 4.7
PERC99.50	92.6 ± 1.7	79.6 ± 5.7	84.7 ± 4.5	82.0 ± 4.1
PERC99.75	$\mathbf{93.6 \pm 1.8}$	$\mathbf{85.7 \pm 4.4}$	$\mathbf{81.1 \pm 6.0}$	$\mathbf{83.3 \pm 4.8}$
PERC99.95	91.6 ± 0.8	94.0 ± 2.3	61.3 ± 3.5	74.2 ± 2.8

5   SVM Classifier

The problem of annotation errors detection can be viewed as a two-class classification problem: whether a word is misannotated or not. For the purposes of our work described in this paper we utilized support vector machines (SVMs) with radial basis function (RBF) kernel [14], which have proven to be successful in various classification tasks. Experiments with other classifiers and also with other approaches to annotation errors detection (such as novelty and outlier detection), and their comparison with the SVM classifier are described in [15].

5.1   Feature Sets

Basic Features (BAS). Similarly as in Section 4, features based on phone durations and acoustic likelihood of each phone model were taken, describing each word by the following 7 features: mean, minimum, and maximum phone duration within the word, mean, minimum, and maximum phone acoustic likelihood within the word, and the number of phones in the word.

Histogram Related Features (HIST). In order to emphasize outlying durations and acoustic likelihoods, histogram of durations H_D and histogram of acoustic likelihoods H_A with non-uniform bin widths were used to extend the basic feature set. The bins for H_D were defined with edges in msec as $[0, 10, 20, 50, 100, 200, \infty]$, and the bins for H_A with edges in log likelihoods as $[-\infty, -200, -150, -100, -70, -40, 0]$, resulting in 12 features.

Phonetic Features (PHON). Another feature set concerned phonetic properties of each word. The following 28 features, observed to often accompany errors caused by misannotations, were taken into account: *"voicedness" ratio* (the ratio between voiced and unvoiced phones in the word—1 feature), *word boundary "voicedness" match* (whether the beginning/end of the word matches the end/beginning of the previous/next word with respect to "voiceness"—2 features), *sonority ratio* (the ratio between sonorized and noised phones in the word—1 feature), *manner of articulation* (number of phones in various manner-of-articulation related classes—11 features), *place of articulation* (number of phones in various place-of-articulation related classes—12 features), and *syllabic consonants* (whether the word contains a syllabic consonant or not—1 feature).

Positional Features (POS). Positional features include the position of the word in a phrase, the position of the phrase in an utterance, both in forward and reverse order, number of words in the phrase, and number of phrases in the utterance (6 features in total).

Deviation from Duration Model (DEV). To emphasize duration-related features, the deviation of the forced-alignment based duration of each phone from the duration predicted by another duration model was used as an another feature set. The duration model was based on classification and regression trees (CART) and trained on the same forced-aligned speech corpus as used throughout this paper. Various phone-level features like the phonetic contexts (up to 2 phones to the left and to the right) and the categorization of the phones into various phone classes were used. In addition, prosody related features like the number of phones in a word, the number of words in the phrase, the number of phrases in an utterance, and the position of each phonetic element (phone, word, phrase) in the parent structure (word, phrase, utterance) were used as well (172 features in total). Since the training phone durations were based on automatically segmented speech corpus, statistically outlying durations were not used—only durations between 5 and 95 percent fractile (computed for each phone independently) were included into the training data [16]. For each phone an independent CART was trained using EST tool *wagon* [17]. Similarly as for the basic feature set, each word was then assigned 3 features—mean, minimum, and maximum deviation of forced-aligned phone duration from CART-based phone duration.

5.2 Classification Procedure

The SVM-based classification procedure can be summarized in the following steps:

1. The data was split into 10 train/evaluation data pairs with 80% of words being used for training and 20% of words being used for evaluation. The splits were made by preserving the ratio of correctly annotated and misannotated words for each class.
2. For each particular training/evaluation pair, the following steps were carried out:
   (a) The training data was standardized to have zero mean and unity variance.
   (b) A classifier was trained on the training data, and its parameters were optimized by grid search using 5-fold cross-validation.
   (c) The same standardization method as for the training data was applied to the evaluation data.
   (d) The performance of the resulting classifier was evaluated with the metrics described in Section 3.
3. The overall performance of the classifier was computed as an average over the evaluations for each training/evaluation data pairs.

Classifier parameters were optimized with respect to $F1$ score during grid search & cross-validation process. Otherwise, with low recall score the classifier would not be able to detect the misannotated words reasonably, and low precision score would indicate that the classifier falsely detects words that are annotated correctly—in the case of manual correction of speech corpora too many words would be unnecessarily checked reducing thus the efficiency of annotation error detection and correction. For training and evaluation, *scikit-learn* toolkit was employed [18].

6 Results and Discussion

The detection performance for different feature sets is given in Table 2. Bold rows denote that the corresponding feature sets performed better than the other ones according to McNemar's test at the significance level 0.05. "All features" stands for BAS+HIST+DEV+PHON+POS. Results for the cases where no acoustic likelihoods and no duration-related features would be available in the speech corpus are shown in the rows "no likelihoods" or "no lklhd., no dur.", respectively.

Table 2. Annotation error detection results. The numbers denote mean values $\pm$ standard deviation over the 10 training/evaluation data splits.

Set ID	Features	A [%]	P [%]	R [%]	$F1$ [%]
ID1	BAS	93.1 ± 0.9	83.2 ± 4.1	82.3 ± 4.0	82.6 ± 2.3
ID2	BAS+HIST	93.9 ± 1.1	85.4 ± 4.0	84.0 ± 4.0	84.6 ± 2.6
ID3	BAS+HIST+PHON	93.3 ± 1.4	81.4 ± 5.2	86.6 ± 5.0	83.7 ± 3.3
ID4	BAS+HIST+POS	92.9 ± 1.7	79.8 ± 5.8	86.8 ± 4.4	83.0 ± 3.7
ID5	**BAS+HIST+DEV**	**95.2 ± 1.2**	**89.0 ± 4.1**	**86.6 ± 5.1**	**87.6 ± 3.1**
ID6	**All features**	**94.9 ± 1.3**	**89.0 ± 3.4**	**85.1 ± 5.2**	**86.9 ± 3.6**
ID7	No likelihoods	93.9 ± 1.4	86.8 ± 4.7	82.3 ± 6.3	84.3 ± 3.8
ID8	No lklhd., no dur.	63.0 ± 2.2	29.8 ± 2.2	63.8 ± 6.9	40.6 ± 3.3
ID0	PERC99.75	93.6 ± 1.8	85.7 ± 4.4	81.1 ± 6.0	83.3 ± 4.8
RND	Random classifier	68.0	20.0	20.0	20.0

Table 3. Statistical significance of the detection results

Feature sets	ID1	ID2	ID3	ID4	ID5	ID6	ID7	ID8
ID0	0	2	3	0	5	6	7	0
ID1		–	–	–	5	6	–	1
ID2			–	4	5	6	–	2
ID3				–	5	6	–	3
ID4					5	6	–	4
ID5						–	5	5
ID6							6	6
ID7								7

McMenar's test was also carried out to interpret the results. The element $[i, j]$ of Table 3 indicates which one from the feature sets IDi and IDj performs significantly better (at the significance level 0.05, see Section 3.2). The value "–" means that the difference in performance of IDi and IDj is not statistically significant. ID0 stands for the baseline outlier-based detection method described in Section 4, and RND denotes a random classifier considering the number of correctly annotated and misannotated words in our data set. As can be seen, both BAS+HIST+DEV (ID5, a small set of 22 duration and acoustic likelihood related features) and the set of all 53 features (ID6) achieved better results than the other feature sets and also better than the baseline detection method. Reasonably good performance was also achieved for the feature set without acoustic likelihoods. On the other hand, duration-related features appeared to be essential for a good performance.

7 Conclusion

We performed a study on the automatic detection of misannotated words in single-speaker read-speech corpora. We proposed an SVM-based classifier which proved to detect the annotation errors better than the baseline outlier-based detection algorithm. We also experimented with various feature sets and showed that a small set of duration and acoustic likelihood related features gives very good results (in terms of both precision and recall) with $F1$ measure being almost 88%. These results were comparable to those achieved for all features.

Given the very promising results, fine-tuning of the proposed classifier (e.g., contextual classification and/or using different acoustic features [19]) and its application for annotation errors detection in multi-speaker spontaneous-speech corpora typical for ASR systems [20] or in multimedia archives for fast information retrieval [21] is planned to be researched.

References

1. Matoušek, J., Tihelka, D., Šmídl, L.: On the impact of annotation errors on unit-selection speech synthesis. In: Sojka, P., Horák, A., Kopeček, I., Pala, K. (eds.) TSD 2012. LNCS, vol. 7499, pp. 456–463. Springer, Heidelberg (2012)

2. Matoušek, J., Romportl, J.: Recording and Annotation of Speech Corpus for Czech Unit Selection Speech Synthesis. In: Matoušek, V., Mautner, P. (eds.) TSD 2007. LNCS (LNAI), vol. 4629, pp. 326–333. Springer, Heidelberg (2007)
3. Adell, J., Agüero, P.D., Bonafonte, A.: Database pruning for unsupervised building of text-to-speech voices. In: Proc. ICASSP, Toulouse, France, pp. 889–892 (2006)
4. Tachibana, R., Nagano, T., Kurata, G., Nishimura, M., Babaguchi, N.: Preliminary experiments toward automatic generation of new TTS voices from recorded speech alone. In: Proc. INTERSPEECH, Antwerp, Belgium, pp. 1917–1920 (2007)
5. Wei, S., Hu, G., Hu, Y., Wang, R.H.: A new method for mispronunciation detection using support vector machine based on pronunciation space models. Speech Commun. 51(10), 896–905 (2009)
6. Kominek, J., Black, A.: Impact of durational outlier removal from unit selection catalogs. In: Proc. SSW, Pittsburgh, USA, pp. 155–160 (2004)
7. Lu, H., Wei, S., Dai, L., Wang, R.H.: Automatic error detection for unit selection speech synthesis using log likelihood ratio based SVM classifier. In: Proc. INTERSPEECH, Makuhari, Japan, pp. 162–165 (2010)
8. Wang, W.Y., Georgila, K.: Automatic detection of unnatural word-level segments in unit-selection speech synthesis. In: Proc. ASRU, Hawaii, USA, pp. 289–294 (2011)
9. Tihelka, D., Kala, J., Matoušek, J.: Enhancements of Viterbi search for fast unit selection synthesis. In: Proc. INTERSPEECH, Makuhari, Japan, pp. 174–177 (2010)
10. Young, S., Evermann, G., Gales, M., Hain, T., Kershaw, D., Liu, X., Moore, G., Odell, J., Ollason, D., Povey, D., Valtchev, V., Woodland, P.: HTK Book (for HTK Version 3.4). The Cambridge University, Cambridge (2006)
11. Matoušek, J., Tihelka, D., Psutka, J.V.: Experiments with Automatic Segmentation for Czech Speech Synthesis. In: Matoušek, V., Mautner, P. (eds.) TSD 2003. LNCS (LNAI), vol. 2807, pp. 287–294. Springer, Heidelberg (2003)
12. Matoušek, J., Romportl, J.: Automatic pitch-synchronous phonetic segmentation. In: Proc. INTERSPEECH, Brisbane, Australia, pp. 1626–1629 (2008)
13. Dietterich, T.G.: Approximate statistical tests for comparing supervised classification learning algorithms. Neural Comput. 10, 1895–1923 (1998)
14. Cortes, C., Vapnik, V.: Support-vector networks. Machine Learning 20(3), 273–279 (1995)
15. Matoušek, J., Tihelka, D.: Annotation errors detection in TTS corpora. In: Proc. Interspeech, Lyon, France (2013)
16. Romportl, J., Kala, J.: Prosody modelling in Czech text-to-speech synthesis. In: Proc. SSW, Bonn, Germany, pp. 200–205 (2007)
17. Taylor, P., Caley, R., Black, A., King, S.: Edinburgh speech tools library: System documentation (1999), http://www.cstr.ed.ac.uk/projects/speech_tools/manual-1.2.0/
18. Pedregosa, F., Varoquaux, G., Gramfort, A., Thirion, V.M.B., Grisel, O., Blondel, M., Prettenhofer, P., Weiss, R., Dubourg, V., Vanderplas, J., Passos, A., Cournapeau, D., Brucher, M., Perror, M.: Édouard Duchesnay: Scikit-learn: Machine learning in Python. J. Machine Learn. Res. 12, 2825–2830 (2011)
19. Přibil, J., Přibilová, A.: Comparison of spectral and prosodic parameters of male and female emotional speech in Czech and Slovak. In: Proc. ICASSP, Prague, Czech Republic, pp. 4720–4723 (2011)
20. Ircing, P., Psutka, J., Psutka, J.V.: Using morphological information for robust language modeling in Czech ASR system. IEEE Trans. Audio Speech Lang. Process. 17, 840–847 (2009)
21. Psutka, J., Švec, J., Psutka, J.V., Vaněk, J., Pražák, A., Šmídl, L., Ircing, P.: System for fast lexical and phonetic spoken term detection in a Czech cultural heritage archive. EURASIP J. Audio Speech Music Process. 10 (2011)

Scratching the Surface of Possible Translations

Ondřej Bojar, Matouš Macháček, Aleš Tamchyna, and Daniel Zeman

Charles University in Prague, Faculty of Mathematics and Physics
Institute of Formal and Applied Linguistics
`{bojar,machacek,tamchyna,zeman}@ufal.mff.cuni.cz`

Abstract. One of the key obstacles in automatic evaluation of machine translation systems is the reliance on a few (typically just one) human-made reference
translations to which the system output is compared. We propose a method of
capturing millions of possible translations and implement a tool for translators to
specify them using a compact representation. We evaluate this new type of reference set by edit distance and correlation to human judgements of translation
quality.

Keywords: machine translation, evaluation, reference translations.

1 Introduction

The relationship between a sentence in a natural language as written down and its
meaning is a very complex phenomenon. Many variations of the sentence preserve the
meaning while other superficially very small changes can distort or completely reverse
it. In order to process and produce sentences algorithmically, we need to somehow
capture the semantic identity and similarity of sentences.

The issue has been extensively studied from a number of directions, starting with
thesauri and other lexical databases that capture synonymy of individual words (most
notably the WordNet [1], [2]), automatic paraphrasing of longer phrases or even
sentences (e.g. [3], [4], [5]) or textual entailment [6]. We are still far away from a
satisfactory solution.

The field of machine translation (MT) makes the issue tangible in a couple of ways,
most importantly within the task of automatic MT evaluation. Current automatic MT
evaluation techniques rely on the availability of some reference translation, i.e. the
translation as produced by a human translator. Obtaining such reference translations
is relatively costly, so most datasets provide only one reference translation, see e.g. [7].

Figure 1 illustrates the situation: while there are many possible translations of a given
input sentence, only a handful of them are available as reference translations. The sets of
hypotheses considered or finally selected by the MT system can be completely disjoint
from the set of reference translations. Indeed, only about 5–10% of reference translations were *reachable* for Czech-to-English translation [8], and about a third of words
in a system output are not confirmed by the reference despite not containing any errors based on manual evaluation [9]. Relying mostly on unreachable reference translations is detrimental for MT system development. Specifically, automatic MT evaluation
methods perform worse and consequently automatic model optimization suffers.

I. Habernal and V. Matousek (Eds.): TSD 2013, LNAI 8082, pp. 465–474, 2013.

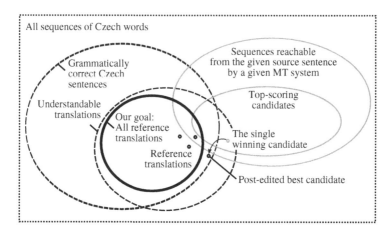

Fig. 1. The space of all considerable translations of a given source sentence. Human-produced sets are denoted using black lines, machine output is in grey.

We would like to bring the sets of acceptable and reachable translations closer to each other, providing more space for optimal hypothesis selection. This paper presents one possible step in that direction, namely significantly enlarging the set of reference translations. As outlined above, the dataset we created could serve well in research well beyond the MT field, e.g. in an analysis of *sentence-level* paraphrases.

In Section 2, we describe our annotation tool for producing large numbers of correct translations and relate it to a similar tool developed for English [10]. Section 3 provides basic statistics about the number of references we collected and Section 4 carefully analyzes and discusses their utility in MT evaluation.

2   Annotation Tools for Producing Many References

The most inspiring work for our experiment was that of Dreyer and Marcu [10]. Their annotators produce "translation networks", a compact representation of many references, to be used in their HyTER evaluation metric.

We experimented with their annotation interface developed primarily for English but found it rather cumbersome for Czech and other languages with richer morphology and a higher degree of word order freedom.

2.1   Recursive Transition Networks

The approach of [10] is based on recursive transition networks (RTN, [11]), a formalism with the power of context-free grammars. The main building block of the annotation tool in [10] is called a "card" and it covers multiple translations of a short phrase. By combining cards, a large network for the whole sentence can be built. Every path through the network represents a new sentence and the annotators construct the networks so that all such sentences are synonymous. See Figure 2 for an example.

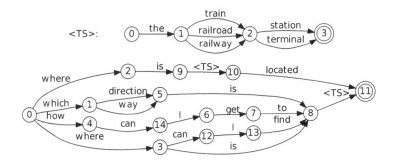

Fig. 2. An example of English recursive transition network from [10]

Reordering of phrases (not discussed in [10]) is possible within the RTN framework by changing the order of cards. The need of such a mechanism in English may seem negligible, yet there are situations such as direct speech where mutual positions of large blocks of text are perfectly interchangeable; similar patterns occur in Czech as well:

- He wanted to step down, he said, "so I could work with more freedom."
- He said: "I want to step down so I could work with more freedom."
- "I want to step down so I could work with more freedom," he said.

More importantly, it is difficult to specify *conditions* under which particular cards can combine in RTN. In morphologically rich languages, we often have to translate a phrase differently based on morpho-syntactic rules. For instance, functions of English verb arguments are determined using word order and prepositions. In Czech, they are determined using prepositions and morphological cases. Verbs subcategorize for noun phrases in particular cases. Consequently, if a verb is replaced by a synonym (or if the verb phrase is passivized or nominalized), the required case for the arguments of the verb may change. The case of the noun must then be reflected by its adjectival modifiers.

Let us illustrate this by an example. Consider the sentence *"The city council approved a new regulation."* The cards that would model pieces of this sentence in English could be combined more or less freely (within the fixed word order):[1]

the (city council / local government) (approved / gave blessing to / agreed with) a new (regulation / directive / decree)

If we ignored morphology, we could get a very similar picture with the Czech equivalents of the phrases:

(městská rada / zastupitelstvo města) (schválila / požehnala / souhlasila s) nový (předpis / směrnici / nařízení)

---

[1] We are aware that certain domains are much more sensitive to meaning distortions (directive vs. regulation) or inappropriate register (approve vs. give blessing). Our definition of meaning equivalence is not strict. We permit slight divergence if it can be reasonably expected that a human translator will pick either of the alternatives.

So far the alternatives within each part of the sentence differ *lexically*. However, the lexical selections have morphological implications and thus we also have to define *morphological* alternatives:

- One subject is feminine *(městská rada)*, the other is neuter *(zastupitelstvo města)*. Their gender must be reflected by the verbs *(schválila / schválilo // požehnala / požehnalo // souhlasila s / souhlasilo s)*.
- On a similar note, the three synonyms for *regulation* differ in gender, which dictates different suffixes for the adjective *new: nový předpis / novou směrnici / nové nařízení*.
- Each of the three verbs subcategorizes for a different case: *schválila* requires object in accusative, *požehnala* in dative and the preposition in *souhlasila s* requires instrumental. Thus we have *nový předpis / novému předpisu / novým předpisem // novou směrnici / nové směrnici / novou směrnicí // nové nařízení / novému nařízení / novým nařízením*.

2.2   Unification-Based Annotation

The RTN framework gives us a powerful tool to *combine* "cards". We would ideally want a tool that also lets us specify the *constraints* that must be fulfilled if two cards are to be combined.

We thus created our own compact representation for languages similar to Czech. Our main building block called *bubble*, comparable to the cards of [10], is defined by:

- the set (possibly discontinuous) of source language tokens it covers;
- the set of conditions it meets;
- the set of translation alternatives in the target language. Every alternative in the set covers the same set of source tokens and meets the same conditions.

A translation alternative is composed of *atoms* (tokens of the target language) and/or *slots* (positions in the translation alternative, specifying properties of other bubbles that are permitted to fill the slot). Where an RTN would refer to a smaller transition network (card) by its name, we refer to a smaller bubble by enumerating the constraints it must meet. For instance, we ask that the bubble covers the source word *regulation* (it may cover more words but this one must be among them) and that its form is in the accusative. Obviously we could achieve the same result in RTN by using more explicative card names, e.g. *regulation-acc*. Our approach is equivalently expressive but it increases annotators' comfort as well as maintainability of the whole system. While RTNs could be rewritten as a context-free grammar, our approach can be thought of as a unification grammar.

Typical creation of a translation network is analogous to traversing the dependency tree structure that models the syntax of the sentence. One starts at the verb, defines its translations and creates slots for its arguments (and adjuncts). Each argument typically receives its own bubble. The bubble can define alternations for a whole noun phrase, or it can again use slots to separate description of a modifier that could be reused elsewhere. Occasionally a bubble represents a subordinated clause and the process is applied recursively.

Unlike in common unification grammars, we are not forced to annotate a full syntactic tree. It is possible e.g. to create one flat bubble for the whole sentence. The only guiding principle in creating nested bubbles is economy: a set of alternations useful at two or more places is a candidate for a new bubble.

Along the same lines, the decomposition of the sentence into bubbles does not need to reflect linguistic constituents. Sometimes it is practical to take punctuation as the root, rather than the verb; high-level word order decisions drive the distribution of commas and quotation marks around clauses; co-ordination could be preferred over dependency etc. The set of possible constraints is not restricted in any way (e.g. to morphological categories and their values). The annotator is free to introduce arbitrary constraints, e.g. for ensuring good co-reference patterns, auxiliaries in coordination, rhematizers and negation interplay or even style and register features.

We developed two annotation environments in which translators create the compact representations. The Prolog programming language appears to be ideally suited for evaluation of constraints and expansion of bubbles. Several translators encode their annotations directly in Prolog. For those less technically capable we also designed a web-based graphical interface.

2.3 Prolog Interface

Roughly 300 lines of pre-programmed Prolog code provide the necessary set of predicates that check constraints (bubble-slot compatibility) and make sure that all tokens of the source sentence are covered. The translator essentially creates a set of clauses for the predicate `option()`, each of those encodes a bubble. Every `option` lists the source words covered, the conditions met and the target sequence consisting of atoms and slots. A slot refers at least to one source word that must be covered by the bubble in the slot; optionally, it also specifies additional conditions that must be met. Example:[2]

```
% option(+SrcWordsCovered, +ConditionsMet, +OutputAtomsAndSlots).
option([the, city, council], [], [městská, rada, [approved, fem]]).
   % "The city council" can be translated as "městská rada" and then
   % it requires the translation of "approved" in feminine gender.
option([approved], [fem], [schválila, [regulation, acc]]).
option([approved], [fem], [souhlasila, s, [regulation, ins]]).
   % Different translations of "approved" require "regulation" in
   % either "acc"usative or "ins"trumental cases.
option([a, new, regulation], [acc], [nový, předpis]).
option([a, new, regulation], [ins], [novým, předpisem]).
option([a, new, regulation], [acc], [novou, směrnici]).
option([a, new, regulation], [ins], [novou, směrnicí]).
```

A few additional constructs such as the logical `or()` and the possibility to simply drop a token further expand the tools the translators have at their disposal; note however that the syntax that the annotators have to grasp is extremely simple and virtually no knowledge of programming in general or Prolog in particular is required.

---

[2] A pre- and post-processor enables using uppercase letters and non-English letters freely in the `option()` predicates.

2.4 Web Interface

Some translators will be scared by any programming language, regardless how easy it may be. Others may face technical issues regarding installation and running a Prolog interpreter on their laptops. In order to accommodate all translators, we also developed a web-based graphical interface. It works as a wrapper for the Prolog engine: bubbles defined in the browser are sent to the server, converted to Prolog clauses and evaluated. The server then sends back either the full list of translations (which is only practical for shorter sentences) or the list of differences against the previous state.

The main problem of the web interface was that we did not anticipate that many bubbles and translations generated. The tool implemented is thus too heavy in terms of processing time, network load and even the required screen space.

3 Collected Data

We selected 50 sentences from the WMT11 test set [12] for our annotation. This particular test set was used in various experiments before and we can thus use:

- manual system rankings of the official WMT11 manual evaluation (see also [13] for a discussion of the rankings)
- the single official reference translation of WMT11 (denoted "O" in the following)
- three more reference translations that come from the German version of the test set [14] (denoted "G" in the following)
- two manual post-edits of a phrase-based MT system similar to those participating in the WMT11 competition (denoted "P"; this set contains only 1997 sentences, each post-edited by two independent annotators.)

Six translators were involved in the task, producing 77 sets of references altogether. Of the 50 sentences, 24 were translated by one annotator, 25 by two and one sentence has three annotations. Each annotator was instructed to spend at most 2 hours translating one sentence. More than 1 hour was needed for a typical sentence.

Utilizing the existing versions, one of the translators used German as the input language; the others used English. All of the annotators had access to one pre-existing human translation in English, German, Spanish and French, and up to four Czech translations, which they could use for inspiration.[3]

We combine networks produced by different translators using simple union of the sets of target sentences (*finite-state union* of [10]). This set is denoted "D" in the following.

Table 1 shows basic statistics of the annotations. In all cases, annotators who used the Prolog interface were more productive, creating over 255 thousand reference translations per sentence on average, compared to roughly 49 thousand references produced by the web users.

---

[3] The texts are news articles of mixed origin. One of the pre-existing "translations" was actually the original text.

Table 1. Basic statistics of our annotations

Annotator	Interface	# of Sents.	Avg. Sent. Length	Avg. Number of Refs.
A		3	14.0	483k
B	Prolog	5	22.5	246k
C		20	24.1	223k
D		25	25.1	54k
E	Web	19	23.0	26k
F		5	15.7	111k

Great when a film has several target groups , but a shame if they are mutually exclusive

Je výborné , když má film větší počet cílových skupin , jen je smutné , že se navzájem vylučují .
Není k zahození , pokud má film vícero cílových skupin , jen je smůla , když se navzájem vylučují .
Skvělé , že má film víc cílových skupin , je ale politováníhodné , pokud se navzájem neslučují .
Prima , pokud má film více cílových skupin , jen je smutné , pokud se vzájemně neslučují .

Je výhodné , když snímek má několik cílových skupin , smutné ale je , pokud jsou navzájem neslučitelné .
Skvělé je , když snímek má několik cílových skupin , škoda ale je , pokud se tyto skupiny vzájemně vylučují .
Dobré je , když má film více cílových skupin , je jen smolné , pokud jsou tyto navzájem neslučitelné .
Výhodou je , pokud se snímek zaměří na několik cílových skupin , je ale smolné , pakliže jsou vzájemně neslučitelné .

Fig. 3. Four random samples from 196k (Web) and 943k (Prolog) variations of one sentence

For the 25 sentences annotated by two translators, we measured the overlap between the sets of references. We created the union and intersection of the two sets (after tokenizing the sentences) and simply measured the ratio. The results provide further evidence of the richness of possible translations – on average, the overlap was only 4.4% of the produced references. With the exception of two very short sentences, the overlap was always below 10%. See Figure 3 for some examples.

4   Analysis

4.1   String Similarity

Our metric of similarity is based on the well-known Levenshtein distance [15] and returns a value between 0 (completely different) and 1 (identical strings):

$$\text{similarity}(x, y) = 1 - \frac{\text{levenshtein}(x, y)}{\max(\text{length}(x), \text{length}(y))}$$

In order to quantify the diversity of the produced manifold translations, we sampled pairs of translations of each sentence, looking for the pair with the smallest similarity. For four sentences, translations with similarity below 0.1 were found in the samples. The minimum similarity averaged over sentences in our dataset was 0.24±0.13 (standard deviation). This result indicates how much the surface realizations differed while preserving identical meaning.

We also measured the string similarity between system outputs and various reference translations. Table 2 summarizes the results. Unsurprisingly, each of the two manually post-edited translations (P_1 and P_2) are much closer to the original system outputs

Table 2. String similarity between system outputs and reference translations

System	String Similarity to the Given Reference			
	D (closest)	O	P$_1$	P$_2$
online-B	0.65	0.55	0.66	0.65
cu-tamchyna	0.65	0.52	0.68	0.69
cu-bojar-contrastive	0.65	0.52	0.64	0.66
cu-bojar	0.65	0.51	0.65	0.68
uedin-contrastive	0.64	0.54	0.64	0.65
uedin	0.64	0.54	0.64	0.65
cu-tamchyna-contrastive	0.64	0.50	0.66	0.66
cu-marecek	0.64	0.52	0.67	0.68
jhu-contrastive	0.62	0.52	0.61	0.61
jhu	0.62	0.51	0.59	0.61
cu-zeman	0.61	0.52	0.61	0.62
cu-popel	0.60	0.51	0.59	0.61
commercial1	0.57	0.48	0.57	0.57
commercial2	0.56	0.46	0.57	0.57

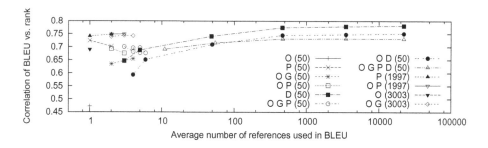

Fig. 4. Correlation of BLEU and manual system rankings with varying sets of references

than the official reference translation (O). However, our annotators managed to produce references (D) which are almost exactly as close as the post-edited translations, even though they did not have access to them or the system outputs (the table shows the similarity with the closest reference translation found).

4.2   Correlation with Manual Ranking

We evaluate the utility of the manifold reference translations by measuring the Pearson correlation between manual MT system evaluation and the common automatic MT evaluation method BLEU [16]. BLEU was originally tested with 4 reference translations and the number of reference translations is known to strongly influence its performance. In spite of that, BLEU is very often used with just a single reference translation, hoping that a larger test set (more sentences) will compensate the deficiency.

Figure 4 plots Pearson correlation of the official WMT11 system rankings and BLEU when varying the size of the test set (50, 1997 or 3003 sentences as noted in the legend) and the average number of references per sentence. The sources O, P, and G provide us with 1, 2 or 3 references respectively. Our new dataset (denoted "D") consists of all the references generated from the web or Prolog annotation interface by any of the annotators. The number of references for each sentence differs. We shuffle them and take up to 5, 50, 500, ..., 50k items from the beginning.

The very baseline correlation is 0.47 ("O 50" in the chart), obtained with only the single official reference translation on the test set reduced to the 50 sentences where we have our extended references. Using the full test set ("O 3003"), the correlation jumps to 0.69. The 1, 2 or 3 additional references coming from German ("O G 3003") indicate how well the "standard" BLEU should fare: around 0.74. These four references (one official and three coming from German) on the small set of 50 sentences lead to quite a low result: 0.65.

A notable result is 0.72 ("P 50") obtained on the 50-sentences test set when we use the post-edited translations instead of the official translations. The official translations are obviously more distant from what the systems are capable of producing given the source. With distant references, large portions of output are not scored and systems may differ greatly in the translation quality of those unscored parts. It thus seems sensible to manually post-edit just 50 sentences coming from a baseline version of an MT system and evaluate modifications of the system on this small but tailor-made test set rather than on a larger less-matching set, perhaps even if it had 4 reference translations. The post-edits may however still suffer some problems: in our case, using not just one of the post-edit versions but both of them, the correlation drops to 0.70.

Lines in the chart extending beyond 10 references include our large reference sets. The limit on the number of references has to be taken with caution: subsampling 50 references from a 100k set constructed in 2 hours of annotation is bound to give better results than stopping the construction as soon as it generates 50 options. We nevertheless see that around 5k references, the correlation curves flatten. This could be attributed both to the inherent limitations of BLEU (evaluating the precision of up to 4-grams of words) as well as still the low coverage of the test set. Dreyer and Marcu mention that even billions of references obtained from two annotators still did not include many of the translations suggested by a third annotator.

The solid baseline 0.69 ("O 3003") is reached shortly after 5 random items from our annotation only (the second point on the line "D 50"), a much smaller set of sentences with many possible translations. Using up to 50k of the possible translations leads to the correlation of 0.78. However, we cannot claim that spending 2 hours × 2 annotators × 50 sentences, i.e. about 200 hours of work, is better than translating 3003 sentences (also about 200 hours of work), because our annotators *did* have access to several versions of each of the sentences.

5   Conclusion

We developed a method and two annotation environments for producing many (possibly all) acceptable translations of a sentence. We performed a small scale experiment in which human translators processed 50 sentences. Obviously this annotation process is very costly (1 to 2 hours per sentence) and we cannot expect anyone to annotate large datasets this way. However, even the small sample provides a completely new point of view for the evaluation of machine translation output (tens or hundreds of thousands of references per average sentence). As expected, automatic evaluation against larger sets of references shows higher correlation with human judgment of translation quality.

A surprising observation is that just 50 post-edited translations serve as an equal or better reference than 3003 independent translations (correlation 0.70–0.72 vs. 0.69).

The annotated data we created is available to the research community. Besides machine translation, it can be also used to evaluate other NLP tasks, ranging from paraphrasing to grammar development or parsing.

Acknowledgements. This work was partially supported by the grants P406/11/1499 of the Grant Agency of the Czech Republic, FP7-ICT-2011-7-288487 (MosesCore) of the European Union and 1356213 of the Grant Agency of the Charles University. We are grateful to Markus Dreyer for a preview of their annotation interface.

References

1. Miller, G.A.: WordNet: A lexical database for English. Commun. ACM 38(11), 39–41 (1995)
2. Pala, K., Čapek, T., Zajíčková, B., et al.: Český WordNet 1.9 PDT (2010)
3. Bannard, C., Callison-Burch, C.: Paraphrasing with bilingual parallel corpora. In: Proc. of ACL, Ann Arbor, Michigan, USA, pp. 597–604 (2005)
4. Kauchak, D., Barzilay, R.: Paraphrasing for Automatic Evaluation. In: Proc. of NAACL/HLT, New York City, USA, pp. 455–462 (2006)
5. Denkowski, M., Lavie, A.: Meteor-next and the meteor paraphrase tables: Improved evaluation support for five target languages. In: Proc. of WMT and MetricsMATR, pp. 339–342. ACL, Uppsala (2010)
6. Androutsopoulos, I., Malakasiotis, P.: A survey of paraphrasing and textual entailment methods. Journal of Artificial Intelligence Research 38, 135–187 (2010)
7. Callison-Burch, C., Koehn, P., Monz, C., et al.: Findings of the 2012 Workshop on Statistical Machine Translation. In: Proc. of WMT, pp. 22–64. ACL, Montréal (2012)
8. Bojar, O., Kos, K.: 2010 Failures in English-Czech Phrase-Based MT. In: Proc. WMT and MetricsMATR, pp. 60–66. ACL, Uppsala (2010)
9. Bojar, O., Kos, K., Mareček, D.: Tackling Sparse Data Issue in Machine Translation Evaluation. In: Proc. of ACL Short Papers, pp. 86–91. ACL, Uppsala (2010)
10. Dreyer, M., Marcu, D.: HyTER: Meaning-Equivalent Semantics for Translation Evaluation. In: Proc. of NAACL/HLT, Montréal, Canada, pp. 162–171 (2012)
11. Woods, W.A.: Transition network grammars for natural language analysis. Commun. ACM 13(10), 591–606 (1970), doi:10.1145/355598.362773
12. Callison-Burch, C., Koehn, P., Monz, C., Zaidan, O.: Findings of the 2011 Workshop on Statistical Machine Translation. In: Proc. of WMT, pp. 22–64. ACL (2011)
13. Bojar, O., Ercegovčević, M., Popel, M., Zaidan, O.: A Grain of Salt for the WMT Manual Evaluation. In: Proc. of WMT, pp. 1–11. ACL, Edinburgh (2011)
14. Bojar, O., Zeman, D., Dušek, O.: Additional German-Czech reference translations of the WMT'11 test set
15. Levenshtein, V.: Binary codes capable of correcting deletions, insertions and reversals. In: Soviet Physics-Doklandy, vol. 10 (1966)
16. Papineni, K., Roukos, S., Ward, T., Zhu, W.J.: BLEU: A Method for Automatic Evaluation of Machine Translation. In: Proc. of ACL, Philadelphia, Pennsylvania, pp. 311–318 (2002)

Selecting and Weighting N-Grams
to Identify 1100 Languages

Ralf D. Brown

Carnegie Mellon University Language Technologies Institute
5000 Forbes Avenue, Pittsburgh, PA 15213, USA
ralf@cs.cmu.edu

Abstract. This paper presents a language identification algorithm using cosine similarity against a filtered and weighted subset of the most frequent n-grams in training data with optional inter-string score smoothing, and its implementation in an open-source program. When applied to a collection of strings in 1100 languages containing at most 65 characters each, an average classification accuracy of over 99.2% is achieved with smoothing and 98.2% without. Compared to three other open-source language identification programs, the new program is both much more accurate and much faster at classifying short strings given such a large collection of languages.

Keywords: language identification, discriminative training, n-grams.

1 Introduction

Language identification (and the related task of encoding identification) is a useful first step in many natural language processing applications. Increasing numbers of languages have significant online presence, but most prior work on language identification focuses on only a handful of (usually European) languages. As in many language identification techniques, the method presented here relies on n-gram statistics trained from text, but it gains much of its accuracy from *which* n-grams are selected for inclusion in a language model and how their statistics are converted into weights.

2 Method

The method used by our whatlang program is a k-nearest neighbor approach using cosine similarity (normalized inner product) of weighted byte n-grams as the metric, which permits each n-gram to be individually weighted according to the strength of the evidence it provides for – or *against* – the hypothesis that the input is in a particular language. One or more language models are trained for each combination of language and character encoding one wishes to identify. The cosine similarity scores are computed incrementally by adding the weighted score of each n-gram match between the input and the models to the overall score for each model, and the languages (and optionally character encodings) corresponding to a user-specified maximum of k highest-scoring models (provided $score \geq 0.85 \times highest$) are output as guesses.

I. Habernal and V. Matousek (Eds.): TSD 2013, LNAI 8082, pp. 475–483, 2013.

Observing that successive strings are more likely to be in the same language in most texts, the scores for the current input string may be smoothed by adding in a portion of the scores from the immediately prior string(s). The smoothed score is an adaptive linear interpolation between the current string's score vector and an exponentially-decaying sum of all previous strings' score vectors. The interpolation takes into account the current string's length and the maximal score of any model; the higher either of these two factors is, the less weight is given to the previous scores. To avoid excessive smoothing, the inter-string score smoothing weights were tuned on held-out data such that the error rate matched the un-smoothed error rate when the language changed after every fourth string, and was better if the same language occurred at least five times consecutively.

Given the task of selecting the most useful n-grams to populate a language model of limited fixed size, one should focus on the most frequent n-grams. However, even among the top n-grams, there are those which are clearly not indicative of a language (e.g. strings of digits or whitespace), and some which are less informative if certain other n-grams are already included in the model.

whatlang treats the input as an untokenized stream of bytes, allowing n-grams to capture multi-word phenomena. The baseline language model extracts from the training data the K highest-frequency three- through N-grams which do not span a linebreak, start with a tab character, two blanks, three digits, or three identical puncation marks. N is set to 6, 10, or 12 for language/encoding pairs which are predominantly one, two, or three bytes per character. Previous experiments [1] showed that 6 is the optimum fixed length limit for K of 3000 and 9000, and the longer limits for multi-byte situations account for the reduced information content of all but the final byte of a character where the script for a language fits into a 256-codepoint block.

The first modification to the baseline model is to filter out n-grams which are substrings of other n-grams also in the model but which contribute little additional information about the language. For example, if the 8-gram "withhold" is in the language model, the 7-gram "withhol" and six-gram "ithhol" would not be informative as all three strings occur the same number of times in the data. The substring need not occur exactly the same number of times in the training data for its removal to be useful, so the removal threshold is a tunable parameter. A threshold of 0.62 (the optimum found for the held-out development set described in Section 4) means that "withhold" would have to occur at least 62% as often as "ithhol" for the latter to be excluded. Removing such uninformative n-grams frees space in the fixed-size model for alternative n-grams.

The second modification to the baseline model is to use discriminative training to add n-grams which provide negative evidence (what we will call "stopgrams"). To compute the appropriate stopgrams for a language model, we first train baseline models for all languages, then select those baseline models whose cosine similarity relative to the baseline model for the language being trained is above some threshold, typically in the range of 0.4 to 0.6. The union of the n-grams in those selected models is formed, and the training text is scanned for occurrences; any which never appear in the training data are added to the baseline model as stopgrams, with an appropriate weight as discussed in the next section. Although this adds n-grams to individual models, the global set of n-grams remains unchanged since the additions were already present in another model.

Weighting the n-grams in a language model is as important as selecting them. A simple uniform weight would not allow an n-gram shared between multiple models to distinguish between them, even though that n-gram is stronger evidence for a language in which it occurs more frequently. On the other hand, using the actual probability of occurrence may over-state differences resulting largely from the relative small amounts of training data. Another factor to be examined in weighting an n-gram is its length, since longer n-grams will typically occur less frequently but are expected to be individually more informative as they are less likely to be encountered by chance.

Empirically, the best weight for an n-gram G was determined to be

$$probability(G)^{0.27} \times length(G)^{0.09}$$

where the probability is simply the frequency of occurrence in the training data divided by the training data's size. These parameters were tuned using the held-out development set but their values proved not to be critical. Varying the exponent for $length(G)$ over the range 0.0 to 1.0 resulted in baseline error rates ranging from 2.414 to 2.421%, a relative change of less than 0.3%. Error rates varied less than 0.5% (relative) for $probability(G)$ exponents ranging from 0.20 to 0.50.

When using discriminative training, stopgrams need to be weighted separately from baseline n-grams; three additional factors contribute to a stopgram's weight. Within the union set of n-grams used as candidate stopgrams, each is weighted by the maximum cosine similarity of any of the individual language models containing it times the maximum baseline score within those models. However, when there is only a small amount of training data available for a language, an n-gram may fail to occur in the training data not because it is not permitted by the language, but simply due to lack of data. Thus, for training sizes less than 2,000,000 bytes, the weight of each stopgram is discounted by a factor of

$$\left(\frac{\max(0, ||t|| - 15000)}{2000000 - 15000} \right)^{0.7}$$

The exponent of 0.7 compensates for the power-law distribution of the most frequent n-grams by making the weight increase more rapidly for small training sizes than a simple linear discount. Finally, an overall weight is given to stopgrams. Their occurrence in the input is strong negative evidence; empirical tuning on the held-out set confirms this with an optimal global weight of -9.0.

3   Related Work

One of the earliest uses of n-grams for language identification, Cavnar and Trenkle's [2] rank-order statistics of the (usually 400) most frequent 1- through 5-grams in the training and test data, has become quite popular; numerous implementations are available, typically including models for between 70 and 100 language/encoding pairs. On a 14-language collection, they reported 98.6% accuracy for documents of 300 or fewer bytes using models with 300 n-grams, trained on 20K to 120K of text.

Ljubesic *et al* [3] use a form of discriminative training to distinguish among Croatian, Slovenian, and Slovak. The discriminative training consisted of stop-word lists for

Croatian and Serbian containing words which appear at least five times in the training data for one language but never in the training data for the other. The occurrence of *any* word on a stop-word list in a document tentatively classified as that language would switch the classification if no stop-words for the other language were present. The authors report a final document-level accuracy among the three languages above 99%.

Ahmed *et al* [4] use an incremental inner product. For 50-byte test strings over a twelve-language collection, accuracy was 88.66% for a Naive Bayes classifier, 96.56% for rank-order statistics, and 97.59% for inner product.

Carter *et al* [5] use semi-supervised priors on Twitter messages to bias language identification based on the assumption that a particular user will only post in a limited number of languages, that conversations will remain in the same language and that pages linked from tweets will be in the same language as the tweet. On a five-language collection, they found that applying these priors improved overall accuracy across the five languages from 91.0% to 93.2%.

Very few published results for language identification involve more than 20 languages. Damashek [6] reported an experiment visualizing the similarities of 31 languages, and Shuyo [7] reported 99.8% average accuracy at the document level for news articles in 49 languages. Xia *et al* [8] used information from the containing document (primarily the occurrence of the language name) to help classify Interlinear Glossed Text examples in 638 languages with an accuracy of 84.7 to 85.1%. Some closed-source offerings now provide identification for large numbers of languages; the most comprehensive we have found to date is Likasoft's Polyglot 3000 [9], which claims to support more than 400 languages.

4   Training and Test Data

The greatest difficulty in performing language identification for more than 1000 languages is in actually obtaining text in that many languages, properly labeled by language. Wikipedia is available in 285 language versions as of this writing, of which some 200 have sufficient text to be useful for our purposes. Translations of the full Christian Bible have been made into at least 475 languages and the New Testament has been translated into a further 1240 [10]; hundreds of them have become available as e-texts since early 2010. The New Testament is large enough at around one million characters to be useful as training data. Bible translations were obtained from web sites such as bible.is, ScriptureEarth.org, PNGScriptures.org, and GospelGo.com; they include languages from all continents except Antarctica, ranging from millions of speakers to nearly extinct. Scripts used by these languages include, among others, Arabic, Cyrillic, Canadian Aboriginal Syllabics, Devanagari, Han, Khmer, Lao, Sinhala, Tamil, Telugu, Thai, and Tibetan.

Wikipedia data was obtained by downloading pages linked via the
{langcode}.wikipedia.org/Special:PrefixIndex
search page and extracting the main entry from each such page. The resulting lines of text were sorted, removing duplicates and lines which were unambiguously English,

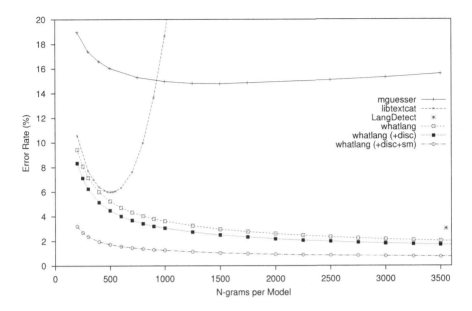

Fig. 1. Average accuracy over 1100 languages when using 200 to 3500 n-grams per language model. whatlang was tested in a baseline configuration and with discriminative training and inter-string smoothing enabled. LangDetect averaged 5642 n-grams per model.

French, or German as well as long sequences of ASCII in non-Latin scripts, and given some manual cleaning such as eliminating most occurrences of templates like "X is a city in Y with a population of Z" to avoid skewing the statistics. Despite this cleaning, there is still a fair amount of pollution from other languages in each language's data.

The available data for each language was split into training and test sets in one of two ways. For Bible translations organized as one file per chapter, the first verse of each chapter was held out as test data and the remainder used for training. In all other cases, a uniform sample of every 30th line was held out as test data. The held-out text was converted into test strings by word-wrapping the lines to 65 characters or less and then filtering out any wrapped lines containing fewer than 25 bytes as well as any resulting lines in excess of 1000 (using a New Testament as training data results in approximately 700 test strings).

For the 153 languages for which the above train/test split produced more than 4 million bytes of training data, the training file was split again, reserving a uniform sample of every 30th line as a development set for parameter tuning.

The final collection of languages for whatlang contains a total of 1119 languages. Nineteen of the languages (kept as potential confounders) do not have enough data to form a useful test set of at least 100 test strings, leaving 1100 languages for testing. Because a number of languages use multiple scripts, there were 1129 test files containing a total of 824,171 lines.

5   Experiments and Results

Multiple sets of language models were trained with differing numbers of n-grams per model, differing amounts of training data per model, varying substring filtering thresholds, and varying stopgram weights. We evaluated whatlang by identifying the language of each of the test strings in each of the 1100 languages using each set of models, and computing micro-average and macro-average classification error rates (total errors divided by total classifications, and average of per-language error rates, respectively) under multiple conditions, including with and without substring filtering applied to the language models and with and without inter-string score smoothing.

Three other open-source language identification programs were trained and run on the same data. libtextcat, version 2.2-9 [11], is a C reimplementation of the Cavnar and Trenkle approach. mguesser version 0.4 [12] is part of the mnoGoSearch search engine; its similarity computation is an inner product between 4096-element hash arrays for the input and each trained language, where the elements are normalized to a mean of zero and standard deviation of 1.0. LangDetect [13] version 2011-09-13 is a Java library using the Naive Bayes approach. All three packages' identification programs were modified to optionally provide language identification on each line of their input rather than on the entire file, and libtextcat and LangDetect were modified to process the entire input rather than a small initial segment in whole-file mode to match whatlang. LangDetect's limitation of one model per language code was worked around by adding disambiguating digits to the language code during training and removing them from its output prior to scoring.

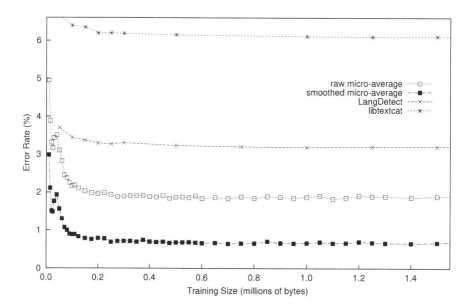

Fig. 2. Average accuracy over the 184 languages with at least 3 million bytes of training data when trained with 15,000 to 1,500,000 bytes of text per language model

Table 1. Comparison of run times in seconds and memory requirements for the programs. Times in parentheses exclude program startup (one invocation per test file) and scoring overhead.

Program	Model Size	Training Time	Error Rate	Evaluation by line		Evaluation by file		Disk Space	Memory
whatlang	500	763s	1.743%	38s	(25s)	35s	(21s)	16 MB	16 MB
whatlang	3500	1622s	0.754%	67s	(51s)	63s	(47s)	97 MB	94 MB
libtextcat	500	583s	5.950%	2138s	(1975s)	125s	(4s)	4.9 MB	18 MB
mguesser	1500	166s	14.783%	15482s	(15218s)	268s	(26s)	19 MB	76 MB
LangDetect	5642	699s	3.035%	9491s	(1158s)	8275s	(7s)	40 MB	8250 MB

Figure 1 compares the accuracies of the programs as the size of the language models is varied from 200 n-grams per model to 3500; `libtextcat` was not evaluated beyond 1100, and `LangDetect` does not provide control over the model size (average 5642 n-grams per model for 1 million bytes training data, with 23 models containing more than 15,000 n-grams). The weighted cosine-similarity approach clearly dominates at a given model size even without the addition of discriminative training and inter-string smoothing. As it appears to asymptotically approach its optimal performance [1], the choice of model size is a compromise between accuracy and size/speed.

Figure 2 shows how the classification error rate varies with the amount of training data. Only the 184 languages for which models could be built from at least 3 million bytes were used. Accuracy improves rapidly up to 300,000 bytes; the slight "jitter" in error rates for larger training sizes is likely due to the selection of different lines of text when subsampling. The lowest error rate of 1.832/0.662% (raw/smoothed micro-average) is achieved at 1.1 million bytes, compared to `libtextcat`'s 6.065% error rate using 500-element models trained on 2.0 million bytes of data and `LangDetect`'s 3.193% on 1.0 million bytes. Substring filtering reduces the micro-average raw error rate on the development set from 2.459% to 2.330% (-5.2% relative), and the micro-average error rate with smoothing is reduced from 1.231% to 1.141% (-7.3%), both at a threshold of 0.62. On the full 1100-language set, discriminative training reduces the micro-average raw error rate from 2.035% to 1.722% (-15.3%) compared to the baseline, while the error rate with inter-string smoothing is reduced to 0.754% (-62.9/-56.2% relative to baseline/discriminative).

Table 1 compares the four programs for training time, evaluation time, and memory requirements on a hex-core Intel i7 processor at 4.3 GHz. Due to the incremental scoring used by `whatlang`, it runs much faster than `libtextcat` and `mguesser` on small inputs such as the 65-character lines used in these experiments; decreasing model sizes further increases its speed. For large inputs, such as the entire test file for each language (e.g. 1129 language identifications instead of 824,171), there is much less disparity in overall speed. Times are compared both with and without startup and scoring overhead, since `LangDetect` has a very long startup time (over seven seconds), but it evaluates large inputs very quickly as a result of randomly sampling a fixed number of n-grams from the input. The given training times are for single-threaded training and include the discriminative training second pass for `whatlang`. Since individual language models can be built independently of each other, training is highly parallelizable.

6   Conclusions

We have shown that a cosine-similarity approach to language identification scales to large numbers of languages and outperforms the popular rank-order method of Cavnar and Trenkle as well as two other programs in both accuracy and speed on short strings. Removing redundant high-frequency n-grams from the language models reduces overall classification error by 5.2 to 7.3%. Adding negative n-grams to the language models results in a further error rate reduction of 15.3% over the basic filtered model, and smoothing model scores reduces errors by more than a factor of two. The resulting error rate of 0.754% across 1100 languages is an accuracy of over 99.2%, and could be further improved (at the cost of greater resource use) by increasing the model size.

The whatlang program is a separately-compilable module within the Language-Aware String Extractor package, available under the terms of the GNU General Public License at http://la-strings.sourceforge.net/. Training data for approximately half of the languages used is redistributable under Creative Commons licenses and may be downloaded from the above URL.

The author thanks the reviewers for their feedback, which improved this paper.

References

1. Brown, R.D.: Finding and Identifying Text in 900+ Languages. Digital Investigation 9, S34–S43 (2012)
2. Cavnar, W.B., Trenkle, J.M.: N-Gram-Based Text Categorization. In: Proceedings of SDAIR 1994, 3rd Annual Symposium on Document Analysis and Information Retrieval, UNLV Publications/Reprographics, pp. 161–175 (April 1994)
3. Ljubešić, N., Mikelić, N., Boras, D.: Language identification: How to distinguish similar languages. In: Lužar-Stifter, V., Hljuz Dobrić, V. (eds.) Proceedings of the 29th International Conference on Information Technology Interfaces, Zagreb, pp. 541–546. SRCE University Computing Centre (2007)
4. Ahmed, B., Cha, S.H., Tappert, C.: Language Identification from Text Using N-gram Based Cumulative Frequency Addition. In: Proceedings of Student/Faculty Research Day, CSIS, Pace University (May 2004)
5. Carter, S., Tsagkias, M., Weerkamp, W.: Semi-Supervised Priors for Microblog Language Identification. In: Proceedings of the Dutch-Belgian Information Retrieval Workshop (DIR 2011), Amsterdam (February 2011)
6. Damashek, M.: Gauging Similarity with n-grams: Language Independent Categorization of Text. Science 267(5199), 843–848 (1995)
7. Shuyo, N.: Language Detection Library - 99% over precision for 49 languages (December 2010), http://www.slideshare.net/shuyo/language-detection-library-for-java (accessed May 30, 2013)
8. Xia, F., Lewis, W.D., Poon, H.: Language ID in the Context of Harvesting Language Data off the Web. In: Proceedings of EACL 2009, pp. 870–878 (2009)
9. Likasoft: Polyglot 3000, http://polyglot3000.com (accessed May 30, 2013)
10. United Bible Societies: Scripture Language Report 2011 (2011), http://www.unitedbiblesocieties.org/wp-content/uploads/uploads/2012/04/-report-TABLE-I-2011-Rec-March-28.doc (accessed May 30, 2013)

11. Hugueney, B.: libtextcat 2.2-9: Faster Unicode-focused C++ reimplementation of libtextcat (2011), https://github.com/scientific-coder/libtextcat (accessed May 30, 2013)
12. Barkov, A.: mguesser version 0.4 (2008),
    http://www.mnogosearch.org/guesser/mguesser-0.4.tar.gz
    (accessed May 30, 2013)
13. Shuyo, N.: Language Detection Library for Java,
    http://code.google.com/p/language-detection/ (accessed May 30, 2013)

Semantic Spaces for Sentiment Analysis

Ivan Habernal[1,2] and Tomáš Brychcín[1,2]

[1] Department of Computer Science and Engineering
Faculty of Applied Sciences, University of West Bohemia
Univerzitní 8, 306 14 Plzeň, Czech Republic
`nlp.kiv.zcu.cz`
[2] NTIS – New Technologies for the Information Society
Faculty of Applied Sciences, University of West Bohemia
Univerzitní 22, 306 14 Plzeň, Czech Republic
`{habernal,brychcin}@kiv.zcu.cz`

Abstract. This article presents a new semi-supervised method for document-level sentiment analysis. We employ a supervised state-of-the-art classification approach and enrich the feature set by adding word cluster features. These features exploit clusters of words represented in semantic spaces computed on unlabeled data. We test our method on three large sentiment datasets (Czech movie and product reviews, and English movie reviews) and outperform the current state of the art. To the best of our knowledge, this article reports the first successful incorporation of semantic spaces based on local word co-occurrence in the sentiment analysis task.

Keywords: document-level sentiment analysis, semantic spaces.

1 Introduction

Supervised document-level sentiment analysis belongs to a very popular branch of opinion mining research [1]. Although the mainstream target language of this research area is English, other languages have been recently gaining attention. However, the lack of linguistic resources (e.g., sentiment lexicons) or publicly-available labeled datasets represents the main obstacle in research for many non-mainstream languages. The datasets for supervised sentiment analysis research usually deal with product or movie reviews, as the labeled data can be easily obtained from the web.

Many current approaches rely on bag-of-word (or bag-of-n-gram) document representation and features based on various weighting metrics of word frequencies [2]. However, for languages with high flection, such as Czech, the feature vectors are very sparse due to a large vocabulary; even after stemming or lemmatization [3].

To tackle the issue of data sparsity in sentiment analysis of Czech, we investigate the possibilities of clustering semantically similar words.[1] Words with similar meanings are clustered according to their distance in semantic spaces that are computed on unlabeled data. We enrich the baseline feature set by additional word cluster features.

---

[1] Recent research on language modeling of inflectional languages revealed that clustering semantically similar words can rapidly improve performance [4].

I. Habernal and V. Matousek (Eds.): TSD 2013, LNAI 8082, pp. 484–491, 2013.
© Springer-Verlag Berlin Heidelberg 2013

By employing Maximum Entropy classifier and the extended set of features, we can outperform the state of the art.

2   Related Work

Recent research in document-level sentiment analysis of English texts has shifted towards semi-supervised methods that exploit small labeled data or highly precise lexicons enriched with large unlabeled datasets [1].

Authors of [5] investigate learning word vectors that contain information about the word's semantics as well as sentiment. They use a graphical model for inferring word vectors and rely on words' global context. Document vectors are obtained as a vector product of words' vectors and document's bag-of-word vector (using TF-IDF weighing scheme). The SVM classifier is used do decide about sentiment of document.

Authors of [6] present a new probabilistic graphical model called JST (Joint Sentiment Topic) based on LDA, but extended with one latent node for word sentiment. The model trained by unsupervised manner performs poorly (accuracy about 60% on a binary classification task). By incorporating sentiment lexicons into the model, the accuracy increased to 82.8%.

Although the above-mentioned approaches give reasonable performance, authors of [7] conclude that it may be hard to beat the baseline word n-grams without additional linguistic processing, if a well-tuned classifier is used.

Sentiment analysis of Czech has gained attention very recently. Authors of [8] presents a pilot study on sentence-level sentiment analysis. They manually annotated 410 sentences from the news, generated a sentiment lexicon using this corpus, and reported only preliminary results using the Naive Bayes classifier.

An in-depth investigation of supervised methods for Czech sentiment analysis on three different domains was introduced in [3]. They tested various preprocessing methods, sets of features, and two classifiers (Maximum Entropy and SVM) on three large corpora (movie reviews, product reviews, and Facebook dataset)[2]. They achieved the best performance (F-measure 0.69 on Facebook data, 0.75 on product reviews, and 0.79 on movie reviews) using Maximum Entropy classifier and unigrams, bigrams, POS (Part-of-Speech) ratio, and emoticon features.

3   Semantic Spaces

The backbone principle of semantic spaces used in this paper is so-called Distributional Hypothesis, saying that *"a word is characterized by the company it keeps"* [9]. The word meaning is thus related to the context in which this word usually occurs, as confirmed in empirical tests carried out on human groups in [10]. In semantic spaces, words are represented as high-dimensional vectors. These vectors are usually derived from statistical properties of the words' contexts.

We briefly describe the most popular methods for building semantic spaces. All presented algorithms are available in an open source package *S-Space* [11].

---

[2] http://liks.fav.zcu.cz/sentiment

HAL — Hyperspace Analogue to Language [12] is a simple method for building a semantic space. HAL iterates over the corpus and records the co-occurring words (in some fixed window—typically 4 words) of each token, resulting into a co-occurrence matrix $\mathbb{M} = |W| \times |W|$, where $|W|$ is the vocabulary size. Finally, the row and column vectors of matrix $\mathbb{M}$ represent co-occurrence information of words appeared before and after, respectively.

COALS — Correlated Occurrence Analogue to Lexical Semantics [13] is an extension of HAL model as it starts with building a similar co-occurrence matrix $\mathbb{M}$. After this step, the raw counts are converted into the Pearson's correlations. Negative values are set to zero, other values are replaced by their square roots. The optional final step, inspired by LSA [14], reduces the dimension of matrix $\mathbb{M}$ using SVD (Singular Value Decomposition) and can also discover latent semantic relationship between words.

RI — Random Indexing [15] use a completely different approach from HAL and COALS. For each word in vocabulary, RI starts by creating random high-dimensional sparse vectors filled with few 1 and -1; the dimension is typically in order of thousands. Such vectors are very unlikely to overlap. Then the algorithm iterates over the corpus and for each token it sums up all the co-occurring words' vectors into the appropriate word vector in the final matrix $\mathbb{M}$.

RRI — Reflective Random Indexing [16] is an iterative extension of RI that focuses on modeling transitive relations between words. This approach is similar to the SVD reduction used in LSA and COALS, but it is less computationally expensive.

BEAGLE — Bound Encoding of the AggreGate Language Environment [17] shares some ideas with RI as it starts with generating high-dimensional vectors for each vocabulary word. The values are, however, drawn from a Gaussian distribution with 0 mean value and $1/D$ variance, where D is the vector dimension (in order of thousands). The final matrix $\mathbb{M}$ contains the co-occurrence information of vectors within a certain window (typically 5) as well as information about word order given by the convolution of n-gram vectors that contain the processed word.

4   Our Model

Our method combines supervised machine learning, which employs a classifier on sentiment-labeled training data, and unsupervised feature space extension, which relies on clustering words represented in semantic spaces.

As a baseline, we adapt the supervised classification approach which has proven to be successful in the related work [3]. This approach requires a set of documents with their appropriate sentiment labels where each document is represented as a set of features. We rely on two kinds of binary features, namely the presence of word unigrams and bigrams. The feature space is pruned by filtering out unigrams and bigrams that occurred less than 10 times in the training data.

For classification we used the Maximum Entropy classifier [18] in the following form:

$$p(s|x) = \frac{1}{Z(x)} \prod_{i=1}^{n} e^{\lambda_i f_i(x,s)}, \tag{1}$$

where s is a sentiment label for a document, x is our knowledge about document, $f_i(x, s)$ is an i-th feature function, λ_i is corresponding weight and $Z(x)$ is a normalization factor. For estimating parameters of Maximum Entropy model we used limited memory BFGS (L-BFGS) method [19].

4.1 Word Clusters as Features

Since words in semantic space are represented as real-valued vectors, we can apply clustering methods. The main assumption is that words in the same cluster are semantically substitutable.

The selection of a suitable clustering algorithm is crucial for such a task. According to the study in [20] we selected Repeated Bisection algorithm because of its efficiency and acceptable computational requirements. We use the implementation from CLUTO software package.[3] As a similarity measure between two words we use cosine similarity of word vectors, calculated as the cosine of the angle between corresponding vectors: $S_{\cos}(a, b) = \frac{a \cdot b}{\|a\| \cdot \|b\|}$.

We extend the baseline feature set (word unigram and bigram features) by adding binary features capturing the presence of cluster unigrams and bigrams in the document represented as a sequence of word clusters.[4] Again, we ignore word cluster unigrams and bigrams that occur less than 10 times in the corpus.

5 Datasets

We perform our experiments on two Czech datasets and one English dataset labeled with their sentiment.

The Czech datasets, provided by [3], contain a product review dataset (102,977 positive, 31,943 neutral, and 10,387 negative reviews) and a movie review dataset (30,897 positive, 30,768 neutral, and 29,716 negative reviews). We pre-process the data in the same way as in the original paper, namely using Ark-tweet tokenizer [21], Czech HPS stemmer,[5] and lowercasing.

The English movie review dataset, provided by [5], consists of 25,000 positive and 25,000 negative reviews for the training part, the same number of reviews is used for testing. There are additional 50,000 unlabeled reviews which are suitable for unsupervised methods.

6 Results

The experiments on the Czech corpora were performed using 10-fold cross-validation. The English corpus was already separated into training and test data. The vocabulary size $|V|$ for building semantic spaces and clustering was limited to the 10,000 most

---

[3] http://www.cs.umn.edu/~karypis/cluto

[4] According to our experiments, incorporating only the cluster features (without words unigrams and bigrams), in order to reduce the feature space, leads to worse performance than a baseline.

[5] http://liks.fav.zcu.cz/HPS/

frequent words (excluding stopwords). Semantic spaces were constructed using training data (Czech) or training+unlabeled data (English). We used default settings for all semantic spaces according to their original papers.

For each semantic space (HAL, COALS, RI, RRI, and BEAGLE) we conducted the experiments with seven different numbers of clusters (50, 100, 200, 500, 1,000, 2,000, and 5,000 clusters, respectively).

Table 1. F-measure results for the Czech product reviews dataset (best results are bold-faced)

State of the art [3]				0.75

Number of clusters	Semantic space				
	COALS	HAL	RI	RRI	BEAGLE
50	0.72	0.72	0.71	0.70	0.75
100	0.75	0.77	0.75	0.70	0.71
200	0.75	0.73	0.73	0.74	0.74
500	0.74	0.72	0.71	0.70	0.69
1000	0.75	0.75	0.72	0.73	0.71
2000	0.73	0.73	0.73	0.73	0.73
5000	0.72	0.73	0.76	**0.78**	0.76

95% confidence interval = ± 0.002.

Table 2. Example of clusters obtained from the RRI model on the Czech product review dataset. To make this example easy to read, we transformed the stemmed words back into their nominative cases. Note that the Czech word *spokojennost* contains a typo (double *n*), however, is was clustered into the same cluster together with the correct word *spokojenost*.

cluster 1	cluster 2	cluster 3	cluster 4
pěkný (pretty)	složitější (more complex)	šmejd (junk)	silnější (stronger)
líbivý (pleasing)	komplikovaný (complicated)	spokojennost (satisfaction)	širší (wider)
moderní (modern)	nepřehledný (confusing)	geniální (ingenious)	užší (tighter)
nadčasový (timeless)	nepraktický (unpractical)	úchvatný (fascinating)	kratší (shorter)
milý (nice)	nešikovný (inept)	spokojenost (satisfaction)	menší (smaller)
atraktivní (attractive)	problematický (problematic)	vyhovujicí (suitable)	větší (bigger)
hezký (handsome)	zmatený (chaotic)		méně (less)

Czech Product Reviews — The best result (F-measure: 0.78) was achieved using RRI and 5,000 clusters, as shown in Table 1. We assume that clustering into 5,000 clusters produces small and precise clusters, where the words tend to have the same meaning as well as the same sentiment. See Table 2 for examples.

Czech Movie Reviews — The best result (0.80) was obtained from COALS with a small number of clusters (50 and 100), as shown in Table 3. Also BEAGLE with 1,000 clusters performs well.

English Movie Reviews — Table 4 shows that the best results (0.895) on the English movie review corpus were achieved using small clusters—50 for RRI and 100 for COALS. However, both semantic spaces yield similar results for larger clusters as well.

Table 3. F-measure results for the Czech movie reviews dataset (best results are bold-faced)

State of the art [3]					0.79
Number of	Semantic space				
clusters	COALS	HAL	RI	RRI	BEAGLE
50	**0.80**	0.78	0.79	0.79	0.78
100	**0.80**	0.78	0.78	0.77	0.79
200	0.79	0.78	0.79	0.78	0.79
500	0.79	0.79	0.78	0.79	0.79
1000	0.78	0.77	0.78	0.78	**0.80**
2000	0.79	0.79	0.78	0.79	0.77
5000	0.79	0.79	0.79	0.79	0.79

95% confidence interval = ±0.003.

Table 4. F-measure results for the English movie reviews dataset; best results are bold-faced. Note that in a two-class balanced-data scenario, F-measure equals accuracy.

State of the art [5]					0.889
No. of	Semantic space				
clusters	COALS	HAL	RI	RRI	BEAGLE
50	0.891	0.890	0.890	**0.895**	0.892
100	**0.895**	0.892	0.891	0.893	0.892
200	0.893	0.891	0.894	0.894	0.892
500	0.893	0.890	0.892	0.893	0.889
1000	0.894	0.889	0.889	0.893	0.889
2000	0.893	0.890	0.889	0.893	0.889
5000	0.893	0.893	0.890	0.894	0.892

95% confidence interval = ±0.004,
90% confidence interval = ±0.003,

In our experiments, RRI and COALS tend to perform consistently best for both languages. On the other hand, HAL and RI give no satisfactory results, regardless of the cluster size. Surprisingly, COALS with 100 clusters gives the best results in the movie review domain in both languages. On both Czech datasets, the current state of the art was significantly outperformed by our method (95% confidence interval). On the English dataset, our results are significantly better than the state of the art on 90% confidence interval.

Since RRI extends RI by modeling transitive relations between words, authors of RRI claim that RRI surpasses RI [16]. Similarly, the COALS model extends HAL (see section 3). We suppose that this is the reason why these two models perform best in our task. We also think that the improvement on Czech movie review dataset using BEAGLE with 1000 clusters is caused only by chance, because BEAGLE performed consistently worst in all other experiments.

7   Conclusion and Future Work

This article presented a promising semi-supervised method for document-level sentiment analysis using semantic spaces and word cluster features. We outperformed the current state of the art on two datasets in Czech and one dataset in English. Our method benefits from its independence of any additional labeled data as it improves the existing methods in a fully-unsupervised manner. We also prove that the method is language independent. Furthermore, it can significantly help in morphologically rich languages.

In the future work, we plan to investigate a combined model that incorporates the best-performing semantic space for each cluster size.

Acknowledgments. This work was supported by grant no. SGS-2013-029 Advanced computing and information systems, by the European Regional Development Fund (ERDF) and by project "NTIS - New Technologies for Information Society", European Centre of Excellence, CZ.1.05/1.1.00/02.0090. Access to the MetaCentrum computing facilities provided under the program "Projects of Large Infrastructure for Research, Development, and Innovations" LM2010005, funded by the Ministry of Education, Youth, and Sports of the Czech Republic, is highly appreciated. The access to the CERIT-SC computing and storage facilities provided under the programme Center CERIT Scientific Cloud, part of the Operational Program Research and Development for Innovations, reg. no. CZ. 1.05/3.2.00/08.0144 is acknowledged.

References

1. Liu, B., Zhang, L.: A survey of opinion mining and sentiment analysis. In: Mining Text Data, pp. 415–463. Springer (2012)
2. Martineau, J., Finin, T.: Delta TFIDF: An improved feature space for sentiment analysis. In: Proceedings of the Third International Conference on Weblogs and Social Media, ICWSM 2009, The AAAI Press, San Jose (2009)
3. Habernal, I., Ptáček, T., Steinberger, J.: Sentiment analysis in czech social media using supervised machine learning. In: Proceedings of the 4th Workshop on Computational Approaches to Subjectivity, Sentiment and Social Media Analysis, pp. 65–74. Association for Computational Linguistics, Atlanta (2013)
4. Brychcín, T., Konopík, M.: Semantic spaces for improving language modeling. In: Computer Speech and Language (2013), doi:10.1016/j.csl.2013.05.001
5. Maas, A.L., Daly, R.E., Pham, P.T., Huang, D., Ng, A.Y., Potts, C.: Learning word vectors for sentiment analysis. In: Proceedings of the 49th Annual Meeting of the Association for Computational Linguistics: Human Language Technologies, pp. 142–150. Association for Computational Linguistics, Portland (2011)
6. Lin, C., He, Y.: Joint sentiment/topic model for sentiment analysis. In: Proceedings of the 18th ACM Conference on Information and Knowledge Management, CIKM 2009, pp. 375–384. ACM, New York (2009)
7. Wang, S., Manning, C.D.: Baselines and bigrams: simple, good sentiment and topic classification. In: Proceedings of the 50th Annual Meeting of the Association for Computational Linguistics: Short Papers, ACL 2012, vol. 2, pp. 90–94. Association for Computational Linguistics, Stroudsburg (2012)
8. Veselovská, K., Hajič Jr., J., Šindlerová, J.: Creating annotated resources for polarity classification in Czech. In: Proceedings of KONVENS 2012, PATHOS 2012 Workshop, ÖGAI, pp. 296–304 (2012)
9. Firth, J.R.: A Synopsis of Linguistic Theory, 1930-1955. Studies in Linguistic Analysis, pp. 1–32 (1957)

10. Charles, W.G.: Contextual correlates of meaning. Applied Psycholinguistics 21, 505–524 (2000)
11. Jurgens, D., Stevens, K.: The s-space package: An open source package for word space models. In System Papers of the Association of Computational Linguistics (2010)
12. Lund, K., Burgess, C.: Producing high-dimensional semantic spaces from lexical co-occurrence. Behavior Research Methods Instruments and Computers 28, 203–208 (1996)
13. Rohde, D.L.T., Gonnerman, L.M., Plaut, D.C.: An improved method for deriving word meaning from lexical co-occurrence. Cognitive Psychology 7, 573–605 (2004)
14. Landauer, T.K., Foltz, P., Laham, D.: An Introduction to Latent Semantic Analysis. Discourse Processes, 259–284 (1998)
15. Sahlgren, M.: An Introduction to Random Indexing. In: Methods and Applications of Semantic Indexing Workshop at the 7th International Conference on Terminology and Knowledge Engineering, TKE 2005 (2005)
16. Cohen, T., Schvaneveldt, R., Widdows, D.: Reflective random indexing and indirect inference: a scalable method for discovery of implicit connections. Journal of Biomedical Informatics 43, 240–256 (2010)
17. Jones, M.N., Mewhort, D.J.K.: Representing word meaning and order information in a composite holographic lexicon. Psychological Review 114, 1–37 (2007)
18. Berger, A.L., Pietra, V.J.D., Pietra, S.A.D.: A maximum entropy approach to natural language processing. Computational Linguistics 22, 39–71 (1996)
19. Nocedal, J.: Updating Quasi-Newton Matrices with Limited Storage. Mathematics of Computation 35, 773–782 (1980)
20. Zhao, Y., Karypis, G.: Criterion functions for document clustering: Experiments and analysis. Technical report, Department of Computer Science, University of Minnesota, Minneapolis (2002)
21. Gimpel, K., Schneider, N., O'Connor, B., Das, D., Mills, D., Eisenstein, J., Heilman, M., Yogatama, D., Flanigan, J., Smith, N.A.: Part-of-speech tagging for twitter: annotation, features, and experiments. In: Proceedings of the 49th Annual Meeting of the Association for Computational Linguistics: Human Language Technologies: Short Papers, HLT 2011, vol. 2, pp. 42–47. Association for Computational Linguistics, Stroudsburg (2011)

Semi-automatic Verb Valence Frame Assignment through VerbNet Classification

Nives Mikelić Preradović and Damir Boras

Faculty of Humanities and Social Sciences, University of Zagreb
Ivana Lučića 3, 10000 Zagreb, Croatia
www.ffzg.unizg.hr
nmikelic, dboras@ffzg.hr

Abstract. An approach to cross-linguistic transfer of verb valence frames for Czech and Croatian language using VerbNet syntactic-semantic classification is presented. Confronting Czech-Croatian pairs of valence frames for each syntactic-semantic class, we extracted the semantic part of the Czech valence frames and applied the subcategorization features to Croatian with minor (mainly syntactic) modifications. The encouraging results suggest that other similar (Slavic) languages could benefit from the same approach. It is shown that even for languages that do not have sufficiently developed NLP resources that are highly relevant for verb subcategorization (such as WordNet), it is possible to obtain valence lexicon applying small number of rules for cross-linguistic syntactic transfer and using Levin's verb classes.

Keywords: VerbNet classification, verb subcategorization, frame assignement, Croatian language, Czech language.

1   Introduction

In this paper we address the question of leveraging NLP resources across closely-related languages. The semi-automatic assignment of Czech verb valence frames to verbs in Croatian language is described. This is the first attempt to develop a Croatian verb valence lexicon through cross-linguistic comparison, with Levin's syntactic-semantic verb classification being used for comparison and evaluation of two valence lexicons (VerbaLex for Czech language and CROVALLEX for Croatian language).

Languages such as English currently possess high-quality and high-coverage lexical resources (such as tagged corpora, parsers, WordNets and verb subcategorization lexicons) and therefore have an excellent starting point for automatic verb classification and acquisition of subcategorization frames (verb valence frames) through statistical approaches. Unfortunately, languages such as Croatian fall behind in development of such prerequisite resources. However, the need for high-quality NLP resources for Croatian is sometimes underestimated with the belief that they can be at least partially replaced by pure statistical techniques. Since state-of-the-art NLP resources for Czech language (including two verb valence lexicons – Vallex [1] and VerbaLex [2]) have already achieved a high level of quality and coverage, the decision was made to implement and evaluate the approach that could connect and compare NLP resources for these two syntactically and semantically quite similar languages.

I. Habernal and V. Matousek (Eds.): TSD 2013, LNAI 8082, pp. 492–500, 2013.

2   Method of Building Czech-Croatian Verb Valence Lexicon

Although there are studies that extend an existing valence lexicon by using information from plain bilingual dictionaries [3], very few studies have examined the cross-linguistic potential of Levin's verb classification for building a valence lexicon. A supervised cross-linguistic classification approach from English to Italian with 59 verbs and 3 classes [4] obtained accuracy of 86.3%. Others [5], [15] revealed similarities between the English and French classification, supporting earlier results [4] that Levin's classification has a cross-linguistic basis, and that the general methodology and the best performing features are transferable between the languages.

Our method aims to build the Czech-Croatian bilingual valence lexicon, enriching Croatian verb arguments in CROVALLEX lexicon with finer grained subcategorization features based on the Czech VerbaLex lexicon, using Croatian and Czech semantic classes inspired by the Levin's classification in both lexicons. Czech and Croatian language, both belonging to the same family of Slavic languages, tend to be morphologically and syntactically similar. The main difference between languages lies in the surface realization of obligatory verb complements (their number and function remains similar or even the same). CROVALLEX 2.008 is the first Croatian verb lexicon containing valence frames of Croatian verbs [6]. It contains 1739 verbs associated with 5118 valence frames (which makes an average of 3 valence frames per verb). These 1739 verbs are the most frequent Croatian verbs selected from the Croatian frequency dictionary [7], according to their number of occurrences. Regarding the syntactic-semantic classification in CROVALLEX 2.008, it contains 72 syntactic-semantic verb classes and 6774 Croatian verb lemmas, but only 1739 verbs have complete valence description. These classes have been originally adopted from VerbNet verb lexicon [8], which is based on Levin's verb classes [9]. VerbNet verb classes were adapted for the Croatian language (semantic criteria were given priority over the syntactic alternations), since most of the alternations used by Levin do not apply for Croatian. Croatian classes were enriched with synonym verbs, aspectual counterparts and prefixed verbs. The motivation for using such classification was to interpret the relation between the syntax and semantics of Croatian verbs, capture generalizations over some linguistic properties and explore possibilities for cross-linguistic development of verb valence lexicons. VerbaLex is a valence lexicon of Czech verbs developed at the University of Brno [1] that includes approx. 10500 Czech verbs with semantic classes inspired by the Levin's classification (similar to CROVALLEX). CROVALLEX describes explicit syntactic and semantic features of each verb lemma, while VerbaLex describes features of the whole verb synset that represents verbs in a synonymic relation followed by their sense numbers (standard Princeton WordNet notation). Synsets are based on the ability of each verb to appear in pairs of frames that are in some sense semantically preserved. Every synset is also defined by valence frames of its verbs, since they contain a set of syntactic descriptions. Since such verb synsets contain all the relevant characteristics of the individual verb, they allow generalization over verb features and also act as a compensation for the lack of necessary information. These synsets create a semantically and syntactically coherent group where members of the group share a range of features, starting with the implementation and interpretation of certain complements up to the existence of morphologically related forms. VerbaLex valence frames are enriched with pronouns (e.g. kdo – who, co – what), apart from the

Table 1. Comparison of a single valence frame in VerbaLex and CROVALLEX

Czech frame	frame: AGENT_kdo1 (person:1) PATIENT_koho4 (person:1)
Czech example	adoptovali chlapečka (pf) – they adopted a boy
Croatian frame	frame: AGENT_1 PATIENT_4
Croatian example	usvojili su dječaka (pf) - they adopted a boy

morphological case number that exists in CROVALLEX. This notation is important to differentiate animate from inanimate agent semantic case and therefore it would be useful to introduce this notation in CROVALLEX as well. An example of two verb frames in CROVALLEX and VerbaLex is displayed in the Table 1.

CROVALLEX and VerbaLex share approx. the same number of first-level semantic roles of verb arguments and have the similar role inventory. But, VerbaLex also distinguishes approx. 1000 second-level roles, each of them being mapped to Princeton WordNet and Czech WordNet (e.g. person:1; location:1, etc). The 1st level roles in VerbaLex are based on the 1stOrder-Entity and 2ndOrderEntity from EuroWordNet Top Ontology [10].

Therefore, the notation of the first-level roles in these lexicons and absence of the second-level roles in CROVALLEX represents the main differences between valence frames in CROVALLEX and VerbaLex. The basic comparison of the first level roles in both lexicons brought us to the conclusion that the most frequent verb meaning shares the same or very similar semantic roles in both lexicons. On the second level of semantics VerbaLex contains around 1000 specific lexical units from the set of EuroWordNet Base Concepts that represent a notation with large degree of sense differentiability. Such notation for the most frequent 1st semantic role AGENT differentiates whether it is a person, an animal, a group of people, an institution or a machine.

Comparing Czech-Croatian verb valence frames pairs for all Levin's classes, we aim to capture the semantic part of the Czech valence frames and apply the subcategorization features to Croatian frames with minor modifications.

Such improvements could solve the problems arising from the polysemous nature of the verb that CROVALLEX 2.008 does not account for. The change in the verb meaning in Croatian does not always affect the verb valence, but CROVALLEX 2.008 does not capture these distinctions (examples 1-4). All sentence constructions (1-4) are syntactically and morphologically valid, but sentences(2) and (3) are semantically possible only in the poetic context, since the agent is not the living entity (i.e. sea as a body of water cannot miss someone, i.e. suffer from the lack of something or someone). This type of problem could be solved introducing the verb subcategorization features that distinguish different verb meanings very well (examples 5 and 6).

(1) Nives nedostaje more [AGT3:obl PAT1:obl] (Nives misses the sea)
(2) *Nives nedostaje moru [PAT1:obl AGT3:obl] (The sea misses Nives)*
(3) *Moru nedostaje Nives [AGT3:obl PAT1:obl] (The sea misses Nives)*
(4) More nedostaje Nives [PAT1:obl AGT3:obl] (Nives misses the sea)
(5) Nives nedostaje more [AGT3:obl(**person:1**) ENT1:obl(**body of water:1**)] (Nives misses the sea)

(6) More nedostaje Nives [ENT1:obl(*body of water:1*) AGT3:obl(*person:1*)] (The sea misses Nives)

Enriching obligatory arguments with subcategorization features will ensure the valence notation with higher degree of sense differentiability. Therefore, CROVALLEX should introduce the semantic typing based on the VerbaLex lexicon in order to get a finer grained semantic classification, while building a bilingual valence lexicon.

3   Experiment: The Semi-automatic Transition of Valence Frames

Our experiment involved seven following steps: (**1**) 19 fine-grained Levin's classes were taken as a gold standard to compare verb semantics in Czech and Croatian (Table 2); (**2**) 7 classes (out of 19) that contain verbs in both languages were selected for cross-linguistic comparison; (**3**) morpho-syntactic rules for cross-linguistic transfer were designed manually; (**4**) Czech verbs belonging to each of these 7 classes were matched to their Croatian equivalents or translated to Croatian if not found in the existing list; (**5**) valence frames of Czech verb synsets were applied to Croatian verbs on a verb-to-verb basis, taking into account morpho-syntactic differences and applying morpho-syntactic rules; (**6**) resulting valence frames for 312 Croatian verbs were manually analysed for the range of features; (**7**) since 2 classes (*18.1 hit* and *37.3 manner of speaking*) contained more Croatian verbs than Czech, valence frames of translated Czech verb synsets were applied to all Croatian verbs in the class in order to test a possibility of obtaining a richer version of Croatian verb valence lexicon in the automatic way. The entire work is described in the following sections.

The classification of verbs in both Croatian and Czech valence lexicon is based on VerbNet lexicon [8]. VerbNet is a hierarchically organized lexicon of verbs that provides detailed syntactic-semantic descriptions of Levin's classes [9] and that proved to be useful in various fields of natural language processing [11], [12], [13], [14].

Regarding the implementation of the VerbNet classification, 17 classes from Levin's classification were used as English gold standard [14] in automatic verb clustering of English verbs. They were also used for investigation of the cross-linguistic applicability for French language [15]. Therefore, we used the same gold standard to compare verb semantics in Czech and Croatian language, since this classification was already implemented in their valence lexicons (Table 2). Out of 17 classes in English and French gold standard, only 5 classes contained verbs in both languages, since Czech verbs for 12 classes were not provided in VerbaLex. It was decided to keep this gold standard for the future comparison and analysis of the valence frames in Croatian with verb valence frames in French and English. Two more classes were added to the gold standard (*judgement-33* and *17.1-throw*) to obtain a reasonable number of Czech and Croatian verbs for the comparison. The verb coverage in Croatian classes is better than in Czech. But, although Croatian has 3460 classified verbs, only 1739 have been described by valence frames, compared to 10.500 verbs in VerbaLex, which are all described by valence frames. Since a significant number of Czech verbs have not yet been assigned to any semantic class, it would be interesting to test whether the morpho-syntactic rules proposed in this paper could be used as reversed rules to systematize the classification of these Czech verbs in a more efficient manner.

Table 2. Test classes for Czech and Croatian based on Levin's original classes

Levin Class	English verb examples	Czech verbs	Croat. Verbs
9.1 Put	*bury, place, install, mount, put, deposit, position, set*	75	46
10.1 Remove	*remove, abolish, eject, extract, deduct, eradicate, sever, evict*	327	97
11.1 Send	*ship, post, send, mail, transmit, transfer, deliver, slip*	86	62
13.5.1 Get	*win, gain, earn, buy, get, book, reserve, fetch*	-	97
18.1 Hit	*beat, slap, bang, knock, pound, batter, hammer, lash*	6	12
22.2 Amalgamate	*contrast, match, overlap, unite, unify, unite, contrast, affiliate*	-	119
29.2 Characterize	*envisage, portray, regard, treat, enlist, define, depict, diagnose*	-	73
30.3 Peer	*listen, stare, look, glance, gaze, peer, peek, squint*	-	88
31.1 Amuse	*delight, scare, shock, confuse, upset, overwhelm, scare, disappoint*	-	193
36.1 Correspond	*cooperate, collide, concur, mate, flirt, interact, dissent, mate*	-	36
37.3 Manner of speaking	*shout, yell, moan, mutter, murmur, snarl, moan, wail*	9	93
37.7 Say	*say, reply, mention, state, report, respond, announce, recount*	-	16
40.2 Nonverbal express.	*smile, laugh, grin, sigh, gas, chuckle, frown, giggle*	-	24
43.1 Light emission	*shine, flash, flare, glow, blaze, flicker, gleam, sparkle*	-	39
45.4 Change of state	*soften, weaken, melt, narrow, deepen, dampen, melt, multiply*	-	249
47.3 Modes with motion	*quake, falter, sway, swirl, teeter, flutter, wobble, waft*	-	23
51.3.2 Run	*swim, fly, walk, slide, run, travel, stroll, glide*	-	89
33. Judgement	***compliment, congratulate, bless, salute, thank, stigmatize***	233	70
17.1 Throw	***throw, toss, catapult, launch, kick, tip***	23	22

Although Czech-Croatian bilingual dictionary offers only the option for mapping individual verbs, not synsets, the mapping between verbs in the source and target language was quite straightforward. Each Czech verb in the dictionary has 3 Croatian translations in average. If the meaning of a verb in Czech synset (expressed as a synset definition) does not equal the meaning of Croatian verb, the verb is labeled as ambiguous and omitted from the test class. On the other hand, polysemic verbs, belonging to multiple Czech synsets, also belong to multiple (different) semantic classes, which facilitates the translation process. Inasmuch as we had to analyze the results manually, the small set of classes and verbs belonging to each class enabled us to thoroughly evaluate our results. These classes were manually compared in order to extract a range of comparable features (morphological cases, syntactic positions and semantic meanings). The manual generation of rules resulted in 103 morpho-syntactic transition rules for prepositional cases, 7 for adverbs, 3 for adjectives and 18 for subordinating conjunctions. Phraseme transition rules were also created, which is especially interesting for foreign language

Table 3. Example of morpho-syntactic transition rules for direct cases

Semantic role	Surface form in Czech	Surface form in Croatian	Example
LOC (location)	čemu3 (to what)	čim7 (with what)	královna kraluje Anglii/kraljevna vlada Engleskom/the Queen reigns the England
PAT (patient)	komu3 (to whom)	kim7 (with whom)	král kraluje svému lidu/kralj kraljuje svim narodima/the king reigns over all people
FEEL (feeling:1)	čim7 (with what)	od+čega2 (from what)	srdce mu plesalo radostí/srce mu je plesalo od radosti/his heart was dancing with joy
EVEN (event:1)	čeho2 (of what)	čemu3 (to what)	nadál se spìchu/nadao se uspjehu/he longed for success
PAT (person:1)	koho2 (of whom)	komu3 (to whom)	nadál se bratra/nadao se bratu/he longed for his brother
PAT (person:1)	kym7 (with whom)	koga4 (whom)	mrštil svim sokem na zem/oborio je protivnika na tlo/he threw his rival to the ground
OBJ (object:1)	čim7 (with what)	koga4 (whom)	pohyboval nábytkem/pomicao je namje- štaj/he moved the furniture

learning. Some of the rules that give the best insight into syntactical differences between two languages are presented in Table 3.

4 Evaluation of the Verb Valence Frames Obtained through Semi-automatic Valence Assignment

A preliminary study was carried out mapping each of the Czech verb synsets that belong to each of the 7 Levin's classes to Croatian using Czech-Croatian bilingual dictionary [16]. Valence frames of translated verbs in Czech verb synsets were applied to Croatian verbs on a verb-to-verb basis, taking into account morpho-syntactic differences and applying morpho-syntactic rules when needed. We obtained results for 312 Croatian verbs and manually evaluated their valence frames (only primary (literal) and metaphorical verb meanings were examined, idioms were omitted). The phrase "identical frame" refers to a verb valence frame that has the same number of complements (semantic roles) with the same 1st and 2nd level of semantics in both Czech and Croatian language. Regarding the morpho-syntactic realization of these roles, they might share the same surface forms or morpho-syntactic rules need to be applied. The results of evaluation are presented in Table 4.

Verb valence accuracy (ACCvv) was calculated for each verb in the class as well as class valence accuracy (ACCcv) for all 7 classes. ACCvv represents the proportion of accurate valence frames associated with a verb, compared to all the frames assigned to a verb. It is a sum of viable valence frames divided by the number of all valence frames transferred from Czech to a Croatian verb. ACCcv represents the sum of ACCvv's from all verbs in each class divided by the number of verbs in class. Although we proved the

Table 4. Evaluation results for experiments on all seven comparable classses

Class (C)	No. of verbs compared	No. of Verbs with identical frames	ACCcv value
9.1 Put	46	41	0.96
10.1 Remove	97	89	0.97
11.1 Send	62	60	0.99
18.1 Hit	6	6	1
37.3 Manner of speaking	9	9	1
33. Judgement	70	68	0.98
17.1 Throw	22	20	0.97

high accuracy in verb valence frame assignment: 293 out of 312 Croatian verbs share the identical frames with their Czech equivalents, with the only difference being different morpho-syntactic behaviour of semantic roles, the transfer method was performed on the same set of verbs for which the manual rules were developed, so the reported accuracy can be only treated as the upper bound that can be achieved with this type of rules.

It can be concluded that Levin's classification helps to resolve the ambiguities in application of the morpho-syntactic rules. For example, Czech phrase *"z+čeho"* can be translated to English and Croatian as *"what from"* and *"where from"*. But, if belonging to the class *build-26.1.4* or *create-26.4.2*, it is translated only as *"what from"*.If belonging to *remove-10.1*or *clear-10.3*class, it is translated only as *"where from"*.

As a result of our experiment with cross-linguistic transition rules and semi-automatic valence frame assignment, the following conclusion was drawn: the most frequent deep cases/semantic roles (such as AGENT or PATIENT) and valence frames containing these roles can be automatically transferred from one language to another if their verbs belong to the same semantic class (since they consequently share the same meaning and the valence frame and can be translated one –to –one). *Example:* Czech verbs "hodit and hazet" (in Croatian "baciti and bacati") belonging to the Levin's class throw-17.1 have the following meaning propel through the air in a certain direction with a sudden motion". Their frame in Czech has obligatory semantic roles AGENT (person:1) and OBJECT (object:1) with the syntactic forms AG(kdo1), OBJ(co4, čim7). In Croatian, the number and the order of the roles is the same, with the one difference: OBJ can only be realized as *što4* (Croatian equivalent for *co4*), not as *čim7*, which is specific for Czech. Also, verbs in both languages can share the meaning and the same number of semantic roles which are syntactically realized with different morphological cases, so we need morpho-syntactic rules to be able to apply the Czech valence frames to Croatian verbs and vice versa. *Example:* Czech verbs "blahopřát, gratulovat, poblahopřát, pogratulovat, popřát, přát" belonging to the Levin's class judgement-33 are translated to Croatian as čestitati, poželjeti, željeti, zaželjeti" with the following meaning "to communicate pleasure, approval, or praise to a person". Their frames in both Czech and Croatian consist of identical obligatory semantic roles AGENT (person:1) and PAT (person:1) with the same syntactic forms, but the optional role TIME is realized as *k+čemu3* or *na+co4* in Czech, while in Croatian it can only be realized as *što4*.

5   Conclusion

Our research on the cross-linguistic properties of Levin's classes and valence frame assignment in CROVALLEX and VerbaLex revealed high similarities between the Czech and Croatian verb classification, which only supports the results obtained by other researchers for French [5] and for Italian [4]. It also revealed the possibility to enrich the Croatian verb valence lexicon in a semi-automatic way with 2-level semantic roles, establishing links to Princeton WordNet and CzechWordNet hierarchy and introducing highly relevant differentiation of animate/inanimate constituents. Our experiment shows that even for languages that do not have sufficiently developed NLP resources that are highly relevant for verb subcategorization, it is possible to obtain valence lexicon applying small number of rules for cross-linguistic syntactic transfer and using Levin's verb classes. As a next step we will try to improve the quality of transitions introducing conditional rules based on co-occurence of each slot with other slots in a valence frame. Finally, we plan to explore the possibility of building the verb part of the Croatian WordNet, linking Croatian verb synset valence frames to Czech verb synsets.

References

1. Lopatková, M., Žabokrtsky, Z., Skwarska, K., Benešová, V.: VALLEX 1.0 Valency Lexicon of Czech Verbs. CKL/UFAL Technical Report TR-2003-18 (2003)
2. Hlaváčková, D., Horák, A.: Verbalex - new comprehensive lexicon of verb valencies for czech. In: Proceedings of the Slovko Conference, Brno, Czech Republic (2005)
3. Fujita, S., Bond, F.: A method of creating new valency entries. Machine Translation 21(1), 1–28 (2007)
4. Merlo, P., Stevenson, S., Tsang, V., Allaria, G.: A multilingual paradigm for automatic verb classification. In: Proceedings of the 40th Meeting of the Association for Computational Linguistics (2002)
5. Sun, L., Poibeau, T., Korhonen, A., Messiant, C.: Investigating the cross-linguistic potential of VerbNet-style classification. In: Proceedings of COLING, pp. 1056–1064 (2010)
6. Mikelić Preradović, N., Boras, D., Kišiček, S.: CROVALLEX: Croatian Verb Valence Lexicon. In: Proceedings of the 31st International Conference on Information Technology Interfaces (ITI), pp. 533–538 (2009)
7. Moguš, M., Bratanić, M., Tadić, M.: Croatian Frequency Dictionary (Hrvatski čestotni rječnik). Zavod za lingvistiku i Školska knjiga. Zagreb (1999)
8. Kipper-Schuler, K.: VerbNet: A broad-coverage, comprehensive verb lexicon. University of Pennsylvania, PA (2005)
9. Levin, B.: English verb classes and alternations: A preliminary investigation. University of Chicago Press, Chicago (1993)
10. Vossen, P., Bloksma, L., Rodriquez, H., Climent, S., Calzolari, N., Roventini, A., Bertagna, F., Alonge, A., Peters, W.: The EuroWordNet base concepts and top ontology. EuroWordNet (LE-4003) Deliverable D017D034D036, University of Amsterdam (1998)
11. Crouch, D., King, H.: Unifying lexical resources. In: Proceedings of the Interdisciplinary Workshop on the Identification and Representation of Verb Features and Verb Classes, Saarbrücken, Germany, pp. 32–37 (2005)
12. Hensman, S., Dunnion, J.: Automatically building conceptual graphs using VerbNet and WordNet. In: Proceedings of the 3rd International Symposium on Information and Communication Technologies (ISICT), Las Vegas, Nevada, pp. 115–120 (2004)

13. Swier, R., Stevenson, S.: Unsupervised semantic role labelling. In: Proceedings of the Conference on Empirical Methods in Natural Language Processing, pp. 95–102 (2004)
14. Swift, M.: Towards automatic verb acquisition from VerbNet for spoken dialog processing. In: Proceedings of Interdisciplinary Workshop on the Identification and Representation of Verb Features and Verb Classes (2005)
15. Sun, L., Korhonen, A., Krymolowski, Y.: Verb class discovery from rich syntactic data. In: Gelbukh, A. (ed.) CICLing 2008. LNCS, vol. 4919, pp. 16–27. Springer, Heidelberg (2008)
16. Merhaut, J.: Czech-Croatian dictionary (Češko-hrvatski rječnik), Zagreb (1998)

Speaker-Specific Pronunciation for Speech Synthesis*

Lukas Latacz[1,2], Wesley Mattheyses[1], and Werner Verhelst[1,2]

[1] Vrije Universiteit Brussel, Dept. ETRO-DSSP, Brussels, Belgium
[2] iMinds, Dept. of Future Media and Imaging, Ghent, Belgium
{llatacz,wmatthey,wverhels}@etro.vub.ac.be

Abstract. A pronunciation lexicon for speech synthesis is a key component of a modern speech synthesizer, containing the orthography and phonemic transcriptions of a large number of words. A lexicon may contain words with multiple pronunciations, such as reduced and full versions of (function) words, homographs, or other types of words with multiple acceptable pronunciations such as foreign words or names. Pronunciation variants should therefore be taken into account during voice-building (e.g. segmentation and labeling of a speech database), as well as during synthesis.

In this paper we outline a strategy to automatically deal with these variants, resulting in a speaker-specific pronunciation. Based on a labeled speech database, the pronunciation lexicon is pruned in order to remove as much as possible pronunciation variation from the lexicon. This pruned lexicon can be used to train speaker-specific letter-to-sound rules. If the speaker has uttered a word in different ways, then these variants are not pruned. Instead, decision trees are trained for each of those words, which are used to select the most suitable pronunciation during synthesis. We tested our approach on five speech databases, and two lexicons per speech database. The automatic selection of pronunciation variants yielded a small improvement over the baseline (selecting always the most common variant).

Keywords: speech synthesis, lexicon, pronunciation variants, speaker-specific.

1 Introduction

In a typical modern speech synthesizer, the input text needs to be converted in a phonemic or phonetic sequence before the actual speech can be produced. This conversion is usually done using a pronunciation lexicon in which the phonemic transcription of each word can be looked up.

Ideally, the transcriptions in a pronunciation lexicon are tailor-made for a specific speaker. In practice, they are made for a specific regional accent of a language. Some lexicons are so-called meta-lexicons, allowing to generate pronunciations in different regional accents (e.g. Unisyn [1]), or different speaking styles (e.g. FONILEX [2]). These rules could also be adapted for a particular speaker, but this requires a significant amount of (manual) work. Another approach to generate speaker-specific pronunciation

---

* The research reported in this paper was partly supported by the projects IWT-SPACE, iMinds-RAILS, iMinds-SEGA and EC FP7 ALIZ-E (FP7-ICT-248116).

I. Habernal and V. Matousek (Eds.): TSD 2013, LNAI 8082, pp. 501–508, 2013.
© Springer-Verlag Berlin Heidelberg 2013

is to use a phone-recognizer to detect variants uttered by the speaker as proposed in [3]. This approach also requires some additional manual work, as the accuracy of the recognizer is not high enough for a fully automatic approach.

Taking pronunciation variations into account during synthesis can improve synthesis quality as mentioned in [4], but their approach cannot easily be applied in most speech synthesizers as multiple pronunciations for the same word need to be supported. Most speech synthesizers use a single pronunciation for each word during synthesis.

In case multiple pronunciation variants of a certain word are present in the lexicon, the speech synthesizer should have a strategy to select the best pronunciation variant. Sometimes the pronunciation variant depends on the part-of-speech (e.g. in Dutch: een (article) versus een (numeral)) or on the meaning of the word (e.g. in English: bass (music) versus bass (fish)). This information can be added to the pronunciation lexicon, and allows selecting the correct variant. This approach can not be used in all cases, as some variants have the same meaning or part-of-speech tag (e.g. the full and reduced versions of function words). An option in the Multisyn synthesizer [5] allows setting whether full or reduced vowels should be chosen, but this affects the whole utterance. Bennett and Black proposed to train decision trees to select the correct pronunciation variant, but applied their approach only to predict reduced or full function words [6].

Our strategy to generate a speaker-specific pronunciation is fully automatic and is also able to deal with pronunciation variants beyond function words. The most suitable pronunciation is selected during voice-building and synthesis. Our approach requires that the same pronunciation lexicon is being used for both labeling the speech database and synthesis, and that post-lexical rules are not used. The latter has the advantage that the risk of a potential mismatch between the segmentation labels and the output of the front-end is reduced.

Instead of modifying existing lexicon entries or adding new entries to the lexicon, we assume that the lexicon contains enough pronunciation variants to match the speech of the speech database sufficiently. We realize that such assumption is difficult to test, but - based on our own experience with building voices - adding additional pronunciations for known words is not often needed if a pronunciation lexicon suitable for speech synthesis is being used. We therefore propose to remove entries (unnecessary variants) from the lexicon, and to automatically select the correct pronunciation variant out of multiple options which were uttered by the speaker.

Table 1. Pronunciation variants in analyzed lexicons. Words are words with a unique orthography (disregarding capitalization). The *unresolved* column refers to the number of unique words which have multiple pronunciation variants which could not be disambiguated using information present in the lexicon. Lexical stress and syllable boundaries are not taken into account.

Lexicon	Entries	Words with multiple pronunciations	Unresolved words
CMUDict	105901	168	6
OALD	72301	428	111
Unisyn-RP	167617	1003	268
Unisyn-GA	167617	1084	330
KUNLEX	312232	100	64
FONILEX	188550	7785	7760

This paper continues as follows. In section 2, we have analyzed the pronunciation variants in several pronunciation lexicons. Even though the number of variants in these lexicons is quite limited in comparison to the total number of lexicon entries, we still need to cope with these variants during labeling and synthesis. In section 3 we describe how we perform labeling and segmentation. In section 4 we describe how we have generated a speaker-specific pronunciation lexicon. Our approach to deal with pronunciation variation during synthesis is described in section 5 and evaluated in section 6 using several speech synthesis databases and lexicons.

2   Lexicon Analysis

We have analyzed the following pronunciation lexicons in order to find out how many words with multiple pronunciations are present in each lexicon. The CMU lexicon [7] is a relatively widely-used American English lexicon. We have used CMUDict 0.4 included in the Festival speech synthesis system. It contains part-of-speech tags for disambiguation. The Unisyn lexicon [1] allows generating lexicons for many English accents. It contains part-of-speech tags for each entry, and an optional field for disambiguation. This optional field contains either the meaning of the word, or indicates whether the word is (un)reduced. We have used both the (British English) Received Pronunciation and General American accent in our experiments. The (un)reduced option was ignored in our experiments. OALD [8] is the computer-useable version of the Oxford Advanced Learner Dictionary and contains British English pronunciation. We have used the version included in Festival, which contains part-of-speech tags for each entry. KUNLEX is a Dutch lexicon which covers standard Northern-Dutch (i.e. the standard Dutch spoken in The Netherlands). It is part of NeXTeNS [9], a Dutch extension for Festival. FONILEX [2] is a Flemish pronunciation lexicon. We are using a modified version of this lexicon, which contains a high-level phonemic transcription (the pronunciation of each morpheme is transcribed as it would be pronounced in isolation) and the default speaking style (normal pronunciation). FONILEX does not contain part-of-speech tags, but these can be obtained from the CELEX lexicon [10] as each FONILEX entry contains a reference to its corresponding CELEX entry.

All these pronunciation lexicons contain some pronunciation variants that cannot be disambiguated using part-of-speech or other information in the lexicon, as shown in table 1. The amount of these variants varies between the lexicons. The largest number of variants can be found in the FONILEX lexicon. We assume that this is because this lexicon was originally constructed for speech recognition purposes. The other lexicons contain far fewer pronunciation variants.

Some of the words with pronunciation variants are homographs, proper names or foreign words with multiple acceptable pronunciations. Reduced and full versions of common function words can also be frequently found. The use of other variants is more questionable. We assume that these could be speaker- or region-depended, or simply exist because of the difficulty of accurate transcription. The lexicon developer needed to make hard decision during the construction of the lexicon, and sometimes none of the options was entirely satisfying: the phonemic transcription is intrinsically an approximation of the continuous nature of speech sounds.

3 Segmentation and Labeling

A large speech database is required for high-quality speech synthesis, and needs to be segmented into phonemes. The word sequence of each utterance of the database is normally known beforehand, as the speaker is instructed to read a recording script. Currently, the most popular segmentation technique is to use hidden Markov models: a large vocabulary continuous speech recognizer is run in forced-alignment mode - the most likely sequence of states is selected for each utterance of the database, given input speech signals and phonemic sequences. If multiple phonemic sequences are possible (due to optional silence insertions or multiple pronunciations), the best phonemic sequence - in terms of likelihood - is automatically chosen using the Viterbi algorithm. Our approach for segmenting and labeling is based on the open source HMM-based speech recognizer SPRAAK [11], but can probably also be extended to other recognizers such as HTK or Sphinx.

The output of the (HMM-based) segmentation algorithm does not include which variant is actually chosen, as only a phoneme sequence including timing information is typically given. We have solved this problem by constructing a word lattice. The correct sequence of pronunciation variants as selected by the segmentation algorithm can be obtained by traversing through the lattice using the output phoneme sequence. If at some point the output sequence does not match the selected variant, a different variant needs to be chosen and backtracking might be necessary. Optional silences (i.e. silences between words) are also taken care off. Since no additional post-lexical rules are being used in our case, a matching word sequence will always be found. Words with the same pronunciation but different lexical stress or syllable boundaries cannot be distinguished using this approach, as the segmentation algorithm is not able to take these differences into account. In our current implementation, the first suitable variant is always chosen.

4 Lexicon Pruning

If a word has multiple possible pronunciations in the lexicon, we need to select the most suitable option during synthesis. As discussed previously, several such words can be found in the lexicons we examined. Part of the variation can be reduced by using part-of-speech. Our approach to deal with the remaining variation is to base the selection of the pronunciation variants on the speech database, and focus on these variants that have been uttered by the speaker. Using this information, we can create a new version of the pronunciation lexicon which contains as few pronunciation variants as possible, resulting in a speaker-specific lexicon.

Ideally, each word in the pronunciation lexicon should have exactly one pronunciation in the lexicon, unless additional information is present in the lexicon to disambiguate the variants. In this section, we refer to a word and this additional information as an unambiguous word in the lexicon. These unambiguous words could also have multiple pronunciations, as previously demonstrated. The goal of the lexicon pruning is to keep a single pronunciation variant for each unambiguous word. If a word is present in the speech database, then all variants which are pronounced by the speaker are kept. Words with multiple pronunciation variants can still be present in the pruned lexicon.

Unambiguous words that are not present in the speech database need to be dealt with differently. We do not know which variant will give the best synthesis quality before actually synthesizing these variants. Choosing the best variants by hand though iterative listening is a potential solution, but not very practical for texts longer than a few utterances. We therefore propose the following solution.

Words that are covered by common diphones can typically be synthesized better than words which contain rare diphones. A very simple quality estimation is to calculate the average diphone coverage of a word, using the diphones present in the speech database. We therefore keep only the variant that has the highest average diphone coverage. If multiple variants have the same coverage, we select the first variant. A probably better solution would be then to look at larger elements such as triphones.

The resulting pruned pronunciation lexicon is a speaker-specific lexicon, tailored to a specific speech database. As such, it becomes possible to train speaker-specific letter-to-sound rules using the pruned lexicon.

5   Dealing with Multiple Pronunciation Variants

The most straight-forward approach is to always select the most common pronunciation for each so-called unambiguous word, i.e. the variant that is chosen most frequently. An alternative is to select a variant based on the average diphone frequency, in order to maximize diphone coverage (as explained previously). Both approaches ignore all non-selected variants from the lexicon. As such, they are less suitable if two or more variants of the same word have similar frequency in the speech database.

Decision trees can also be used to select the most suitable pronunciation. Our approach is quite similar to the approach proposed by Bennett and Black [6] They used decision trees to select either the full or the reduced versions of function words. The trees were trained using linguistic data extracted from an annotated speech database. We have generalized their approach for all words with pronunciation variants in a speech database. We have used wagon part of the Edinburgh Speech Tools to train the decision trees. A separate tree is constructed for each distinct word. The training data is based on

Table 2. Results of the lexicon pruning. *Unresolved words* are words which could not be disambiguated using information in the lexicon.

Database	Lexicon	Entries in pruned lexicon	Unresolved words
SLT	CMUDict	105894 (-0,01%)	0
	Unisyn GA	167167 (-0,27%)	26
AWB	CMUDict	105894 (-0,01%)	0
	Unisyn GA	167172 (-0,27%)	29
RJS	OALD	72002 (-0,41%)	3
	Unisyn RP	167172 (-0,27%)	26
AVKH	FONILEX	178262 (-5,46%)	21
	KUNLEX	311696 (-0,17%)	2
AWDC	FONILEX	178251 (-5,46%)	11
	KUNLEX	311694 (-0,17%)	0

the instances uttered by the speaker. Linguistic features are extracted for each instance. Our current features take part-of-speech, phrase breaks, and phonemes and graphemes of the surrounding words into account. If the pronunciation of each word is processed sequentially (which is normally the case), then it is not possible to use features below the word level for words which succeed the current word, as this information is not yet known. The best stop size (a parameter to control the size of the tree) can be selected using n-fold cross-validation (e.g. n=5).

6 Evaluation

6.1 Databases

The following databases were used in our experiments: SLT and AWB [12], RJS [13], the audio part of the AVKH [14] database, and AWDC [15]. A subset of the AWDC database was used, containing about 2.5 hours of speech and consisting of sentences and paragraphs selected from childrens stories. Each database was segmented twice with SPRAAK, with two different lexicons. The SLT and AWB databases were segmented with the CMUDict and Unisyn (General American) lexicons. We chose to use US English lexicons for the Scottish-accented AWB database, because the CMUDict lexicon is the default lexicon for this database in the AWB voices supplied with Festival. The RJS database was segmented with the OALD and Unisyn (Received Pronunciation) lexicons. The Flemish databases were segmented with the KUNLEX and FONILEX lexicons. All databases contained some out-of-vocabulary words. These words were manually transcribed, based on the existing entries of the lexicon. These additional transcriptions were needed during segmentation. As we are not expert English phoneticians, words were only transcribed if the transcription was straight-forward, such as in case of compound words. Utterances which contained non-transcribed out-of-vocabulary words were not used in our experiments.

6.2 Lexicon Pruning

Result of the lexicon pruning can be seen in table 2. As expected from table 1, the impact of the lexicon pruning is quite small for all lexicons except FONILEX. As FONILEX was originally constructed for speech recognition, many alternative pronunciations were added by the lexicon developers in order to improve recognition.

6.3 Prediction of Pronunciation Variants

We evaluated the algorithms to select the best pronunciation variants using 5-fold cross-validation. Words with less than 6 instances in the database are ignored in this evaluation. The results in table 3 indicate that the best approach is to use decision trees. This indicates that the selection of the best variant can be generalized up to a certain extend - from data in the speech database. The worst approach is to select a variant based on diphone frequency. This is not unexpected, as this does not directly use information about which of the variants have been uttered by the speaker. Its performance is

Table 3. Evaluation of the automatic selection of pronunciation variants. Accuracy (% correct) is shown. The results of other voices are not included because not enough instances were available for evaluation.

Database	Lexicon	Instances in evaluation	Decision tree	Majority voting	Diphone selection
SLT	Unisyn GA	1660	**81%**	80%	60%
AWB	Unisyn GA	1914	**76%**	74%	58%
RJS	Unisyn RP	6962	**86%**	83%	75%
AVKH	FONILEX	2192	**90%**	82%	75%
AWDC	FONILEX	1273	**86%**	85%	52%

still better than randomly selecting a variant. This is an indication that our approach to prune words which do not occur in the speech database, is justified.

If we compare the results across speech databases, we can see that in general larger databases yield better performance. Performance across the RJS, AWDC and AVKH databases is quite similar for the decision tree-based method. The difference in accuracy between the AWB and SLT database might be explained by the Scottish accent of the speaker in the AWB speech database. We have not examined whether using the Scottish variant of the Unisyn lexicon would yield better performance.

7 Conclusion

In this paper we presented an approach to automatically generate a speaker-specific pronunciation for speech synthesis based on lexicon pruning and decision tree-based variant selection. Our approach allows the fully automatic prediction of these variants, without the need of the synthetic voice developer to know the language in particular. The impact of the lexicon pruning is quite limited if the original lexicon contains few words with multiple pronunciations, and seems more appropriate for lexicons with a relatively large number of variants, such as the FONILEX or Unisyn lexicons. Our approach may lead to fewer pronunciation differences between the front-end and back-end of the synthesizer, as the speaker-specific pronunciation is generated using the actual segmentation labels in the speech database. A lower amount of these differences can result in an improved speech synthesis quality [4].

Several improvements are still possible. As relatively standard features were used to construct the decision trees for prediction, a potentially larger improvement can be obtained by the use of more advanced linguistic features. Our current segmentation algorithm could be extended to take lexical stress (e.g. [16]) and syllable boundaries into account. Furthermore, the success of the lexicon pruning and pronunciation variant selection depends partly on the accuracy of the segmentation (i.e. whether the correct variant is actually chosen in the speech database). We have not yet examined the influence of the segmentation accuracy in our experiments. Our approach can also be implemented differently: the existing lexicon is kept and decision trees are constructed for each word with multiple pronunciations in the original lexicon. This has the advantage that the same lexicon can be reused for multiple speakers, while still allowing a speaker-specific pronunciation.

References

1. Fitt, S.: Unisyn multi-accent lexicon, version 1.3,
   `http://www.cstr.ed.ac.uk/projects/unisyn`
2. Mertens, P., Vercammen, F.: FONILEX manual. Technical report, K.U.Leuven CCL (1998)
3. Kim, Y.J., Syrdal, A., Conkie, A.: Pronunciation lexicon adaptation for TTS voice building. In: Proceedings Interspeech 2004, Jeju Island, Korea, pp. 2569–2572 (2004)
4. Hamza, W., Eide, E., Bakis, R.: Reconciling pronunciation differences between the front-end and the back-end in the IBM speech synthesis system. In: Proceedings Interspeech 2004, Jeju Island, Korea, pp. 2561–2564 (2004)
5. Clark, R.A.J., Richmond, K., King, S.: Multisyn: Open-domain unit selection for the festival speech synthesis system. Speech Communication 49, 317–330 (2007)
6. Bennett, C., Black, A.: Prediction of pronunciation variations for speech synthesis: A data-driven approach. In: Proceedings of the IEEE International Conference on Acoustics, Speech, and Signal Processing 2005 (ICASSP 2005), Philadelphia, PA, USA, vol. 1, pp. 297–300 (2005)
7. Weide, R.L.: The carnegie mellon university pronouncing dictionary, version 0.4 (1995)
8. Mitton, R.: A description of a computer-usable dictionary file based on the oxford advanced learner's dictionary of current english. Technical report, Oxford Text Archive (1992)
9. Kerkhoff, J., Marsi, E.: NeXTeNS: a new open source text-to-speech system for dutch. In: 13th Meeting of Computational Linguistics in the Netherlands (2002)
10. Baayen, R.H., Piepenbrock, R., Gulikers, L.: The CELEX lexical database (CD-ROM). Technical report, Linguistic Data Consortium, University of Pennsylvania, Philadelphia, PA (1995)
11. Demuynck, K., Roelens, J., Compernolle, D.V., Wambacq, P.: SPRAAK: an open source "SPeech recognition and automatic annotation kit". In: Proceedings Interspeech 2008, Brisbane, Australia, p. 495 (2008)
12. Kominek, J., Black, A.W.: The CMU arctic speech databases. In: Proceedings Fifth ISCA Workshop on Speech Synthesis (SSW5). ISCA (2004)
13. King, S., Karaiskos, V.: The blizzard challenge 2010. In: Blizzard Challenge Workshop 2010 (2010)
14. Mattheyses, W., Latacz, L., Verhelst, W.: Auditory and photo-realistic audiovisual speech synthesis for dutch. In: Proceedings International Conference on Auditory-Visual Speech Processing 2011 (AVSP 2011), Volterra, Italy, pp. 55–60 (2011)
15. Duchateau, J., Kong, Y.O., Cleuren, L., Latacz, L., Roelens, J., Samir, A., Demuynck, K., Ghesquière, P., Verhelst, W., et al.: Developing a reading tutor: Design and evaluation of dedicated speech recognition and synthesis modules. Speech Communication 51, 985–994 (2009)
16. Van Dalen, R.C., Wiggers, P., Rothkrantz, L.J.M.: Lexical stress in continuous speech recognition. In: Proceedings Interspeech 2006, Pittsburgh, PA, USA (2006)

Structuring a Multimedia Tri-Dialectal Dictionary[*]

Nikitas N. Karanikolas[1], Eleni Galiotou[1], George J. Xydopoulos[2], Angela Ralli[2],
Konstantinos Athanasakos[1], and George Koronakis[1]

[1] Dept. of Informatics, Technological Educational Institute of Athens, 12210 Aigaleo, Greece
{nnk,egali}@teiath.gr, {k.athanasakos,gkoronakis}@gmail.com
[2] Dept. of Philology, University of Patras, 26504 Rio, Greece
{gjxydo,ralli}@upatras.gr

Abstract. This paper deals with the problem of designing and implementing a
multimedia electronic dictionary of three Greek dialects in Asia Minor (Pontic,
Cappadocian, Aivaliot). At first, we present the linguistic and lexicographic approach adopted, as well as the principles for designing the macro/microstructure
of the dictionary. Next, we present and describe the conceptual model of the tri-
dialectal dictionary. Finally, we discuss the relational schema and implementation
issues. Although, in its current state, the system hosts three dialects, it is designed
so as to be able to incorporate other dialectal dictionaries in the future.

Keywords: Computational Dialectology, Dialectal Lexicography, Electronic
Dictionaries, Macrostructure, Microstructure, Canonical Forms, Lemma, Modern
Greek Dialects, Asia Minor Greek.

1 Introduction

Structuring and implementing an electronic dialectal dictionary gives rise to a number
of issues as for the lexicographic approach adopted and its design and implementation
procedures. Existing dialectal databases are usually concerned with a single dialect and
are structured as monolingual dictionaries. Therefore, natural language processing tools
developed for monolingual dictionaries can be used for the creation and update of dialectal lexica. Yet, the creation of an electronic dictionary having as a starting point a
printed dialectal lexicon can come across various problems. For example, the transfer
of Joseph Wright's English Dialect Dictionary (EDD) into a digitized databank, on one
hand has enabled the availability of data for dialectology and historical lexicology but,
on the other hand, has come across different problems mainly due to its lexicographic
structure ([11,12]).

Recent approaches to the creation of electronic dialectal dictionaries tend to adopt a
multilingual framework. For example, in an attempt to document and archive an endangered linguistic system such as the Formosan language, Yang et al. ([23]) propose the
creation of an online system for the archiving and processing of shared resources based

[*] This research has been co-financed by the European Union (European Social Fund – ESF) and
Greek national funds through the Operational Program "Education.and Life-long Learning"
of the National Strategic Reference Framework (NSRF) – Research Funding Program: Thalis.
Investing in knowledge society through the European Social Fund.

I. Habernal and V. Matousek (Eds.): TSD 2013, LNAI 8082, pp. 509–518, 2013.

on a participatory process. Their system is based on the Web 2.0 platform and provides tribal teachers in Taiwan the possibility to develop their own languages resources and online dictionaries. In a totally different context, de Vriend et al. ([22]) propose a data model which enables the unification of data and classifications from different dictionaries of Dutch dialects. Existing and new dialectal resources are converted to XML and access to data is provided through a web-based interface. In addition, they propose the development of an infrastructure for electronic access to all dialect dictionaries. In order to facilitate archiving and interoperability, they follow the Lexical Markup Framework (LMF) which was developed in order to deal with problems posed by search, merging, linking and comparison at a cross lexicon level ([5]). To this end, all core data are imported to LMF using the LEXUS lexicon tool which is described in [8].

Electronic lexicography for Modern Greek was not concerned with the creation of dialectal dictionaries until very recently. The online dictionaries developed at the Center for the Greek Language ([3]) comprise the computerized versions of Georgacas' Greek-English Dictionary, Triandafyllides' Dictionary of Standard Modern Greek and Anastasiadi-Symeonidi's Reverse Dictionary. In addition, the portal of the Center provides access to the computerised version of Kriaras' Concise Dictionary of Medieval Vulgar Greek Literature. The Institute for Language and Speech Processing has developed online bilingual dictionaries (Greek-English, Greek-German, Greek-Russian, Greek-Turkish, Greek-Arabic). The dictionaries are under continuous development and enhancement and they are available from [7]. In addition, NLP tools in order to support lexicographic applications have been developed. Indicatively, In [18] infrastructure tools which are used for encoding morphological, syntactic and semantic information are reported as well as proofing tools such as a spelling checker, a hyphenator etc. As far as Greek dialects are concerned, the only computerized dictionary to our knowledge is the online lexical database of Cypriot Greek ([17]). The online dictionary environment provides an enhanced searching mechanism as well as text to speech feature for the pronunciation of Cypriot Greek words.

In this paper, we discuss the design and implementation of a multimedia tridialectal dictionary of Greek dialects in Asia Minor (Pontic, Cappadocian, Aivaliot) within the framework of the THALIS program "Pontus, Cappadocia, Aivali: in search of Asia Minor Greek" (AMiGre) which aims at: (a) providing a systematic and comprehensive study of these three linguistic systems of common origin and of parallel evolution that are faced with the threat of extinction and, (b) digitizing, archiving and processing a wide range of oral and written data, thus providing the sustainability of this longwinded cultural heritage. In section 2, we present the linguistic approach and the lexicographic principles for the design of the tridialectal dictionary. In section 3, we discuss design and implementation issues and we present the current state of the electronic lexicon. Finally, in section 4, we draw the necessary conclusions and point to future research.

2   Linguistic Structure and Information

The three Asia Minor dialects, Pontic, Cappadocian and Aivaliot are not sufficiently documented, although they are on the way to extinction. With the exception of the old Papadopoulos' (1958) historical dictionary of Pontic ([15]), there are only glossaries containing words and idiomatic phrases accompanied by their meaning in Standard Modern Greek. In most of these glossaries, lemmas are stored in a very unsystematic way and crucial information, such as pronunciation or usages, is missing. For instance, some verbs are listed in their past tense form while others appear in the present tense (see for example, [9]), and there is no distinction made between words and phrases.

In this project, we aim to build TDGDAM (a Tri-Dialectal Greek Dictionary of Asia Minor), a linguistically-sound tri-dialectal dictionary in electronic form. In this dictionary, among other things (e.g. the dialectal area or the source where the lemma has been extracted from), users can have access to a graphic representation of each lemma in a conventionally-adopted character set, pronunciation, meaning, usages and other possible related lemmas. This type of dictionary constitutes an innovation not only for the Greek language and its dialects, but also for the international standards, as will be explained below.

Dialectal dictionaries are usually treated as monolingual synchronic dictionaries, due to limits in macrostructure ([10,24]). However, we have decided to treat and design TDGDAM as a trilingual dictionary, given that its macrostructure is in a different system/variety from that of microstructure (Three Asia Minor dialects vs. Standard Modern Greek) ([2,6,21]). Regarding its geographic and time scope, TDGDAM is a local / microareal dialectal dictionary of non-synchronic nature as it includes entries from different areas and time periods ([14]). Lemmas in TDGDAM are drawn from oral speech and written material of the particular dialectal varieties, either directly or indirectly ([19]).

TDGDAM's projected macrostructure includes *ca.* 2,500 entries from each of the three dialects of Asia Minor Greek (a total of *ca.* 7,500 entries). These entries are drawn from collected vocabulary solely from the three dialects concerned and exclude all vocabulary found in Standard Greek (unless differently used). Their listing is based on alphabetical, and not onomasiological, organization, accessed via dynamic searching options ([20]).

TDGDAM's microstructure includes formal information about pronunciation (phonetic form), grammar (categorial and morphological information), origin (etymology), meaning (synonymic and/or descriptive definitions), usage (thematic and register labels) and it is linked to multimedia information resources (internal or external to TDGDAM) in order to enrich the semantics and pragmatics of lemmas ([1,16,21]). In order to avoid different and arbitrary spelling codes for the same dialect ([4,20]), orthographic forms (as headwords) do not appear in a "semi-phonetic" transcription but in (capitalized) orthographic form. This is so in order to avoid different and arbitrary spelling codes for the same dialect ([4,20]). In particular, the capitalized orthographic form departs from the spelling form in the standard dialect; it does not prescribe spelling rules in the dialect and allows for any alternative orthographic forms to appear in microstructure ([13,20]). Cross-reference is made to other entries, related either through derivational processes or through semantic relations. Finally, authentic examples of use have been considered to constitute essential information in entries which will appear in non-standard spelling, reflecting pronunciation as closely as possible with the use of diacritics, but avoiding a "semi-phonetic" transcription ([16]).

3   Design and Implementation

3.1   Design

In this project, we aim at building a dialectal lexicon of three dialects that can host diverse realizations of lemmas which depend on the geographic area where the dialects are spoken. The fields that we include are: headword, dialect (dialectal region), morphological information/process, etymology, realizations (usually slightly different phonetic realizations), meanings, usage examples, and related lemmas (e.g. synonyms). We address the following issues:

- headword, dialect (dialectal region), morphological information/process and etymology are primary information with single values that together define and depend on the lemma;
- each lemma can have many different realizations and each one of them is characterized by a slightly different phonetic realization depending on the microdialectal region it originates from (the specific area within the wider dialectal region where the lemma's realization occurs);
- each lemma can possibly have different meanings (i.e. polysemy), or be homonymous with other, semantically distinct, lemmas;
- for each meaning, different usage examples are essential.

Regarding, the relations between lemmas and meanings we decided to set three relations we deemed valuable for a tri-dialectal lexicon:

- Cross reference ("See also") links can be available for connecting lemmas that are semantically / pragmatically / morphologically / etymologically related to each other.
- Synonyms (words with similar meanings) and Antonyms (words with opposite meanings) are two semantic relations that apply between lemmas. Since both relations relate a lemma meaning with a lemma (the referenced one), we could merge them in a single relation (named "Thesaurus") that would act either as a synonym or as an antonym. Thus, the distinction between a synonym and an antonym could be an attribute of the "Thesaurus" link. In our case (tri-dialectal lexicon), the "Thesaurus" link is restricted between a lemma meaning and a lemma from the same dialect.
- In some cases, there are meanings of different lemmas from different dialects that share the same definition (have the same meaning). This is the third relation that our system supports. This relation can be shortly named (labelled) "Other Dialect". In contrast with the previous two relations, "Other Dialect" is a symmetrical relation.

The above principles are formally presented in the Entity Relationship Diagram (ERD) in Fig. 1.

The principles regarding the 3-dialectal lexicon could be applicable to any multidialectal lexicon. In our ERD diagram (Fig. 1), we can directly locate (pinpoint) the basic entity ("Lemma", depicted as a parallelogram with single contour in the ERD), the relations ("has", "See Also", "Thesaurus" and "Other Dialect", as rhombus), the

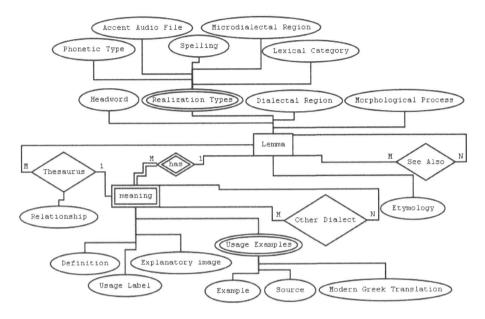

Fig. 1. Conceptual Model of our 3-Dialectal Lexicon

composite attributes with multiple values ("Usage Examples" and "Realization Types" as ellipses with double contour) and the weak entity ("meaning", as parallelogram with double contour). However, more detailed information (simple and single attributes of entities and relations and sub-attributes of composite attributes) needs further explanation regarding definition, data format and example values. Tables 1 to 5 contain this information.

Table 1. Lemma's Attributes

Attribute	Definition	Data format	Example
Headword	The canonical form of the word	String containing only capital letters of the Greek alphabet	ΑΛΛΟΥΓΥΡΙΣΤΡΑ
Etymology	Basic information about the origin of the word.	String written in Greek with accents (polytonal)	Από το ρήμα αλλουγυ-ρίζου (from the verb aluji'rizu)
Morphological Process	Different processes involved in word-formation.	A value from a predefined list of morphological processes	Σύνθετο (Compound noun)
Dialectal Region	The region/dialect in which the lemma is found	A value from a predefined list of Dialects	Αϊβαλί (AIVALI)

3.2 Implementation Issues

One possible implementation of the conceptual model (ERD) using a relational database is depicted in the relational schema of figure 2 which has thirteen tables. However, only seven tables are important. The rest six tables are lookup tables (listing the set of available values existing) for some fields of the important tables. The important tables are

Table 2. Realization Types' subAttributes

sub Attribute	Definition	Data format	Example
Phonetic Type	Phonetic transcription of (the examined) pronunciation of the word.	String containing letters of the International Phonetic Alphabet (IPA).	aluji'ristra
Accent Audio	Audio file of the authentic pronunciation of the word	String containing a file path	`http://amigre.gr/ xyzR1.wav`
Spelling	Non-standard graphic representation of pronunciation according to the orthographic rules of Standard Greek, combined with diacritics to annotate any phonological alternations.	String containing the letters of the Greek alphabet and other diacritic symbols (accent, hyphens, parentheses and apostrophes)	αλλουγυρίστρα
Microdialectical Region	Name of a specific area within the wider dialectal region of the lemma in which the realization form is found	Value from a predefined list of microdialectal regions	
Lexical Category	Part of Speech & Gender	Value from a predefined list of lexical categories	Ουσιαστικό Θηλυκό (noun feminine)

Table 3. Meaning's Attributes

Attribute	Definition	Data format	Example
Definition	Short description of the meaning of a lemma	String in Standard-Modern Greek	Γυναίκα που περιφέρεται εδώ χι εχεί ('woman who goes around')
Explanatory Image File	Image illustrating the meaning of a lemma	String containing a file path	`http://amigre.gr/xyzM1. png`
Usage Label	Formal indication of the context (stylistic/register/other) in which the lemma is used	A value from a predefined list of domains	ΥΠΟΤΙΜΗΤΙΚΟ (pejorative)

Table 4. Usage examples' subAttributes

sub Attribute	Definition	Data format	Example
Usage example	Example (phrase or sentence) demonstrating the usage of the lemma under one specific meaning, in the original dialect	The whole example (the whole string) is written with the letters of the Greek alphabet and other diacritic symbols (accent, hyphens, parentheses and apostrophes)	Ξιπόρτσι πάλ'-η-γ'-αλλουγυρίστρα
Standard Modern Greek Translation	Translation of the usage example into Standard Modern Greek	String in Standard Modern Greek	Πάλι βγήκε η αλλουγυρίστρα.
Source	Reference to the source from which the usage example was extracted	String (can be a book, a URL, etc)	

Table 5. Thesaurus' Attributes

Attribute	Definition	Data format	Example
Relationship	Distinction between Synonyms and Antonyms	'Synonym' or 'Antonym'	Synonym

highlighted (in figure 2) with thicker border and larger font in their title. Four (out of seven) important tables are the relational equivalents of the main conceptual entity ("Lemma"), the weak entity ("meaning") and the two multiple-valued composite attributes ("Realization Types" and "Usage Examples"). The rest three (out of seven) important tables are the relational equivalents of the conceptual relations ("See Also", "Thesaurus" and "Other Dialect").

Only table MeaningSets (the implementation of the conceptual relation "Other Dialect") needs more explanation. This relation is symmetrical by nature. That means that whenever one meaning of some lemma from one dialect is declared as being the equivalent of the meaning of some other lemma from a different dialect, then the reverse is implied. It is the structure of table MeaningSets (and the application's logic) that assures this symmetry. The other two relations ("See Also" and "Thesaurus") are not symmetrical by nature. This is reflected by the relational schema (and the application logic). Consequently, the user has to define the relation in both directions, in case that an instance of them (the "See Also" or the "Thesaurus" relation) is symmetrical.

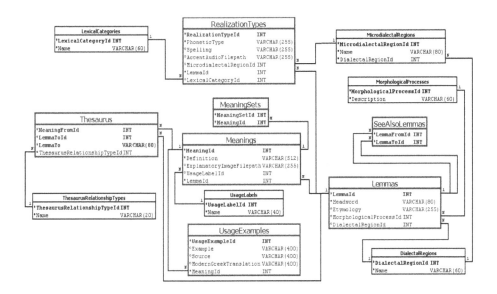

Fig. 2. Relational Schema (Implementation of ERD)

The 3-dialectal lexicon is implemented using the java language and the MySQL DBMS. Figures (screenshots) 3, 4 and 5 illustrate the functioning of the system.

4   Conclusions

In this paper, we have described the design and implementation of a multimedia electronic dictionary of three Greek dialects in Asia Minor. Available information to the users of such a lexicon, includes the dialectal area, pronunciation, meaning, usages and other possible related lemmas. We claim that the type of the dictionary constitutes an innovation not only for the Greek language and its dialects but for international standards as well. Although in its present state, the dictionary contains three dialects, it is

Fig. 3. The realization types of lemma ΑΛΛΟΥΓΥΡΙΣΤΡΑ

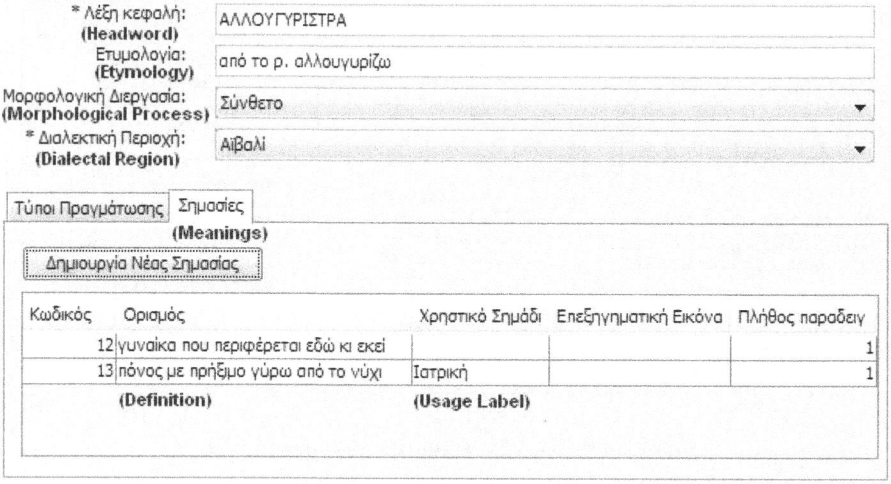

Fig. 4. The (two) meanings of lemma ΑΛΛΟΥΓΥΡΙΣΤΡΑ

designed so as to be able to incorporate more dialects in the future. Future work on the electronic dictionary includes the implementation of an advanced retrieval component and the introduction of a innovate module for automatic (or semi-automatic) identification of the "other dialect" relations. The latter, will be based on the similarity of lemma meanings' definitions.

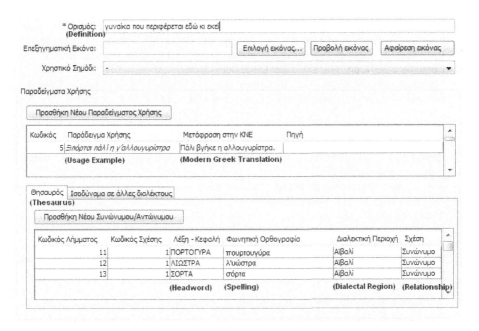

Fig. 5. Usage Examples and Synonyms of the first meaning of lemma ΑΛΛΟΥΓΥΡΙΣΤΡΑ

References

1. Barbato, M., Varvaro, A.: Dialect dictionaries. International Journal of Lexicography 17(4), 429–439 (2004)
2. Béjoint, H.: Modern lexicography: An introduction. Oxford University Press, Oxford (2000)
3. Center for the Greek Language: Online Dictionaries,
   http://www.greek-language.gr/greekLang/modern_greek/tools/
   lexica/index.html
4. Durkin, P.: Assessing non-standard writing in lexicography. In: Hickey, R. (ed.) Varieties of English in Writing: The Written Word as Linguistic Evidence, pp. 43–60. John Benjamins, Amsterdam (2010)
5. Frangopoulo, G., Bel, N., Monte, G., Calzolari, N., Monochini, M., Pet, M., Soria, C.: Lexical Markup Framework: ISO Standard for Semantic Information in NLP Lexicons. In: GLDV, Tuebingen (2007)
6. Geeraerts, D.: Principles in monolingual lexicography. In: Hausmann, F.J., Reichmann, O.O., Wiegand, H.E., Zgusta, L. (eds.) Wörterbücher / Dictionaries / Dictionnaires. Ein Internationales Handbuch zur Lexikographie. An International Encyclopedia of Lexicography. Encyclopédie Internationale de Lexicographie, pp. 287–296. Walter de Gruyter, Berlin (1989)
7. Institute for Language and Speech Processing: The Dictionaries,
   http://www.xanthi.ilsp.gr/dictionaries/ (in Greek)
8. Kemps-Snijders, M., Wittenburg, P.: LEXUS – A Web-based Tool for Manipulating Lexical Resources. In: Proceedings of the 5th LREC, Genoa (2006)
9. Kotsanidis, L.: The Cappadocian Dialect of Misti. Gnomi, Kilkis (2009)
10. Landau, S.I.: Dictionaries: The art and craft of lexicography, 2nd edn. (2001)
11. Markus, M.: Wright's English Dialect Dictionary Computerised: towards a new source of information. In: Proceedings of the 27th International ICAME Conference, May 24-26 (2006)

12. Markus, M.: Joseph Wright's English Dialect Dictionary computerised: a platform for a new historical English dialect geography. In: Amano, M., Ogura, M., Ohkado, M. (eds.) Historical Englishes Varieties of Texts and Contexts. The Global COE Programme, International Conference 2007, pp. 335–353. Peter Lang, Bern (2008)

13. Markus, M., Heuberger, R.: The architecture of Joseph Wright's English Dialect Dictionary: preparing the computerized version. International Journal of Lexicography 20(4), 355–368 (2007)

14. Penhallurik, R.: Dialect dictionaries. In: Cowie, A.P. (ed.) The Oxford History of English Lexicography, vol. II, pp. 290–313. Oxford University Press, Oxford (2009)

15. Papadopoulos, A.: Historical Dictionary of the Pontic Dialect. Epitropi Pontiakon Meleton. Annex 3, Athens (1958)

16. Rys, K., Van Keymeulen, J.: Intersystemic correspondence rules and headwords in Dutch dialect lexicography. International Journal of Lexicography 22, 129–150 (2009)

17. Themistocleous, C., Katsogiannou, M., Armosti, S., Christodoulou, K.: Cypriot Greek Lexicography: An Online Lexical Database. In: Proceedings of Euralex 2012, pp. 889–891 (2012)

18. Tsalidis, C., Pantazara, M., Minos, P., Mantzari, E.: NLP Tools for Lexicographic Applications in Modern Greek. In: Granger, S., Paquot, M. (eds.) eLexicography in the 21st Century: New Challenges, New Applications (Proceedings of eLex2009), Louvain-la-Neuve, pp. 457–462 (2010)

19. Van Keymeulen, J., De Tier, V.: Pilot project: A Dictionary of the Dutch Dialects. In: Proceedings of the 14th Euralex International Congress, pp. 754–763. Fryske Akademy, Ljouwert/Leeuwarden (2010)

20. Xydopoulos, G.J.: Metalexikografikes paratiriseis sta leksika Benardi ke Syrkou (Metalexicographic comments to Benardi and Syrkou (dialectal) dictionaries). In Patras Working Papers in Linguistics 2.1, pp. 96–113 (2011)

21. Xydopoulos, G.J., Ralli, A.: Greek dialects in Asia Minor: Setting lexicographic principles for a tridialectal dictionary. Paper read at the 5th MGDLT Conference (September 2012), Ghent, Belgium (2012)

22. de Vriend, F., Boves, L., van den Heuvel, H., van Hout, R., Kruijsen, J., Swanenberg, J.: A Unified Structure for Dutch Dialect Dictionary Data. In: Proceedings of the 5th LREC, Genoa (2006)

23. Yang, M.-C., Chou, H.-T., Guo, H.-S., Chen, G.-P.: A Formosan Multimedia Dictionary Designed Via a Participatory Process. In: Rau, D.V., Florey, M. (eds.) Documenting and Revitalizing Austronesian Languages, pp. 202–218. LD&C Special Publication (2007)

24. Zgusta, L.: Manual of lexicography. Mouton, The Hague (1971)

Stylistic Changes for Temporal Text Classification

Sanja Štajner[1] and Marcos Zampieri[2]

[1] Research Group in Computational Linguistics, University of Wolverhampton, UK
[2] Romance Philology Department, University of Cologne, Germany
sanjastajner@wlv.ac.uk,
mzampier@uni-koeln.de

Abstract. This paper investigates stylistic changes in a set of Portuguese historical texts ranging from the 17th to the early 20th century and presents a supervised
method to classify them per century. Four stylistic features – average sentence
length (ASL), average word length (AWL), lexical density (LD), and lexical richness (LR) – were automatically extracted for each sub-corpus. The initial analysis
of diachronic changes in these four features revealed that the texts written in the
17th and 18th centuries have similar AWL, LD and LR, which differ significantly
from those in the texts written in the 19th and 20th centuries. This information was
later used in automatic classification of texts per century, leading to an F-Measure
of 0.92.

Keywords: text classification, stylistic changes, historical corpora, Portuguese.

1 Introduction

It is well known that language changes over time. These changes occur in all aspects
of language: phonetics, lexicon, grammar, and discourse, as well as in its style. While
reading a text dating from a previous century, the reader can often spot that the text contains features that are not common to contemporary language, even if not being aware
of its publication date. As it can be seen in [1], studies on lexical and syntactic change
are abundant for most languages. The interest of philologists and historical linguists
in tracking language change is long-standing, and it exists prior to the development of
the first electronic corpora, which are the fundamental resource for current studies in
language change.

To the best of our knowledge, very little has been said regarding the stylistic changes
of texts. Studies on lexical richness, density and other stylistic aspects of historical
texts have mostly been neglected. This is mainly due to the difficulty in quantifying this
information before the development of electronic corpora and reliable NLP tools. Only
recently have a couple of experiments applied NLP techniques to quantify changes in
diachronic corpora [2,3].

In this paper we investigate stylistic changes of historical texts and use this information to train machine learning algorithms to classify texts automatically. Historical
manuscripts are sometimes unidentified regarding its geographical source and/or date
of publication, and classification methods can be trained to estimate this information.
The methods presented here were applied to a Portuguese historical corpus [4], but they

I. Habernal and V. Matousek (Eds.): TSD 2013, LNAI 8082, pp. 519–526, 2013.

can be replicated to any language. This study is of interest to researchers in text classification and NLP in general, and historical linguists as well as scholars in the digital humanities who deal with historical manuscripts.

2   Related Work

The vast majority of corpus-based studies on language change focus on grammatical changes (e.g. Leech et al. [5] for English and Galves et al. [6] for Portuguese). A number of English diachronic corpora are available for this kind of study which makes it possible for scholars to use NLP and quantitative methods to examine language change. For Portuguese, only a few resources are available, including: Tycho Brahe [7], and Colonia [4]. On stylistic diachronic changes in 20th-century English language, Štajner and Mitkov [3] report significant changes in several features. Among them – the most relevant for this study – is a significant increase in lexical density and lexical richness in 20th-century British and American English general prose.

Regarding temporal text classification, a couple of studies are worth mentioning. Dalli and Wilks [8] present a computational model to date texts from a time span of nine years. The method is aided by lexical items which increase their frequency at some point of time (e.g. *Bin Laden* in September, 2001 or *World Cup* in June, 2010). The experiments described by Abe and Tsumoto [9] work under a similar assumption. The authors proposed the use of similarity measures to categorise texts based on keywords that are calculated by indexes such as the popular tf-idf. The method obtains document clusters based on temporal differences in the usages of terms.

Mohkov [10] presented one of the systems that participate in the DEFT2010[1] shared task. In this shared task, systems aimed to classify short French journalistic texts of up to 300 words not only with respect to their geographical location, but also regarding the decade in which they were published. Trieschnigg et al. [11] describe a classification experiment using the Dutch Folktale Database. This database includes texts from different dialects and varieties of Dutch, but also historical texts written in middle and 17th-century Dutch. Researchers report a micro average F-measure of 0.799 with the highest F-measure reaching 0.987 for one of the classes.

To the best of our knowledge, the idea of using stylistic features for temporal text classification is new to Portuguese and not substantially explored to most languages. Most studies use lexical and orthographic features to identify the date of publication of a text.

3   Methods

The study consists of two main parts: (1) quantitative analysis of four stylistic features automatically extracted from the corpus; and (2) five text classification experiments.

---

[1] http://www.groupes.polymtl.ca/taln2010/deft.php

3.1 Corpus

We used the aforementioned Colonia[2] [4], a diachronic collection of historical Portuguese containing texts ranging from the 16th to the early 20th century. The corpus is annotated with lemma and part-of-speech (POS) information, using TreeTagger [12], which is regarded to achieve performance of over 95% accuracy using coarse-grained tags. According to the authors, spelling variation was not systematically normalised, but they acknowledge that some texts presented edited orthography prior to their compilation. At its compilation stage, authors addressed solely the question of unknown lemmas caused by non-standard spelling.

The original Colonia corpus contains 100 texts spanning from 16th to 20th century, balanced between European and Brazilian Portuguese (it contains 52 Brazilian texts and 48 European texts). The time span covered in our experiments comprises the period from 17th to the 20th century and a total of 87 texts. As to the size of the articles, the original corpus contains complete manuscripts of up to 90,000 tokens each. For our experiments, we decided to work with samples of up to 2,000 tokens per text, which were retrieved automatically, starting from a random point in the text (Table 1).

Table 1. Corpora

Century	Texts	Sentences	Tokens
17th	18	1,667	31,635
18th	14	2,566	23,175
19th	38	5,217	63,950
20th	17	2,602	28,569
Total	87	12,052	147,329

We decided to use this sample size in order to obtain results which could be compared with a similar study in English language [3] based on the 'Brown family' of corpora (which also has approx. 2000 tokens per text).

3.2 Experimental Settings

Four stylistic features – average sentence length (ASL), average word length (AWL), lexical density (LD), and lexical richness (LR) – were automatically extracted from the corpora (Table 2). Based on the initial analysis of the distribution of these four features across the four centuries (17th–20th), we decided to conduct five text classification experiments:

1. Classification across all four centuries (17th–20th);
2. Classification between (17th–18th) and (19th–20th) centuries;
3. Classification between the 17th and 18th centuries;

---

[2] http://corporavm.uni-koeln.de/colonia/

4. Classification between the 18th and 19th centuries;
5. Classification between the 19th and 20th centuries.

Table 2. Features

Feature	Code	Formula
Average sentence length	ASL	ASL = words/sentences
Average word length	AWL	AWL = characters/words
Lexical density	LD	LD = (unique tokens)/tokens
Lexical richness	LR	LR = (unique lemmas)/tokens

All classification experiments were conducted in Weka[3] Experimenter [13], employing four different classification algorithms – Naive Bayes [14]; SMO (Weka implementation of Support Vector Machines) [15,16] with normalisation and using poly kernels; JRip [17], and J48 (Weka implementation of C4.5) [18] – in 5-fold cross-validation setup with 10 repetitions. In all experiments, we considered the majority class as the baseline.

4   Results and Discussion

The averaged values for each of the four investigated features (ASL, AWL, LD, and LR) in each of the sub-corpora, together with their standard deviations, are presented in Table 3. Statistically significant differences between adjacent centuries are presented in bold. The difference in ASL between the 18th and 19th centuries was reported as statistically significant at a 0.05 level of significance, while all other statistically significant differences were significant at a 0.001 level of significance. Statistical significance was calculated using the two-independent samples t-test in SPSS (in cases where both compared sets followed approximately normal distribution) and using the two-sample Kolmogorov-Smirnov test (in cases where at least one of the sets did not follow approximately normal distribution).

Table 3. Statistics of the corpora (Key: ASL = average sentence length (in words); AWL = average word length (in characters); LD = lexical density; LR = lexical richness)

Century	ASL	AWL	LD	LR
17th	**20.53 ± 6.29**	4.48 ± 0.16	0.38 ± 0.04	0.14 ± 0.02
18th	**11.73 ± 6.42**	**4.52 ± 0.16**	**0.39 ± 0.03**	**0.15 ± 0.02**
19th	**13.73 ± 5.55**	**4.80 ± 0.18**	**0.46 ± 0.03**	**0.19 ± 0.02**
20th	12.79 ± 6.24	4.89 ± 0.32	0.47 ± 0.04	0.18 ± 0.02

The skewness and the existence of outliers can be observed from the box-plots presented in Figure 4. The height of the rectangle indicates the spread of the values

---
[3] http://www.cs.waikato.ac.nz/ml/weka/

for the variable, the horizontal line inside the rectangle indicates the mean, while the "whiskers" outside the rectangle indicate the smallest and largest observations which are not outliers. Outliers are presented as numbered cases beyond the whiskers. If the rectangle is not equally distributed on both sides of the mean line, then the data is skewed (not normal).

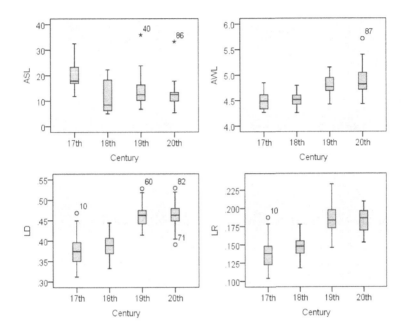

Fig. 1. Distribution of features across the corpora

The results presented in Table 3 and Figure 4 indicate that the average sentence length (ASL) was significantly higher in the 17th than in the 18th century, but then significantly lower in the 19th than in the 18th century. More interestingly, it revealed that the texts written in the 17th and 18th centuries have similar AWL, LD and LR, which were significantly lower than those in the texts written in the 19th and 20th centuries.

These results motivated us to conduct the second text classification experiment (where the texts from the 17th and 18th centuries were grouped together in one class, and those from the 19th and 20th centuries in the other class), in addition to the first classification experiment across all four centuries (17th–20th) and the other three classification experiments between each pair of adjacent centuries (17th and 18th, 18th and 19th, and 19th and 20th).

The results of all classification experiments are presented in Table 4. Columns 'NB', 'SMO', 'JRip', and 'J48' contain weighted average F-measures of the four classification algorithms (Section 3.2), while the column 'baseline' contains the classification accuracy if for each text we select the majority class. Figure 2 contains the rules of the JRip classifier which were used in each of the five experiments.

Table 4. Classification results

Exp.	Classes	NB	SMO	JRip	J48	Baseline
(1)	17th, 18th, 19th, 20th	0.59	0.54	0.52	0.56	0.44
(2)	17th+18th, 19th+20th	0.92	0.92	0.87	0.87	0.63
(3)	17th, 18th	0.64	0.67	0.63	0.73	0.56
(4)	18th, 19th	0.91	0.86	0.88	0.86	0.73
(5)	19th, 20th	0.59	0.57	0.57	0.55	0.69

Experiment I – Classification between the 17th, 18th, 19th, and the 20th century texts:

```
(LD <= 0.421742) and (ASL <= 8.787234) => text=18th (9.0/1.0)
(LD <= 0.407346) and (AWL <= 4.675958) => text=17th (16.0/0.0)
 => text=19th (62.0/24.0)
```

Experiment II – Classification between the 17th+18th and the 19th+20th century texts:

```
(LD <= 0.421742) => text=17th (34.0/5.0)
 => text=19th (53.0/3.0)
```

Experiment III – Classification between the 17th and the 18th century texts:

```
(ASL <= 8.787234) => century=18th (8.0/0.0)
(AWL >= 4.548444) and (AWL <= 4.601594) => century=18th (3.0/0.0)
(AWL >= 4.6875) => century=18th (4.0/1.0)
 => century=17th (17.0/0.0)
```

Experiment IV – Classification between the 18th and the 19th century texts:

```
(LD <= 0.406519) => century=18th (11.0/0.0)
 => century=19th (41.0/3.0)
```

Experiment V – Classification between the 19th and the 20th century texts:

```
(AWL >= 5.186118) => century=20th (3.0/0.0)
 => century=19th (52.0/14.0)
```

Fig. 2. JRip rules for the classification experiments

From the results presented in Table 4, it can be noted that classification accuracies were significantly higher in the second than in the first experiment for all four algorithms, achieving the weighted average F-measure up to 0.92. This is not a surprise given that initial analysis revealed a statistically significant difference in all four features (ASL, AWL, LD, and LR) between the 18th- and the 19th-century texts (Table 3), and the classification between the texts from the 18th and the 19th centuries (experiment 4) achieved almost equally good results. The results of the first and the third experiment, although being significantly lower than those of the second and the fourth experiments, still outperformed the baseline. The results of the classification of texts between 19th

and 20th century, however, did not even reach the performance of the baseline. One possible explanation for this difficulty in classifying texts from these two centuries is that the 20th century class contains only texts published in the first half of the century. The newest text was published in 1948. The style of the texts are therefore very similar to those published in the end of the previous century and this has direct impact on the classifiers' performance.

The difference in the results achieved in the first and the third experiments, and those achieved in the fifth experiment, could also be explained by the fact that the initial analysis of the corpora revealed that there was a significant difference in one of the features (ASL) between the texts from the 17th and the 18th centuries, and there was no significant differences in any of the four investigated features between the texts from the 19th and the 20th centuries. The presented results (Table 3 and Table 4) thus indicate a high correlation between the classification accuracy and the number of features reported to be significantly different between two classes.

5   Conclusions and Future Work

This study was, to the best of our knowledge, a first attempt of comparing the style of historical Portuguese texts in a purely automatic manner. The results indicated similarities between texts from the 17th and 18th as well as the 19th and 20th centuries, and a great dissimilarity between the 18th- and the 19th-century texts. It was also observed that the lexical density (LD) and lexical richness (LR) were substantially higher in the 19th- and 20th-century texts than in the 17th- and 18th- century texts.

As a practical application of our initial analysis of the corpora, we carried out five automatic classification experiments. The first setting containing four classes (one class for each century) achieved a modest 0.59 F-measure which outperformed the baseline. The second setting, binary classification (17th and 18th centuries; and 19th and 20th centuries grouped together), achieved a 0.92 F-measure, thus reflecting the already reported significant differences in all four features (ASL, AWL, LD, and LR) between the texts from the 18th and the 19th century. The lowest classification performances were reported for the classification between the texts from the 19th and the 20th century, again reflecting the fact that the initial analysis of the corpora did not report any significant differences in any of the four investigated features between those two sets of texts.

We continue to experiment with historical texts in different directions. As Portuguese is a pluricentric language, it would be interesting to investigate whether there are significant stylistic differences between these two varieties (both synchronic and diachronic). Previous studies [19] suggest that classification methods are able to distinguish Brazilian and European current texts with 99.8% accuracy when using lexical and orthographic features. It would be worth exploring whether a similar classification accuracy could be achieved by using some language-independent features, thus enabling the use of the same methodology for other languages with their regional varieties.

References

1. Joseph, B., Janda, R.: The Handbook of Historical Linguistics. Blackwell Publishing (2003)
2. Smith, J., Kelly, C.: Stylistic constancy and change across literary corpora: Using measures of lexical richness to date works. Computers and the Humanities 36, 411–430 (2002)
3. Štajner, S., Mitkov, R.: Diachronic stylistic changes in british and american varieties of 20th century written english language. In: Proceedings of the Workshop on Language Technologies for Digital Humanities and Cultural Heritage, Hissar, Bulgaria, pp. 78–85 (2011)
4. Zampieri, M., Becker, M.: Colonia: Corpus of historical portuguese. ZSM Studien, Special Volume on Non-Standard Data Sources in Corpus-Based Research 5 (2013)
5. Leech, G., Hundt, M., Mair, C., Smith, N.: Change in Contemporary English: A Grammatical Study. Cambridge University Press, Cambridge (2009)
6. Galves, C., Sandalo, F.: Clitic-placement in modern and classical European Portuguese. MIT Working Papers in Linguistics 47, 115–128 (2004)
7. Britto, H., Finger, M., Galves, C.: Computational and linguistic aspects of the Tycho Brahe parsed corpus of historical portuguese. In: Proceedings of the First Freiburg Workshop on Romance Corpus Linguistics, Freiburg, Germany (2000)
8. Dalli, A., Wilks, Y.: Automatic dating of documents and temporal text classification. In: Proceedings of the Workshop on Annotating and Reasoning about Time and Events, Sidney, Australia, pp. 17–22 (2006)
9. Abe, H., Tsumoto, S.: Text categorization with considering temporal patterns of term usages. In: Proceedings of ICDM Workshops, pp. 800–807. IEEE (2010)
10. Mokhov, S.: A marf approach to deft 2010. In: Proceedings of TALN 2010, Montreal, Canada (2010)
11. Trieschnigg, D., Hiemstra, D., Theune, M., de Jong, F., Meder, T.: An exploration of language identification techniques for the dutch folktale database. In: Proceedings of LREC 2012 (2012)
12. Schmid, H.: Probabilistic part-of-speech tagging using decision trees. In: Proceedings of International Conference on New Methods in Language Processing, Manchester, UK (1994)
13. Witten, I., Frank, E.: Data mining: Practical machine learning tools and techniques. Morgan Kaufmann Publishers (2005)
14. John, G.H., Langley, P.: Estimating Continuous Distributions in Bayesian Classifiers. In: Proceedings of the Eleventh Conference on Uncertainty in Artificial Intelligence, pp. 338–345 (1995)
15. Keerthi, S.S., Shevade, S.K., Bhattacharyya, C., Murthy, K.R.K.: Improvements to Platt's SMO Algorithm for SVM Classifier Design. Neural Computation 13, 637–649 (2001)
16. Platt, J.C.: Fast Training of Support Vector Machines using Sequential Minimal Optimization. In: Schoelkopf, B., Burges, C., Smola, A. (eds.) Advances in Kernel Methods – Support Vector Learning (1998)
17. Cohen, W.: Fast Effective Rule Induction. In: Proceedings of the Twelfth International Conference on Machine Learning, pp. 115–123 (1995)
18. Quinlan, R.: C4.5: Programs for Machine Learning. Morgan Kaufmann Publishers, San Mateo (1993)
19. Zampieri, M., Gebre, B.G.: Automatic identification of language varieties: The case of Portuguese. In: Proceedings of KONVENS 2012, Vienna, Austria, pp. 233–237 (2012)

SummEC: A Summarization Engine for Czech

Michal Rott and Petr Červa

Institute of Information Technology and Electronics, Technical University of Liberec
Studentska 2, 461 17, Liberec, Czech Republic
{michal.rott,petr.cerva}@tul.cz
https://www.ite.tul.cz/itee/

Abstract. This paper describes a summarization engine developed primarily for the Czech language. Therefore, the engine takes advantage of language-dependent preprocessing modules performing segmentation of the input document into sentences, lemmatization and substitution of synonyms. Our system is also implemented as a dynamic library which can be employed in either a web or a desktop application, and supports a variety of summarization methods. To evaluate the performance of the system, several experiments are conducted in this paper using a set of manually created summaries. The obtained results show that our engine yields an outcome for Czech which is better or at least comparable to other online summarization systems. The above-mentioned reference summaries and the presented summarization engine are available online at http://summec.ite.tul.cz.

Keywords: automatic summarization, single-document summarization, latent semantic analysis, term frequency, inverse document frequency.

1 Introduction

Automatic summarization [1] is a wide scientific discipline which makes use of linguistic, mathematical and computer science knowledge and skills. The primary goal of the discipline is to instruct computers how to distill the most important information from a source (single-document summary) or set of sources (multi-document summary). It is also possible to distinguish between summaries which are created as extracts and those which represent abstracts and contain new sentences or phrases generated by the automatic summarizer [2]. The next traditional distinction is between informative summaries, which can be used instead of the source, and indicative summaries, which allow selection of documents for further, more detailed analysis.

All aforementioned types of summaries can be useful in a wide range of applications. For example, informative summaries of lectures allow the review of course materials [3]. On the other hand, indicative summaries can simulate the work of an intelligence analyst or enable the analyst to more accurately find the most relevant documents. It would also be interesting to have a system providing a multi-document information summary of a story from various newspapers or even directly from TV channels.

It is therefore not surprising that systems for automatic summarization have attracted a lot of attention recently and several evaluations [4] of these systems have been conducted within the Text Analysis Conferences (TAC)[1]. Several summarization

---

[1] http://www.nist.gov/tac

I. Habernal and V. Matousek (Eds.): TSD 2013, LNAI 8082, pp. 527–535, 2013.

systems also exist that are available online, such as the Open Text Summarizer[2], Free Summarizer[3] or Text Compactor[4].

Unfortunately, all of these systems, as well as others, are language-dependent and their use for the highly inflective Czech language is limited (as proven by results presented in Section 4.6) Note that only the Open Text Summarizer provides direct support for the Czech language. There is also the Almus[5] summarizer developed in the Czech republic, but the data files released on Almus web pages only allow the creation of summaries from English texts.

Therefore, we decided to develop a new Summarization Engine for Czech (SummEC). The motivation for this development also stems from the fact that we cooperate with a media-monitoring company which operates in the Czech, Slovak and Polish market and delivers news digests particularly from the print sources. These digests could be created automatically by a summarization engine.

Hence, the current version of our system produces informative summaries, which are created on a single-document basis. The multi-document extension and support for other (Slavic) languages is under development.

Note that our summaries are created as extracts and with the use of a text-preprocessing module, which performs language-dependent operations such as substitution of synonyms and lemmatization, that are important for the resulting accuracy of the system (as proven in Section 4.5).

The rest of the paper is structured as follows: The next section describes the architecture and modules of the SummEC system. Section 3 then reviews the principles of summarization methods that are supported by SummEC. Experimental evaluation of these methods is then given in Section 4, where we also compare our results with those obtained using several online summarization tools. The last section 5 then concludes this paper.

2   Description of the SummEC System

2.1   System Architecture

The overall scheme of SummEC is depicted in Figure 1. The figure shows that the system is composed of a preprocessing and summarization modules and that the output from the former module serves as the input to the latter one.

The figure also demonstrates that SummEC is implemented as a dynamic library which can be employed in either a web or a desktop application. The output summary (i.e., the sequence of the most important sentences) can be converted to the xml format for evaluation and further processing, or to the format appropriate for display on a web page.

The preprocessing module is detailed in the following subsection. The summarization module provides support for several summarization methods. Their principles are described in Section 3.

---

[2] http://libots.sourceforge.net
[3] http://freesummarizer.com
[4] http://textcompactor.com
[5] http://textmining.zcu.cz/?lang=en&section=download

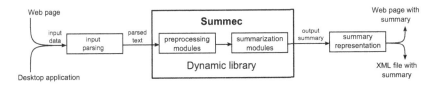

Fig. 1. The scheme of the developed Summarization Engine for Czech (SummEC)

2.2 The Preprocessing Module

The preprocessing module converts the input text to its normalized form. This process is carried out in three consecutive phases as follows:

Sentence Segmentation: The text is split into sentences in the first step. The splitting routine takes into account Czech words ending with the symbol '.' such as academic titles (Ing.), military titles (gen.) or abbreviations.

Lemmatization: In the next step, every sentence is lemmatized using an external morphological tool Fmorph[6]. This analyzer was chosen as it has a higher latency than newer Czech taggers Morče[7] and Compost[8]. This factor is important, because SummEC also performs summarization in the online mode (over the web interface).

Note that, after lemmatization, the resulting text does not just contain lemmas, but also word forms, such as numbers or typing errors, which cannot be lemmatized. We further call all of these items of the lemmatized text as terms.

Substitution of Synonyms: The goal of the third step is to substitute all synonyms of every lemma using one preferred form. The substitution is based on the use of a lemmatized dictionary of synonyms, which contains 7443 different groups of synonyms with a total of 22856 lemmas. These items are compiled from two sources. The first is the Czech version of the project Wiktionary[9]. The second is the Thesaurus project[10].

3 Supported Summarization Methods

3.1 The Heuristic Method

The heuristic method [1] is based on a natural idea that a word which occurs frequently in the input document is important and should therefore be presented in the resulting

---

[6] http://ufal.mff.cuni.cz/pdt/Morphology_and_Tagging/Morphology/index.html
[7] http://ufal.mff.cuni.cz/morce/
[8] http://ufal.mff.cuni.cz/compost/
[9] http://cs.wiktionary.org
[10] http://packages.debian.org/sid/myspell-cs

summary of the document. However, common words exist for every language, e.g., prepositions or conjunctions, which are generally considered irrelevant to the meaning or topic of the given document. These words should be included in a stop list.

An advantage of the heuristic method is that it is simple to implement, because it works in three simple steps, as follows:

At first, the frequency of occurrence, i.e., the term frequency (TF), is calculated for every term from the input document, excluding those in the stop list. In the second step, the score of each sentence is given as a sum over TFs of all words in the given sentence. Finally, the sentences with the highest scores are included in the resulting summary.

3.2 The TF-IDF Method

The TF-IDF method [5,6] represents a modification of the heuristic method. It does not rely on a stop list; instead, it assigns a weight to the frequency of each term in the sentence by its inverse document frequency (IDF). For this purpose, it is necessary to first create an IDF dictionary containing IDF values for all terms t from the set of training documents D.

These IDF values can be expressed as:

$$IDF(t) = log \frac{|D|}{|\{d \in D : t \in d\}|} \tag{1}$$

where $|D|$ is the total number of training documents and $|\{d \in D : t \in d\}|$ is the number of documents containing the term t.

Note that the previous equation demonstrates that the resulting IDF value is low for common terms. This is why the IDF method does not require a stop list.

The score of each sentence s from the input document is then simply given as:

$$score(s) = \sum_{t \in s} TF(t) \times IDF(t) \tag{2}$$

Similar to the heuristic method, the output summary is created from sentences with the best scores. However, the resulting summary then usually contains very similar sentences, which are composed of the same terms with a high value of the product $TF(t) \times IDF(t)$. To eliminate these redundant sentences, an enhanced $TF(t) \times IDF(t)$ approach was proposed in [7]. This method is carried out in the following steps:

1. The TF values are determined for all terms in the input document.
2. The TF x IDF score is calculated for each sentence.
3. The sentence with the highest score is added to the output summary.
4. The TF values for all terms from this sentence are set to zero.
5. Steps 2-4 are repeated until the summary has the required number of sentences.

3.3 Latent Semantic Analysis

The Latent Semantic Analysis (LSA) [8] is inspired by latent semantic indexing. Therefore, it employs the singular value decomposition (SVD) to the matrix A, in which each column vector represents the weighted TF vector of one sentence of the input document. That means that, when the document contains m terms and n sentences, the matrix A has a size of $m \times n$. The SVD of A is then defined as:

$$A = U\Sigma V^T \tag{3}$$

where U is an $m \times n$ column-orthonormal matrix of left singular vectors, $\Sigma = diag(\sigma_1, \sigma_2, ..., \sigma_n)$ is an $n \times n$ diagonal matrix whose diagonal elements represent non-negative singular values sorted in descending order, and V is an $n \times n$ orthonormal matrix of right singular vectors.

The authors in [8] show that the matrix V describes an importance degree for each topic of the document in each sentence. Hence, the resulting summary is created by choosing the most important sentences as follows:

1. The summary does not contain any sentence, $k = 1$.
2. The matrix A is constructed and the SVD of this matrix is performed. Each sentence of the document is then represented in the singular vector space by the column vector of V^T.
3. The k'th vector from the matrix V^T is selected.
4. The sentence that has the largest index value with this vector is included in the summary
5. Similar to the TF x IDF method, the value of k is incremented and steps 3 and 4 are repeated until the summary has the required number of sentences.

As mentioned above, each column vector of the matrix A represents the weighted TF vector of one sentence. The weighting can be performed using several different methods [9]. The recent version of SummEC supports two basic approaches. The first takes advantage of the IDF values defined above. In the second approach, the frequency of each term in the sentence is normalized by its frequency over the whole document.

3.4 Modified LSA

The approach described above was modified in [10] using elements of the diagonal matrix Σ. The score of each sentence s is then expressed as:

$$score(s) = \sqrt{\sum_{i=1}^{p} v_{k,i}^2 \cdot \sigma_i^2} \tag{4}$$

where p is the number of chosen dimensions of the new space, $v_{k,i}$ is the i'th element of the k'th column vector of the matrix V, and σ_i is the corresponding element of the matrix Σ. The authors in [10] also suggest choosing only the dimensions whose singular values were smaller than half of the highest singular value.

4 Experimental Evaluation

4.1 Data for Evaluation

Unfortunately, no publicly available reference data exist for evaluation of automatic summarization in the Czech language. Therefore, we had to create our own test set:

we asked 15 persons to produce informative extracts of 50 different newspaper articles. The articles contained 92089 words in total and were selected from columns on local and international news, economics and culture. The resulting extracts contain an average of six sentences from each article (i.e., 20 % of the sentences from each article) and are made public on the SummEC web pages.

4.2 Tools and Metrics Used

We used the toolkit ROUGE [11] for evaluation, which supports various metrics to compare automatically-produced summaries against manually-produced references. In this paper, we chose the metrics ROUGE-1 and ROUGE-W. The former is based on co-occurrence of unigrams. The latter represents statistics based on Weighted Longest Common Subsequence (WLCS) [11]. The weight was set on the value of 1.2 in our case.

In following subsections, the results obtained using these metrics are presented in terms of Recall, Precision and F-score. These are defined for ROUGE-1 as:

$$Precision = \frac{TP}{TP + FP} \quad Recall = \frac{TP}{TP + FN} \quad F - score = \frac{2RP}{R + P} \quad (5)$$

where TP, FP and FN are explained in Table 1.

Table 1. The meaning of variables in equation (5) for ROUGE-1

# unigrams	selected by anotators	not selected by annotators
selected by the system	TP	FN
not selected by the system	FP	TN

A more complex definition of Precision, Recall and F-score for ROUGE-W can be found in [11].

4.3 Experimental Setup

The summarization methods reviewed in Section 3 make use of a stop list and an IDF dictionary. In this work, both of these components are created using 2.2M newspaper articles. The resulting stop list contains 283 items, including the most frequent Czech words and Czech prepositions, conjunctions and particles. The IDF dictionary has 491k items. They represent all terms from the lemmatized articles whose frequency of occurrence was higher than five.

4.4 Comparison of Supported Summarization Methods

The aim of the first experiment performed was to compare the results of individual summarization methods that are supported by SummEC. The experiment was carried out using all preprocessing modules and the stop list. The obtained results are presented in Table 2. In this table, the method denoted as LSA-IDF corresponds to the enhanced version of LSA with weighting based on IDF. In contrast, LSA-TF stands for the enhanced LSA using TF normalization.

It also should be noted that in this experiment, we take advantage of the approach based on increasing the frequency of terms that are important for the topic of the document. We suppose that these topic terms are those included in the title of each document and we multiply their TF values by two.

The presented results show that the TFxIDF method yielded the highest F-score for the metric ROUGE-1 (57.3 %) as well as ROUGE-W (30.0 %). This approach also led to the best Recalls of 62.6 % and 35.5 % respectively. In contrast, the highest Precisions of 55.2 % and 28.9 % were reached by LSA-IDF. It is also evident that the worst F-scores were obtained by using LSA-TF.

Table 2. Comparison of results of individual summarization methods supported by SummEC

method	ROUGE-1			ROUGE-W		
	Recall [%]	Prec. [%]	F-score [%]	Recall [%]	Prec. [%]	F-score [%]
Heuristic	57.2	54.3	55.3	30.3	27.9	28.3
TFxIDF	62.6	53.3	57.3	35.5	26.6	30.0
LSA-IDF	55.4	55.2	55.1	28.6	28.9	28.4
LSA-TF	57.6	50.1	53.3	28.6	22.2	24.6

4.5 Performance of SummEC's Components

In the second series of experiments, individual components of our system were activated gradually to show their contribution to the system's overall accuracy. We employed the TFxIDF method, which yielded the best results in the previous experiment. The obtained results are presented in Table 3.

This shows that the highest absolute increase in F-score was reached for both metrics by using lemmatization. The other components and approaches yielded only a small additional improvement of this measure. The exception is the stop list, which increased the F-score of ROUGE-W from 29.4 % to 30.3 %.

4.6 Comparison with other Online Systems

The final experiment (see Table 4) compares the results yielded by SummEC (using TFxIDF) with several online summarization systems. We can see that SummEC outperformed not only the systems without explicit support for Czech, as we expected, but the OTS system as well. This proves that SummEC is a useful tool for automatic summarization of documents in the Czech language.

Note that the worst results reached by the Free Summarizer are caused by the fact that this tool does not correctly accept the Czech set of characters. For that reason, some other online systems were not evaluated at all.

Table 3. Contribution of individual components to the overall performance of our engine

component	ROUGE-1			ROUGE-W		
	Recall [%]	Prec. [%]	F-score [%]	Recall [%]	Prec. [%]	F-score [%]
no component	58.3	50.6	53.9	31.0	23.8	26.6
lemmatization	63.9	51.6	56.9	37.1	24.9	29.5
+ stop list	63.0	52.5	57.0	35.7	25.5	29.4
+ topic terms	61.6	53.7	57.1	35.2	27.3	30.3
+ synonyms	62.6	53.3	57.3	35.5	26.6	30.0

Table 4. Comparison of results yielded by SummEC and several online summarizers

system	ROUGE-1			ROUGE-W		
	Recall [%]	Prec. [%]	F-score [%]	Recall [%]	Prec. [%]	F-score [%]
SummEC	62.6	53.3	57.3	35.5	26.6	30.0
Text Compactor	56.1	51.6	53.2	32.0	27.6	28.5
Free Summarizer	25.9	36.0	29.3	7.4	14.2	9.0
Open Text Summ.	50.8	54.7	52.4	27.1	31.8	28.8

5 Conclusion

In this paper, we presented our summarization engine developed for the Czech language and evaluated its performance on the set of manually created reference summaries. This evaluation demonstrated that a) the TFxIDF method is capable of producing the best automatic summaries and b) the lemmatization module is an important component of the system, because Czech is a highly inflective language. The comparison of SummEC's results with those yielded by several online summarization systems showed that our engine produces summaries of high accuracy. As previously mentioned, the online version of our engine and the reference summaries are available for free at http://summec.ite.tul.cz. Support for other Slavic languages, particularly for Slovak, Polish and Croatian, is under development.

Acknowledgments. This paper was supported by the Technology Agency of the Czech Republic (project no. TA01011204) and by the Student Grant Scheme (SGS) at the Technical University of Liberec.

References

1. Luhn, H.P.: The automatic creation of literature abstracts. IBM J. Res. Dev. 2, 159–165 (1958)
2. Jing, H., McKeown, K.R.: Cut and paste based text summarization. In: Proceedings of the 1st North American Chapter of the Association for Computational Linguistics Conference, NAACL 2000, pp. 178–185. Association for Computational Linguistics, Stroudsburg (2000)
3. Fujii, Y., Kitaoka, N., Nakagawa, S., Nakagawa, S.: Automatic extraction of cue phrases for important sentences in lecture speech and automatic lecture speech summarization. In: INTERSPEECH, pp. 2801–2804 (2007)

4. Mani, I., Klein, G., House, D., Hirschman, L., Firmin, T., Sundheim, B.: Summac: a text summarization evaluation. Nat. Lang. Eng. 8, 43–68 (2002)
5. Manning, C., Schütze, H.: Foundations of Statistical Natural Language Processing. MIT Press (1999)
6. Skorkovská, L.: Application of lemmatization and summarization methods in topic identification module for large scale language modeling data filtering. In: Sojka, P., Horák, A., Kopeček, I., Pala, K. (eds.) TSD 2012. LNCS, vol. 7499, pp. 191–198. Springer, Heidelberg (2012)
7. Vanderwende, L., Suzuki, H., Brockett, C., Nenkova, A.: Beyond sumbasic: Task-focused summarization with sentence simplification and lexical expansion. Inf. Process. Manage. 43, 1606–1618 (2007)
8. Gong, Y., Liu, X.: Generic text summarization using relevance measure and latent semantic analysis. In: Proceedings of the 24th Annual International ACM SIGIR Conference on Research and Development in Information Retrieval (2001)
9. Berry, M., Browne, M.: Understanding Search Engines. Society for Industrial and Applied Mathematics, Philadephia (2005)
10. Steinberger, J., Ježek, K.: Text summarization and singular value decomposition. In: Yakhno, T. (ed.) ADVIS 2004. LNCS, vol. 3261, pp. 245–254. Springer, Heidelberg (2004)
11. Lin, C.Y.: Rouge: A package for automatic evaluation of summaries. In: Proceedings ACL Workshop on Text Summarization Branches Out (2004)

Text-to-Speech Alignment for Imperfect Transcriptions

Marek Boháč and Karel Blavka

SpeechLab, Faculty of Mechatronics, Technical University of Liberec
Studentská 2, 461 17 Liberec, Czech Republic
{marek.bohac,karel.blavka}@tul.cz

Abstract. In this paper we propose a method for text-to-speech alignment intended for imperfect (text) transcriptions. We designed an ASR-based (automatic speech recognition) tool complemented with a special post-processing layer that finds anchor points in the transcription and then aligns the data between these anchor points. As the system is not dependent on usually employed keyword-spotter it is not as vulnerable to the noisy recordings as some other approaches. We also present other features of the system (e.g. keeping of the document structure and processing of the numbers) that allow us to use it in many other specific tasks. The performance is evaluated over a challenging set of recordings containing spontaneous speech with many hesitations, repetitions etc. as well as over noisy recordings.

Keywords: unsupervised, text-to-speech alignment, inaccurate transcription, automatic speech recognition.

1 Introduction

Text-to-speech alignment is a task to assign time stamps to every word in the input text transcription. The task is not as demanding as the continuous speech recognition but still it is very useful in many applications [1,2]. It can be used when indexing a spoken-word document for searching [3,4]; the time stamps can be used for automatic timing of subtitles or for automatic mining and sorting of the training data [5].

If the input transcription is correct the solution is very straightforward and it is sufficient to use the forced alignment method (e.g. the one implemented in HTK Toolkit [6]). The complication begins when there are some inaccuracies in the input transcription or if there are noisy segments without speech in the recording. The accuracy of the alignment suffers great losses in such cases. Another problem of the baseline forced alignment method is that its real time factor (the time needed to process the data) is strongly non-linear and rapidly increases with the increasing length of the processed recording (approximate complexity of the computation is N^3).

In this paper we present a solution coming from our foregoing work related to this topic [3] that was similar to [7]. We propose a solution based on automatic speech recognition (ASR) module with especially built language model complemented with a layer of ASR output post-processing. As the task of the alignment between a good recording (recorded with a small amount of noises) and its correct transcription is well solved we aim our experiments on a more challenging task. We test our method in the

I. Habernal and V. Matousek (Eds.): TSD 2013, LNAI 8082, pp. 536–543, 2013.

task of aligning inaccurate transcriptions (we synthesized a set of differently impaired transcriptions). We align these transcriptions to high-quality recordings as well as to recordings with a large amount of background speech and noise. The recordings contain unprepared (not read) speech so there are many hesitations, repetitions, reparations and silent segments in the recordings that make the task even more challenging. We also test the system's ability to handle abbreviations, titles, numbers and other phenomena for which there are more alternative phonetic transcriptions.

2   Proposed Method

The main principles of our new method were derived from the foregoing approaches [3,7]. The older one used the key-word spotter to find sequences of words that were unique enough to be used as anchor points. The recording and the transcription were split into shorter segments by these anchor points. The searching for anchor points and splitting was iteratively repeated until the segments were short enough or there could not be found any relevant anchor point. When the iteration was over we used HTK-implemented forced alignment module to align the short segments. The approach is vulnerable when a segments of speech (a word sequence) repeats and the anchor point can be assigned to the wrong repetition of the sequence.

From this original approach we wanted to use the most advantageous features - the splitting of the document by finding the confident anchor points. We also wanted to minimize the number of loops in the algorithm and we wanted to get rid of the dependency on the key-word spotter (because of its uncertainty when processing noisy or poorly pronounced recordings). All these demands were fulfilled by creating a special language model for one-pass automatic speech recognizer (ASR) that is followed by a post-processing layer that performs the searching for the anchor points directly in the ASR output.

Figure 1 depicts the overall scheme of the proposed method. The base algorithm is encapsulated into two layers that remember and reconstruct the original document structure. This encapsulation passes on the plain text that is used to formulate the corresponding language model for the ASR module together with a phonetic dictionary. The output of the ASR is post-processed to reduce possible transcription errors and the results are produced.

2.1   Remembering and Reconstruction of the Document Structure

The proposed method supports two input file types a plain text or a structured XML file format denoted as '.trsx' that contains not only the transcription text but is also able to keep the time stamps and pronunciation of the words together with the speaker information. It can also handle the detected non-speech events and keeps the structure of the document in the terms of chapters and paragraphs. The same file format is used for the saving of the results. During the alignment of the document all the available structure information is stored and used in the final step of document structure reconstruction. The same format is used by the NanoTrans - a tool that allows us to simply see and check the results [8].

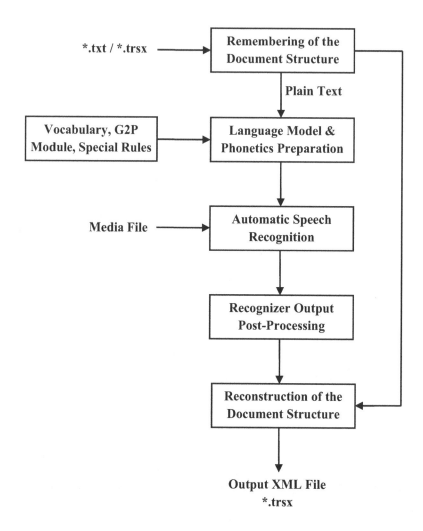

Fig. 1. The overall scheme of the proposed method

2.2 Preparation of the ASR Input Data

The ASR is run with a user-specified language model (tailored for the solved task) and with a corresponding phonetic dictionary. The language model comes directly from the input text and enables the recognizer to pass between two subsequent words directly or through a set of non-speech events. The word can be omitted also.

When preparing appropriate phonetic dictionary we can utilize an existing phonetic dictionary as well as any rule-based G2P module or some task-specific dictionary. The system is able to use all these information sources to create sets of alternative pronunciations for the aligned words. This is important mainly for abbreviations and titles that have many forms (the gender, inflection and other grammatical categories generate tens of word forms in the case of Slavic languages). Special category is the alignment of numbers. For the Czech we prepared a set of all possible pronunciations of the numerals and a combination logic that allows us to generate all possible pronunciations of numbers. This option is useful when aligning the numbers as well as it enables to detect which number form was pronounced in the recording - the recognizer uses the corresponding pronunciation that is determined from the result.

2.3 Automatic Speech Recognizer

The ASR system we use in our experiments was originally built for large vocabulary applications of the recognition of highly inflective Slavic languages [9]. It processes audio documents converted into 16 kHz PCM format (via FFmpeg codec) so it can process audio as well as audio-video inputs. The parameterization uses standard 39-dimensional MFCC vectors computed on 20-ms frames with 10-ms overlap. The CMS (Cepstral Mean Subtraction) normalization is applied. The acoustic model uses 42 phonemes and 8 types of non-speech events trained on more than 320 hours of annotated recordings.

The output of the ASR provides rich information about the recognized recording. It provides the orthographic and phonetic forms of the recognized items together with its start and end times (with 10-ms resolution). It also transcribes the non-speech events (e.g. silence, click, breathe, hesitation) detected in the recording and the time of its occurrence.

2.4 ASR Output Post-processing

The first step of the ASR output post-processing is the searching for the anchor points. An anchor point is defined as a sequence of subsequent words (words are for the needs of the system replaced by their index; the words occurring more times are listed as possible substitutions). This sequence must fulfill the constraint of a minimal length (in our case 20 characters) and new anchor point is represented by the beginning of the first word forming it. When all the anchor points are found they are checked if they appear in a right order along the recording (some specific transcription errors may lead to detection of an expected sequence in an incorrect time). Detected wrong anchor points are removed.

The second step is the alignment of segments bordered by the found anchor points. The Minimum Edit Distance[1] (MED) is computed between the phonetics of the segment transcription and the phonetics of the ASR output. For our needs the MED works over whole words phonetics and takes into account possible alternative pronunciations. The MED produces a sequence of hits, substitutions, insertions and deletions. If the result for the word pair (first one from the transcription, second one from the ASR output) is a hit or a substitution the time stamps of the result are the same as those in the ASR output. If there is an insertion the recognizer output is excluded from the final result and if there is a deletion the time stamps are derived from the borders of the events

---

[1] The original Minimum Edit Distance [10] is defined as the minimum number of editing operations (hit, substitution, deletion or insertion) needed to transform one string into the other.

(words or non-speech events) surrounding the aligned word. When this step is finished the results are passed to the module reconstructing the document structure.

3 Experiments

3.1 Experimental Data

To perform the experiments we gathered a set of recordings containing some recorded lectures and few public speeches. Even if this type of speech should be prepared the recordings are partly spontaneous, it contains repetitions of single words as well as of shorter sentences. There are also mispronunciations, unfinished words, silence parts (up to 5 seconds), slang and dialect pronunciation distortions and different amount of noise (the background speech or just many types of noise). As the presence of the noise in the recording may be significant for the system accuracy we chose 167 minutes of not too noisy recordings and 181 minutes of very noisy recordings for our tests. As we had accurate transcriptions for the recordings we created the references by checking (and correcting) the first round alignment of the data. Then we aligned it again with the corrected transcription and checked the references again (using NanoTrans tool [8]). As the reference transcription was correct the automatic results are almost perfect for both - noisy and normal data.

3.2 Proposed Experiments and Evaluation References

To test the different types of transcription inaccuracies we impaired the reference transcriptions in 6 different ways. We added/removed randomly placed single words, we added/removed 1-minute continuous segments and we added/removed 5-minute continuous segments. We prepared a script that performed the required impairment and prepared the ideal result for the impaired test data so we could automatize the accuracy evaluation.

In the case of adding/removing single words we made 4 experiments specified by the ratio of added/removed words to the number of words contained in the reference transcription (5 %, 10 %, 15 % and 20 %). In the case of adding/removing whole segments we impaired the transcription until the impairment reached 10 %, 20 %, or 30 % of the total length of the impaired recording. For the needs of adding words we estimated the average number of words per minute - 115 words per minute. The words were chosen randomly from a Czech dictionary (containing approx. 430,000 items). The test transcription preparation was the same for the noisy recordings and for those with a small amount of noise.

4 Evaluation and Results

As we mentioned in the previous chapter, the references for the evaluation were assembled together with the impaired test transcriptions. Thank to this we could automatically evaluate all the words in the results. We assumed the word as correctly aligned (OK) if its and reference time stamps (beginning and ending times) did not differ more than

Table 1. The alignment accuracy affected by single-word impairments

Experiment	Normal recordings			Noisy recordings		
	OK	ERR	ACC [%]	OK	ERR	ACC [%]
add words 5 %	12,424	3,724	76.94	5,137	6,718	43.30
add words 10 %	13,184	3,800	77.63	5,347	7,156	42.77
add words 15 %	11,054	6,099	64.44	8,630	5,171	62.53
add words 20 %	14,747	3,818	79.43	6,376	7,305	46.60
rem words 5 %	11,742	1,996	85.47	5,594	6,827	45.04
rem words 10 %	13,443	1,095	92.47	7,303	5,462	57.21
rem words 15 %	12,326	1,412	89.72	5,847	6,574	47.07
rem words 20 %	11,147	1,768	86.31	4,162	7,565	35.49

Table 2. The alignment accuracy affected by continuous transcription segment impairments

Experiment	Normal recordings			Noisy recordings		
	OK	ERR	ACC [%]	OK	ERR	ACC [%]
add1min 10 %	16,103	40	99.75	13,498	481	96.56
add1min 20 %	16,059	110	99.32	13,230	768	94.51
add1min 30 %	16,090	103	99.36	12,628	1,299	90.67
add5min 10 %	16,117	13	99.92	13,741	230	98.35
add5min 20 %	16,075	60	99.63	13,189	707	94.91
add5min 30 %	16,080	59	99.63	12,819	1,044	92.47
rem1min 10 %	11,429	188	98.38	6,940	3,237	68.19
rem1min 20 %	11,203	177	98.44	7,025	3,269	68.24
rem1min 30 %	11,428	157	98.64	6,774	3,660	64.92
rem5min 10 %	8,239	109	98.69	7,787	2,161	78.28
rem5min 20 %	11,422	103	99.11	7,649	2,283	77.01
rem5min 30 %	11,239	132	98.84	7,187	2,934	71.01

30ms. Other words were classified as errors (ERR). The alignment accuracy (ACC) is given by (1).

$$ACC = \frac{OK}{OK + ERR} \cdot 100[\%] \tag{1}$$

4.1   Aligner Accuracy

The reached results are summarized in Tables 1 and 2. The first one deals with added (add) and removed (rem) single words. The second one deals with the impairment of continuous segments of transcription. The percentage in the experiment name marks the amount of the impairment relative to the reference transcription.

As can be seen in Table 1, the impacts of single missing or added words in the transcription can be very significant. There are two main reasons i) the anchor-point finding is more difficult due to the fact that many potential anchor segments are interrupted by single-word errors; ii) the time borders of words around the added/removed word are vulnerable to inaccuracies. Table 2 illustrates the system ability to handle longer continuous errors because such segments are delimited by anchor points around.

4.2 Processing-Time Requirements

As we wanted to evaluate the time requirements of the proposed system and to compare the results with our previous work [3] we used the same computer for our experiments (CPU Intel Core i7 (920), 3GB RAM - DDR3, 1066 MHz). For the recordings marked as normal (those with adequate amount of noise) the real-time factor (the ratio of the processing time and the recording length) takes the values 0.23 - 0.39. For the noisy recordings the RT increases in the range 0.66 - 0.72. This slowdown is caused by the usual behavior of any speech recognizer - the noisy recording prevents it from the application of acceleration optimizations (e.g. hypothesis pruning during the decoding phase).

5   Conclusions and Future Work

We introduced a system for the alignment between a text transcription and an audio recording. The method was proposed to reduce impacts of incorrect transcriptions. We tested the system on a challenging set of spontaneous speech recordings. The test recordings contained two groups - normal recordings and noisy ones. We performed a set of tests that evaluated impacts of different types of transcription inaccuracies. The results prove the system ability to eliminate the impact of longer continuous transcription errors while the impact of single word errors is still to be solved. We also evaluated the impact of strong noise present in the recording. When the transcription was correct the alignment was almost as perfect as for normal recording but the impacts of transcription inaccuracies are much stronger.

If we compare this system to our foregoing work [3] the accuracy of the new system outperforms the old one (it reached 96.2 % accuracy for transcriptions with continuous error transcriptions while the new one reaches approx. 99 % accuracy see Table 2). The computational demands of the new method are also reduced - the old method reached RT factor 0.5 - 0.6 while the new one reaches RT factor 0.23 - 0.39 while processing the data of comparable quality of recordings.

The proposed system also incorporates some new features - it keeps the structure of the input text document, it is able to handle numbers and it is able to reliably process very long recordings (we successfully aligned a 10 hours long document).

For our future work we want to solve the single-word transcription errors. We plan to replace the MED-based alignment of the words between the anchor points with a more sophisticated pronunciation-based multi-word alignment (similar to the method used in [11]).

We also plan to make the pre-processing and post-processing phases more modular. Such modularity will allow us to adapt the system for specific tasks and moreover the cross-lingual adaptation will be much easier [12].

It is reasonable to ask why we did not evaluate our system on existing aligned data (e.g. movie subtitles or transcriptions created for people with hearing handicap). The point is that such alignments align whole phrases only so it contains just a fraction of information needed for the complete evaluation. We are also working on a system for automatic timing of subtitles so we decided to address these tasks in our future publications.

Acknowledgments. This work was supported by the Technology Agency of the Czech Republic (project no. TA01011204).

References

1. Zhang, J., Pan, F., Yan, Y.: An LVCSR Based Automatic Scoring Method in English Reading Tests. In: 4th International Conference on Intelligent Human-Machine Systems and Cybernetics, IHMSC 2012, Nanchang, pp. 34–37 (2012)
2. Córdova Lucero, D.P., Toledano, D.T.: Preliminary Results of Alignment of Text and Audio in News and Songs. In: Joint 7th Spanish Speech Technology Workshop and the Iberian SLTech Workshop, Madrid, pp. 59–68 (2012)
3. Bohac, M., Blavka, K.: Automatic Segmentation and Annotation of Audio Archive Documents. In: 10th International Workshop on Electronics, Control, Measurement and Signals, ECMS 2011, Liberec, pp. 1–6 (2011)
4. Nouza, J., Zdansky, J., Cerva, P.: System for automatic collection, annotation and indexing of Czech broadcast speech with full-text search. In: Proc. of 15th IEEE MELECON Conference, Malta, pp. 202–205 (2010)
5. Stanislav, P., Švec, J., Šmídl, L.: Unsupervised Synchronization of Hidden Subtitles with Audio Track Using Keyword Spotting Algorithm. In: Sojka, P., Horák, A., Kopeček, I., Pala, K. (eds.) TSD 2012. LNCS, vol. 7499, pp. 422–430. Springer, Heidelberg (2012)
6. HTK Toolkit (March 2013), http://htk.eng.cam.ac.uk
7. Moreno, P.J.: Joerg, Ch. F., Van Thong, J.-M. and Glickman, O.: A recursive algorithm for the forced alignment of very long audio segments. In: The 5th International Conference on Spoken Language Processing - ICSLP, Sydney (1998)
8. Seps, L.: NanoTrans Editor for Orthographic and Phonetic Transcriptions. In: The 36th International Conference on Telecommunications and Signal Processing (TSP), Rome (in press, 2013)
9. Nouza, J., Zdansky, J., Cerva, P., Silovsky, J.: Challenges in Speech Processing of Slavic Languages (Case Studies in Speech Recognition of Czech and Slovak). In: Esposito, A., Campbell, N., Vogel, C., Hussain, A., Nijholt, A. (eds.) Second COST 2102. LNCS, vol. 5967, pp. 225–241. Springer, Heidelberg (2010)
10. Wagner, R.A., Fischer, M.J.: The String-to-String Correction Problem. Journal of the ACM 21(1), 168–173 (1974)
11. Boháč, M., Nouza, J., Blavka, K.: Investigation on Most Frequent Errors in Large-Scale Speech Recognition Applications. In: Sojka, P., Horák, A., Kopeček, I., Pala, K. (eds.) TSD 2012. LNCS, vol. 7499, pp. 520–527. Springer, Heidelberg (2012)
12. Nouza, J., Cerva, P., Zdansky, J., Kucharova, M.: A Study on Adapting Czech Automatic Speech Recognition System to Croatian Language. In: 54th International Symposium ELMAR 2012, Zadar, pp. 227–230 (2012)

The CNG Corpus of European Portuguese Children's Speech

Annika Hämäläinen[1,2], Silvia Rodrigues[1], Ana Júdice[1], Sandra Morgado Silva[3],
António Calado[1], Fernando Miguel Pinto[1], and Miguel Sales Dias[1,2]

[1] Microsoft Language Development Center, Lisbon, Portugal
[2] ADETTI – ISCTE, IUL, Lisbon, Portugal
{t-anhama,v-antonc,a-fpinto,Miguel.Dias}@microsoft.com
[3] Diferente Jogo, Caldas da Rainha, Portugal
sandrasilva@diferencas.net

Abstract. Speech recognisers trained with adults' speech do not work well with
children's speech because of the inherent acoustic and linguistic differences in
the speech of these two populations. To develop speech-driven applications capable of successfully recognising children's speech, a sufficient amount of children's speech is needed for training acoustic models from scratch or for adapting
acoustic models trained with adults' speech. However, the availability of suitable
children's speech corpora is still limited, especially in the case of less-spoken languages. This paper describes the design, collection, transcription and annotation
of a 21-hour corpus of prompted European Portuguese children's speech collected
from 510 children aged 3-10. Before the development of this corpus, European
Portuguese children's speech data have not been available at all for parts of this
age range.

Keywords: automatic speech recognition, children's speech, corpus, European
Portuguese, prompted speech.

1 Introduction

Speech interfaces have tremendous potential in educational applications, speech therapy
systems and games targeted at children. However, it is well known that automatically
recognising children's speech is a very challenging task. Recognisers trained on adults'
speech tend to suffer from a substantial deterioration in recognition performance when
used to recognise children's speech [1–4]. Moreover, word error rates on children's
speech are usually much higher than those on adults' speech even when using a recog-
niser trained on age-specific speech – although they do show a gradual decrease as the
children get older [1–6].

The difficulty of automatically recognising children's speech can be attributed to it
being acoustically and linguistically very different from adults' speech [1, 2]. For instance, due to their vocal tracts being smaller, the fundamental and formant frequencies
of children's speech are higher [1, 2, 7–9]. What is particularly characteristic of children's speech is its higher variability as compared with adults' speech, both within and
across speakers [1, 2, 8]. This variability is caused by rapid developmental changes in

I. Habernal and V. Matousek (Eds.): TSD 2013, LNAI 8082, pp. 544–551, 2013.

their anatomy, speech production et cetera, and manifests itself, for instance, in speech rate, the degree of spontaneity, the frequency of disfluencies, fundamental and formant frequencies, and pronunciation quality [1, 2, 8–11].

To develop speech-driven applications capable of successfully recognising children's speech, a sufficient amount of children's speech is needed for training acoustic models from scratch or for adapting acoustic models trained with adults' speech. Suitable children's speech corpora are available, for instance, for British [12] and American [13] English, Dutch [14], German [12], Italian [12], and Swedish [12]. However, for many other languages, especially less-spoken ones, the availability of children's speech corpora is still limited. This has, for instance, been the case for European Portuguese (pt-PT). Before the development of the corpus presented in this paper, only two corpora containing children's speech have been available for pt-PT: a 158-minute corpus of isolated words collected from 111 children aged 5-6 for developing a computer-aided speech therapy system [15], and the Portuguese Speecon Database, which contains speech from 52 children aged 8-14 [16]. The Portuguese Speecon Database contains both spontaneous speech and read speech, the read speech ranging from phonetically rich sentences to number expressions, proper names, command words et cetera. In this paper, we describe the design, collection, transcription and annotation of the CNG Corpus of pt-PT Children's Speech, a 21-hour corpus of prompted speech collected from 3-10-year-old children. The corpus was developed in the Contents for Next Generation Networks (CNG) project, whose end product will be a multimodal educational game for 3-10-year-old Portuguese children.

2   The CNG Project and Educational Game

The CNG project is an ongoing Portuguese industry-academia collaboration with seven partners: Association CCG/ZGDV, Diferente Jogo, INESC Porto, I.Zone, Microsoft Language Development Center (MLDC), and the Universities of Aveiro and Porto. The project studies speech and gesture as natural alternatives for child-computer interaction in the education of 3-10-year-old children. To this end, the project partners are developing a multimodal educational game that can be played in a cave automatic virtual environment (CAVE) or with a desktop computer using Kinect for Windows, a motion sensing input device by Microsoft [17]. The game addresses two main areas of development: motor skill development (physical coordination skills) and cognitive development (attention, problem solving, mathematical and musical skills). In the game, the children will use their voice and gestures to complete different kinds of puzzles and tasks in themed 3D and 2D scenarios. M.A.T., an intelligent virtual assistant with a synthesised voice, will guide and help them throughout the game. In one of the themed scenarios, the Age of Discovery, M.A.T., might, for instance, say, *"Cheer up the sailors by counting from one to five and moving your body. With each number you say, you will need to move a part of your body."* Speech input will be enabled for tasks related to mathematics (e.g. counting objects, simple mathematical operations) and music (e.g. completing musical note sequences). The expected speech input includes isolated cardinal numbers, sequences of cardinal numbers and musical notes.

3   The CNG Corpus

When it comes to children's speech data, the goal of the CNG project was to develop a corpus of about 20 hours of children's speech suitable for training and testing acoustic models (AMs) for the speech-driven parts of the CNG game. The data collection was a collaboration between MLDC and Diferente Jogo, with MLDC being responsible for the technical aspects of the data collection, the design of the prompts to record, as well as the transcription and annotation of the recorded data, and Diferente Jogo being responsible for collecting most of the data. The following subsections describe the design, collection, transcription and annotation of the corpus, as well as the details of the corpus and the datasets proposed for automatic speech recognition (ASR) experiments.

3.1   Corpus Design

Speaker Selection. As the CNG game is aimed at 3-10-year-old children, we only collected speech from speakers in that age range. Based on the children's capabilities (see below), we split them into two age groups that are considered homogenous populations for the purposes of the CNG project: 3-6-year-old and 7-10-year-old children.

We collected speech from children attending nurseries and schools in and around the cities of Lisbon, Leiria and Aveiro. The cities were chosen for practical reasons; the time and budget available for the data collection campaign was tight, so we had to concentrate on places where the project partners had existing contacts at nurseries and schools and were able to easily attend recording sessions. We tried to keep the ratio of girls and boys as even as possible but were not, for instance, able to aim at a specific ratio of speakers from the different areas.

Prompt Design. Collecting speech from children poses some special challenges. First, children's attention span depends on their age [18]; they might get distracted from a prolonged recording task. Second, they may have difficulty reading or repeating long, complex words and sentences. Taking these challenges and the requirements of the CNG game into account, we designed four types of prompts to record: 292 phonetically rich sentences, musical notes (e.g. *dó*), isolated cardinals (e.g. *44*), and sequences of cardinals (e.g. *28, 29, 30, 31*). The phonetically rich sentences originated from the CETEMPúblico corpus of Portuguese newspaper language [19]. They were short (about 4 words/sentence) and did not include any difficult words. In the case of the 3-6-year-olds, the cardinals ranged from 0 to 30, and the sequences of cardinals consisted of 2-3 numbers. In case of the 7-10-year-olds, the numbers ranged from 0 to 999, and the sequences of cardinals consisted of 4 numbers.

The younger children produced a set of 30 prompts selected across the different types of prompts in a balanced way. This resulted in a bit more than one minute of speech per speaker. The older children read out a set of 50 prompts resulting in about 3 minutes of speech per speaker.

The differences in the contents and the targeted number of prompts between the two age groups were designed based on our experiences from pilot recording sessions with children of different ages. They take into account the differences in the attention span and linguistic capabilities between the two age groups.

3.2   Data Collection

We used the *Your Speech* online speech data collection platform [20] for collecting the speech and some biographical information (age group, gender and region of origin) about the speakers. The platform's web interface presented the speakers with each of the prompts to record (see Fig. 1). The recording sessions were supervised by six people trained for the task, and took place in a quiet room. In the case of the 3-6-year-olds and the 7-10-year-olds who had problems reading the prompts, the recording supervisor read the prompts out first and the children then repeated them. Each utterance was recorded separately using a noise-cancelling Life Chat LX 3000 USB headset and digitized at 16 bits and 22 kHz.

The recording of an utterance started when the recording supervisor clicked the *Record* button (see Fig. 1) and ended when (s)he clicked the *Stop* button, or when no more speech was detected by the system. The recorded utterance was then uploaded to the web backend of the system and checked for the presence of speech and clipping; if speech was indeed detected and if the utterance did not contain any samples of clipping, the recording supervisor could proceed to the next prompt by clicking the *Next Phrase* button or, if unhappy with the utterance, rerecord the utterance by clicking the *Rerecord* button. If the automatic quality control was not passed, the speaker was requested to rerecord the utterance.

Fig. 1. The *Your Speech* data collection platform displaying a prompt to record

3.3   Transcriptions, Annotations and Quality Control

Using an in-house transcription tool, a native speaker trained for the task transcribed the corpus orthographically, and annotated it for filled pauses, noises, damaged words (e.g. mispronunciations) and other speakers' speech using the tags presented in Table 1 The annotation scheme was designed to be compatible with the requirements of our in-house AM training tool.

Table 1. The tags available for annotating the recordings

Tag	Meaning
<FILL/>	Filled pauses (e.g. umm, er, ah)
<NON/>	Non-human noises (e.g. mouse or keyboard noises, radio, TV, music)
<SPN/>	Human noises (e.g. coughs, audible breath, utterance echoes)
<UNKNOWN/>	False starts; mispronounced, unintelligible or truncated words; or words with considerable background noise
<NPS/>	Speech from non-primary speakers (e.g. the recording supervisors)

The transcriber discarded sessions that contained recordings with consistently poor audio quality, or speech from children who were not native speakers of pt-PT or had consistent problems repeating or reading the prompts out. After the transcription and annotation work, we identified typographical errors in the transcriptions by checking them against a large Portuguese lexicon. In addition, we used forced alignment to identify potentially problematic utterances; utterances that cannot be aligned are more likely to have problems in the quality of the speech, the transcriptions and/or the audio. We cast these utterances (0.2% of all the recorded utterances) aside. As the proportion of utterances annotated with the <NPS/> tag was low (2.3% of all the recorded utterances), we did not include them in the final version of the corpus, either.

3.4   Overview of the Corpus and the Datasets for ASR Experiments

We collected a total of 21 hours of speech from 510 children – 30% of them aged 3-6, and 70% of them aged 7-10. 56% of the children were girls and 44% of them boys. The vast majority of them (84%) were from the Leiria area. Table 2 presents further details about the age and gender distribution of the speakers, as well as their regional origin, whereas Table 3 details the amount of speech collected per prompt category.

For the purpose of ASR experiments, we are proposing three speaker-independent datasets that respect the proportions of speaker age and gender in the full corpus: a training set used for training AMs (85% of the data), a development test set for optimisation purposes (5% of the data), and an evaluation test set for the final testing of the AMs (10% of the data). Due to the limited number and type of prompts recorded, it is not possible to rule out the same prompts appearing in the three different datasets; with hindsight, this could have been avoided by using a corpus design that splits the speakers and prompts between the three datasets before the recordings. The main statistics of the corpus and the proposed datasets are presented in Table 4.

4   Discussion

The CNG corpus was specifically designed for training and testing AMs for the speech-driven parts of the CNG game, which expects cardinals, sequences of cardinals and

Table 2. The number of speakers per age group and region of origin

	Lisbon	Leiria	Aveiro	Total
Ages 3-6	46	100	7	153
Girls	25	55	2	82
Boys	21	45	5	71
Ages 7-10	11	326	20	357
Girls	4	191	8	203
Boys	7	135	12	154
Total	57	426	27	510
Girls	29	246	10	285
Boys	28	180	17	225

Table 3. The amount of speech (hh:mm:ss) collected per prompt category

	Ages 3-6	Ages 7-10	Total
Phonetically rich sentences	01:29:03	05:59:07	07:28:10
Musical notes	00:17:22	01:28:10	01:45:32
Isolated cardinal numbers	00:22:55	02:12:49	02:35:44
Sequences of cardinal numbers	00:49:57	08:14:59	09:04:56
Total	02:59:17	17:55:05	20:54:22

musical notes as the speech input. In addition to those types of data, the corpus also contains phonetically rich sentences, making it useful for developing other speech-driven applications targeted at children, as well.

AMs trained using speech from the targeted age or age group are expected to lead to the best speech recognition performance [1]. However, when a limited amount of children's speech is available, children are often treated as a homogenous population and AMs are trained with speech collected from children of all ages [1]. In the CNG corpus, the children are split into two age groups (3-6 and 7-10) that are considered homogenous populations; their exact age is not included in the corpus. Considering the

Table 4. The main statistics of the final corpus and the datasets proposed for ASR experiments

	Training set	Devel Test Set	Eval Test Set	Total
#Speakers	432	26	52	510
#Word Types	605	482	521	614
Ages 3-6	557	218	319	560
Ages 7-10	585	458	494	591
#Word Tokens	102,537	6229	12,029	121,046
Ages 3-6	9553	676	1148	11,424
Ages 7-10	92,984	5553	10,881	109,622
hh:mm:ss	17:42:22	01:06:26	02:05:34	20:54:22
Ages 3-6	02:30:24	00:10:22	00:18:31	02:59:17
Ages 7-10	15:11:58	00:56:04	01:47:03	17:55:05

rapid development of children, this choice somewhat limits the usefulness of the corpus in speech research. However, it can still be used, for instance, for studying the acoustic characteristics of children belonging to the two age groups.

The majority of the children were from the Leiria area, and the other speakers were from the Lisbon and Aveiro areas. Therefore, the collected speech data is hardly representative of all the regional accents of Portugal. This shortcoming of the corpus is due to the limited resources that were available for the data collection campaign; we could not afford to be too selective about the speakers. We did, however, manage to collect speech from a much larger slice of the Portuguese population of children (510 children) than the other pt-PT children's speech data collection efforts that we are aware of [15, 16]. In addition, the CNG corpus contains speech from 3-4-year-old and 7-year-old children – ages that the other existing resources of pt-PT children's speech do not cover.

In addition to 21 hours of children's speech data, the speech data collection campaign gave us valuable experience recording speech from an age group that is considered particularly challenging to record speech from: children under six years of age [1]. The prompts designed for this age group were shorter and simpler and the targeted number of prompts per speaker lower than those for 7-10-year-old children. However, many children still had difficulty repeating the prompts correctly, especially in the case of prompts containing several words to memorise. The youngest children, the 3-year-olds, were almost impossible to record speech from due to their shyness and difficulties repeating the prompts. Because of the difficulty of finding very young speakers and recording speech from them, as well as the limited amount of speech collected from each of them, the majority of the speech in the CNG corpus is from 7-10-year-olds.

5 Summary

In this paper, we presented a European Portuguese children's speech corpus containing 21 hours of prompted speech from 510 children aged 3-10. The corpus comes with manual orthographic transcriptions and annotations indicating filled pauses, noises and damaged words (e.g. mispronunciations). It was specifically designed for training and testing acoustic models for an educational game that teaches children basic mathematical and musical skills. However, it could also prove useful for developing other speech-driven applications for children and for use in children's speech research. It is currently the largest available corpus of European Portuguese children's speech, and covers some ages that other resources of European Portuguese children's speech do not cover.

The corpus is available at request for R&D activities. Please contact Miguel Sales Dias (Miguel.Dias@microsoft.com) for further information.

Acknowledgements. The QREN 7943 CNG – Contents for Next Generation Networks project is co-funded by Microsoft, the Portuguese Government, and the European Structural Funds for Portugal (FEDER) through the Operational Program for Competitiveness Factors (COMPETE) and the National Strategic Reference Framework (QREN). The authors are indebted to the children, recording supervisors, nurseries and schools that took part in the data collection.

References

1. Gerosa, M., Giuliani, D., Narayanan, S., Potamianos, A.: A Review of ASR Technologies for Children's Speech. In: Proc. Workshop on Child, Computer and Interaction, Cambridge, MA (2009)
2. Russell, M., D'Arcy, S.: Challenges for Computer Recognition of Children's Speech. In: Proc. SLaTE 2007, Farmington, PA (2007)
3. Potamianos, A., Narayanan, S.: Robust Recognition of Children's Speech. IEEE Speech Audio Process. 11(6), 603–615 (2003)
4. Wilpon, J.G., Jacobsen, C.N.: A Study of Speech Recognition for Children and the Elderly. In: Proc. ICASSP, Atlanta, GA (1996)
5. Elenius, D., Blomberg, M.: Adaptation and Normalization Experiments in Speech Recognition for 4 to 8 Year Old Children. In: Proc. Interspeech, Lisbon (2005)
6. Gerosa, M., Giuliani, D., Brugnara, F.: Speaker Adaptive Acoustic Modeling with Mixture of Adult and Children's Speech. In: Proc. Interspeech, Lisbon (2005)
7. Narayanan, S., Potamianos, A.: Creating Conversational Interfaces for Children. IEEE Speech Audio Process. 10(2), 65–78 (2002)
8. Gerosa, M., Giuliani, D., Brugnara, F.: Acoustic Variability and Automatic Recognition of Children's Speech. Speech Commun. 49(10-11), 847–860 (2007)
9. Huber, J.E., Stathopoulos, E.T., Curione, G.M., Ash, T.A., Johnson, K.: Formants of Children, Women and Men: The Effects of Vocal Intensity Variation. J. Acoust. Soc. Am. 106(3), 1532–1542 (1999)
10. Lee, S., Potamianos, A., Narayanan, S.: Acoustics of Children's Speech: Developmental Changes of Temporal and Spectral Parameters. J. Acoust. Soc. Am. 10, 1455–1468 (1999)
11. Eguchi, S., Hirsh, I.J.: Development of Speech Sounds in Children. Acta Otolaryngol. Suppl. 257, 1–51 (1969)
12. Batliner, A., Blomberg, M., D'Arcy, S., Elenius, D., Giuliani, D., Gerosa, M., Hacker, C., Russell, M., Steidl, S., Wong, M.: The PF_STAR Children's Speech Corpus. In: Proc. Interspeech, Lisbon (2005)
13. Eskernazi, M.: KIDS: A Database of Children's Speech. J. Acoust. Soc. Am. 100(4), 2759–2759 (1996)
14. Cucchiarini, C., Van Hamme, H., van Herwijnen, O., Smits, F.: JASMIN-CGN: Extension of the Spoken Dutch Corpus with Speech of Elderly People, Children and Non-natives in the Human-Machine Interaction Modality. In: Proc. LREC, Genoa (2006)
15. Lopes, C., Veiga, A., Perdigão, F.: A European Portuguese Children Speech Database for Computer Aided Speech Therapy. In: Caseli, H., Villavicencio, A., Teixeira, A., Perdigão, F. (eds.) PROPOR 2012. LNCS, vol. 7243, pp. 368–374. Springer, Heidelberg (2012)
16. The Portuguese Speecon Database,
    http://catalog.elra.info/product_info.php?products_id=798
17. Kinect for Windows,
    http://www.microsoft.com/en-us/kinectforwindows/
18. Unger, H.G.: Encyclopedia of American Education, 3rd edn. Facts on File Inc., New York (2007)
19. CETEMPúblico, http://www.linguateca.pt/cetempublico/
20. Freitas, J., Calado, A., Braga, D., Silva, P., Dias, M.: Crowd-Sourcing Platform for Large-Scale Speech Data Collection. In: Proc. FALA 2010, Vigo (2010)

The Joint Optimization of Spectro-Temporal Features and Neural Net Classifiers

György Kovács[1] and László Tóth[2,*]

[1] Department of Informatics, University of Szeged, Szeged, Hungary
[2] Research Group on Artificial Intelligence, Hungarian Academy of Sciences, Szeged, Hungary
{gykovacs,tothl}@inf.u-szeged.hu

Abstract. In speech recognition, spectro-temporal feature extraction and the training of the acoustical model are usually performed separately. To improve recognition performance, we present a combined model which allows the training of the feature extraction filters along with a neural net classifier. Besides expecting that this joint training will result in a better recognition performance, we also expect that such a neural net can generate coefficients for spectro-temporal filters and also enhance preexisting ones, such as those obtained with the two-dimensional Discrete Cosine Transform (2D DCT) and Gabor filters. We tested these assumptions on the TIMIT phone recognition task. The results show that while the initialization based on the 2D DCT or Gabor coefficients is better in some cases than with simple random initialization, the joint model in practice always outperforms the standard two-step method. Furthermore, the results can be significantly improved by using a convolutional version of the network.

Keywords: spectro-temporal features, Neural Net, phone recognition, TIMIT.

1   Introduction

Neurophysiological and biological studies (e.g. [1]) suggest that filters responsive to spectro-temporal modulations can be used for feature extraction in automatic speech recognition. Standard techniques for the extraction of modulation features like these include the application of the 2D DCT [2–4] or a set of Gabor filters [5–7] on the spectro-temporal representation of the speech signal. These features then form the input for some statistical modelling technique such as a hidden Markov model (HMM) or an artificial neural net (ANN). The feature extraction and the statistical modelling steps are usually separate, which is convenient, but suboptimal. Here, we propose to combine these steps in a specially designed neural net, and use this net not just to optimize new feature extraction filter sets, but also to improve the standard 2D DCT and Gabor filters. We will compare the 2D DCT, Gabor and the ANN-based optimized filter sets by evaluating their performance on the TIMIT speech database.

In Section 2, we describe the standard spectro-temporal feature extraction methods. Then in Section 3 we present the concept behind the joint handling of both the feature

---

* This publication is supported by the European Union and co-funded by the European Social Fund. Project title: Telemedicine-focused research activities in the fields of mathematics, informatics and medical sciences. Project number: TÁMOP-4.2.2.A-11/1/KONV-2012-0073.

I. Habernal and V. Matousek (Eds.): TSD 2013, LNAI 8082, pp. 552–559, 2013.

extraction filters and the neural net classifier. A further refinement – the application of convolutional neural nets – is also elucidated. After, in sections 4 and 5 we present the experiments and discuss the results. Lastly, in Section 6, we draw some brief conclusions about our study, and make a suggestion about future work.

2   Spectro-Temporal Filters

Localized spectro-temporal analysis is a neurophysiologically motivated feature extraction method for speech recognition [6] that has received much attention over the past few years. In this approach we extract spectro-temporally localized patches from the spectrogram of the speech signal, and create features for ASR purposes by processing them using standard filtering methods. Formally, a spectro-temporal feature can be described by the formula

$$o = \sum_{f=0}^{N} \sum_{t=0}^{M} P(f,t)F(f,t), \tag{1}$$

where N and M are the height and width of patch P and filter F, which have to be the same size. There are many different methods for getting the proper coefficients for the filter $F(f,t)$. Below, we describe two well-known methods, then in Section 3 we present a new method.

2.1   2D DCT

A common approach is to process the patches using a 2D DCT, which works with the following filter coefficients:

$$F_{pq}(f,t) = \cos \frac{\pi \cdot (f+0.5) \cdot p}{N} \cos \frac{\pi \cdot (t+0.5) \cdot q}{M}, \quad \begin{matrix} 0 \le q \le N-1 \\ 0 \le p \le M-1 \end{matrix} \tag{2}$$

where N and M are the respective height and width of the filters for f and t, while p and q specify the modulation frequencies of the filter along the frequency and time axis. Using all possible values of p and q would result in as many features as the number of inputs. However, it is common practice [2] to retain just the output of the filters corresponding to the lowest-order coefficients. This is motivated by research suggesting that *"the auditory system may extract [...] relational information through computation of the low-frequency modulation spectrum in the auditory cortex"* [8]. For example, by keeping only 9 coefficients we achieved a performance competitive with the widely used MFCC features [3]. It should be mentioned, however, that though this approach works well in practice, the filters defined by the 2D DCT coefficients are not necessarily the optimal choice.

2.2   Gabor Filters

Another family of filters that has been used for feature extraction in speech recognition is Gabor filters [7]. Their application is motivated by their similarity with the spatio-temporal receptive fields of the auditory cortex. These filters are defined [9] as a product of a two-dimensional Gaussian (3)

$$W(f,t) = \frac{1}{2\pi\sigma_f\sigma_t}e^{-\frac{1}{2}\left(\frac{(f-f_0)^2}{\sigma_f^2}+\frac{(t-t_0)^2}{\sigma_t^2}\right)},$$
(3)

and an oriented sinusoid (4)

$$S_{p,q}(f,t) = e^{j\left(\frac{\pi \cdot f \cdot p}{N}+\frac{\pi \cdot t \cdot q}{M}\right)},$$
(4)

where we iterate f and t over the frequency and time intervals of the patch, and σ_f and σ_t specify the respective bandwiths of the filters. Again, N and M specify the transform size, while p and q specify the slanting of the sinusoid as well as its periodicity. These parameters allow many different filters, and unlike in the case of 2D DCT (where there is an assumption about which filters should be kept), the selection of the right Gabor filters for ASR is a question yet to be answered [9, 10].

3   Joint Optimization of Neural Net Classifiers and Spectro-Temporal Filters

The spectro-temporal features extracted by the filters form the input of a machine learning algorithm, which is usually a hidden Markov model (HMM), though the artficial neural net (ANN) algorithm is also a feasible alternative. The feature extraction and the classification steps are conventionally performed in two distinct steps. Our proposal here is to treat the feature extraction filters as the lowest layer of a neural net, and let the training algorithm tune the filter coefficients as well. To explain how our approach works, let us examine the operation of a simple perceptron model. In general, its output can be obtained using the formula.

$$o = a\left(\sum_{i=1}^{L} x_i \cdot w_i + b\right),$$
(5)

where $\mathbf{x}$ is the input of the neuron, L is the length of the input, $\mathbf{w}$ is the weight vector, and b is a bias corresponding to that neuron. For the activation function a we usually apply the sigmoid function, but it is also possible to create a linear neuron by setting a to the identity function. In that case, setting $b = 0$ and $L = N \cdot M$, and representing filter F and patch P in (1) in vector form (which is actually just a notational change), we see that (1) is just a special case of (5). This means that the spectro-temporal filters can be integrated into an ANN classifier system as special neurons, with the filter coefficients corresponding to the weights of the given neuron.

3.1   Structure of the ANN for Combined Feature Extraction and Posterior Estimation

Fig. 1 shows the proposed structure of the ANN that can perform spectro-temporal feature extraction and classification (phone posterior estimation) in one step. When compared to a conventional neural net, the main difference is the introduction of what we

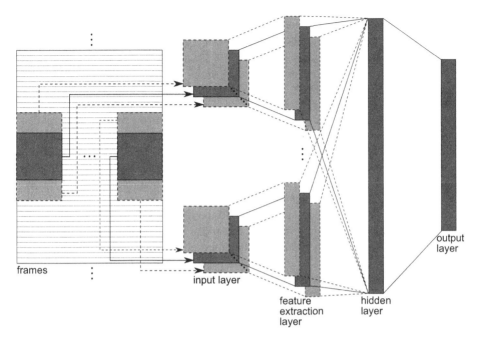

Fig. 1. Structure of the ANN for joint feature extraction and classification. The boxes in light grey correspond to additional units used by the convolutional version of the network.

call the feature extraction layer. Here, the dark grey areas in Fig. 1 mean that the spectro-temporal patches of the speech signal are concatenated to form the input data for the input layer. Then the linear neurons in the feature extraction layer perform the spectro-temporal filtering of (1). The output of this layer is channelled into the hidden layer, and from this point on the system behaves just like a conventional neural net. Hence, if the weights of the feature extraction layer were initialized with 2D DCT or Gabor filter coefficients, and only the weights of the hidden and output layers were tuned during training, then the model would be equivalent to a more traditional system, and incorporating the feature extraction step into the system would be just an implementational detail.

3.2 Fine-Tuning the Spectro-Temporal Filters

The structure in Fig. 1 allows the algorithm to evaluate the spectro-temporal features and the ANN in one step. However, our main goal here was to extend the scope of the backpropagation algorithm to the feature extraction layer as well. This way, we could also train the weights associated with the spectro-temporal filters, and hence fine-tune the initial coefficients. Of course, we had the option to initialize these coeffcents randomly (just as we do with all the other weights of the network), but it was also possible to initialize them with the 2D DCT or Gabor coefficients. Usually, as the backpropagation algorithm guarantees only a locally optimal solution, initializing the model with weights that already provide a good solution may help the backpropagation algorithm find a better local optimum than the one found using random initial values.

3.3 Convolutional Neural Nets

It is well known that integrating a longer temporal context into the acoustic features can significantly improve recognition performance. In HMM-based recognition the Δ and $\Delta\Delta$ features are used for this purpose, while in ANN/HMM hybrids a common technique is to use several neighbouring acoustic vectors [11]. Although spectro-temporal features process longer time intervals than tradional techniques (such as MFCC), we observed that adding the delta features to the feature set improves the results [4]. Unfortunately, incorporating the delta features into the joint model would be technically challenging. However, training the network on several neighbouring feature vectors instead of just one is possible by modifying the proposed structure and creating a convolutional neural net [12, 13]. This modification is shown in Fig. 1 by the boxes drawn in light grey. As can be seen, in convolutional networks the feature extraction layer performs its operation on several input patches instead of just one. We should add that the same weights are applied on each input block, so the number of weights will not change in this layer. Obviously, the number of feature vectors processed by the hidden layer increases, but in other respects the hidden and output layers work just as before. Note also that the patches used do not necessarily have to be immediate neighbours, but here we chose this simplest scenario.

4   Experimental Setup

All the experiments reported here were conducted on the TIMIT speech corpus. In the train-test partitioning, we followed the widely accepted standard of having 3696 train sentences and a core test set of 192 sentences. The phonetic labels of the database were "fused" into 39 categories, as is standard practice [14]. To create a phone recognizer from the frame-level phone posterior estimates of the neural net, we utilized a modified version of the Hidden Markov Model toolkit (HTK) [15] with a simple bigram language model.

4.1   Time-Frequency Processing

We chose the log mel-scaled spectrogram as the initial time-frequency representation of the signal. We computed the spectrograms using 400 samples (25 ms) per frame at 160 sample (10ms) hops, and applied a 1024-point FFT on the frames. They were then transformed to a log mel-scale with 26 channels, and each sentence was normalized so as to give zero mean and unit variance. After, a copy of the lower four channels were mirrored in order to avoid artificially down-weighting low frequency bins near the lower edge of the spectrogram.

Table 1. Phone recognition correctness/accuracy scores (the average of 20 independently trained neural nets)

Initial	filter weights	
filter weights	unaltered	trained
Random	73.95% / 67.04%	76.58% / 69.73%
2D DCT	75.29% / 68.81%	76.64% / 69.79%
Gabor	75.25% / 67.59%	76.56% / 69.71%

4.2   Initialization of Filter Coefficients

In an earlier paper, we performed an extensive search to get the optimal size of the time-frequency patches [3]. Based on these findings, the patches – and consequently the filters used here – had a size of 9x9, which corresponded to 9 mels in height and 90 ms in width (9 frames). The filters were applied with a step size of 4 mels (4 channels) in frequency. We tried out three different initialization schemes for the filter coefficients (i.e. the feature extraction layer of the network). In the first case, they were initialized with random numbers, as they usually are with neural nets. In the second case, they were initialized using the 2D DCT filter coefficients we utilized in our studies [3, 4]. And in the third case, the coefficients were initialized based on the Gabor filter coefficients that we found in earlier studies and had given us good results.

4.3   Neural Net Classifier

In the experiments, the classifier we applied was a multilayer neural net modified for this purpose. It consisted of a hidden feature extraction layer with a linear activation function, a hidden layer (with 1000 neurons) with the sigmoid activation function, and an output layer containing softmax units. The number of output neurons was set to the number of classes (39), while the number of neurons in the input and feature extraction layers varied, depending on how many neighbouring patches were actually used. The neural net was trained with random initial weights in the hidden and output layers, using standard backpropagation on 90% of the training data in semi-batch mode, while cross-validation on the remaining, randomly selected 10% of the training set was used as the stopping criterion.

5   Results and Discussion

The phone recognition results we got on the TIMIT corpus using a non-convolutional network are listed in Table 1. The rows of the table correspond to the various filter initialization schemes. The first column shows what we got when the filter coefficients were not trained, while in the second column they were also modified by backpropagation. The first thing we notice is that the joint training method always gives better scores than those obtained with fixed filter coefficients (with significance $p < 10^{-11}$). Second, the initialization techniques gave practically the same results, so starting from the 2D DCT or Gabor filters did not help the optimization process compared to the case with

Table 2. Phone recognition correctness/accuracy scores obtained with the convolutional network (taking the average score of 20 networks)

Initial	filter weights	
filter weights	unaltered	trained
Random	77.65% / 72.02%	78.24% / 72.61%
2D DCT	77.52% / 71.17%	78.14% / 72.52%
Gabor	78.32% / 71.74%	78.46% / 72.83%

random initialization. However, we also see that when there is no fine-tuning of filters involved, the 2D DCT and Gabor filter sets clearly outperform the randomly initialized ones ($p < 10^{-5}$). This sounds reasonable and, in fact, one might expect much worse results from random filters. Interestingly, there are studies which show that in many cases a large set of random base functions can give a representation that is just as good as a carefully selected function set. Recently, a similar study was published for the case of dictionary learning for speech feature extraction [16]. The 'extreme learning machine' of Huang et al. also exploits this suprising fact: this learning model is practically a two-layer network, where both layers are initialized randomly, and the lowest layer is not trained at all [17].

Table 2 shows the phone recognition scores on the TIMIT speech corpus using the convolutive version of the network with 4 neighbouring patches. We see that the difference between the performance of the fine-tuned and the untrained filter sets is smaller than that for Table 1. As regard the initialization methods of the trained filters, Gabor filters gave slightly better results in this case ($p < 10^{-2}$). However, the convolutional network seems to work just as well with random filters as with 2D DCT coefficients. This is an interesting observation that needs to be examined further. But it is already quite clear that a convolutional structure brings about a large improvement to the network. The superior performance of a convolutional network is in accordance with findings in similar studies [12, 13].

6   Conclusions

Here, we presented a method for the joint training of spectro-temporal filters and acoustic models using a special neural network structure. The proposed algorithm was tested in a phone recognition task for the TIMIT speech database. Our results confirmed that joint optimization does indeed result in a better recognition performance than that got by the standard, separate feature extraction and acoustic modelling approach. We also found that further significant improvements could be attained with a convolutional neural network structure. However, starting the training using a filter coefficient set like the 2D DCT set or Gabor set did not always result in better recognition accuracy scores compared to those using simple random initialization. In the future, we would like to study the behaviour of the new network in more detail, so as to learn more about its properties and limitations.

References

1. Aertsen, A.M., Johannesma, P.I.: The spectro-temporal receptive field. A functional characteristic of auditory neurons. Biological Cybernetics 42(2), 133–143 (1981)
2. Bouvrie, J., Ezzat, T., Poggio, T.: Localized Spectro-Temporal Cepstral Analysis of Speech. In: Proc. ICASSP, pp. 4733–4736 (2008)
3. Kovács, G., Tóth, L.: Localized Spectro-Temporal Features for Noise-Robust Speech Recognition. In: Proc. ICCC-CONTI 2010, pp. 481–485 (2010)
4. Kovács, G., Tóth, L.: Phone Recognition Experiments with 2D-DCT Spectro-Temporal Features. In: Proc. SACI 2011, pp. 143–146 (2011)
5. Meyer, B.T., Kollmeier, B.: Optimization and evaluation of Gabor feature sets for ASR. In: Proc. Interspeech 2008, pp. 906–909 (2008)
6. Kleinschmidt, M.: Localized Spectro-Temporal Features for Automatic Speech Recognition. In: Proc. EuroSpeech 2003, pp. 2573–2576 (2003)
7. Kleinschmidt, M.: Methods for capturing spectrotemporal modulations in automatic speech recognition. Acta Acustica United With Acustica 88(3), 416–422 (2002)
8. Greenberg, S.: Understanding Speech Understanding: Towards A Unified Theory Of Speech Perception. In: Proceedings of the ESCA Tutorial and Advanced Research Workshop on the Auditory Basis of Speech Perception, pp. 1–8 (1996)
9. Ezzat, T., Bouvrie, J., Poggio, T.: Spectro-Temporal Analysis of Speech Using 2-D Gabor Filters. In: Proc. Interspeech 2007, pp. 506–509 (2007)
10. Kleinschmidt, M., Gelbart, D.: Improving Word Accuracy with Gabor Feature Extraction. In: Proc. ICSLP 2002, pp. 25–28 (2002)
11. Bourlard, H., Morgan, N.: Connectionist speech recognition: A hybrid approach. Kluwer Academic Pub. (1994)
12. Abdel-Hamid, O., Mohamed, A., Jiang, H., Penn, G.: Applying Convolutional Neural Networks concepts to hybrid NN-HMM model for speech recognition. In: Proc. ICASSP 2012, pp. 4277–4280 (2012)
13. Vesely, K., Karafiat, M., Grezl, F.: Convolutive Bottleneck Network features for LVCSR. In: Proc. ASRU 2011, pp. 42–47 (2011)
14. Lee, K.-F., Hon, H.-W.: Speaker-independent phone recognition using Hidden Markov models. IEEE Trans. Acoust., Speech Signal Processing 37, 1641–1648 (1989)
15. Young, S., et al.: PC The HTK book version 3.4. Cambridge University Engineering Department, Cambridge (2006)
16. Vinyals, O., Deng, L.: Are sparse representations enough for acoustic modeling? In: Proc. INTERSPEECH (2012)
17. Huang, G.-B., Wang, D.H., Lan, Y.: Extreme learning machines: A survey. International Journal of Machine Learning and Cybernetics 2(2), 107–122 (2011)

Three Syntactic Formalisms
for Data-Driven Dependency Parsing of Croatian

Željko Agić and Danijela Merkler

Faculty of Humanities and Social Sciences, University of Zagreb
Ivana Lučića 3, 10000 Zagreb, Croatia
{zagic,dmerkler}@ffzg.hr

Abstract. A new syntactic formalism for dependency parsing of Croatian and its implementation in the SETimes Dependency Treebank of Croatian – the SE-TIMES.HR Treebank – is presented. Its new syntactic tagset is targeted towards improving dependency parsing accuracy, with special emphasis on the main syntactic categories such as predicates, subjects and objects. It is compared with two versions of Croatian Dependency Treebank (HOBS): one with explicit encoding of subordinate syntactic conjunctions and one without. Manual annotation quality and dependency parsing accuracy were inspected. An improvement in inter-annotator agreement was observed, as Cohen's kappa coefficient for label attachment $\kappa(LA)$ peaked at 0.92, topping the two HOBS instances by 0.036 and 0.081 points. Overall dependency parsing accuracy reached 77.49 in labeled attachment (LAS), 2.99 and 5.78 points over HOBS, using a standard graph-based dependency parser.

Keywords: dependency treebank, dependency parsing, Croatian language.

1 Introduction

While from one viewpoint, the two CoNLL shared tasks in multilingual – or, in terms of the underlying paradigm, mostly data-driven – dependency parsing [3,10] have marked a checkpoint in the general resurgence of interest in dependency parsing, from another viewpoint, they have also emphasized the sparseness of treebanks as language resources, especially for under-resourced languages. For example, out of the total sum of 2 million tokens, the Prague Dependency Treebank (PDT) [7] accounted for 500 thousand tokens in the CoNLL 2007 shared task [10], while certain treebanks allocated 1.5% of the sum, contributing with less than 30 thousand tokens.

Croatian language was not included in the CoNLL shared tasks in dependency parsing as no treebanks of Croatian existed at that time. Just recently, the first experiments in data-driven dependency parsing of Croatian were conducted [1] by using a draft version of Croatian Dependency Treebank (HOBS) [11] with approximately 90 thousand tokens in 3 500 sentences. Similar to parsing experiments with Slovene Dependency Treebank (SDT) [4] – both based on the PDT syntactic formalism [7] – overall parsing accuracy peaked at somewhat over 70% in labeled attachment score using standard data-driven, i.e. transition-based and graph-based dependency parsers. This in turn opened a line of research in improving parsing quality, e.g., by combining parsers with language

I. Habernal and V. Matousek (Eds.): TSD 2013, LNAI 8082, pp. 560–567, 2013.

resources such as valency lexicons [1], as well as by reducing existing and designing new syntactic tagsets for more efficient treebank annotation and parsing [9].

Drawing from the experience of PDT-based SDT and subsequent development of a new simplistic syntactic formalism for Slovene dependency syntax – encompassing only 10 syntactic functions in comparison with 70 in the PDT formalism – and its implementation in JOS corpus of Slovene [6], this line of research presents a new dependency formalism for Croatian and its implementation in form of a 2 500 sentence prototype SETimes Dependency Treebank of Croatian (SETIMES.HR Treebank). Latest versions of HOBS and SETIMES.HR are presented and compared in terms of underlying formalisms, annotation choices and inter-annotator agreement. The treebanks are utilized in an experiment with data-driven dependency parsing and compared regarding overall parsing accuracy, learning rates and accuracy for specific syntactic functions.

2   Treebanks

HOBS is developed by implementing an adaptation of the PDT analytical layer formalism for Croatian, i.e., by directly applying the PDT-style syntactic functions and a set of additional rules accounting for Croatian language specifics, e.g., annotation of complex predicates [2]. It is built on top of the CW100 subcorpus of Croatian National Corpus,[1] which was manually lemmatized and MSD-tagged by using an adaptation of MTE v4 morphosyntactic tagset [5]. This version of HOBS is publicly available via META-SHARE,[2] but its syntactic tags were stripped and only the barebone dependency tree structure is provided. HOBS was further adapted by introducing a new set of syntactic functions (also called analytical functions – *afuns*) for explicit annotation of syntactic conjunctions that introduce subordinate clauses. This 11-function set (called *Sub**, as the new tags are prefixed with *Sub*, e.g., *Sub_Adv* for adverbial clause and *Sub_Obj* for objective clause) implements explicit subordinate clause classification and consequently also alters the PDT formalism by explicitly encoding subordinate clause predicates. The stats for the two versions of HOBS are given in Table 1. Both are built on top of the same sentence set, they slightly differ in MSD annotation and the number of syntactic functions reflects the annotation formalism. Basic and full HOBS syntactic tagset is exposed: all sub-classification tags are stripped from the full tagset – e.g., all predicate tags (*Pred*, *Pred_Co*, *Pred_Pa*) collapse into a single tag *Pred* – to create the basic tagset.

SETimes Dependency Treebank of Croatian (SETIMES.HR Treebank) is built on top of the Croatian part of the SETimes parallel corpus.[3] It was sentence split and tokenized, lemmatized and MSD-tagged by human annotators. The treebank is fully compliant with MTE v4 morphosyntactic tagset. This preprocessed dataset of approximately 4 000 sentences was then used for manual syntactic annotation of approximately 2 500 sentences to create the treebank.

A new model for syntactic annotation was developed for SETIMES.HR. It consists of 15 analytical functions, aiming at simplicity and easier manual annotation, but still

---

[1] http://hnk.ffzg.hr/

[2] http://meta-share.ffzg.hr/

[3] http://www.nljubesic.net/resources/corpora/setimes/

Table 1. Treebank stats

treebank	features	sent's	tokens	types	lemmas	MSDs	afuns
HOBS	full without *Sub**	4 626	117 369	25 038	12 388	914	70
	full with *Sub**	4 626	117 369	25 038	12 388	911	81
	basic without *Sub*	4 626	117 369	25 038	12 388	914	27
	basic with *Sub*	4 626	117 369	25 038	12 388	911	28
SETIMES.HR	full MSD	2 488	56 334	13 409	6 901	804	15
	reduced MSD	2 488	56 334	13 374	6 943	665	15
	POS	2 488	56 334	13 374	6 943	12	15

Table 2. Inter-annotator agreement

treebank	features	LAS	UAS	LA	κ(LA)
HOBS	full with *Sub**	78.89	89.16	84.07	0.839
HOBS	basic with *Sub*	82.05	89.16	88.83	0.884
SETIMES.HR	full MSD	86.11	91.29	92.51	0.920

hoping to enable clear and distinct annotation of the most important syntactic features. The main categories – predicate, subject, object, adverb, attribute and apposition – are annotated with functions *Pred*, *Sb*, *Obj*, *Adv*, *Atr* and *Ap*, in direct conformance with grammatical rules of Croatian. *Pnom* is used for annotation of nominal predicates. Predicates in Croatian can be very complex and cover many words – most often, a combination of modal or phase verb and infinitive, or modal or phase verb, infinitive and nominal predicate. Infinitives in these examples are annotated with function *Atv*, which is also assigned to verbal adverbs and predicate attributes – predicate expansions which occur with nominal words. *Aux* denotes auxiliary elements that are typically parts of predicates, e.g., auxiliary verbs, negation, reflexive pronoun *se*, etc. Functions *Co* and *Sub* are used to annotate elements that link coordinate or subordinate clauses, and *Co* also links coordinate elements on the phrase level. Function *Prep* denotes all prepositions. Function *Oth* is used for annotating various other types of elements, e.g., emphasizing words, redundant or emotional items, parts of multiword syntactic elements like complex conjunctions, etc. *Elp* is used both for elements whose head is missing from the syntactic structure and for elements and clauses which are parenthetical in syntactic structure. Function *Punc* is assigned to all graphic symbols in the sentence when they do not have a coordinating function.

SETIMES.HR stats are also given in Table 1. At 2 488 sentences and 56 334 tokens, it is at roughly 50% size of HOBS. Two additional versions of SETIMES.HR are introduced for the experiment – one using a reduced MTE v4 based MSD tagset and one using only POS. They also introduce corrections in lemmatization and thus these figures differ slightly.

Inter-annotator agreement was estimated by assigning two annotators with 100 sentences from HOBS with explicit subordination encoding (HOBS with *Sub**) and 100 sentences from SETIMES.HR and measuring the standard parsing accuracy scores.

Table 3. Inter-annotator confusion matrices

HOBS basic with *Sub*			SETIMES.HR		
afun pair	frequency	pct	afun pair	frequency	pct
Obj Adv	48	17.65	*Obj Atr*	24	15.09
Obj Atr	18	6.62	*Adv Oth*	19	11.95
Sb ExD	11	4.04	*Obj Adv*	16	10.06
AuxG ExD	8	2.94	*Adv Atr*	11	6.92
Sb Atr	8	2.94	*Pnom Pred*	8	5.03
Adv Atr	7	2.57	*Pred Aux*	8	5.03
Atr Sb	7	2.57	*Pnom Sb*	4	2.52
other	165	60.67	other	69	43.40

Cohen's kappa was calculated for linear label attachment (LA). The scores are given in Table 2. Being that the unlabeled attachment scores (UAS) are expectedly comparable as head attachment is not tagset-dependent, the increase in labeled attachment score (LAS) can be seen as influenced exclusively by attachment of token labels. LAS increases by 3.16 points from full HOBS with *Sub** to basic HOBS with *Sub* and 4.06 from basic HOBS with *Sub* to SETIMES.HR, amounting to an overall increase of 7.22 in LAS from full HOBS with *Sub** to SETIMES.HR. The ratio is maintained in LA scores, while the complexity of syntactic tagsets reflects in kappa coefficients – all label attachment agreements are considered very good, with SETIMES.HR topping full HOBS with *Sub** by 0.081 and basic HOBS with *Sub* by 0.036. The improvements in agreement moving from HOBS to SETIMES.HR are thus shown to be induced by syntactic tagset design, with the simpler tagset enabling more consistent annotation.

Confusion matrices for inter-annotator agreement in HOBS and SETIMES.HR are given in Table 3. The data indicates that choosing label assignments between *Adv*, *Atr* and *Obj* is the most frequent source of confusion for both treebanks. Being that these are inherently ambiguous categories, the larger spread and size of the confusion matrix for HOBS further supports the choice of SETIMES.HR tagset design, as the chance for disagreement increases with tagset size. From another viewpoint, introducing one additional syntactic tag to HOBS basic (*Sub*) and 11 additional tags to HOBS full (*Sub**) yielded a substantial improvement in overall inter-annotator agreement over original HOBS, indicating that tagset design choices may not be as simple as *the smaller the better* rule of thumb and that achieving a balance in explicitness and informativeness of syntactic formalisms is a non-trivial task, even though the (over)simplistic implication of quality by simplicity might generally still be supported by parsing experiments [9]. As this experiment deals with treebank design for improved parsing accuracy, syntactic tagset design remains an interesting future research topic.

3   Experiment and Results

In this experiment, three syntactic formalisms are inspected for impact on dependency parsing of Croatian:

1. HOBS without *Sub** – the initial PDT-based formalism, somewhat adapted to Croatian regarding annotation of complex predicates [2], implemented in original HOBS,
2. HOBS with *Sub** – its further adaptation by introduction of an additional set of syntactic functions (*Sub**) for explicit annotation of subordinating syntactic conjunctions, implemented on top of original HOBS and
3. SETIMES.HR – a new, simplified and parsing-oriented 15-tag syntactic formalism, implemented in the SETIMES.HR Treebank.

Experiment setup follows the guidelines of CoNLL dependency parsing shared tasks [3,10]. The treebanks are split into tenfold training and testing sets, the latter consisting of approximately 5.000 tokens or 200 sentences each. As it was previously shown – for Slovene in the CoNLL 2006 shared task [3] and for Croatian in [1] – that a statistically significant preference exists between standard data-driven parsers in favor of the graph-based parsing paradigm, MSTParser parser generator [8] was used in the experiment. The MSTParser training setup was `decode-type:non-proj order:2 training-k:5 iters:10 loss-type:punc` as it was shown [1] to be the top performing setting for Croatian.[4] Parsing output was evaluated by observing labeled and unlabeled attachment score (LAS, UAS), attachment of labels to tokens (LA) and treebank LAS learning curves. Statistical significance of differences in results was tested pairwise by using two-tailed paired t-test.

Seven different treebank instances were used in the experiment: four versions of HOBS (full tagset or basic tagset, with *Sub** or without *Sub** annotation of subordinating syntactic conjunctions) and three versions of SETIMES.HR (full MSD tagset, reduced MSD tagset and POS tagset). The first set of results is given in Table 4. It shows that parser models using HOBS versions with explicit *Sub** tagging outperform HOBS versions without *Sub** with statistical significance, in full and basic tagset instances. Even the full tagset version with *Sub** outperforms the basic tagset without *Sub*. SETIMES.HR instances trained with MSD features all significantly outperform HOBS treebanks, while the one trained with POS features remains comparable to the top performing HOBS model (basic with *Sub*). The top performing model is trained on reduced MSD SETIMES.HR dataset, scoring 77.49% LAS (2.99 points over the best HOBS model, 5.78 and 5.56 over the no *Sub** models, 4.45 over HOBS full with *Sub**). These differences are statistically significant, while the difference between reduced MSD and full MSD SETIMES.HR models is not. Similar to inter-annotator agreement, the LAS differences are supported by label attachment (LA) more substantially than by dependency tree structure, i.e., unlabeled attachment (UAS). It should be noted that the difference between HOBS and SETIMES.HR models is obtained by using completed versions of HOBS with 117 369 tokens and draft versions of SETIMES.HR with 56 334 tokens (48% size of HOBS).

Learning curves in Figure 1 might serve as standpoints for comparisons between HOBS and SETIMES.HR regarding treebank sizes, as well as additional illustrations of overall scores in Table 4. The patterns of differences within the two treebank groups are maintained, while the third diagram indicates the LAS difference between top performing HOBS and SETIMES.HR at training set sizes of approximately 2 500 sentences.

---

[4] This MSTParser configuration indicates using the Chu-Liu/Edmonds non-projective maximum spanning tree parsing algorithm with second order features.

Table 4. Overall parsing accuracy

treebank	features	LAS	UAS	LA
HOBS	full without *Sub**	71.71	80.34	81.75
	full with *Sub**	73.04	81.10	82.85
	basic without *Sub*	71.93	79.98	84.65
	basic with *Sub*	74.50	81.41	86.87
SETIMES.HR	full MSD	77.13	83.08	88.82
	reduced MSD	77.49	83.58	89.00
	POS	74.56	81.59	85.87

Table 5. Parsing accuracy and test set frequencies for matching syntactic functions

afun	HOBS without *Sub*			HOBS with *Sub*			SETIMES.HR		
	LAS	UAS	pct	LAS	UAS	pct	LAS	UAS	pct
Adv	65.88	84.81	9.98	68.33	88.33	8.99	61.57	84.72	4.72
Ap (Apos)	38.10	47.62	0.64	36.84	42.11	0.64	89.60	92.00	3.05
Atr	81.61	88.29	28.7	83.06	89.18	25.8	80.75	88.39	26.5
Co (Coord)	48.21	49.23	4.15	56.85	59.39	4.18	46.00	48.00	2.87
Obj	62.81	79.40	8.39	70.06	87.65	6.53	74.10	89.76	7.25
Pnom	58.73	80.95	1.51	60.61	77.27	1.74	65.75	73.97	2.03
Pred	65.89	72.87	4.76	80.69	82.19	9.29	86.58	88.10	9.32
Prep (AuxP)	69.85	70.50	9.28	71.54	71.94	9.99	74.04	75.11	9.44
Sb	68.85	81.26	7.84	73.99	82.37	7.01	75.56	82.87	6.61
Sub	–	–	–	72.91	73.89	4.04	65.22	65.76	3.81

LAS of this specific HOBS instance is measured at 73.18%, i.e., 4.31 LAS or 1.32 points more than in overall evaluation. Projecting the learning curves indicates that this ratio would be maintained when increasing the size of SETIMES.HR.

Table 5 compares basic tagset HOBS to the reduced MSD SETIMES.HR model regarding LAS and UAS on matching syntactic functions. Frequencies of these functions are also indicated to provide a more weighted comparison. SETIMES.HR parser outperforms HOBS parsers on basic syntactic categories – predicates (*Pred*, at least 5.89 LAS points), subjects (*Sb*, 1.57) and objects (*Obj*, 4.04), with nominal predicate (*Pnom*) scores spread between the three models. Basic tagset HOBS with *Sub* significantly outperforms SETIMES.HR on coordinating (*Co*) and subordinating (*Sub*) syntactic conjunctions. This is most likely due to their absolute frequency difference in the training sets. Other differences are due to relative frequency disproportion. It should be noted that HOBS with *Sub* significantly outperforms HOBS without *Sub* in almost all categories, particularly for predicates (*Pred*, 9.32, with frequency almost doubling in HOBS with *Sub* due to explicit encoding) and objects (*Obj*, 8.25, with a frequency decrease due to encoding object clause introduction using syntactic conjunctions *Sub* rather than by denoting subordinate clause predicates as *Obj*).

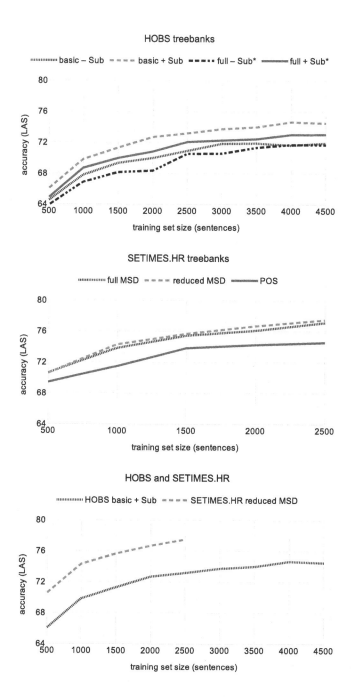

Fig. 1. Learning curves for HOBS and SETIMES.HR treebanks

4   Conclusions and Future Work

Three formalisms for dependency-syntactic representation and parsing of Croatian were presented, two of them implemented in Croatian Dependency Treebank (HOBS) and one in newly-developed SETIMES.HR Treebank. In SETIMES.HR a new syntactic formalism for Croatian was introduced and compared with HOBS for inter-annotator agreement and parsing performance. SETIMES.HR and respective parsing models were shown to be significantly more efficient than HOBS in manual and automatic annotation of Croatian texts. Results are made publicly available.[5] Future work plans include further development of HOBS and SETIMES.HR regarding size and consistency and implementing more advanced parsing approaches. Influence of syntactic tagset design on annotation quality and parsing accuracy should be carefully investigated.

References

1. Agić, Ž.: K-Best Spanning Tree Dependency Parsing With Verb Valency Lexicon Reranking. In: Proceedings of COLING 2012: Posters, COLING 2012 Organizing Committee, pp. 1–12 (2012)
2. Berović, D., Agić, Ž., Tadić, M.: Croatian Dependency Treebank: Recent Development and Initial Experiments. In: Proceedings of LREC 2012, pp. 1902–1906. ELRA (2012)
3. Buchholz, S., Marsi, E.: CoNLL-X Shared Task on Multilingual Dependency Parsing. In: Proceedings of CoNLL-X, pp. 149–164. ACL (2006)
4. Džeroski, S., Erjavec, T., Ledinek, N., Pajas, P., Žabokrtský, Z., Žele, A.: Towards a Slovene Dependency Treebank. In: Proceedings of LREC 2006, pp. 1388–1391. ELRA (2006)
5. Erjavec, T.: MULTEXT-East: Morphosyntactic Resources for Central and Eastern European Languages. Language Resources and Evaluation 46(1), 131–142 (2012)
6. Erjavec, T., Fišer, D., Krek, S., Ledinek, N.: The JOS Linguistically Tagged Corpus of Slovene. In: Proceedings of LREC 2010, pp. 1806–1809. ELRA (2010)
7. Böhmová, A., Hajič, J., Hajičová, E., Hladká, B.: The Prague Dependency Treebank: A Three-Level Annotation Scenario. In: Abeillé, A. (ed.) Treebanks: Building and Using Parsed Corpora. Springer (2003)
8. McDonald, R., Lerman, K., Pereira, F.: Multilingual Dependency Parsing With a Two-Stage Discriminative Parser. In: Proceedings of CoNLL-X, pp. 216–220. ACL (2006)
9. Mille, S., Burga, A., Ferraro, G., Wanner, L.: How Does the Granularity of an Annotation Scheme Influence Dependency Parsing Performance? In: Proceedings of COLING 2012: Posters, COLING 2012 Organizing Committee, pp. 839–852 (2012)
10. Nivre, J., Hall, J., Kübler, S., McDonald, R., Nilsson, J., Riedel, S., Yuret, D.: The CoNLL 2007 Shared Task on Dependency Parsing. In: Proceedings of the CoNLL Shared Task Session of EMNLP-CoNLL 2007, pp. 915–932. ACL (2007)
11. Tadić, M.: Building the Croatian Dependency Treebank: The Initial Stages. Suvremena lingvistika 63(1), 85–92 (2007)

---

[5] http://zeljko.agic.me/resources/

Topic Models for Comparative Summarization

Michal Campr and Karel Ježek

Department of Computer Science and Engineering, FAV,
University of West Bohemia,
20 March 2013, 301 00, Plzen, Czech Republic
{mcampr,jezek_ka}@kiv.zcu.cz
http://textmining.zcu.cz

Abstract. This paper aims to sum up our work in the area of comparative summarization and to present our results. The focus of comparative summarization is the analysis of input documents and the creation of summaries which depict the most significant differences in them. We experiment with two well known methods – Latent Semantic Analysis and Latent Dirichlet Allocation – to obtain the latent topics of documents. These topics can be compared and thus we can learn the main factual differences and select the most significant sentences into the output summaries. Our algorithms are briefly explained in section 2 and their evaluation on the TAC 2011 dataset with the ROUGE toolkit is then presented in section 3.

Keywords: comparative summarization, latent semantic analysis, latent dirichlet allocation, topic model, rouge.

1 Introduction

With the continual growth of the internet as an information source, where a great amount of data is being uploaded every minute, the need for data compression is obvious. This necessity is not important only for audio or video files, but also for textual documents. As the amount of textual data grows, the probability of appearance of documents with very similar features is increasing, e.g. political programs or descriptions of university courses. In any application, when facing a set of documents sharing a similar topic, people are interested to know what are their differences. This problem can be addressed by comparative summarization and it is the primary focus of this paper.

Several papers addressing the comparison of text mining methods [1] and the problem of document comparison have already been published (e.g. [2]), as well as papers covering the area of summarization using a variety of methods, e.g. discriminative sentence selection [3] or linear programming [4]. In this paper, we explore the possibilities of utilizing two very well known methods for acquiring latent semantic topics of documents for comparing them, extracting the most characteristic sentences and forming summaries which depict the main differences of the documents. These two methods are Latent Semantic Analysis (LSA) and Latent Dirichlet Allocation (LDA). We have already published papers which discuss the algorithms that use LSA or LDA for basic summarization or for topic comparison so in this paper we focus solely on algorithms for comparative summarization and their evaluation.

I. Habernal and V. Matousek (Eds.): TSD 2013, LNAI 8082, pp. 568–574, 2013.

2   Topic Models for Comparative Summarization

This section briefly explains the use of LSA and LDA for comparative summarization. Both algorithms operate under the same assumptions and have several similar features, e.g. the method of lemmatization and removing stop-words.

Both of these methods are based on a specialized dictionary. Our stop-word dictionary contains 578 strings which are compared to every word in the input documents to decide if the given word should be removed. Similarly, our lemma dictionary contains over 22248 words, each paired with its lemmatized form. Each word of the input document is compared to those in the dictionary and its corresponding lemma is used. If the input word is not found, the original form is used.

The principle of comparative summarization is loosely based on update summarization (e.g. [5]) but with some important differences. Its goal is the comparison of two different documents or two sets of documents D_1 and D_2, where we do not assume any familiarity with any of the documents. We just assume, that those two sets of documents refer to a similar topic, but contain some different information about it. The aim is finding those differences.

2.1   Latent Semantic Analysis

LSA is an algebraic method which analyses relations between terms and sentences of a given document. For decomposing matrices, it uses Singular Value Decomposition (SVD) which is a numerical process used for data dimensionality reduction, classification, searching in documents and also for text summarization.

The algorithm starts with creating two matrices A_1 and A_2 for each of the document sets. Column vectors of matrix A_1 or A_2 contain term frequencies in the given sentences. However, both matrices must be created with the same term set (terms combined from both document sets) to avoid inconsistencies with lengths of singular vectors during their comparison. Matrix A_1 has $t \times s_1$ dimensions and matrix A_2 $t \times s_2$ dimensions, where t is the number of terms in both document sets, s_1 is the number of sentences in D_1 and s_2 is the number of sentences in D_2. The values of the matrix elements are computed as $a_{ij} = L(t_{ij}) \cdot G(t_{ij})$, where $L(t_{ij})$ is a boolean value (1 if term i is present in sentence j, 0 otherwise) and $G(t_{ij})$ is the global weight for term i in the whole document set:

$$G(t_{ij}) = 1 - \sum_j \frac{p_{ij} \log(p_{ij})}{\log(n)}, p_{ij} = \frac{t_{ij}}{g_i}, \tag{1}$$

where t_{ij} is the frequency of term i in sentence j, g_i is the total number of times that term i occurs in the whole document set and n is the number of sentences.

The Singular Value Decomposition of A, constructed over one document set with m terms and n sentences, is defined as $A = U\Sigma V^T$, where $U = [u_{ij}]$ is an $m \times n$ matrix and its column vectors are called left singular vectors. Σ is a square diagonal $n \times n$ matrix and contains the singular values. $V^T = [v_{ij}]$ is an $n \times n$ matrix and its columns are called right singular vectors. This decomposition provides latent semantic structure of the input documents represented by A. This means, that it provides a decomposition of documents into n linearly independent vectors, which represent the main topics contained in the documents. If a specific combination of terms is often present within the document set, it is represented by one of the singular vectors. And furthermore, the singular values contained in the matrix Σ represent the significance of these topics. Matrix U provides mapping of terms into topics and V^T provides mapping of sentences into topics.

By applying SVD on both matrices A_1 and A_2 separately, we get matrices U_1 and U_2, Σ_1 and Σ_2, V_1^T and V_2^T, which provide the mapping of terms/sentences to topics contained in both document sets. We can then start comparing those topics in matrices U_1 and U_2.

For each topic (left singular vector) from U_2, we want to find the most similar topic in U_1. The redundancy between two vectors is computed as a cosine similarity:

$$red(t) = \frac{\sum_{j=1}^m U_1[j,i] * U_2[j,t]}{\sqrt{\sum_{j=1}^m U_1[j,i]^2} * \sqrt{\sum_{j=1}^m U_2[j,t]^2}}, \tag{2}$$

where t is the index of the topic from U_2, j is the index of topic from U_1. With redundancy computed, we can get the dissimilarity of the given topic as $dis(t) = 1 - red(t)$. We store the values $dis(t)$ in a diagonal matrix DS_1 (Dissimilarity Score) and create the final matrix $F_1 = DS_1 * \Sigma_2 * V_2^T$ which contains the dissimilarity, as well as the importance of individual topics mapped on sentences.

From matrix F_1, we can start selecting sentences into the final extract. This selection is based on finding the longest sentence vectors, i.e. the length s_r of a sentence r is defined as:

$$s_r = \sqrt{\sum_{i=1}^t F_1[r,i]^2}. \tag{3}$$

When a vector is selected, we need to make sure that a sentence with similar information will not be selected. We have tested three different solutions (the first one provided the best results):

– Set all values of the selected vector to 0. This is the simplest solution and it guarantees that once a sentence was selected, a similar one will not be selected again.
– Subtract the selected vector from matrix F. This removes the selected sentence (and information it contains) from the whole matrix.
– Use cosine similarity to detect possible resemblance between the candidate sentence and any of the already selected sentences. This does not make any alterations to matrix F.

This process is run in both directions, i.e. we create matrices F_1 and F_2 which contain the differences of topics in both document sets and search for the most suitable sentences until the resulting summary reaches a predefined length.

2.2   Latent Dirichlet Allocation

LDA [6] can be viewed as a model which breaks down the collection of documents into topics. The collection is represented as a mixture of documents, where the probability (importance) of the document D_B for the collection is denoted as $P(D_B)$. It also represents each document as a mixture of topics with the probability distribution representing the importance of j-th topic for the document D_B denoted as $P(T_j|D_B)$. The topics are then represented as a mixture of words with their probability representing the importance of the i-th word for the j-th topic (denoted as $P(W_i|T_j)$).

The first step of the algorithm is to load the input data from two document sets A and B. The important thing here is that from the perspective of LDA, we treat every sentence as one document. Before we run the Gibbs sampler (we used the JGibbLDA implementation [7]) to obtain the LDA distributions, we have to remove the stop-words and perform term lemmatization. This way we are sure that there are no words that carry no useful information.

The obtained topic-word distributions for each document set are stored in matrices T_A for the document set A and T_B for B, where rows represent topics and columns represent words. A very important aspect of saving the distributions into matrices is matching their dimensions, i.e. to include words that appear only in one set into both matrices (with zero probability).

After this, we can compute topic-sentence matrices U_A and U_B with sentence probabilities :

$$P(S_r|T_j) = \frac{\sum_{W_i \in S_r} P(W_i|T_j) * P(T_j|D_r)}{length(S_r)^l}, \tag{4}$$

where $l \in < 0, 1 >$ is an optional parameter to configure the handicap of long sentences. The row vectors of U_A and U_B represent topics and the columns are sentences. Next step includes creating a symmetrical diagonal matrix SIM which contains the similarities of topics from both sets. This is accomplished as follows:

$T_A = [T_{A1}, T_{A2}, ..., T_{An}]^T, T_B = [T_{B1}, T_{B2}, ..., T_{Bn}]^T$, where T_{Ai} and T_{Bi} are row vectors representing topics and n is the number of topics. For each T_{Ai} find red_i (redundancy of i-th topic) by computing the largest cosine similarity between T_{Ai} and T_{Bj}, where $j \in < 1..n >$ and storing value $1 - red_i$ representing the dissimilarity of i-th topic into matrix SIM.

Finally, we construct matrices $F_A = SIM * U_A$ and $F_B = SIM^T * U_B$ combining the probabilities of sentences with the dissimilarity of topics. Then, it is a simple matter to find sentences with the best score and including them in the summary. For better results, it is essential to compare (using cosine similarity) the candidate sentence with already selected sentences to avoid information redundancy. If a sentence is selected, the respective vector in F_A or F_B is set to 0 in order to remove the information from the matrix.

The output of this algorithm consists of two summaries of predefined length depicting the most significant information in which the documents differ.

3   Evaluation

For the evaluation of our algorithms we used a very well known tool – ROUGE which stands for Recall-Oriented Understudy for Gisting Evaluation. This family of measures, which are based on the similarity of overlapping units such as n-grams[1], word

---

[1] An n-gram is a subsequence of n words from a given text.

sequences, and word pairs between the computer-generated summary and the ideal summaries (created by humans), was firstly introduced in 2003 [8]. The ROUGE scores have been widely used for the evaluation of summarization algorithms since then and so we have also decided to compute six different ROUGE scores:

- ROUGE-1, ROUGE-2, ROUGE-3 and ROUGE-4 are N-gram based metrics.
- ROUGE-W is a weighted longest common subsequence measure.
- ROUGE-SU4 is a bigram measure that enables at most 4 unigrams inside of bigram components to be skipped.

3.1   The Experiment

Due to the lack of unified testing data for evaluating comparative summarization, we utilized data from the TAC 2011 conference and created our own dataset to find out if the proposed methods brings the expected results.

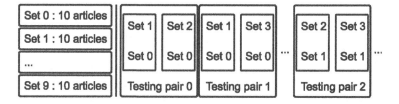

Fig. 1. Arrangement of testing data

The available data consist of 100 news articles, divided into 10 topics, 10 articles each. We have created pairs of sets of articles by combining different topics (Figure 1). In every pair, there is one identical topic present in both sets and one topic for each of the sets that is different. The purpose is to simulate two sets of documents which have something in common, but also some differences. This arrangement allows us also to easily compute the precision of selecting sentences because we know from which topic the sentences should be.

The reason for using the TAC 2011 dataset is also the fact, that it contains three human-created summaries for each of the 10 topics. This allows us to further evaluate our method with the ROUGE package. Note that this dataset only provides an approximation of documents that share some similar topics. We would like to test our algorithms on a dataset specifically designed for evaluating the task of comparative summarization, but to our knowledge, there is no such corpus.

3.2   The Results

With the use of the method, that was described in the previous section, we obtained 360 combinations of topics. Each of these combinations was then used as an input set of documents for our algorithms (producing two summaries for each input set),

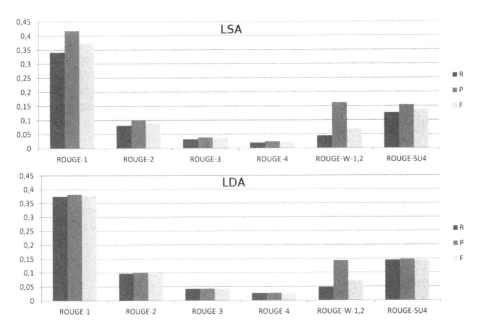

Fig. 2. ROUGE scores (P – precision, R – recall, F – F-score). LSA-based algorithm on the top, LDA-based algorithm on the bottom.

so in the end, we acquired 720 summaries. These summaries were then compared with the human-made summaries (from the TAC 2011 dataset) with ROUGE.

The Figure 2 shows the average scores for both our algorithms. Evidently, both techniques (LSA and LDA) show comparable results, however, LDA performs slightly better considering recall scores, but its precision is lower.

The results are also comparable with the ROUGE scores of other summarization methods where the recall scores ranged from 0.117 to 0.19 (i.e. [9]). It is important to note, that the used reference summaries were primarily meant to be used for basic summarization, however, due to the construction of our experiment, it is possible to use them for our evaluation. Because we extracted the information from a larger set of documents which contained "an interference" in the form of additional information, we obtained lower ROUGE scores, which was expected. The highest average ROUGE-2 recall score for LDA we achieved was 0.097 and 0.081 for LSA.

4   Conclusion and Further Research

This paper is focused on our experiments in the area of comparative summarization. We experimented with two well known topic models – LSA and LDA – from which the LDA provided better recall results but LSA higher precision. However, for summarizing very large data, it is worth also speculating about the aspect of computation time. LSA performs significantly faster for smaller data, but with the increasing length of documents, LSA starts to slow down due to the computations of large sparse matrices.

For improving the performance of our algorithms, we would like to try utilizing some other techniques of natural language processing, such as the analysis of synonymy, named entities or compositional expressions. For the full comparison of document content is necessary to consider not only its variety of topics, but also the polarity of sentences (their sentiment). Sentiment analysis and contrastive summarization will be the focus of our future work, i.e. comparing and summarizing the differences in opinions expressed in one or multiple documents like product reviews or political debates.

Acknowledgement. This project was supported by grant SGS-2013-029 Advanced computing and information systems.

The access to computing and storage facilities owned by parties and projects contributing to the National Grid Infrastructure MetaCentrum, provided under the programme "Projects of Large Infrastructure for Research, Development, and Innovations" (LM2010005) is highly appreciated.

References

1. Lee, S., Baker, J., Song, J., Wetherbe, J.C.: An Empirical Comparison of Four Text Mining Methods. In: Proceedings of the 2010 43rd Hawaii International Conference on System Sciences, pp. 1–10 (2010)
2. Gelbukh, A.F., Sidorov, G., Guzman-Arenas, A.: Document Comparison with a Weighted Topic Hierarchy. In: Proceedings of the 10th International Workshop on Database & Expert Systems Applications, pp. 566–570 (1999)
3. Wang, D., Zhu, S., Li, T., Gong, Y.: Comparative document summarization via discriminative sentence selection. In: Proceedings of the 18th ACM Conference on Information and Knowledge Management, pp. 1963–1966 (2009)
4. Huang, X., Wan, X., Xiao, J.: Comparative news summarization using linear programming. In: Proceedings of the 49th Annual Meeting of the Association for Computational Linguistics: Human Language Technologies: short papers, vol. 2, pp. 648–653 (2011)
5. Steinberger, J., Ježek, K.: Update Summarization Based on Latent Semantic Analysis. In: Matoušek, V., Mautner, P. (eds.) TSD 2009. LNCS, vol. 5729, pp. 77–84. Springer, Heidelberg (2009)
6. Blei, D.M., Ng, A.Y., Jordan, M.I.: Latent dirichlet allocation. The Journal of Machine Learning Research, 993–1022 (2003)
7. Phan, X.-H., Nguyen, C.-T.: http://jgibblda.sourceforge.net/
8. Lin, C.-Y., Hovy, E.: Automatic evaluation of summaries using N-gram co-occurrence statistics. In: Proceedings of the 2003 Conference of the North American Chapter of the Association for Computational Linguistics on Human Language Technology, pp. 71–78 (2003)
9. Steinberger, J., Jezek, K.: Update summarization based on novel topic distribution. In: Proceedings of the 9th ACM Symposium on Document Engineering, Munich, Germany, pp. 205–213 (2009)

Using Low-Cost Annotation to Train a Reliable Czech Shallow Parser

Adam Radziszewski[1] and Marek Grác[2]

[1] Institute of Informatics
Wrocław University of Technology, Wrocław, Poland
adam.radziszewski@pwr.wroc.pl
[2] Computational Linguistics Centre
Department of Czech Language, Faculty of Arts, Masaryk University, Brno, Czech Republic
grac@fi.muni.cz

Abstract. *Bushbank* is a relatively new concept — a type of annotated corpus where annotation is driven by use of automatic tools and the task of human annotators is limited to accepting or rejecting parts of their output. This creates a possibility to obtain annotated corpora of considerable size at relatively low cost.

In this paper we ask the question if the Czech Bushbank is reliable enough to be used for a NLP task instead of a traditional corpus with high annotation rigour. We perform evaluation of three different parsers using its shallow syntactic annotation, including a CRF chunker made originally for Polish. The results are very promising, showing that many practical applications could benefit from low-cost annotation.

Keywords: corpus annotation, shallow parsing, Czech.

Text corpora are the most widely used language resources for NLP applications. It is possible to gather a multi-million corpus automatically from the Internet. However, in order to fully utilize such a resource, it is important to add additional annotation layers. Manual annotation is believed to guarantee high quality and reliability. Unfortunately, it is expensive to hire qualified annotators who will work according to a manual. What is more, it is extremely hard to find experienced annotators, especially for less-resourced languages. This results in a necessity to devote substantial amounts of time to annotator training and supervision. These costs are further multiplied if we want to annotate several layers.

The number of large corpora that are manually annotated on levels other than morphosyntax is quite limited, since their development was a huge enterprise. One of the typical examples is the Prague Dependency Treebank [1]. In order to obtain high quality unambiguous annotation of natural language (which is ambiguous on every level), trained annotators and hundreds of pages of manual [2] were needed.

While treebanks are indispensable for carrying out syntactic research, many practical NLP applications do not need such a level of detail. Areas such as information extraction, text summarisation and question answering typically require recognition of noun and verb phrases in text without the need of full parsing [3]. Unfortunately, switching

I. Habernal and V. Matousek (Eds.): TSD 2013, LNAI 8082, pp. 575–582, 2013.

to shallow syntactic annotation does not guarantee obtaining the desired level of simplicity, e.g. the guidelines for shallow syntactic annotation of a Polish corpus counts as much as 70 pages [4].

A possible solution is to use an automatic parser and have the annotators correct its output [5,6]. This idea in its most radical form was embodied in a new type of an annotated corpus named *bushbank* and implemented in the Czech Bushbank [7]. The most conspicuous features of a bushbank are the crucial role of automatic parsers in its development (the annotators may only accept or reject whole annotations from their output) and relying entirely on untrained staff as annotators.

In this study we ask the question if the Czech Bushbank is reliable enough to be used for a NLP task instead of a traditional corpus. We use its shallow syntactic annotation to perform parser evaluation. We test two parsers which have already been applied for Czech — Collins' statistical parser [8] and SET rule-based parser [9], but also IOB-BER, a chunker based on Conditional Random Fields developed originally for Polish [10]. The obtained results are very interesting for a couple of reasons. First, the performance of IOBBER for the task of NP chunking is surprisingly high, exceeding the achievements of two other parsers. Secondly, its performance is on par with the figures reported for the very same parser tested against 1-million-token part of the National Corpus of Polish, whose manual annotation was carried out under very strict rigour [6]. These two observations confirm that the annotation of the bushbank is consistent enough to obtain a high quality automatic chunker. Incidentally, this is the first reported attempt at shallow parsing of Czech language using a Machine Learning Method (to the best of our knowledge).

In the rest of the paper we present the idea of bushbank, the Czech Bushbank and we describe the parsers we used. Then we discuss the problems related to evaluation of parsers against the incomplete data that are to be found in a bushbank. Finally, we present our results and closing conclusions.

1   Czech Bushbank

The observations summarised in the previous section provided motivation for development of a new corpus type — a bushbank [7]. A bushbank is a essentially an annotated corpus, where:

- simplicity and usability is preferred to consistency and completeness,
- automatic tools are used to provide initial annotation, possibly several conflicting versions of it,
- human annotation is limited to deciding which parts of the automatic annotation is to be accepted and which is to be rejected,
- no measures are taken to resolve ambiguity and the information on every annotator's decision is preserved,
- the annotation manual is reduced to minimum so that an untrained person given a time span of a couple of hours may already contribute practical value.

Along these lines, a proof-of-concept bushbank was developed, namely the Czech Bushbank. Its annotation currently contains two layers: morphosyntactic tagging performed using desamb tagger [11] and shallow syntactic annotation performed by untrained annotators. Each of the annotators was working for 10–20 hours, including one hour of training [7].

In this paper we make use of a part of the Czech Bushbank, namely *CBB.blog*. This is a subcorpus, created from ten blogs of experienced authors. This part is annotated with two types of syntactic chunks: verb phrase (VP) and noun phrases (NP). Verb phrases, according to annotation manual, can contain only tokens that belong to verbal classes, which makes the annotation pretty straightforward. E.g., the annotators do not have to decide if reflexive pronouns should or should not be part of VPs. An example VP is given as example (1).

The term *noun phrases* is here understood rather broadly: it covers actual noun phrases, but also prepositional phrases, adjective phrases and numeral phrases. The employed definition of NP forbids nested structures, so NPs must be disjoint. In order to improve consistency and inter-annotator agreement, prepositional phrase attachment is not included in this annotation layer. Instead, it belongs to an upper layer where relations between phrases are defined (we do not use this layer here). This annotation principle forces to split NPs on every preposition, even if it is not intuitive (e.g. city names such as *[Jablonec] [nad Nisou]* are split). Example (3) presents an NP, example (2) presents a PP (also annotated as NP here).

(1) *nemusí   kupovat*
     doesn't have to buy

(2) *typická ukázka pitoreskního   českého humoru*
     typical example of picturesque Czech   humour

(3) *na českém koncertě americké   kapely*
     on a Czech concert   of an American band

Annotation of CBB.blog is in progress. We were given access to a part containing more than 800 000 tokens annotated by at least single annotator. The goal of this project is to reach one million tokens and have this part annotated by at least three different annotators.

2   Three Parsers

We decided to evaluate three parsers. Two of them (Collins and SET) were also used for the development of the bushbank. This influences the possibilities of performing evaluation: the bushbank annotation already contains all the chunks that were returned by the two parsers and each of the chunks is labelled by an annotator whether it was accepted or rejected.

Besides those two, we also evaluated a trainable chunker (IOBBER) that so far had been evaluated only for Polish. Note that its output is not present in the bushbank in any form, which will influence the evaluation procedure (Sec. 3).

2.1   Collins' Parser Adapted for PDT

The *Collins' parser adapted for PDT* [8] is a representative of dependency parsers that was developed at the Institute of Formal and Applied Linguistics in Prague. It uses Prague Dependency Treebank (PDT, [1]) as training data. The parser outputs dependency trees. To obtain NP chunks compliant with their bushbank definition, additional conversion was needed. This was performed by a conversion script developed earlier for parser evaluation [12]. Conversion is based on identifying constituents in dependency tress according to morphological categories of their syntactic heads. Unfortunately, the definition of VPs in the bushbank data is considerably different to the one used in PDT. This is why we could not use Collins' parser to recognise VP chunks.

2.2   SET

The *SET parser* [9] is an open source[1] rule-based system that comes from NLP Centre at Masaryk University. This parser is based on pattern matching linking rules that are used for incremental segmentation of the sentence. SET parser results its output in several forms including one that contains noun and verb phrases according to bushbank annotation manual. Besides this generic grammar, we were also given access to a specialized grammar that was written by Iveta Beranová for the purpose of accurate recognition of verb phrases.

2.3   IOBBER

IOBBER is an open source[2] chunker based on Conditional Random Fields (CRF) developed originally for Polish [10]. The chunker is very flexible. Most importantly, features used for classification are written in an expressive domain language (WCCL) which allows referring to particular grammatical attributes inferred from positional tags and construct features as complex as tests for syntactic agreement on number, gender and case between given stretch of tokens [13].

IOBBER comes with a configuration tailored for Polish, which also includes a set of features. The feature set includes the following items:

- the wordforms of tokens occupying a local window $(-2, \ldots, +2)$,
- grammatical class of tokens in the window,
- values for the following grammatical categories: number, gender and case in the window,
- a couple of tests for morphosyntactic agreement on the values of number, gender and case,
- two tests for orthographic form: if it starts with an upper-case letter, if it starts with a lower-case letter.

---

[1] Source codes are available at http://nlp.fi.muni.cz/trac/set/ under GNU GPL 3.0.

[2] Source codes are available at http://nlp.pwr.wroc.pl/redmine/projects/iobber/wiki under GNU LGPL 3.0.

These features seem equally valid for Polish as for Czech, thus we decided to keep them intact. The only necessary adaptation to process Czech language was the need to transform the tags' textual representation in the corpus and provide a configuration file with formal definition of the Czech tagset.

3 Experimental Design

Quality evaluation of NLP applications is a very difficult task because usually there are no clear boundaries between good and bad results. Manually annotated data are mostly considered as good enough in this respect. This is also the case of syntactic analysis.

The standard parser evaluation procedure assumes comparison of parser output with syntactic trees available in a treebank. Treebank data are used as the gold standard and basic metrics, such as precision, recall and F-score, are used to measure discrepancies between parser output and the gold standard [14]. This approach is based on the assumption that the gold standard is complete — it contains all correct syntactic elements with their grammatical relations.

In order to use bushbank for the same purpose, we need to reformulate these metrics in terms of syntactic elements annotated as accepted and rejected. The bushbank is inherently incomplete — we know which of the automatically generated phrases were accepted, but we cannot be sure there are no correct phrases that are not present in our data. To account for this, parser output could be divided into three categories: *true positives* (present in parser output and accepted by annotator), *false positives* (present in parser output but rejected) and unknown elements (not found in the bushbank at all, that is not present as accepted nor rejected phrases).

The initial annotation was made using a couple of parsers. The set of parsers used includes the Collins' parser and the SET parser, which makes their evaluation straightforward. We calculate precision simply as the ratio of true positives to the number of chunks output by the parser. Recall is the ratio of true positives to the number of all chunks accepted in the bushbank.

Evaluation of a parser that was not used in initial annotation phrase (i.e., IOBBER) is more difficult, since it may generate unknown elements — phrases that are not included in the positive nor negative data. For practical reasons, we decided to treat all unknown elements as false positives. In other words, we take a false assumption that the bushbank annotation is complete. This way we know that the real value of precision will be greater or equal to the value we will obtain. Computing recall is more complicated because we do not know the number of false positives. Thus, our value of recall is approximated and its reliability depends on the quality of the parsers involved in the initial annotation.

The part of the bushbank we had access to (CBB.blog) is annotated with NP and VP chunks. We treat the problem as a chunking task, that is, we assess only the parsers' capability to recognise chunk boundaries correctly. SET is a rule-based parser, hence it requires no training set. Collins' parser is trainable, but the adapted version we used was trained elsewhere (on PDT), thus the whole CBB.blog could be used as testing data for both of them. To evaluate IOBBER, we had to allocate a part of the data for training set. We decided to use standard ten-fold cross-validation. The figures reported for IOBBER are values of respective measures averaged over ten runs.

Also note that the morphosyntactic tagging was performed automatically, so the results will already include the influence of tagging error on the observed chunking performance. This is desirable, since under such conditions the results are a better approximation of the performance of the tested tools against unseen data, which anyway needs to be tagged first.

4   Results

The results of parser evaluation are presented in Tab. 1. SET parser is tested in two variants: the generic Czech grammar (*SET std*) and the new specialized grammar written for VP (*SET VP*).

Table 1. Performance of the three parsers

	NP			VP		
	P	**R**	**F**	**P**	**R**	**F**
Collins	73.23%	72.84%	73.02%			
SET std	74.70%	89.50%	81.42%	80.18%	85.01%	82.52%
SET VP				89.37%	95.54%	92.35%
IOBBER	85.91%	90.31%	88.05%	81.61%	90.74%	85.93%

The first conclusion is that IOBBER clearly outperforms the remaining parsers in the task of NP chunking. Both precision and recall are substantially higher. IOBBER's precision (almost 86%) corresponds to 44% smaller error rate than in the case of the SET parser (almost 75% precision). The difference in recall is less impressive, but still visible.

The results for VP chunking are somewhat surprising: given the simple definition of VP chunks one could expect higher values of precision. The specialized VP grammar outperforms the other solutions, which corresponds to 42% smaller error rate than in the case of IOBBER, measured in terms of precision alone. Similar ratio holds between values of recall.

The particularly good results of IOBBER tested against Czech data are very encouraging. So far, most effort related to shallow parsing of Slavic languages was carried out using hand-written grammars, while it seems that Machine Learning methods may be better suited. Note that a similar observation was made for Polish [10]: using the very same chunker allowed to achieve 86% of both precision and recall, while a parser based on a hand-written grammar performed significantly worse on the same data ($P = 78\%$, $R = 81\%$). The cited work is especially relevant, since those experiments were carried out using 1-million-token part of the National Corpus of Polish, whose manual annotation was carried out under very strict rigour [6]. The very fact that we obtained slightly higher values of the metrics using the same chunker confirms the practical value of the bushbank: while the annotation was carried out in fast and economic manner, it is reliable enough to train a reliable chunker.

5 Conclusion and Further Work

In this paper we discussed some theoretical and practical issues related to the syntactic annotation of the Czech Bushbank and performed parser evaluation using this data set. The results of our experiment are surprisingly good, leading to the conclusion that low-cost annotation may be sufficient for many practical NLP projects. What is more, we showed that a Machine Learning method, namely CRF, may outperform a hand-written grammar also in the case of Czech language.

The experiment also confirms that similar languages may be processed using the same tools, hence it is worth putting more effort towards reusing existing tools made for Slavic languages to process other Slavic languages rather than always starting with adaptation of those tools which were made for English.

As described in Sec. 3, our evaluation was simplified and it would be interesting to gain more insight into the data that are missing from the bushbank. Also, the annotation of the bushbank is in progress and we hope that the next version will contain decisions of multiple annotators, which would allow for a more thorough analysis of its consistency.

Acknowledgements. This work was financed by Innovative Economy Programme project POIG.01.01.02-14-013/09.

References

1. Böhmová, A., Hajič, J., Hajičová, E., Hladká, B.: The Prague Dependency Treebank. In: Treebanks, pp. 103–127. Springer (2003)
2. Hajič, J., Panevová, J., Buráňová, E., Urešová, Z., Bémová, A., Štěpánek, J., Pajas, P., Kárník, J.: Anotace na analytické rovině. Návod pro anotátory (2004)
3. Shen, H.: Voting between multiple data representations for text chunking. Master's thesis, Simon Fraser University, Canada (2004)
4. Radziszewski, A., Maziarz, M., Wieczorek, J.: Shallow syntactic annotation in the Corpus of Wroclaw University of Technology. Cognitive Studies 12 (2012)
5. Kordoni, V., Zhang, Y.: Annotating Wall Street Journal texts using a hand-crafted deep linguistic grammar. In: Proceedings of the Third Linguistic Annotation Workshop, ACL-IJCNLP 2009, pp. 170–173. Association for Computational Linguistics, Stroudsburg (2009)
6. Waszczuk, J., Glowińska, K., Savary, A., Przepiówski, A.: Tools and methodologies for annotating syntax and named entities in the National Corpus of Polish. In: Proceedings of the International Multiconference on Computer Science and Information Technology (IMCSIT 2010): Computational Linguistics – Applications (CLA 2010), pp. 531–539. PTI, Wisla (2010)
7. Grác, M.: Case study of bushbank concept. In: Proceedings of the 25th Pacific Asia Conference on Language, Information and Computation, pp. 353–361. Institute of Digital Enhancement of Cognitive Processing, Waseda University, Singapore (2011)
8. Collins, M., Ramshaw, L., Hajič, J., Tillmann, C.: A statistical parser for Czech. In: Proceedings of the 37th Annual Meeting of the Association for Computational Linguistics on Computational Linguistics, pp. 505–512. Association for Computational Linguistics (1999)
9. Kovář, V., Horák, A., Jakubíček, M.: Syntactic analysis using finite patterns: A new parsing system for Czech. In: Vetulani, Z. (ed.) LTC 2009. LNCS, vol. 6562, pp. 161–171. Springer, Heidelberg (2011)

10. Radziszewski, A., Pawlaczek, A.: Large-scale experiments with NP chunking of Polish. In: Sojka, P., Horák, A., Kopeček, I., Pala, K. (eds.) TSD 2012. LNCS, vol. 7499, pp. 143–149. Springer, Heidelberg (2012)
11. Šmerk, P.: K morfologické desambiguaci češtiny (2008)
12. Grác, M., Jakubíček, M., Kovář, V.: Through low-cost annotation to reliable parsing evaluation. In: Proceedings of the 24th Pacific Asia Conference on Language, Information and Computation, pp. 555–562. Waseda University, Tokio (2010)
13. Radziszewski, A., Wardyński, A., Śniatowski, T.: WCCL: A morpho-syntactic feature toolkit. In: Habernal, I., Matoušek, V. (eds.) TSD 2011. LNCS, vol. 6836, pp. 434–441. Springer, Heidelberg (2011)
14. Grishman, R., Macleod, C., Sterling, J.: Evaluating parsing strategies using standardized parse files. In: Proceedings of the 3rd ACL Conference on Applied Natural Language Processing, pp. 156–161 (1992)

Verb Subcategorisation Acquisition
for Estonian Based on Morphological Information

Siim Orasmaa*

Institute of Computer Science, University of Tartu
J. Liivi Str 2, 50409 Tartu, Estonia
`{siim.orasmaa}@ut.ee`

Abstract. A method for automatic acquisition of verb subcategorisation information for Estonian is presented. The method focuses on detection of subcategorisation relations between verbs and nominal phrases. Simple comparison of verb-specific argument candidate's frequency ranking against a global frequency ranking of the candidate is used to decide whether the argument candidate is likely governed by the verb. The method also requires only limited linguistic resources from the input corpora: morphological annotations and clause boundary annotations. The results obtained are evaluated against a manually built valency lexicon.

Keywords: verb subcategorisation acquisition, morphological information, frequency ranking comparisons, Estonian.

1 Introduction

Verb subcategorisation information is important in Natural Language Processing, as it specifies morphosyntactic forms of verb arguments and therefore supports tasks related to further syntactic analysis of texts (e.g parsing, grammar building). Many existing methods for automatic verb subcategorisation acquisition require predefined subcategorisation structure (a set of subcategorisation frames) ([1],[2]) or require that the input corpus is in treebank form or fully/partially parsed [3]. However, large machine-readable lists of subcategorisation frames and robust parsers are not available for many languages. The goal of current work is to explore a possibility of subcategorisation acquisition with limited linguistic resources: a corpus having only morphological annotations (word part-of-speech and grammatical categories coded in word form) and clause boundary annotations is taken as the input for the task. The proposed method focuses on acquisition of subcategorisation relations between verbs and nominal phrases (NPs) morphologically marked with semantic cases[1].

In contrast to previous approaches, which often have employed sophisticated statistical methods for identifying verb arguments/subcategorisation frames, the current

---

* This work is supported by the European Regional Development Fund through the Estonian Centre of Excellence in Computer Science (EXCS) and by European Social Funds Doctoral Studies and Internationalisation Programme DoRa, which is carried out by Foundation Archimedes.

[1] What is meant by 'semantic cases' is further explained in Section 2.

I. Habernal and V. Matousek (Eds.): TSD 2013, LNAI 8082, pp. 583–590, 2013.

method uses simple comparison of verb-specific argument candidate's ranked list against a globally calculated ranked list of potential candidates to decide whether the verb tends to attract one or some of the candidates. Verb subcategorisation is represented as a ranked list of morphological cases (argument markers), where ranking of a case indicates how likely the case is governed by the verb.

The current study focuses on subcategorisation acquisition for Estonian verbs. Estonian has limited linguistic resources regarding verb valency information and the task has not been attempted for the language before.

This paper has following structure: first, relevant properties of Estonian language in the context of verb subcategorisation acquisition are described, followed by presentation of the related work. Then, the acquisition method is described and achieved results are presented and discussed. Finally, conclusions are drawn and future work is pointed out.

2   Properties of Estonian

Estonian is a language belonging to the Finnic group of the Finno-Ugric language family. As it is characterised by free word order, position of an argument in a clause cannot be taken as an important clue in subcategorisation acquisition (like it has been done in English).

In Estonian, a verb can subcategorise for following constituent types:

1. NP with a specific case-marking. For example, in sentence *Ma hoolin sinust* 'I care about you', the verb *hoolima* 'to care' requires that the object of caring is marked by elative case (word suffix *-st* indicates the elative in *sinust* '(about) you').
2. Adpositional phrase. Example: *Ametnikud vastutavad andmete eest* 'Officials are responsible for the data', the verb *vastutama* 'to be responsible' requires adpositional phrase (*andmete eest* 'for the data') headed by the adposition *eest* 'for'.
3. Infinite verb. Example: *Anu proovis laulda* 'Anu attempted to sing', the verb *proovima* 'to attempt' requires that the action attempted is marked as an infinite verb (*laulda* 'to sing').
4. Specific subclause type. Example: *Ta teatas, et kohtumine jääb ära* 'He announced that the meeting was cancelled', the verb *teatama* 'to announce' requires that-clause (started by subordinating conjunction *et*).

Current work focuses on relations where verb subcategorises for NP with a case marking. In Estonian, nouns and adjectives decline in 14 morphological cases. Traditionally, 3 cases are considered grammatical cases (nominative, genitive and partitive) and 11 cases semantic ones. Because grammatical cases have multiple syntactic functions (marking subject, object and genitive attribute), current work leaves these out and concentrates on syntactically less ambiguous semantic cases.

3   Related Work

For languages with available syntactic resources such as computational valency lexicons, grammars or annotated treebanks, subcategorisation acquisition task is often viewed as task of finding verb subcategorisation frames (SCFs). Methods proposed by Manning [1], and Briscoe and Carroll [2] for verb SCF acquisition in English assumed that SCFs were known in advance (e.g from valence dictionaries). They collected evidence of verb co-occurrence with SCFs from corpora and used statistical hypothesis testing to decide whether a particular verb subcategorises for a certain SCF. Both approaches relied on automatic corpus preprocessing: Manning used partial parsing (part-of-speech tagging and limited chunking), and Briscoe and Carrol applied full parsing.

More recently, Lippincott et al. [4] showed for English that state-of-the-art verb subcategorisation acquisition can be done without the parsed input, just by learning grammatical relations from POS tags within a close proximity of a verb. They used an unsupervised probabilistic model for the task and their model did not need a predefined subcategorisation frame inventory.

Several authors also consider languages with limited syntactic resources. Aldezabal et al. [3] addressed the task on Basque. They noted that adjuncts are a substantial source of noise in SCF acquisition, especially in the context of limited resources, and therefore they focused on the argument/adjunct distinction task. They used partial parsing of the input corpora to obtain instances of verbs together with their dependents and applied statistical filtering methods to distinguish arguments from adjuncts. Kermanidis et al. [5] experimented on subcategorisation acquisition on Modern Greek and English, using only limited preprocessing of input corpora (morphological and part-of-speech tagging, phrase chunking). While these preprocessing settings are similar to the settings in current work, notable difference is that Kermanidis et al. [5] still used phrase chunking (which is not readily available for Estonian), while they did not use information about clause boundaries to limit the set of possible argument candidates.

4 The Method

The method used in current work is based on the empirical notion that the total frequency of a morphological case in a corpus is the sum of verb-specific case frequencies (case occurrences in verb contexts). Because occurrences of a case do not distribute evenly across all verbs, verb-specific case frequency can be used as an indicator of subcategorisation relation with the verb. To confirm this indication, only cases that are more frequent in context of a verb than would be expected by their total frequency are brought out as cases possibly governed by the verb.

The method requires that the input corpus has been annotated for basic linguistic information: sentence boundaries, morphological information (word lemmas, part of speech tags, morphological case and conjugation information) and clause boundaries inside sentences.

In principle, if one tries to find words having possible subcategorisation relation with the main verb in sentence, one could include all the words co-occurring in same sentence with the verb, or one could use a fixed-size window (e.g take N words from the left and the right context of the verb). However, such approaches will be problematic in case of complex sentences consisting of more than one clause, as words from other clause could not have subcategorisation relation with the verb and will add noise to the co-occurrence counts. Therefore the method proposed in this work is based upon linguistically motivated clause boundary annotation, introduced by Kaalep and Muischnek [6]. In addition to separating different clauses, the annotation also marks embedded clauses and thus allows uniting clause parts that have been cut by an embedded clause. For example, from sentence *The house, in which we lived with Piret, belonged to a childless old couple*, two separate clauses must be extracted: *The house belonged to a childless old couple* and *in which we lived with Piret*, so the verbs *lived* and *belonged* can be associated only with the words belonging to their clause-context.

In the first processing step, the method extracts from corpus clauses that contain a finite verb (belonging to the grammatical category of indicative mood active voice). Clauses containing more than one verb will be discarded as they would require additional analysis to determine, which verb governs which nouns. Clauses that contain potential phrasal verbs are also left out. It is done because phrasal verbs (constructions verb + adverb, such as *üle ajama* 'spill over') change subcategorisation structure of a clause, so the resulting structure is different than the structure for a single verb. Clauses with phrasal verbs are filtered out using a list of phrasal verb expressions compiled from the Explanatory Dictionary of Estonian (EKSS) [7] and from the Estonian-Russian dictionary (EVS) [8].

Next, the following information is extracted from each clause: the lemma of the finite verb and the morphological cases of declinable words (nouns, adjectives, numerals and pronouns). While counting morphological cases co-occurring with a verb in a clause, only word types are counted. E.g if there are 5 clauses where verb *haarama* 'to grab' co-occurs with declinable word *käega* 'by hand', then the word *käega* increases counts of its morphological case, comitative, only by 1. This way of counting aims at reducing the bias introduced by lexical items that co-occur frequently with the verb and form idiomatic expressions.

After the counting phase, each verb is associated with a list of case frequencies. Relatively high frequency of a case in context of a verb indicates that it can be in a subcategorisation relation with the verb. However, there can be other reasons for relatively high frequency:

A) The case could indicate a subcategorisation relation with some non-predicate clause member (a noun, an adjective or an adverb) co-occurring with the predicate (finite verb). For example, in clause *see võimaldab disketile salvestamist* 'this enables saving to disk', the verb *võimaldama* 'to enable' only governs nominalisation *salvestamist* '(of) saving' and the noun *disketile* '(to) disk' is governed by the nominalisation.

B) The case could be subcategorised not by the verb alone, but by a multiword verb construction. For example, verb *hakkama* 'to begin' with noun *silma* '(into) eye' forms an idiomatic expression (*silma hakkama* 'meet the eye') which has different subcategorisation structure than the verb alone. Thus, a case frequently co-occurring with *silma hakkama* does not reflect subcategorisation structure of the verb *hakkama*.

C) The case could have overall high frequency in the corpus, so the high frequency in the context of the verb does not necessarily indicate a subcategorisation relation.

For example, the inessive case is the most frequent semantic case in the corpus[2]; however, it often indicates location of action, and because many verbs can optionally specify location of action, the inessive can indicate an adjunct rather than an argument of a verb.

Current work does not address the situations of type A, as these situations would require syntactic analysis. Filtering out phrasal verbs and counting only unique declinable words in the verb's context should reduce the number of situations of type B. In order to address the effects of overall high frequency (C), the list of frequency-sorted cases associated with a verb is compared to the list of frequency-sorted cases from the whole corpus, and only important ranking changes are brought out.

To get the list of total case frequencies, cases are counted in all obtained clauses. To be in accordance with verb specific case counting, here also only word form types contribute to the count of their respective case.

In the following example, (1) is a list of frequency sorted cases from all clauses and (2) is a list of frequency sorted cases associated with verb *tutvustama* 'to introduce'. Both lists are sorted in case frequency descending order. Asterisk and number following a case denote the increase in rank, when compared to list (1).

(1) *total*: nom; gen; part; in; el; ad; all; com; ill; tr; abl; es; ter; ab;

(2) tutvustama: nom; part*1; gen; all*3; es*7; in; com*1; ad; ill; el; abl

In the final step, a list of frequency sorted cases associated with a verb is further filtered: only semantic cases that had their rank increased are kept in the list. After the final step, the list of cases associated with the verb *tutvustama* 'to introduce' is:

(3) tutvustama: all; es; com

This result shows 3 subcategorisation possibilities of the verb *tutvustama*. The allative case (word suffix -*le*) marks a person to whom someone/something is introduced, e.g *Ta tutvustas sind meile* 'She introduced you to us'. The essive case (word suffix -*na*) marks the role in which someone/something is introduced, e.g *Ta tutvustas end arstina* 'She introduced herself as a doctor'. The comitative case (word suffix -*ga*) has two roles: it can mark a person/group to whom someone/something is introduced (*Ta tutvustas sind kõigiga* 'She introduced you to everyone'), and if that role is already occupied by allative case, it marks a manner of introducing (*Ta tutvustas sind meile uhkusega* 'She introduced you to us proudly').

5   Evaluation

5.1   Corpus

For subcategorisation acquisition, a 5.8 million word fiction subcorpus of the Reference Corpus of Estonian [9] was chosen, because lexicographers often take examples of verb

---

[2] Assuming the corpus introduced in the next section. Other examples in the current section are also based on this corpus.

usage (including subcategorisation examples) from fiction texts, and so the results of the system can be more easily compared to examples listed in dictionaries.

After clauses containing multiple verbs and potential phrasal verbs had been filtered out, total 486,192 clauses were obtained, associated with 4677 different verbs. 4542 (97 %) of these verbs co-occurred with at least one declinable word, and 3534 verbs (76 %) co-occurred with at least one declinable word in a semantic case.

5.2   Automatic Evaluation

In order to evaluate performance of the method, the obtained verb subcategorisation information is compared to subcategorisation information in manually built valency lexicon of syntactic analyser for Estonian [10]. The valency lexicon specifies in detail the morphological case alternation related to object of a clause, and also brings out the semantic cases that are subcategorised by the main verb. However, cases listed in the lexicon do not form a complete subcategorisation frame of a verb and it is not specified, whether a case marks an obligatory or an optional argument of the verb.

Because the cases having higher ranking in the results list are interpreted as being more likely governed by the verb, the evaluation method must take this into account. So, each case that occurs in the lexicon, but has a low ranking in the results list or does not appear in the results at all, must be penalised. Similar situation appears in the evaluation of information retrieval systems, where one typically obtains a list of documents as a result of a query and wants to ensure that all the documents relevant to the query appear at the top of the document list. One of the frequently used measures in such setting is mean average precision (MAP), which aggregates results across multiple recall levels and queries to provide a single-figure precision measure [11]. This measure is also used here.

Calculating MAP requires first finding an average precision (AP) for each verb and then taking the mean of all found APs. For a single verb, a precision is calculated at each position in the results list where some case from the lexicon appears, and these precisions are then averaged over all the cases in the lexicon to get the AP. If some case from the lexicon is missing in the results list, the precision is taken 0 at that point. For example, the semantic cases associated with the verb *helistama* 'to phone' in the lexicon are $L = \{all, ill\}$ and the list acquired from the corpus is $C = \{all, ad, ill, abl\}$. The precision for $all \in L$ is 1.0 (as *all* has top ranking in C) and the precision for $ill \in L$ is $\frac{2}{3}$ (because *ill* is the 3rd case in C and one redundant case appears before *ill*). The average precision on detection of semantic cases governed by the verb *helistama* is $\frac{1+2/3}{2} = 0.83$. In order to find the MAP score, such calculations are done for each verb and then the mean of all verb specific APs is taken.

Only verbs governing at least one semantic case were taken from the lexicon for evaluation. Also, lexicon verbs that did not appear in the clauses extracted from the corpus were discarded from the evaluation. This gave a total of 413 verbs for evaluation. These verbs were split into 3 similar size groups by their occurrence frequency (high, medium and low), and MAP scores were calculated for each group separately and for all verbs together.

Results in Table 1 show that the method is sensitive to verb frequency: for verbs occurring less than 17 times, governed cases are detected only with mean average

Table 1. Mean Average Precision on detection of semantic cases governed by verbs. Verbs are divided into groups by their frequency in the corpus.

Group	Verbs in group	Verb frequency range	MAP
Low frequency verbs	140	1–16	56.0%
Medium frequency verbs	136	17–95	79.6%
High frequency verbs	137	98–5415	82.9%
All verbs	413	1–5415	72.7%

precision 56.0%. However, considering medium and high frequency verbs, the method has rather promising mean average precisions (79.6% and 82.9% respectively).

These results support previous research, which has found that simple co-occurrence frequency can be an effective indicator for subcategorisation relations. Kermanidis et al. [5] compare different statistical filtering methods (log likelihood ratio, T-score, binomial hypothesis testing, and filtering by relative frequency threshold) and report that filtering by relative frequency threshold, despite its simplicity, nearly outperforms the other statistical methods.

As lexicons listing the complete subcategorisation frames are not available for Estonian, it is not possible to estimate which is the percentage of subcategorization frames covered by the proposed method.

However, the case lists obtained in this work can be used to aid valency lexicon building: from cases acquired with the method, lexicographer can choose cases for further studying and for including into the lexicon. Case ranking (which can be made more informative by bringing out exact occurrence counts) supports this, as one can have higher confidence about high ranked cases being governed by the verb.

6   Conclusions

In this paper, a method for automatic acquisition of verb subcategorisation information for Estonian has been presented. The focus of the method is on detection of subcategorisation relations between verbs and NPs. The method requires only minimal linguistic annotation (morphological and clause boundary annotations) of the input corpus, and uses simple comparison of verb-specific argument candidate's frequency ranking against total frequency ranking of the candidate to decide whether the candidate is possibly governed by the verb. Verb subcategorisation is represented as a ranked list of morphological cases (argument markers), where ranking of a case indicates the likelihood of the case being governed by the verb. Ranking performance of the method was evaluated against a manually built valency lexicon and mean average precision 72.7% was measured. In future work, the plan is to extend the set of argument types used in the method and also to experiment with other statistical filtering methods used in literature to see whether these methods will produce comparable rankings.

Appendix: List of Case Abbreviations

ab	abessive	*ill*	illative
abl	ablative	*in*	inessive
ad	adessive	*com*	comitative
all	allative	*nom*	nominative
el	elative	*part*	partitive
es	essive	*ter*	terminative
gen	genitive	*tr*	translative

References

1. Manning, C.: Automatic Acquisition of a Large Subcategorization Dictionary from Corpora. In: Proceedings of 31st Meeting of the Association of Computational Linguistics, Columbus, Ohio, pp. 235–242 (1993)
2. Briscoe, T., Carroll, J.: Automatic extraction of subcategorization from corpora. In: Proceedings of the 5th ACL Conference on Applied Natural Language Processing, Washington, DC, pp. 356–363 (1997)
3. Aldezabal, I., Aranzabe, M., Gojenola, K., Sarasola, K., Atutxa, A.: Learning Argument/Adjunct Distinction for Basque. In: Proceedings of the ACL 2002 Workshop on Unsupervised Lexical Acquisition, ULA 2002, Philadelphia, Pennsylvania, vol. 9, pp. 42–50 (2002)
4. Lippincott, T., ÓSéaghdha, D., Korhonen, A.: Learning Syntactic Verb Frames Using Graphical Models. In: Proceedings of the 50th Annual Meeting of the Association for Computational Linguistics (ACL 2012), Jeju, Korea (2012)
5. Kermanidis, K., Fakotakis, N., Kokkinakis, G.: Automatic acquisition of verb subcategorization information by exploiting minimal linguistic resources. Corpus Linguistics 9(1), 1–28 (2004)
6. Kaalep, H.-J., Muischnek, K.: Robust clause boundary identification for corpus annotation. In: Proceedings of the Eight International Conference on Language Resources and Evaluation (LREC 2012), Istanbul, Turkey (2012)
7. EKSS: Eesti kirjakeele seletussõnaraamat. ETA KKI, Tallinn (1988–2000)
8. EVS: Eesti-venesõnaraamat I. Eesti Keele Instituut, Tallinn (1997)
9. Kaalep, H.-J., Muischnek, K., Uiboaed, K., Veskis, K.: The Estonian Reference Corpus: Its Composition and Morphology-aware User Interface. In: Proceedings of the 2010 Conference on Human Language Technologies – The Baltic Perspective: Proceedings of the Fourth International Conference Baltic HLT, pp. 143–146 (2010)
10. Müürisep, K.: Parsing Estonian with Constraint Grammar. In: Online proceedings of NODALIDA 2001, Uppsala (2001),
http://stp.ling.uu.se/nodalida01/pdf/myyrisep.pdf
11. Manning, C.D., Raghavan, P., Schütze, H.: Introduction to Information Retrieval. Cambridge University Press (2008)

Whispered Speech Database: Design, Processing and Application

Branko Marković[1], Slobodan T. Jovičić[2,3], Jovan Galić[4], and Đorđe Grozdić[2,3]

[1] Čačak Technical College, Computing and Information Technology Department, Čačak, Serbia
brankomarko@yahoo.com
[2] School of Electrical Engineering*,
University of Belgrade,Telecommunications Department, Belgrade, Serbia
jovicic@etf.rs
[3] Life Activities Advancement Center, Laboratory for Psychoacoustics and Speech Perception,
Belgrade, Serbia
djordjegrozdic@gmail.com
[4] Faculty of Electrical Engineering, University of Banja Luka,
Department of Electronics and Telecommunications, Banja Luka, Bosnia and Herzegovina
jgalic@etfbl.net

Abstract. This paper presents creation of a whispered speech database Whi-Spe for Serbian language. The database has been collected in order to investigate how well the whisper is used by humans in intelligible verbal communication and how well whispered information can be used in human-computer communication. The database consists of 50 isolated words. They are generated by ten speakers (five male and five female). Each of them pronounced this vocabulary ten times in two modes: normal and whispered. So, the database contains 5.000 pairs of normal/whispered pronunciations. Database evaluation was performed by an analysis of specific manifestations in whispered articulation. Finally, the preliminary results in whispering recognition by using of HMM, ANN and DTW techniques are presented.

Keywords: database, whisper, speech recognition, HMM, ANN, DTW.

1 Introduction

Whispering is a common mode of communication when someone speaks quietly or privately. Nowadays the mobile phone communication is present everywhere and people often uses whisper to speak privately in a public place [1]. It is interesting that this type of speech communication is perfectly understandable.

The whisper has a lot of specific characteristics. The sound source in the production of the whisper is aperiodic turbulent flow of the air current between the vocal cords that do not vibrate [2]; the measurement of the three-dimensional shape of the vocal tract by magnetic resonance imaging (MRI) showed a narrowing of the vocal tract around the ventricular folds [3]. The consequence of this configuration of the glottal system

---

* This work was supported by the Ministry of Education and Science of the Republic of Serbia under Grant numbers OI-178027 and TR-32032.

I. Habernal and V. Matousek (Eds.): TSD 2013, LNAI 8082, pp. 591–598, 2013.

is a much weaker acoustic coupling of the subglottal and supraglottal systems. Due to the absence of the glottal vibrations, whispering lacks the fundamental frequency of the voice and much prosodic information. In addition, whispered speech has a significantly lower energy as compared to the normal speech [4], the slope of the spectrum being much flatter than in the normal speech [5], and the vowel formants at lower frequencies that are shifted to higher frequencies [6].

Therefore, the whisper is a great challenge for modern human-computer communication. Contemporary investigations are focused on recognition of whispered speech [1, 7], on detecting whisper-islands embedded within normally phonated speech [8], and on possible speaker recognition with whispered speech [9]. But, other scientific fields are involved in whisper research. For instance, recent research dealt with the aerodynamics of the glottal system in a whisper [10], brain activity in aphonia [11], whisper improvements in patients after laryngectomy [12], laryngeal hyperfunction during whispering [13], acoustic analysis of consonants in a whisper [4], and so on.

Any investigation of the whisper requires a speech/whisper database. This paper presents how a database for whispered speech in Serbian language was created and intended for application in the investigation of whispered speech recognition. The paper is structured in the following way. Firstly, the design, recording and processing of whispered speech database is described. Then the specific manifestations in a whispering are described and discussed. Finally, the preliminary results in whisper recognition using HMM, ANN and DTW techniques are presented and the conclusion is drawn.

2   Design of the Database Whi-Spe

The database is named Whi-Spe (using the first letters of Whispered Speech) and it is designed to contain the two parts: the first one contains the speech patterns of a whispered speech, and the second one contains the speech patterns of the normal speech. Both modes of speech were collected from the five female and five male speakers and each speaker during the session of recording read 50 words. This was repeated 10 times with a pause of a few days between recordings. Finally, the database collection grew to 10.000 utterances, half in the whispered speech and half in the normal speech. The speakers with ages between 20 and 30 years, were Serbian native volunteers from the Čačak Technical College. All of them had good articulation in speech and whisper production and correct hearing.

The words stored in the database were divided in three sub-corpora: basic colors (6 words), numbers (14 words) and specific words earlier defined in the GEES database (30 words) [14]. The specific words were carefully chosen to cover the basic linguistic criteria of Serbian language (phonemes distribution, accentual structure, syllable composition and consonants clusters). The appendix provides exactly what these words are. All words in the database are uniquely labeled, so it is easy to use them. They are organized using a tree of directories. The normal and whispered speech patterns are separated and organized in forms of sub-corpora. The name convention of speech patterns allows quick and easy access to data and it is suitable for applications, especially when massive comparison is required.

The database is open for upcoming upgrades and it is free for use by others.

3   Recording and Processing

The Whi-Spe database was recorded in a quiet laboratory room by using an Optimus omni-directional microphone with good frequency response up to 16 kHz. For normal speech the microphone was at a distance of about 25 cm from the mouth of a speaker, while for whispered speech the distance was about 5 cm. By using these scenarios we tried to obtain the speech patterns, especially for the whisper, as good as possible. The speech was digitized by using the sampling frequency of 22.050 Hz, with 16 bits per sample, and stored in the form of Windows PCM wave files.

The sessions of recording were organized more than ten times so as to collect a suf-ficient number of good quality representatives. During a single session speakers had read 50 words in two modes: whispered and natural. Then the whole set of 100 words per speaker was segmented manually and the quality control applied to it. Special attention was applied to the segmentation of whispered recordings where two experts in speech signal processing and one phonetician were involved in this process. If the examined word was satisfactory, it was labeled and stored in the Whi-Spe database; otherwise, it was eliminated. It is on this basis that a collection of more than 10.000 words was generated, but only 10.000 of them were stored in the Whi-Spe database.

The quality control of recordings found various type of error. Some of them were re-lated to an incorrect articulation, a wrong pronunciation, but most of them were related to the whispered speech (they are explained in detail below). One of the major problems of whispered recordings was the low level of the signal in relation to the ambient noise, especially with female speakers. Multiple new recordings were sometimes required to solve this problem.

4   Specific Manifestations in Whispering

During the creation of the database, two types of mistakes related to whisper came up. The first type are controlled mistakes like incorrect articulation of some phonemes that speaker randomly produced or an irregular way of recording. These mistakes are cor-rectable with recording repetitions and quality control. Figure 1 shows some examples of mistakes in comparison to the correct pronunciation of the particular word given in Fig. 1a.

The most frequent mistakes during the whisper recording were penetrations of sonor-ity in whisper, Fig. 1b, and an excessive friction of fricatives and affricates articulation, Fig. 1c. Typical example of a wrong recording due to the position of the microphone near the mouth of a speaker is given in Fig. 1d. In that case a blown microphone caused an intensive acoustic pulse. Some other mistakes were present, for example: a whisper that was too low was masked by ambient noise, there were cases of omission of phones in some word pronunciations, or wrong phone pronunciation, etc. These mistakes can be omitted with a proper speaker training for whisper.

The second type of mistakes is uncontrolled mistakes. These mistakes are produced by the articulation action of a specific speaker. These types of deviations can be random or systematic. If deviations are pathological in nature then that speaker is eliminated from the experiment.

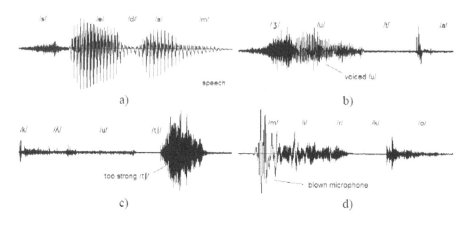

Fig. 1. Mistakes in pronunciation: a) the normal speech, b) voiced segment in whisper, c) excessively pronounced affricate, and d) blown microphone

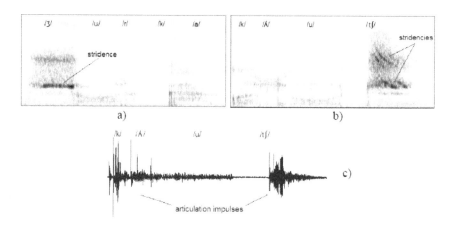

Fig. 2. Examples of specific manifestations in articulation: a) excessive stridence in voiced fricative, b) multiple stridencies in affricate, and c) effects of a tongue contacts with the palate

Fig. 2 shows two examples of intensive resonances known as a stridence [15] which appear when fricative segments are articulated. Fig. 2a and 2b shows the spectrograms of very strong and stable stridence which is caused by the voiced fricative /ʒ/, and strong and unstable multiple stridence caused by affricate /ʧ/, respectively. The stridence is initiated by an unusual tongue position in relation to the palate and it produces a strong and an unpleasant whistling sound or more sounds.

Another interesting case is related to palatal phones like /k/, /ʎ/ and /ʧ/. It is depicted in Fig. 2c and shows a series of very short pulses. They are produced by the contact of parts of the tongue body with the palate during the creation of these phones. This process cannot be controlled and it is specific for some speakers.

Fig. 2c is giving additional information about the whisper and it is useful for further analysis. This is a case of an appearance of many fine features in articulation during whisper production in both temporal and spectral domains. These features are masked in normal speech production, but in whispered speech they can be used for whisper segmentation, phonemes classification or recognition.

5   Database Application in Whisper Recognition

The preliminary results of whisper recognition using three different techniques (the HMM, the ANN and the DTW) are presented in this section. We have used the existing software solutions [16, 17, 18] and our main goal was to investigate how they can be applied to the Whi-Spe database. The results of recognition are not directly comparable in all cases, but they are useful for finding out a path for the further research.

In order to perform these experiments we got feature vectors based on the MFCC coefficients. On the digitized speech signal, represented by wave file, the pre-emphasis, segmentation into frames overlapped 50%, and the Hamming windowing are applied. The Fast Fourier transform is then applied and calculation of log energy based on the Mel-frequency scale gives the MFCC coefficients. Also we calculated the first and the second derivations and finally obtained vectors which consist of 36 coefficients (12 cepstral, 12 delta cepstral and 12 delta-delta cepstral). We have used these vectors for comparison.

The main focus of these experiments was to find out the recognition rate for different scenarios (i.e. normal/normal, whisper/whisper, whisper/normal and normal/whisper) depending on that which is possible at this level of developed ASRs.

5.1   Whisper Recognition with the HMM

In this experiment we used software package AlfaNumCASR [16] for continuous Serbian speech recognition. It is based on the semi continuous HMM and is using the recognition of phonemes in context. This system is independent of the speaker and is trained for normal speech recognition with small vocabulary and recognition rate over 98%. In this experiment we used it for the whisper recognition. The results are given in Table 1 where S1-S5 are male, while S6-S10 are female speakers.

Table 1. Recognition rate for male and female speakers using the HMM (in %)

Speaker	S1	S2	S3	S4	S5	S6	S7	S8	S9	S10
Recognition	55.0	55.4	45.6	66.2	49.2	51.4	30.4	40.6	30.8	25.8
Average			54.3					35.8		

The recognition rate for the whisper is twice less than for the normal speech. The results are expected and further investigation will find out the cause for the difference between male and female speakers.

5.2   Whisper Recognition with the ANN

In order to verify how the whisper can be recognized by the ANN, the two speaker-dependent ASRs were developed with MATLAB Neural Network Toolbox [17]. The structures of these ANNs were: 396 input nodes, 140 hidden neurons and 50 output neurons. The first one is trained with the normal speech, and the second one with the whisper. Then the subset of Whe-Spi (two male and two female speakers) is used for testing. For each speaker all words are divided into three parts: 60% of them are used for training, 20% for validation and 20% for testing. We tested four different scenarios (normal/normal, normal/whisper, whisper/whisper and whisper/normal) and the averaged results are given in Table 2.

Table 2. Recognition rate for male and female speakers using the ANN (in %)

Mode of speech/Gender	Nor/Nor	Nor/Whi	Whi/Whi	Whi/Nor
Male speakers	100	65.6	99.7	78.9
Female speakers	99.7	68.2	98.9	79.2

Based on the results we can conclude that the system which is trained for the whisper and then tested with normal speech gives better results than vice versa. This fact is a point for the further investigation.

5.3   Whisper Recognition with the DTW

The DTW system is developed to see how whispered speech can be recognized by this technique. The DP algorithm uses the symmetric form [18] and the warping paths are not constrained. The system was not trained; it used a randomly chosen pattern and compared it with other patterns. The experiments were conducted for four speakers from the Whi-Spe (two male and two female) and the results are given in Table 3.

Table 3. Recognition rate for male and female speakers using the DTW (in %)

Mode of speech/Gender	Nor/Nor	Nor/Whi	Whi/Whi	Whi/Nor
Male speakers	100	35.7	92.1	27.1
Female speakers	99.3	32.4	96.0	25.2

Based on these results we concluded the normal/normal scenario gives very good recognition, the whisper/whisper also gives a solid score, but the normal/whisper and the whisper/normal scenarios give poor matching. Further investigation will focus on finding out how to improve these results (using robust training, changing type of slope weighting etc.).

6   Conclusion

In this paper we explained how the Whi-Spe database is created, how we recorded the whispering and normal speech, what the specific manifestations of whispering were. We also provided the preliminary results of whisper recognition using this database. Three classic techniques that we have used (HMM, ANN and DTW) give diverse views allowing us to get deeper into this type of recognition. The results indicate the ANN as a promising technique for whisper recognition.

The database is open for new upgrades and it is free for use by others. Our objectives in future are to expand the Whi-Spe database with more subjects and different linguistic contexts and to apply different applications on it.

References

1. Ito, T., Takeda, K., Itakura, F.: Analysis and Recognition of Whispered speech. Speech Communication 45, 129–152 (2005)
2. Catford, J.C.: Fundamental problems in phonetics. Edinburgh University Press, Edinburgh (1977)
3. Matsuda, M., Kasuya, H.: Acoustic nature of the whisper. In: Proc. Eurospeech 1999, vol. 1, pp. 137–140 (1999)
4. Jovičić, S.T., Šarić, Z.M.: Acoustic analysis of consonants in whispered speech. Journal of Voice 22(3), 263–274 (2008)
5. Zhang, C., Hansen, J.H.L.: Analysis and classification of Speech Mode: Whisper through Shouted. In: Interspeech 2007, pp. 2289–2292 (2007)
6. Jovičić, S.T.: Formant feature differences between whispered and voiced sustained vowels. ACUSTICA - Acta Acoustica 84(4), 739–743 (1998)
7. Jou, S.C., Schultz, T., Waibel, A.: Whispery speech recognition using adapted articulatory features. In: ICASSP 2005, Paper SP-P15 (2005)
8. Zhang, C., Hansen, J.H.L.: Whisper-Island Detection Based on Unsupervised Segmentation With Entropy-Based Speech Feature Processing. IEEE Transactions on Audio, Speech, and Language Processing 19(4), 883–894 (2011)
9. Fan, X., Hansen, J.H.L.: Speaker identification within Whispered Speech Audio Stream. IEEE Transactions on Audio, Speech and Language Processing 19(5), 1408–1421 (2011)
10. Sundberg, J., Scherer, R., Hess, M., Müller, F.: Whispering-A Single-Subject Study of Glottal Configuration and Aerodynamics. Journal of Voice 24(5), 574–584 (2010)
11. Tsunoda, K., Sekimoto, S., Baer, T.: Brain Activity in Aphonia After a Coughing Episode: Different Brain Activity in Healthy Whispering and Pathological Aphonic Conditions. Journal of Voice 26(5), 668.e11–668.e13 (2012)
12. Sharifzadeh, H.R., McLoughlin, I.V., Ahamdi, F.: Voiced Speech from Whispers for Post-Laryngectomised Patients. IAENG International Journal of Computer Science, IJCS-36-4-13 (November 19, 2009) (advance online publication)
13. Rubin, A.D., Praneetvatakul, V., Gherson, S., Moyer, C.A., Sataloff, R.: Laryngeal hyperfunction during whispering: reality or myth? Journal of Voice 20, 121–127 (2004)
14. Jovičić, S.T., Kašić, Z., Djordjević, M., Rajković, M.: Serbian emotional speech database: design, processing and evaluation. In: SPECOM-2004, pp. 77–81. St. Petersburg, Russia (2004)
15. Jovičić, S.T., Punišić, S., Šarić, Z.: Time-frequency detection of stridence in fricatives and affricates. In: Int. Conf. Acoustics 2008, Paris, pp. 5137–5141 (2008)

16. Jakovljević, N., Pekar, D.: Description of Training Procedure for AlfaNum Continuous Speech Recognition System. In: EUROCON 2005, pp. 1646–1649 (2005)
17. Demuth, H., Beale, M.: Neural Network Toolbox User's Guide. The MathWorks, Inc. (2002)
18. Marković, B.: Call by voice - the feature of a mobile telephone, MS work, School of Electrical Engineering, Belgrade University (2004) (in Serbian)

Appendix: Vocabulary of the Whi-Spe Database

The vocabulary of the Whi-Spe database is given in Tables 4 and 5.

Table 4. Colors and Numbers

Serbian	English	Serbian	English
/bela/	white	/ʒuta/	yellow
/tsrna/	black	/tsrvena/	red
/plava/	blue	/zelena/	green
/nula/	zero	/jedan/	one
/dva/	two	/tri/	three
/tʃetiri/	four	/pet/	five
/ʃest/	six	/sedam/	seven
/osam/	eight	/devet/	nine
/deset/	ten	/sto/	hundred
/hiλadu/	thousand	/milion/	million

Table 5. Balanced words

Serbian	English	Serbian	English
/Marko/	Marko (name)	/ʒurka/	party
/Petar/	Petar (name)	/demonstratsije/	demonstration
/standard/	standard	/pijatsa/	market place
/padavine/	drops	/ponedeλak/	Monday
/godina/	year	/predstava/	play
/kompjuteri/	computers	/inostranstvo/	abroad
/drvo/	tree	/Mirjana/	Mirajana (name)
/more/	sea	/kiʃa/	rain
/zgrade/	buildings	/klintsi/	kids
/Milan/	Milan (name)	/rezultati/	results
/telefon/	telephone	/svetlo/	light
/prozor/	window	/ruke/	hands
/lokal/	locale	/kλutʃ/	key
/suntse/	sun	/pare/	money
/sef/	treasure	/blok/	block

* Serbian and IPA notation for consonants and vowels are the same except for the following consonants: ʃ(š), h(x), ʒ(ž), ts(c), ʨ(ć), tʃ(č), ʥ(đ), ʤ(dž), ɲ(nj) and λ(lj).

Author Index